Advances in Oceanography

Advances in Oceanography

Edited by Oscar Roberts

MURPHY & MOORE
www.murphy-moorepublishing.com

Published by Murphy & Moore Publishing,
1 Rockefeller Plaza,
New York City, NY 10020, USA
www.murphy-moorepublishing.com

Advances in Oceanography
Edited by Oscar Roberts

International Standard Book Number: 978-1-63987-021-9 (Hardback)

Cataloging-in-Publication Data

Advances in oceanography / edited by Oscar Roberts.
p. cm.
Includes bibliographical references and index.
ISBN 978-1-63987-021-9
1. Oceanography. 2. Ocean. 3. Earth sciences. 4. Marine sciences. I. Roberts, Oscar.
GC11.2 .A38 2022
551.46--dc23

TABLE OF CONTENTS

PREFACE

The study of the oceans is known as oceanography. It encompasses a wide range of topics such as marine life, ecosystems, currents, waves, the movement of sediments, and seafloor geology. Oceanography studies the physical, chemical, and biological characteristics of the ocean along with its history, current condition, and future. Concepts from numerous branches of science such as chemistry, geology, meteorology and biology are applied in oceanography. Some of the branches of oceanography are biological oceanography, chemical oceanography, geological oceanography, physical oceanography, and paleoceanography. This book attempts to understand the multiple branches that fall under the discipline of oceanography and how such concepts have practical applications. It aims to equip students and experts with the advanced topics and upcoming concepts in this area. This book will provide comprehensive knowledge to the readers.

This book is a comprehensive compilation of works of different researchers from varied parts of the world. It includes valuable experiences of the researchers with the sole objective of providing the readers (learners) with a proper knowledge of the concerned field. This book will be beneficial in evoking inspiration and enhancing the knowledge of the interested readers.

In the end, I would like to extend my heartiest thanks to the authors who worked with great determination on their chapters. I also appreciate the publisher's support in the course of the book. I would also like to deeply acknowledge my family who stood by me as a source of inspiration during the project.

Editor

1

Theoretical constraints of physical and chemical properties of hydrothermal fluids on variations in chemolithotrophic microbial communities in seafloor hydrothermal systems

Kentaro Nakamura[1,2*] and Ken Takai[1,3]

Abstract

In the past few decades, chemosynthetic ecosystems at deep-sea hydrothermal vents have received attention as plausible analogues to the early ecosystems of Earth, as well as to extraterrestrial ecosystems. These ecosystems are sustained by chemical energy obtained from inorganic redox substances (e.g., H_2S, CO_2, H_2, CH_4, and O_2) in hydrothermal fluids and ambient seawater. The chemical and isotope compositions of the hydrothermal fluid are, in turn, controlled by subseafloor physical and chemical processes, including fluid–rock interactions, phase separation and partitioning of fluids, and precipitation of minerals. We hypothesized that specific physicochemical principles describe the linkages among the living ecosystems, hydrothermal fluids, and geological background in deep-sea hydrothermal systems. We estimated the metabolic energy potentially available for productivity by chemolithotrophic microorganisms at various hydrothermal vent fields. We used a geochemical model based on hydrothermal fluid chemistry data compiled from 89 globally distributed hydrothermal vent sites. The model estimates were compared to the observed variability in extant microbial communities in seafloor hydrothermal environments. Our calculations clearly show that representative chemolithotrophic metabolisms (e.g., thiotrophic, hydrogenotrophic, and methanotrophic) respond differently to geological and geochemical variations in the hydrothermal systems. Nearly all of the deep-sea hydrothermal systems provide abundant energy for organisms with aerobic thiotrophic metabolisms; observed variations in the H_2S concentrations among the hydrothermal fluids had little effect on the energetics of thiotrophic metabolism. Thus, these organisms form the base of the chemosynthetic microbial community in global deep-sea hydrothermal environments. In contrast, variations in H_2 concentrations in hydrothermal fluids significantly impact organisms with aerobic and anaerobic hydrogenotrophic metabolisms. Particularly in H_2-rich ultramafic rock-hosted hydrothermal systems, anaerobic and aerobic hydrogenotrophy is more energetically significant than thiotrophy. The CH_4 concentration also has a considerable impact on organisms with aerobic and anaerobic methanotrophic metabolisms, particularly in sediment-associated hydrothermal systems. Recently clarified patterns and functions of existing microbial communities and their metabolisms are generally consistent with the results of our thermodynamic modeling of the hydrothermal mixing zones. These relationships provide important directions for future research addressing the origin and early evolution of life on Earth as well as for the search for extraterrestrial life.

Keywords: Deep-sea hydrothermal systems; Chemosynthetic ecosystems; Hydrothermal fluid chemistry; Host rock geochemistry; Geochemical modeling; Bioavailable energy yield

* Correspondence: kentaron@sys.t.u-tokyo.ac.jp
[1]Precambrian Ecosystem Laboratory (PEL), Japan Agency for Marine-Earth Science and Technology (JAMSTEC), 2-15 Natsushima-cho, Yokosuka, Kanagawa 237-0061, Japan
[2]Current address: Department of Systems Innovation, School of Engineering, The University of Tokyo, 7-3-1 Hongo, Bunkyo-ku, Tokyo 113-8656, Japan
Full list of author information is available at the end of the article

Review

Introduction

Deep-sea hydrothermal vents host some of the most diverse microbial communities on Earth (Takai and Nakamura 2011). Since the first discovery of black smoker vents inhabited by dense and unique chemosynthetic macrofaunal communities (Spiess et al. 1980), submarine hydrothermal systems and their associated biota have attracted great interest (e.g., Humphris et al. 1995; Van Dover 2000; Wilcock et al. 2004). Unlike most biological communities, in which photosynthetic organisms are the base of the food web, deep-sea hydrothermal vent ecosystems are dependent on primary production by symbiotic and free-living chemolithoautotrophic microorganisms that obtain energy from inorganic redox substances (e.g., H_2S, CO_2, H_2, CH_4, and O_2) in hydrothermal fluids and ambient seawater (Karl, 1995; Kelley et al. 2002). Because of the unique features of deep-sea hydrothermal vent ecosystems, they are considered plausible analogues to the early ecosystems of Earth and also to extraterrestrial life on other planets and moons (e.g., Jannasch and Mottl 1985; Nealson et al. 2005; Takai et al. 2006a).

To date, more than 300 high-temperature hydrothermal vent systems have been identified at mid-ocean ridges (MOR), island arcs, and back-arc spreading centers (Hannington et al. 2011). Deep-sea hydrothermal fluids vary greatly in their chemical compositions due to subseafloor physical and chemical processes such as fluid-rock interactions, magmatic volatile inputs, and phase separation of hydrothermal fluids (Von Damm 1995; Butterfield et al. 2003; German and Von Damm 2004; Tivey 2007). Compositional variations in hydrothermal fluids (particularly energy and carbon sources) in turn affect biomass production and the diversity of hydrothermal vent-endemic communities. Consequently, clarifying the relationships among the geological background of hydrothermal environments, physical and chemical variations in hydrothermal fluids, and the compositional and functional diversity of chemosynthetic ecosystems has provided important information on the diversification and development of extant deep-sea hydrothermal ecosystems as well as the generation and sustenance of early ecosystems and possible extraterrestrial life forms.

In deep-sea hydrothermal vents, rapid mixing between hot reduced hydrothermal fluids and cold oxidized seawater provides chemical energy for microbial activity and biomass production. To quantify the in situ energetics of chemolithotrophic microorganisms in hydrothermal mixing environments, a thermodynamic model was first proposed and applied to a basalt-hosted hydrothermal system at 21° N on the East Pacific Rise (EPR) (McCollom and Shock 1997). Using batch-mixing models modified from this original model, thermodynamic calculations have been conducted for several hydrothermal vent systems at MOR and arc-backarc (ABA) hydrothermal systems (Shock and Holland 2004; Tivey 2004; McCollom 2007; Amend et al. 2011). These studies have demonstrated differing patterns in the potential energy yields of various in situ metabolic reactions in the mixing zones of these habitats, providing the theoretical basis for relationships between hydrothermal fluid chemistry and the diversity of hydrothermal vent-endemic biological communities. However, the hydrothermal systems studied were quite limited. In addition, the structures and functions of extant chemosynthetic biological communities, which have been characterized in many previous investigations, have not yet been integrated into development of these theoretical relationships.

Many studies have identified high compositional and functional diversity of chemosynthetic ecosystems in geographically and geologically diverse hydrothermal systems (e.g., in reviews by Huber and Holden 2008; Nakagawa and Takai 2008; Takai et al. 2006b). Some of these studies have noted possible relationships between the metabolic abundances and compositions of hydrothermal vent-endemic microbial communities and the chemical characteristics of hydrothermal vent fluids in deep-sea hydrothermal systems (Perner et al. 2007, 2010; Reysenbach and Shock 2002; Takai and Horikoshi 1999; Takai et al. 2001, 2004a). However, most of these studies were qualitative and focused mainly on the genetic and phylogenetic diversity of microbial communities and their constituents. Thus, the relationships between the abundance and composition of chemolithotrophic microbial communities and the geological and geochemical environments of global deep-sea hydrothermal systems remain unclear.

Takai and Nakamura (2010, 2011) first provided clear evidence of biogeochemical relationships among microbiological community development, the chemical composition of hydrothermal fluids, and the geological environment of deep-sea hydrothermal systems through both thermodynamic calculations of the potential energy yields of various in situ metabolic reactions and observed compositional and functional diversity of chemosynthetic ecosystems in the mixing zones of these habitats. However, examples of hydrothermal systems for this comparison were still scarce; thus, only microbial populations in chimney habitats adjacent to high-temperature hydrothermal fluids were characterized by quantitative cultivation techniques and included. In the present study, we conducted a more comprehensive evaluation of the relationships among variations in geology, geochemistry, and microbial metabolisms and the diversity of communities in global deep-sea hydrothermal environments, based on compilation of a substantial hydrothermal fluid chemistry data set and microbial communities in the mixing zones of a wide variety of habitats.

Methods

We compiled end-member fluid chemistry data for 89 hydrothermal vent sites (Additional file 1). Hydrothermal vent sites were included only if the data set contained complete chemical composition data essential for the thermodynamic calculations performed in this study, including H_2, H_2S, CH_4, CO_2, Na, Cl, Ca, K, Fe, Mn, and Si. In addition, the data set included representative geological settings such as MOR hydrothermal systems in the Pacific, Atlantic, and Indian Oceans; ABA hydrothermal systems in the western Pacific region; and the sediment-associated (SED) hydrothermal systems in the eastern Pacific and Okinawa Trough (Figure 1).

The amount of metabolic energy available for production by chemolithotrophic microorganisms was evaluated as in Takai and Nakamura (2010, 2011). Four aerobic and anaerobic reactions were considered representative of chemolithotrophic energy metabolisms (Table 1). To simulate mixing of hydrothermal fluids with seawater in a seafloor hydrothermal system, we employed a thermodynamic reaction path model, following McCollom and Shock (1997), Shock and Holland (2004), and McCollom (2007). The compositions of the hydrothermal solutions in the mixing zones were calculated from those of the end-member vent fluids and seawater (Additional file 1). The model calculation began with 1 kg of vent fluid and continued with addition of successive increments of seawater until a seawater to vent fluid mixing ratio of 1,000:1 was reached. In the mixing calculations, minerals were not allowed to precipitate and all redox reactions were prohibited, while acid-base reactions were allowed to equilibrate. In addition, the temperatures of the calculated mixed solutions were assumed to scale linearly with the temperatures of the end-members, ignoring conductive cooling and/or heating.

The overall Gibbs free energy for the metabolic reactions was calculated using the following equation:

$$\Delta G_r = \Delta G_r^\circ + RT \ln Q_r \quad (1)$$

where ΔG_r is the Gibbs free energy of the reaction, ΔG_r° is the standard-state Gibbs free energy of the reaction, R is the universal gas constant, T is the temperature in Kelvin, and Q_r is the activity quotient of the compounds involved in the reaction. The Q_r term takes into account the contribution of the fluid composition to the Gibbs free energy of each reaction, determined based on the

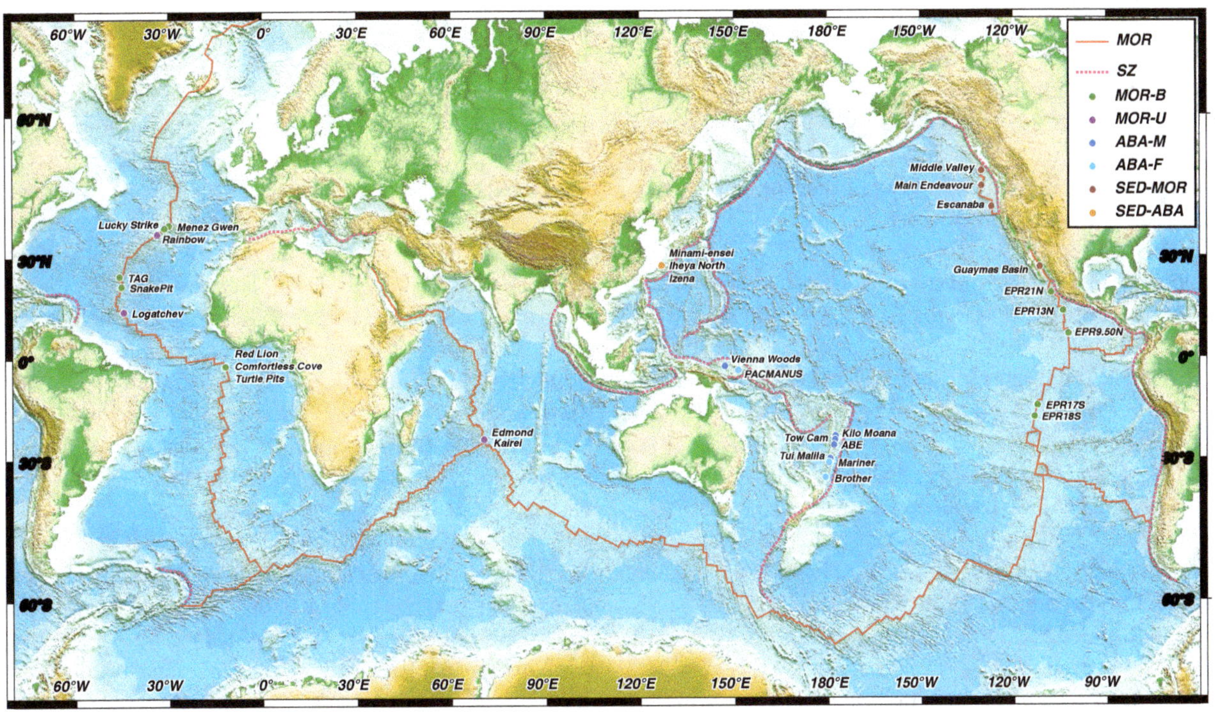

Figure 1 Index map showing mid-ocean ridges and subduction zones with active hydrothermal vents used in this study. Abbreviations: MOR, mid-ocean ridge; SZ, subduction zone; MOR-B, basalt-hosted system in a mid-ocean ridge setting; MOR-U, ultramafic rock-hosted system in a mid-ocean ridge setting; ABA-M, mafic rock-hosted system in an arc-backarc setting; ABA-F, felsic rock-hosted system in an arc-backarc setting; SED-MOR, sediment-associated system in a mid-ocean ridge setting; SED-ABA, sediment-associated system in an arc-backarc setting.

Table 1 Metabolic reactions for chemolithoautotrophy considered in this study

Energy metabolism	Overall chemical reaction	Identified (I)/cultured (C)	$\Delta G_r°_{2,\ 250}$ (kJ)[a]
Aerobic reactions			
Aerobic methanotrophy	$CH_4 + 2O_2 = CO_2 + 2H_2O$	I and C	−860.7
Hydrogenotrophic O_2 reduction	$H_2 + 1/2O_2 = H_2O$	I and C	−264.4
Thiotrophic (H_2S-oxidizing) O_2 reduction	$H_2S + 2O_2 = SO_4^{2-} + 2H^+$	I and C	−758.2
Fe(II)-oxidizing O_2 reduction	$Fe^{2+} + 1/4O_2 + H^+ = Fe^{3+} + 1/2H_2O$	I and C	−52.6
Anaerobic reactions			
Hydrogenotrophic methanogenesis	$H_2 + 1/4CO_2 = 1/4CH_4 + 1/2H_2O$	I and C	−49.2
Hydrogenotrophic SO_4 reduction	$H_2 + 1/4SO_4^{2-} + 1/2H^+ = 1/4H_2S + H_2O$	I and C	−74.9
Hydrogenotrophic Fe(III) reduction	$H_2 + 2Fe^{3+} = 2Fe^{2+} + 2H^+$	I and C	−159.2
Anoxic methanotrophy with SO_4 reduction	$CH_4 + SO_4^{2-} = HCO_3^- + HS^- + H_2O$	I but not yet C	−30.1

[a]Standard-state Gibbs free energy of the metabolic reactions at 2°C, 250 bar.

chemical composition of the mixed fluid estimated from the reaction path calculations. The energy available from the metabolic reactions as a function of temperature (equivalent to the mixing ratio) was calculated by multiplying the calculated Gibbs free energy for the reaction at each temperature by the concentrations of the reactants in the mixed fluid. This method takes into account the stoichiometry of the reaction and the reactants that are limiting, multiplied by the total amount of mixed fluid at that temperature (McCollom and Shock 1997; McCollom 2007).

This calculation yields an estimate of the maximum energy that is potentially available from the metabolic reactions per kilogram of mixed fluid. We used the average ΔG_r values for four temperature ranges: <25°C, 25°C to 45°C, 45°C to 80°C, and 80°C to 125°C, representing the chemolithotrophic metabolisms of psychrophilic, mesophilic, thermophilic, and hyperthermophilic microbial organisms, respectively.

Geological settings and geochemical characteristics of seafloor hydrothermal systems

As noted above, we compiled data for 89 deep-sea hydrothermal vent sites located in various geological settings (Figure 1), including MOR, ABA, and SED hydrothermal systems. These can be further subdivided into the following six categories: basalt-hosted systems in MOR settings (MOR-B), ultramafic rock-associated systems in MOR settings (MOR-U), mafic rock-hosted systems in ABA settings (ABA-M), felsic rock-hosted systems in ABA settings (ABA-F), SED systems in MOR settings (SED-MOR), and SED systems in ABA settings (SED-ABA).

The molecular and elemental compositions of H_2, H_2S, CH_4, CO_2, Na, Ca, K, Fe, Mn, and Si as well as the pH values of the end-member hydrothermal fluids from these sites are plotted against the Cl content in Figure 2. Nearly all substances exhibited some degree of positive or negative correlation to the Cl content, reflecting the influence of phase separation of hydrothermal fluids. Here, we focus on hydrothermal fluids with similar Cl concentrations to seawater in providing a brief overview of the geological settings and corresponding principal characteristics of their hydrothermal environments.

MOR hydrothermal systems

The global MOR system is the largest mountain chain in the world, with a total length of more than 60,000 km. Magma production in the MOR system far exceeds that in any other tectonic environment, accounting for more than 60% by volume of the total annual flux of magma from the mantle to the crust (Fisher and Schmincke 1984; Schmincke 2004). Therefore, the MOR system is the locus of the greatest amount of hydrothermal activity on Earth. Most MOR hydrothermal vents are hosted by basalt (often termed mid-ocean-ridge basalt). Therefore, many studies of deep-sea hydrothermal systems have focused mainly on basalt-hosted hydrothermal systems (Karl 1995; Kelley et al. 2002). However, in recent years, increasing attention has been paid to hydrothermal systems associated with ultramafic rocks present in the MOR (Kelley et al. 2001, 2005; Früh-Green et al. 2004), because the geochemical characteristics of ultramafic rock-hosted hydrothermal fluids are significantly different from those of basalt-hosted ones (Charlou et al. 2002).

Hydrothermal fluids from basalt-hosted systems are characterized by enrichment in base metals, a generally neutral or weakly acidic in situ pH, and strongly reducing conditions compared to seawater, due to subseafloor hydrothermal reactions between basaltic rocks and seawater at high temperature and pressure (Seyfried et al. 1991; Wetzel and Shock 2000). However, the fluids from ultramafic rock-associated systems are often characterized by an abundance of dissolved H_2 and CH_4 compared to basalt-hosted hydrothermal fluids (Figure 2)

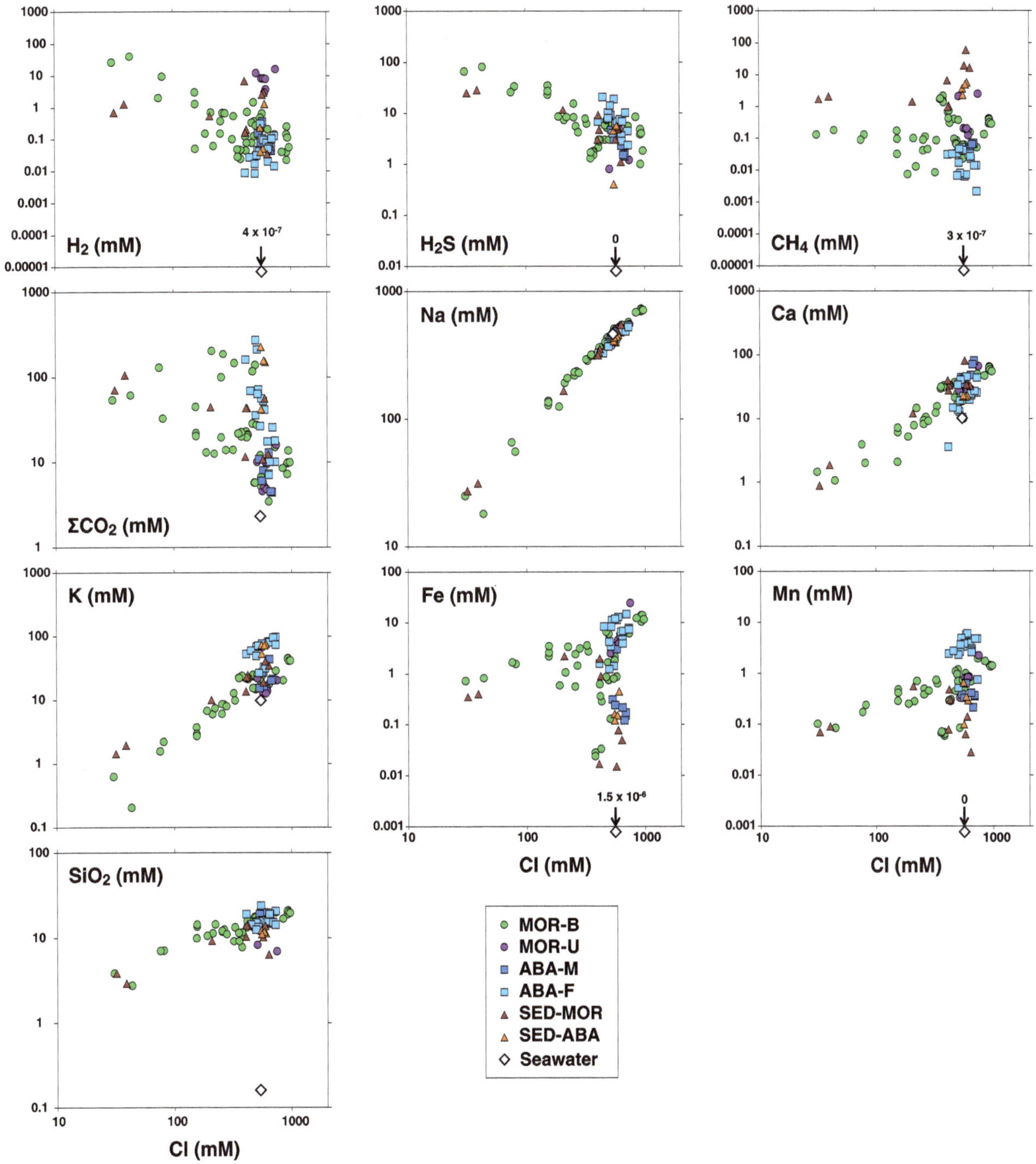

Figure 2 Major elements versus Cl concentration plots for hydrothermal vent fluids. The composition of seawater is also plotted for comparison.

(Charlou et al. 1998, 2002). This is attributed to differences in the mineralogy and bulk chemistry of the rocks, resulting in a substantially different alteration reaction known as serpentinization (Janecky and Seyfried 1986; Wetzel and Shock 2000; Allen and Seyfried 2003). H_2 concentrations of 12 to 16 mmol/kg (roughly 1 to 2 orders of magnitude higher than in basalt-hosted hydrothermal fluids) due to serpentinization have been reported in ultramafic rock-hosted deep-sea hydrothermal vent fluids (Additional file 1) (Charlou et al. 2002; Kelley et al. 2005). In addition, enrichment of CH_4 in ultramafic rock-associated systems is likely associated with reduction of CO_2 to CH_4 under highly reducing (high H_2) conditions.

ABA hydrothermal systems

Although significant global magma production occurs in the MOR, the second most active volcanic area is the subduction zone (Fisher and Schmincke 1984; Schmincke 2004), where ABA volcanic systems develop. Some arc volcanoes are subaerial, particularly in continental arc settings, whereas more than 40% of volcanic arcs are oceanic island arcs where most volcanoes are submarine (Leat and Larter 2003). In oceanic island arcs and associated back-arc basins, many deep-sea hydrothermal vent sites hosted by active submarine ABA volcanisms have been discovered (Ishibashi and Urabe 1995; Gamo et al. 2006). The chemical characteristics of these ABA hydrothermal systems are significantly different from those of MOR hydrothermal systems, although they share many common features derived from the basic reactions and processes of subseafloor hydrothermal circulation. In particular, ABA hydrothermal fluids are characterized by large variations in chemical composition (Ishibashi and Urabe 1995; Gamo et al. 2006). This chemical variability is primarily attributed to a variety of host-rock compositions (from basaltic to rhyolitic), abundant input of volatile elements from magmas (e.g., CO_2 and SO_2), and the variable water depth (shallow to deep) of the submarine volcanoes.

ABA hydrothermal systems hosted by mafic rocks exhibit chemical characteristics similar to those of basalt-hosted hydrothermal systems at the MOR (Figure 2). In contrast, systems hosted by felsic rocks are generally characterized by lower pH, higher metal concentrations, and a variable redox state compared to mafic rock-hosted hydrothermal systems (Figure 2). In addition, because of the subduction of hydrated ocean crust beneath volcanic arcs and back-arc basins as well as the potential involvement of more evolved felsic magmas, the volatile content (H_2O, CO_2, and SO_2) of magmas in ABA settings is much higher than that in MOR settings (e.g., Gamo et al. 1997; Embley et al. 2007). ABA hydrothermal systems are also characterized by wide variations in water depth, and most are significantly shallower than typical MOR hydrothermal systems (Figure 3). This results in variations in phase separation (boiling) and hydrothermal fluid temperatures in the subseafloor reaction zones, which significantly affect the chemical composition of the hydrothermal fluids. The maximum temperatures of the hydrothermal vent fluids in the compiled data were clearly controlled by the two-phase boundary of seawater (Figure 3).

SED hydrothermal systems

Some hydrothermal vent sites are covered with sediments, which influence the chemical composition of their hydrothermal fluids. SED hydrothermal systems are found in both MOR (e.g., Juan de Fuca, Middle Valley, Escanaba Trough, and Guaymas Basin) and ABA (e.g., Okinawa Trough) settings in which hydrothermal vent sites are in close proximity to continental runoff sources. Irrespective of differences in tectonic setting or host-rock composition, SED hydrothermal systems have characteristic chemical compositions, e.g., relatively high pH, low metal content, and high CH_4 and NH_3 concentrations (German and Von Damm 2004). There are only minor differences in chemical composition between SED hydrothermal systems in MOR and ABA settings (Figure 2). Thus, the chemical features of SED hydrothermal vent fluids are more strongly affected by the sediment than by the tectonic setting or host-rock composition.

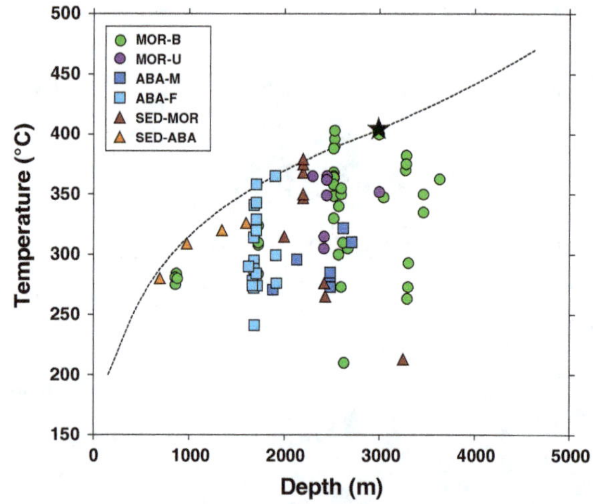

Figure 3 Maximum temperature conditions versus water depth for individual vent areas. Two-phase curve for seawater (dashed line) is also plotted with the liquid field to the lower right of the curve and the liquid + vapor field to the upper left of the curve. The star represents the critical point of seawater at 407°C and 298 bar. The maximum temperature of nearly all of the vent fluids is limited by the two-phase boundary.

Factors controlling hydrothermal fluid chemistry

It is believed that hydrothermal fluid flux is primarily controlled by the crust (magma) production rate, because magmatic heat derived from creation of new oceanic crust (cooling from 1,200°C to 350°C) heats hydrothermal fluids from 2°C to 350°C (e.g., Elderfield and Schultz 1996). Based on this, Kawahata et al. (2001) estimated that the maximum volume of high-temperature hydrothermal fluid is twice that of the upper oceanic crust (the volcanic and sheeted dike sequence) and if 65% to 85% of the rocks are altered to secondary minerals, the volumetric water/rock ratio would be 2.3 to 3.1. This clearly suggests that high-temperature hydrothermal fluids are generated under rock-dominated (low water/rock ratio) conditions (Kawahata et al. 2001). Therefore, the chemical compositions of hydrothermal fluids are predominantly controlled by water-rock reactions and fluid-mineral equilibria (Seyfried et al. 1991; Shock 1992; Seyfried and Ding 1995). Thus, host-rock geochemistry is one of the most important factors controlling hydrothermal fluid chemistry. In addition to the chemical compositions of host rocks, however, other processes such as input of volatiles from magmas, phase separation of hydrothermal fluids, and interactions with sediment also have a significant impact on the observed compositions of hydrothermal fluids.

Chemical composition of host rocks

Differences in the rock types in the reaction zone can lead to large variations in the hydrothermal fluid chemistry, because the bulk chemical composition and primary minerals in the source rock control the reaction sequences and their chemical equilibria.

Mafic rocks Most ocean basins (including MOR and ABA settings) are of mafic composition (mainly basaltic). The chemical systematics of mafic rock-seawater systems (particularly basalt-seawater systems) have been well characterized based on field observations, hydrothermal experiments, and theoretical studies (Bischoff and Dickson 1975; Mottl and Holland 1978; Seyfried and Mottl 1982; Mottl 1983; Reed 1983; Bowers and Taylor 1985; Seyfried 1987; Von Damm 1990, 1995; Seyfried et al. 1988, 1991; Shock 1992; Saccocia et al. 1994; Wetzel and Shock 2000; Butterfield et al. 2003; Nakamura et al. 2007).

Basalt-hosted hydrothermal fluids have relatively low pH at 25°C and 1 atm (generally 3 to 4) and notable enrichment in base metals (e.g., Fe and Mn) and dissolved gases (e.g., H_2, CO_2, and CH_4) compared to seawater (Figure 2). The in situ pH of the basalt-hosted vent fluids are neutral to weakly acidic, significantly different from the pH measured at 25°C and 1 atm in laboratory experiments (German and Von Damm 2004). This discrepancy is mainly caused by dissociation of H^+-bearing aqueous complexes (e.g., HCl^0) due to cooling during sample processing (Shock et al. 1989; Seyfried et al. 1991; Ding and Seyfried 1992; Ding et al. 2005). Another possible mechanism of pH change is precipitation of metal sulfides below the seafloor, producing protons (German and Von Damm 2004).

In high-temperature MOR vent fluids, most metals are enriched by up to 7 to 8 orders of magnitude compared to seawater. The stability of chloride complexes (e.g., $FeCl_2^0$) increases with increasing temperature, and therefore, most metal ions are present in high-temperature fluids as chloride complexes (Helgeson et al. 1981; Ding and Seyfied 1992). The substantial increase in the metal solubility produces the high concentrations in the vent fluids.

Basalt-hosted hydrothermal fluids are also highly reducing, as evidenced by the presence of H_2S rather than SO_4, as well as by significant amounts of H_2, CH_4, Fe^{2+}, and Mn^{2+}. These chemical features are generally consistent with phase-equilibrium calculations involving the observed primary and secondary minerals (feldspar, chlorite, epidote, quartz, magnetite, anhydrite, pyrite, and pyrrhotite) and seawater at in situ hydrothermal fluid temperatures and pressures (Bowers et al. 1988; Seyfried et al. 1988, 1999; Saccocia and Seyfried 1990; Seyfried and Ding 1995; McCollum and Shock 1998; Wetzel and Shock 2000). Time-series measurements of chemical compositions of hydrothermal fluids in MOR regions have indicated steady-state concentrations of dissolved species (Campbell et al. 1988; Bowers et al. 1988; Butterfield et al. 1994; Von Damm 1988, 1995, 2000), reflecting clear solubility control by mineral phases in the subseafloor reaction zones and recharge zones.

Ultramafic rocks Compared with hydrothermal fluids in mafic rock-hosted systems, ultramafic rock-associated hydrothermal fluids are characterized by conspicuous enrichment in H_2 in the presence or absence of phase separation processes (Figure 2). Production of H_2 during serpentinization of ultramafic rocks results from reaction of water with Fe^{2+}-bearing minerals, primarily olivine and pyroxene. In these reactions, Fe^{2+} is partially oxidized to Fe^{3+} by the water, resulting in precipitation of magnetite in the serpentinite. The water reduced by the Fe^{2+} also produces H_2. One of the important chemical features of ultramafic rocks that constrains abundant H_2 production during serpentinization is relatively low Al concentrations. Low Al activity results in formation of alteration minerals (particularly serpentines and brucite) that have a tendency to exclude Fe^{2+} from their structure. This leads to oxidation of Fe^{2+} by water to form magnetite. In contrast, because mafic rocks have higher Al contents than ultramafic rocks, a much greater proportion of the Fe^{2+} is sequestered in Al-bearing alteration minerals (e.g., chlorite), limiting oxidation of Fe^{2+} to Fe^{3+}. Thus, hydrothermal alteration of mafic rocks generates much lower amounts of H_2 and magnetite than serpentinization of ultramafic rocks, even though the Fe content of basaltic rocks is higher than that of ultramafic rocks. Basaltic rock-hosted vent fluids typically exhibit H_2 concentrations of <1 mmol/kg, whereas >10 mmol/kg H_2 has frequently been reported for ultramafic rock-associated fluids (Additional file 1). The potential for even higher H_2 concentrations (>100 mmol/kg) has been suggested by both petrological (Frost 1985; Alt and Shanks 1998) and experimental (Berndt et al. 1996; McCollom and Seewald 2001) studies.

In addition, ultramafic rock-associated hydrothermal fluids are often characterized by CH_4 enrichment (Figure 2). Although the origin of CH_4 in these fluids is controversial, it is generally thought to be derived from reduction of CO_2 by high concentrations of H_2 through abiotic methanogenesis (Nakamura et al. 2009). However, the hydrothermal fluid CH_4 concentrations are always disequilibrated with the concomitant H_2 and CO_2 concentrations, suggesting that the subseafloor hydrothermal circulation system is an open system with respect to CH_4 content. Enriched CH_4 (>0.5 mmol/kg) has also been observed in hydrothermal fluids from certain basalt-hosted systems at the MOR (Figure 2; Lucky Strike and Menez Gwen hydrothermal fields) (Additional file 1). Both of these fields are located in the northern Mid-Atlantic Ridge, where several ultramafic rock-associated hydrothermal systems have been discovered. Based on geological and geochemical lines of evidence, it has been suggested that the high CH_4 concentrations in these basalt-hosted hydrothermal fluids were caused by serpentinization of ultramafic rocks somewhere in the subseafloor (Charlou et al. 2000).

Felsic rocks Hydrothermal systems hosted by felsic rocks have been identified only in ABA settings (Figure 1). Felsic rock-hosted hydrothermal fluids are characterized by relatively low pH and enrichment in H_2S, CO_2, K, Mn, and Fe compared to mafic rock-hosted fluids (Figure 2). The chemical characteristics of felsic rock-hosted hydrothermal fluids are generally consistent with experimental results for seawater-felsic rock interactions (e.g., Hajash and Chandler 1981). For example, felsic rocks contain high concentrations of incompatible elements such as K. Enrichment of K in felsic rock-hosted fluids originates from the bulk rock composition. In addition, the lower pH is attributed to the low ability of felsic rocks to consume H^+ in solution. The pH of a hydrothermal solution is lowered by removal of Mg^{2+} via precipitation of Mg-hydroxysulfate during heating of the seawater (Bischoff and Seyfried 1978) and by formation of Mg-bearing alteration minerals (e.g., smectite and chlorite) during seawater-rock interactions (Mottl 1983). In a basalt–seawater system, the H^+ generated in the fluid is consumed by dissolution of Ca from the reacted rocks (Seyfried and Mottl 1982; Mottl 1983). Mg-Ca exchange reactions control the fluid pH to approximately neutral under in situ conditions (Seyfried and Mottl 1982; Wetzel and Shock 2000). However, because felsic rocks are relatively depleted in Ca, their ability to buffer pH changes is significantly lower than that of their mafic counterparts.

Low pH in hydrothermal fluids promotes leaching of heavy metals through water-rock interactions. This ultimately results in high concentrations of heavy metals, although the initial concentrations of heavy metals in felsic rocks are lower than those in basaltic rocks. The Fe and Mn concentrations in deep-sea hydrothermal fluids are correlated not only with Cl concentrations but also with pH (Figure 4). Moreover, the Fe/Cl and Mn/Cl ratios for hydrothermal fluids exhibit strong correlations with pH (Figure 4). This clearly shows that the concentrations of heavy metals in deep-sea hydrothermal fluids are mainly controlled by the pH of the hydrothermal fluids.

Inputs of magmatic volatiles

Enrichment of H_2S and CO_2 in felsic rock-hosted hydrothermal fluids cannot be explained only by water-rock interactions. Instead, enrichment in these volatiles can be caused by inputs of magmatic volatiles into hydrothermal systems. ABA magmas (particularly those that are felsic in composition) have very high concentrations of volatile elements and molecules, and hydrothermal fluids that are highly enriched in sulfur and/or CO_2 have been observed in ABA hydrothermal systems, mainly hosted by more siliceous rocks, e.g., andesite, calcite, and rhyolite rather than by basalt (Gamo et al. 1997; Sakai et al. 1990; Inagaki et al. 2006; Lupton et al. 2006, 2008). Therefore, enrichment in CO_2 and H_2S of felsic rock-hosted hydrothermal fluids is likely due to significant inputs of magmatic volatiles into these hydrothermal systems.

Dissolution of CO_2 gas into a hydrothermal fluid results in production of H^+ in the fluid via the following reaction:

$$CO_2 + H_2O = HCO_3{}^- + H^+. \qquad (2)$$

Compared to that for SO_2 (see below), the dissociation constant of this reaction is much smaller, particularly under high-temperature conditions. It is therefore believed that the effect of CO_2 on the fluid pH is not very significant. However, segregation of CO_2 from upwelling hydrothermal fluids in the subseafloor can result in consumption of H^+ in the fluid, increasing the pH. Even a small pH increase during ascent of the hydrothermal fluid can cause subseafloor precipitation of metal-sulfide minerals, resulting in low heavy metal concentrations in the hydrothermal fluids. This process would be expected in hydrothermal systems with significant inputs of CO_2 but not SO_2 from magma, such as hot spot-influenced MOR hydrothermal systems and basalt-hosted arc-backarc hydrothermal systems.

Volatile sulfur species in the magma have a more significant effect on deep-sea hydrothermal systems. The predominant gaseous sulfur species in magma are SO_2 and H_2S (Wallace and Edmonds, 2011). Although dissolution of H_2S does not significantly affect hydrothermal fluid chemistry, that of SO_2 into hydrothermal fluids

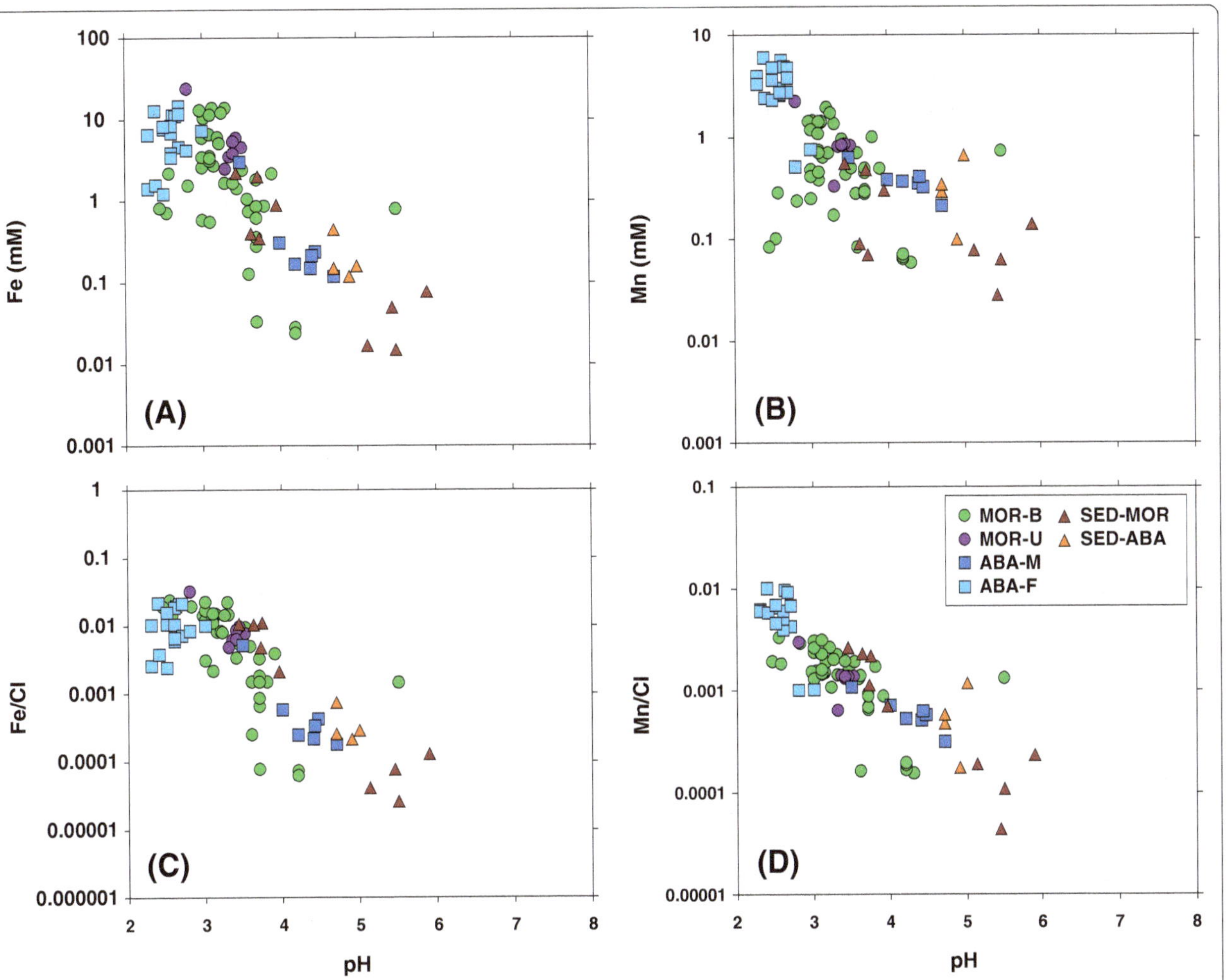

Figure 4 Plots of (A) Fe, (B) Mn, (C) Fe/Cl, and (D) Mn/Cl versus pH for hydrothermal vent fluids. These plots clearly show that concentrations of Fe and Mn in hydrothermal vent fluids are mainly controlled by pH.

increases the f_{O_2} and sulfuric acid in the fluids via the following reactions:

$$2SO_2 + 2H_2O = 2H_2S^0 + 3O_2 \tag{3}$$

$$2SO_2 + O_2 + 2H_2O = 2HSO_4^- + 2H^+. \tag{4}$$

Once the f_{O_2} of the hydrothermal fluid reaches the sulfate-sulfide boundary via reaction (3), sulfuric acid is produced by reaction (4). The presence of sulfuric acid can cause a significant decrease in pH. Because the solubility of heavy metals is quite sensitive to pH, the pH decrease caused by SO_2 promotes dissolution of heavy metals from the reacted rocks. Volatile inputs from magmas occur intermittently, and even during such activity, the chemistry of hydrothermal fluids is controlled by reactions induced by both volatile inputs and water-rock interactions. Therefore, it is likely that the chemical composition of the fluids, as is typical of open and dynamic systems, significantly varies with space and time. Nevertheless, felsic rock-hosted hydrothermal systems affected by magmatic volatile inputs clearly exhibit high concentrations of base metals and low pH (Figure 4).

Phase separation

As described above, the concentrations of all components in hydrothermal fluids exhibited positive or negative correlations with Cl, except for H_2, CO_2, and CH_4 in several samples affected by serpentinization of ultramafic rocks, CO_2 inputs from magma, and CH_4 inputs from sedimentary organic matter and/or microbial processes (Figure 2). Deep-sea hydrothermal fluids often reach temperatures high enough that they separate into vapor and brine phases. The phase separation temperature depends primarily on the pressure conditions (i.e., water depth) of the hydrothermal system. The observed temperatures of hydrothermal fluids from various vent sites were clearly limited by the two-phase boundary of seawater (Bischoff and Pitzer 1989) (Figure 3). Phase separation and subsequent remixing of the vapor and brine phases produce hydrothermal fluids with a wide range of salinities, from 1 order of magnitude lower to several times higher than that of seawater (Figure 2).

The chemical properties of the vapor and brine phases are quite different from each other and from their parent fluid, because each of the chemical species in the parent fluid is distributed preferentially into the vapor and liquid phases according to their physical and chemical properties during phase separation (Butterfield et al. 2003; Foustoukos and Seyfried 2007). The concentrations of gaseous species greatly increased with decreasing Cl concentration, indicating the strong affinity of these volatile components for the vapor phase (Figure 2). In contrast, the positive correlations between the Cl concentration and other dissolved species that are primarily ionic (e.g., Na^+ and Cl^-) indicate the strong affinity of these species for the liquid phase. In addition, the relationships between the Cl concentration and the ratios of elements to chloride clearly showed a strong increase in the ratios of gaseous species to chloride with decreasing Cl concentration (Figure 5). The Si/Cl ratio also slightly increased with decreasing Cl concentration. This indicates that, in addition to gaseous species, neutral species such as SiO_2^0 have some affinity for the vapor phase. On the other hand, there was little change in the ratios of ionic species with Cl concentration as a result of phase separation, indicating that they all partition strictly into the brine phase. These chemical behaviors of the dissolved species during the phase separation and partitioning into hydrothermal fluids result in formation of low-Cl, vapor-dominated fluids enriched in gases and of residual brines enriched in ionic species and depleted in gases.

Presence of sediments

SED hydrothermal systems have chemical compositions distinct from those of other hydrothermal fluids. Compared to those of sediment-starved MOR and ABA hydrothermal systems, SED hydrothermal fluids generally have relatively high pH, lower heavy metal contents, and higher CH_4 and NH_4^+ concentrations (Figure 2) (German and Von Damm, 2004). The very high CH_4 concentrations of hydrothermal fluids in SED systems are attributed to thermal decomposition of organic matter at high temperatures during hydrothermal reactions at discharge zones and/or microbial methanogenesis at relatively low temperatures at sedimentary recharge zones (Lilley et al. 1993; Kawagucci et al. 2011, 2013). Likewise, the source of NH_4^+ in SED hydrothermal systems is considered to be thermal decomposition and microbial ammonification of organic matter (Kawagucci et al. 2011, 2013). High concentrations of NH_4^+ in these hydrothermal fluids provide an NH_3/NH_4^+ buffer that maintains the relatively high pH of the fluid (German and Von Damm, 2004). This greatly decreases the solubility of metal-sulfide minerals, leading to low heavy metal concentrations in the hydrothermal-vent fluids.

Effects of hydrothermal fluid chemistry on the bioavailable energy yield

The chemosynthetic primary producers that sustain deep-sea hydrothermal vent ecosystems utilize inorganic substances (e.g., H_2S, CO_2, H_2, and CH_4) derived from hydrothermal vent fluids as energy and carbon sources. Thus, deep-sea hydrothermal vent ecosystems should be at least partially controlled by the chemical composition of the hydrothermal fluids. The effects of hydrothermal fluid compositions on deep-sea hydrothermal vent ecosystems based on the energy yields available to various chemolithotrophic metabolisms are described below.

Hydrogen sulfide (H_2S)

In all of the deep-sea hydrothermal systems in all settings, the potential energy yields for sulfur-oxidizing chemolithotrophy (thiotrophy) using H_2S in the hydrothermal fluids were uniformly high at >10 J/kg mixed fluid (Figure 6A). Even in ultramafic rock-associated hydrothermal systems with relatively low H_2S concentrations, the metabolic energy of sulfur oxidation (H_2S oxidation) was nearly identical to that in other types of hydrothermal systems, some of which had 2 orders of magnitude higher H_2S concentrations (Figure 6A). This uniformity is attributed to the relatively high concentrations of H_2S (mostly >1 mmol/kg) present in hydrothermal fluids in all of the hydrothermal systems. As a result, the amount of H_2S always exceeded the O_2 concentration throughout the habitable temperature range in the mixing zones, except at very low temperatures (several degrees Celsius). The potential energy yield of sulfur-oxidizing chemolithotrophy is therefore solely controlled by the dissolved O_2 concentration in the seawater (which is globally similar in all deep-sea water). Thus, concentrations of H_2S in end-member hydrothermal fluids may not significantly affect the abundance and composition of sulfur oxidizers.

The exception is in hydrothermal plumes, where much higher seawater mixing ratios (>1000) and low temperatures (up to several degrees Celsius) are found. In hydrothermal plume environments, the concentration of H_2S in the source hydrothermal fluid, rather than the seawater-dissolved O_2, becomes the limiting factor for sulfur-oxidizing chemolithotrophic metabolism.

Hydrogen (H_2)

In contrast to H_2S, variations in the H_2 concentration directly affected the potential energy for the chemolithotrophic microbial population, not only for aerobic H_2 oxidation but also for most anaerobic energy metabolisms

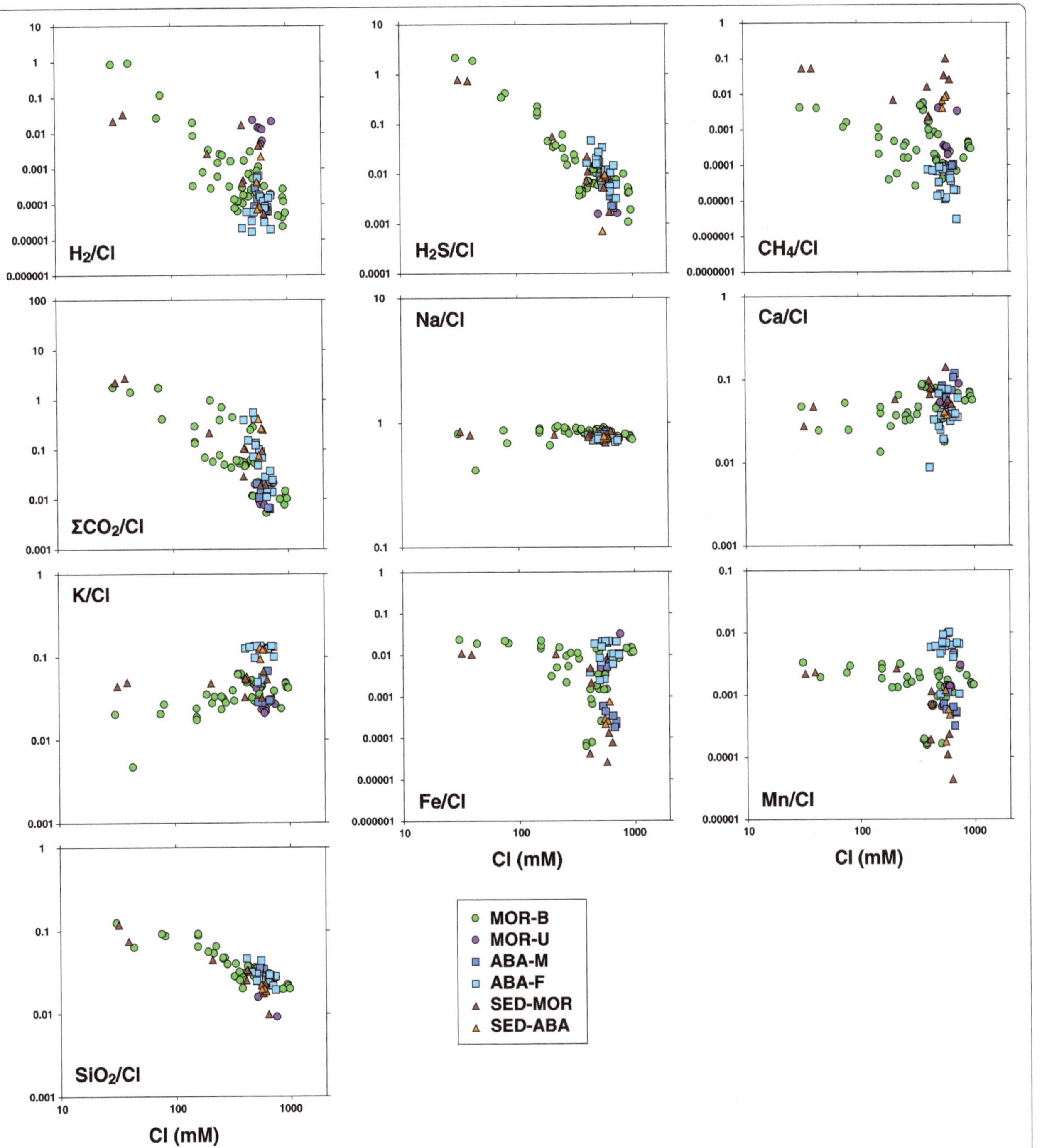

Figure 5 Plots of major element to Cl ratios versus Cl for hydrothermal vent fluids.

other than anaerobic methane oxidation. This feature is partially attributed to the wide range of H_2 concentrations in hydrothermal fluids. For example, typical basalt-hosted hydrothermal fluids had H_2 concentrations of <1 mmol/kg (Figure 2). In contrast, ultramafic rock-hosted hydrothermal fluids had H_2 concentrations 1 to 2 orders of magnitude higher, whereas hydrothermal fluids in ABA settings had H_2 concentrations 1 order of magnitude lower (Figure 2). Thus, potential energy yields for aerobic and anaerobic H_2-oxidizing (H_2-trophic) metabolisms were directly correlated with H_2 concentrations in the end-member hydrothermal fluids (Figure 6B, C, D). For aerobic H_2-oxidizing metabolism, nearly the same amount of metabolic energy was available irrespective of the H_2 concentration in the end-member hydrothermal fluid (Figure 6B), when the H_2 concentration exceeded

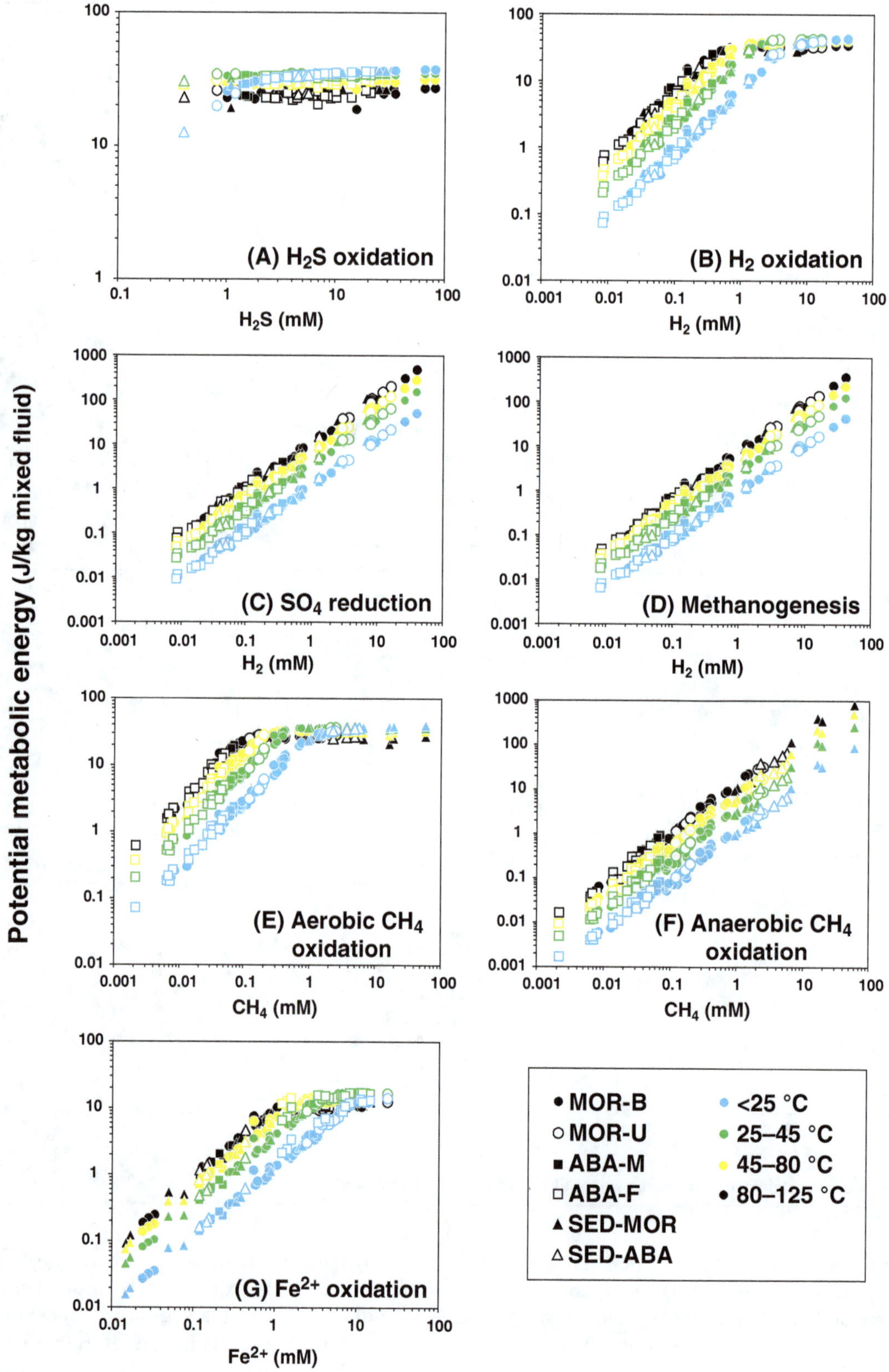

Figure 6 (See legend on next page.)

(See figure on previous page.)
Figure 6 Metabolic energy yields versus H2S, H2, CH4, and Fe2+ concentrations in hydrothermal fluids. Potential metabolic energy available from **(A)** thiotrophic (H_2S-oxidizing) O_2 reduction as a function of the H_2S concentration in the end-member hydrothermal fluid; **(B)** hydrogenotrophic O_2 reduction, **(C)** hydrogenotrophic SO_4 reduction, and **(D)** hydrogenotrophic methanogenesis as a function of the H_2 concentration in the end-member hydrothermal fluid; **(E)** aerobic methanotrophy and **(F)** anoxic methanotrophy as a function of the CH_4 concentration in the end-member hydrothermal fluid; and **(G)** Fe^{2+}-oxidizing O_2 reduction as a function of the Fe^{2+} concentration in the end-member hydrothermal fluid are plotted for each kilogram of mixing water for four temperature ranges. The temperature ranges of <25°C, 25°C to 45°C, 45°C to 80°C, and 80°C to 125°C generally correspond to psychrophilic, mesophilic, thermophilic, and hyperthermophilic microbial components, respectively.

approximately 1 mmol/kg (for hyperthermophiles at relatively high temperatures) to approximately 10 mmol/kg (for psychrophiles to hyperthermophiles at all temperature ranges). As with H_2S oxidation, this occurs because seawater-dissolved O_2 rather than H_2 becomes the limiting factor for aerobic H_2-oxidizing metabolism.

However, because most seafloor hydrothermal fluids had H_2 concentrations <1 mmol/kg, except for ultramafic rock-hosted and highly vapor phase-enriched hydrothermal fluids (Figure 2), variations in the H_2 concentration of the end-member hydrothermal fluid significantly affected the potential energy yields of nearly all H_2-trophic chemolithotrophs (Figure 6B, C, D). This variation is expected to have a strong impact on biomass production and the compositional structure of most H_2-trophic microbial populations as well as the entire microbial community. As described above, the abundance of H_2S in hydrothermal fluids is relatively constant among most hydrothermal systems, tectonic settings, and geological conditions. In addition, H_2S is coupled only to aerobic (both O_2- and NO_3-reducing) sulfur-oxidizing metabolisms. In contrast, the abundance of H_2 in end-member hydrothermal fluids is highly variable and constrains various aerobic and anaerobic chemolithotrophic energy metabolisms for primary production.

Although ultramafic rock-associated hydrothermal fluids are significantly enriched in H_2 (Charlou et al. 2002), in several phase separation-influenced (vapor phase-abundant) hydrothermal fluids hosted by MOR-B, H_2 can also be comparably enriched (Lilley et al. 2003) (Figure 2). In these hydrothermal systems, therefore, the energy yield of H_2-trophic metabolisms was similar to that in ultramafic rock-hosted hydrothermal systems (Figure 6B, C, D). However, the abundant availability of H_2 (as well as other gaseous species) induced by phase separation in MOR-B-hosted hydrothermal fluids is generally transient and lasts several weeks to several months (Lilley et al. 2003). Thus, the effect of temporal enrichment of hydrothermal fluid H_2 on biomass production and compositional structure of the entire microbial community is likely transient and minor.

Methane (CH_4)

The CH_4 concentration in seafloor hydrothermal fluids was also variable, ranging from the nanomole per kilogram level in ABA hydrothermal fluids to the millimole per kilogram level in SED and ultramafic rock-hosted fluids. Variations in the CH_4 concentration in hydrothermal fluids could affect both aerobic and anaerobic methane oxidizers (Figure 6E, F).

The potential metabolic energy from anaerobic CH_4 oxidation (methanotrophy) was directly correlated with the CH_4 concentration in the hydrothermal fluid (Figure 6F), suggesting that the anaerobic methanotrophic population is likely affected by CH_4 concentrations in end-member hydrothermal fluids. This is attributable to the large amount of SO_4^{2-} in seawater; these high levels of SO_4^{2-} are always present in seawater-hydrothermal fluid mixing zones in all habitable temperature ranges. For aerobic CH_4 oxidation reactions, the potential energy yield was also affected by the CH_4 concentration in end-member hydrothermal fluids, except for hydrothermal fluids with high CH_4 concentrations (>0.2 mmol/kg for hyperthermophiles to >2 mmol/kg for psychrophiles) (Figure 6E).

Iron (Fe)

Variation in the Fe^{2+} concentration of the hydrothermal fluids was also large, from the nanomole per kilogram level to the millimole per kilogram level (comparable to the ranges for H_2 and CH_4 concentrations). This variation in the Fe^{2+} concentration of the hydrothermal fluids could affect aerobic iron-oxidizing chemolithotrophy. The potential energy yield from the aerobic Fe oxidation reaction was well correlated with the Fe^{2+} concentration in the hydrothermal fluid (Figure 6G). However, for highly Fe^{2+}-enriched hydrothermal fluids (>0.2 mmol/kg for hyperthermophiles to >2 mmol/kg for psychrophiles), the amount of available metabolic energy was saturated with respect to the Fe^{2+} concentration in the end-member hydrothermal fluid (Figure 6G). The metabolic energy yield potentially obtained from aerobic oxidation of 1 mol of Fe^{2+} was several times smaller than that from aerobic oxidation of 1 mol of H_2S, H_2, or CH_4 (Figure 6). This may affect the relative abundance of aerobic Fe-oxidizer populations in deep-sea hydrothermal vent ecosystems.

In contrast to aerobic Fe-oxidizing chemolithotrophy, there was an essentially negligible energy yield predicted for anaerobic Fe^{3+} reduction using H_2 or CH_4 for all types of deep-sea hydrothermal fluids and systems. This is attributed to an extremely low concentration of Fe^{3+} in both seawater and hydrothermal fluids.

Effect of the geological setting of the hydrothermal system on the bioavailable energy yield

In the MOR-B and all ABA settings (comprising most deep-sea hydrothermal systems), almost all potentially bioavailable energy can be obtained from aerobic metabolism (Figure 7). The end-member hydrothermal fluids in these settings contain only small amounts of H_2 and CH_4 (other than highly vapor-enriched hydrothermal fluids). The most energetically favorable chemolithotrophic metabolism was aerobic sulfur oxidation (Figure 8A). Thus, aerobic sulfur oxidizers are the chemolithoautotrophic population most likely to sustain primary production in these deep-sea hydrothermal vent ecosystems.

In the MOR-B and ABA-M settings, aerobic oxidation of H_2 and CH_4 were the second most available metabolic reactions, particularly at higher temperatures (lower mixing ratios). On the other hand, in the ABA-F setting, aerobic Fe^{2+} oxidation was the second most favorable chemolithotrophic metabolism (Figure 8A). These differences are directly related to the different chemical compositions of the end-member hydrothermal fluids (e.g., H_2, CH_4, and Fe^{2+} concentrations) in these settings, ultimately derived from the oxidation state of the magmas and/or volatile (particularly SO_2) inputs into the hydrothermal fluids. Relatively little energy was predicted to be available from anaerobic chemolithotrophic metabolisms for these settings, except for temporally and spatially limited habitats induced by phase separation of hydrothermal fluids (Figure 8B).

In the MOR-U and SED settings, the potential energy from aerobic oxidation of H_2 or CH_4 exceeded that from aerobic sulfur oxidation at higher temperatures (lower mixing ratios) (Figure 8A). More importantly, in these settings, considerable energy for primary production can be obtained from anaerobic chemolithotrophic metabolisms and populations (Figure 8B). Particularly in high-temperature habitats, anaerobic chemolithotrophs are expected to play a prominent role as primary producers (Figure 8A, B). This represents a marked difference in potential chemolithotrophic microbial communities between the MOR-U and SED settings and the more common MOR-B and ABA settings.

In the MOR-U setting, because of the high H_2 concentrations in end-member fluids, both aerobic and anaerobic H_2-trophic population reducers were energetically dominant primary producers (Figure 8A, B). In addition, the total potentially bioavailable energy yields in the mixing zones were greater in the MOR-U setting than in the typical MOR-B and ABA settings (Figure 8A, B). Thus, the MOR-U deep-sea hydrothermal environments may supply more abundant and diverse energy sources for biological production.

In the SED setting, aerobic methanotrophy could be competitive with aerobic sulfur oxidation in all temperature ranges (Figure 8A). Interestingly, among all possible chemolithotrophic metabolisms, anaerobic (sulfate-reducing) methanotrophy was by far the most favorable energy-generating metabolism, particularly in high-temperature habitats (Figure 8B). The total amount of potentially bioavailable energy yield in the mixing zones of habitats in the SED setting was also greater than in the typical MOR-B and ABA settings.

Comparison of existing chemolithotrophic microbial communities with the results of thermodynamic modeling

Above, we have provided the theoretical basis for relationships between the geological environments of hydrothermal activity (e.g., tectonic settings, basement-rock geochemistry, abundance of sediments, magmatic volatile input, and phase separation related to subseafloor hydrothermal processes), physical and chemical variations in hydrothermal fluids, and the compositional diversity of potentially bioavailable energy for various vent-endemic chemolithotrophic metabolisms estimated using thermodynamic models. We have shown that the abundance and composition of chemolithotrophic energy metabolisms in hydrothermal vent biological communities is directly constrained by the physical and chemical characteristics of the hydrothermal mixing zones of habitats, which are subject to the physical and chemical properties of end-member hydrothermal fluids. Furthermore, the physical and chemical characteristics of the end-member hydrothermal fluids are substantially controlled by the geological settings that host the hydrothermal systems. Thus, it seems likely that the abundance and composition of chemolithotrophic energy metabolisms in microbial communities located in a given deep-sea hydrothermal system could be systematized in terms of geological backgrounds based on the results of the thermodynamic models.

In typical mixing zones in deep-sea hydrothermal habitats, chemolithotrophic microbial communities consist of organisms with three typical lifestyles: surface-attached or biofilm-forming free-living entities, planktonic free-living entities, and symbiotic entities. Here, we discuss potential patterns in chemolithotrophic microbial community development delineated by recent microbiological investigations (biogeochemical, ecological, and molecular approaches) for representative hydrothermal mixing zones of habitats in which the predominant organisms have different lifestyles.

Hydrothermal plumes

Hydrothermal plumes are typical of mixing zone habitats that host planktonic free-living microbial communities

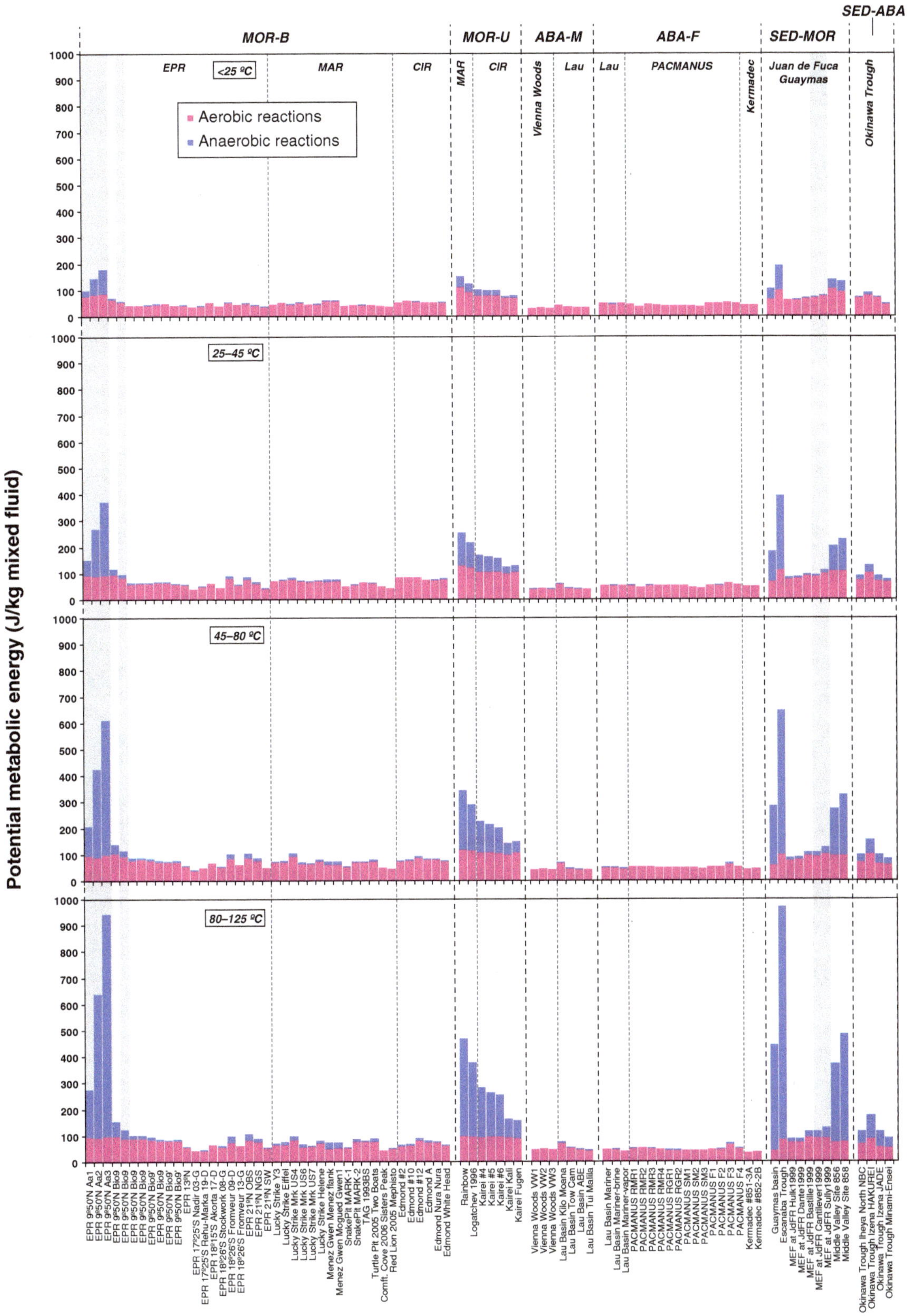

Figure 7 Metabolic energies available from the aerobic and anaerobic reactions shown in Table 1. Calculations were performed for all the 89 hydrothermal vent sites in MOR-B, MOR-U, AMA-M, ABA-F, SED-MOR, and SED-ABA settings. Shaded areas represent hydrothermal vents issuing highly vapor-rich (<100 mmol/kg Cl) fluids.

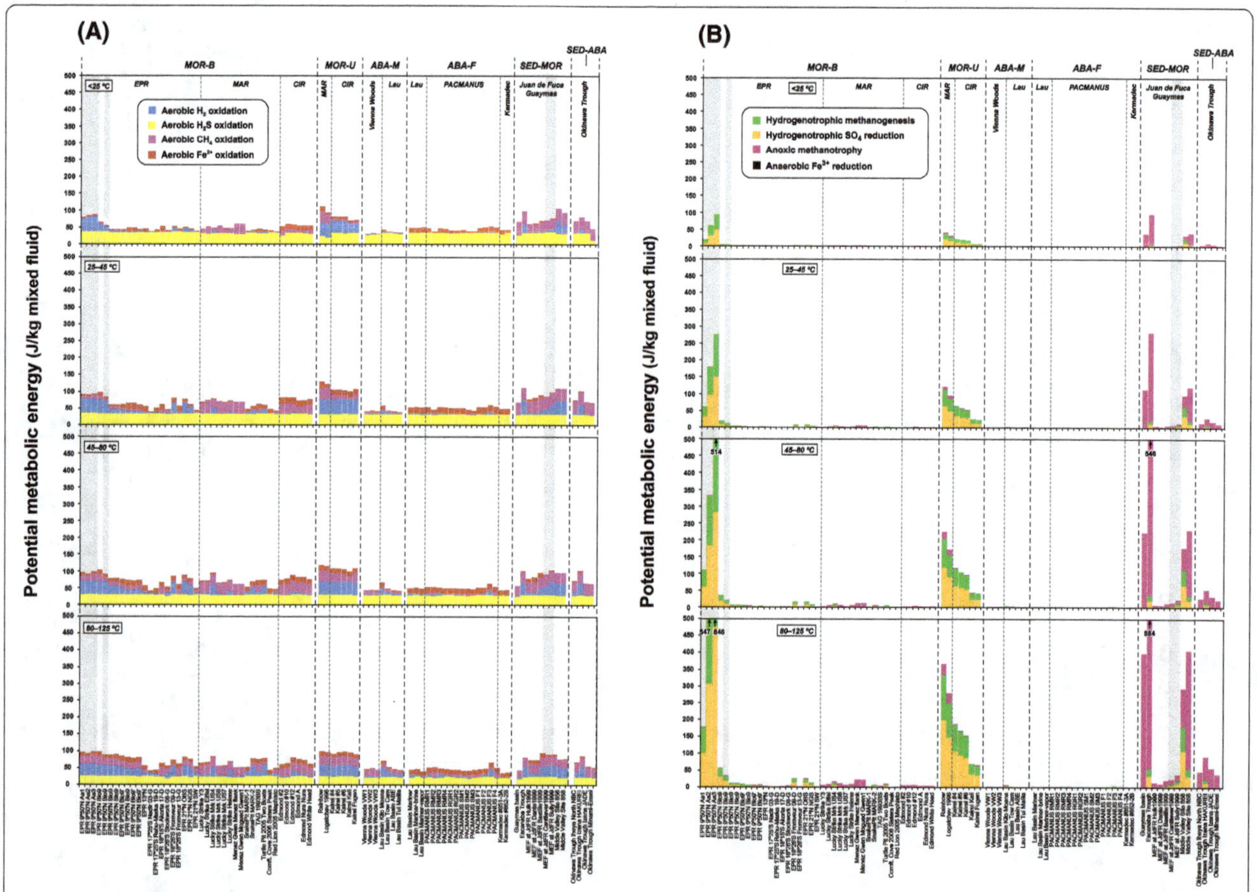

Figure 8 Metabolic energies available from each (A) aerobic and (B) anaerobic reactions shown in Table 1. Calculations were performed for H_2 oxidation, H_2S oxidation, methanotrophy, and Fe^{2+} oxidation (aerobic reactions) and hydrogenotrophic methanogenesis, hydrogenotrophic SO_4 reduction, anoxic methanotrophy, and anaerobic Fe^{3+} reduction (anaerobic reactions) at the 89 hydrothermal vent sites in MOR-B, MOR-U, AMA-M, ABA-F, SED-MOR, and SED-ABA settings. Shaded areas represent hydrothermal vents issuing highly vapor-rich (<100 mmol/kg Cl) fluids.

and are characterized by low temperatures and quite low mixing ratios of hydrothermal fluids (<0.1%). In hydrothermal plumes, the SUP05 group of Gammaproteobacteria is the most abundant and geographically widespread microbial component (Sunamura et al. 2004; Dick and Tebo 2010; German et al. 2010). The deep-sea vent Epsilonproteobacteria (Sunamura et al. 2004; Nakagawa et al. 2005a; German et al. 2010), methanotrophic Gammaproteobacteria (Lam et al. 2008), and ammonia-oxidizing Betaproteobacteria and Thaumarchaeota (Lam et al. 2008; Takai et al. 2004b) have been identified as other dominant microbial populations.

Although microbial communities in hydrothermal plumes have variable abundances and compositions of chemolithotrophic metabolisms, there have been few quantitative studies or functional characterizations of these communities. Thus, we still do not know much about how these microbial communities and their metabolisms respond to the varying physical and chemical states of hydrothermal plume habitats or how they are related to the geological backgrounds that produce the source hydrothermal fluids. Nevertheless, the general landscape of metabolic abundance and composition in hydrothermal plume microbial communities has been established via biogeochemical (Cowen et al. 1986; de Angelis et al. 1993; Mandernack and Tebo, 1993; Dick et al. 2009) and microbial ecology (Sunamura et al. 2004; Takai et al. 2004b; Nakagawa et al. 2005a; Lam et al. 2008; Dick and Tebo 2010; Lesniewski et al. 2012) investigations. For example, dominance of the sulfur-oxidizing SUP05 group of Gammaproteobacteria and the deep-sea vent Epsilonproteobacteria is commonly observed in geographically and geologically diverse deep-sea hydrothermal systems. These organisms primarily sustain themselves via aerobic sulfur oxidation; considerable populations occasionally utilize aerobic H_2 oxidation for additional energy conversion if metabolically available H_2 is present. In addition, much smaller but ubiquitous populations of phylogenetically diverse bacteria also utilize Fe^{2+} and Mn^{2+} oxidation for biomass production. In contrast, in hydrothermal plumes highly enriched in CH_4 and NH_4^+, particularly in SED hydrothermal systems, the methanotrophic Gammaproteobacteria and ammonia-oxidizing Betaproteobacteria and Thaumarchaeota dominate the microbial communities in spatially and temporally varying hydrothermal plume habitats.

An increasing number of recent observations appear consistent with our conclusions based on thermodynamic estimates of bioavailable energy yields in low-temperature mixing zones: (1) aerobic sulfur oxidation is the most common and basic chemolithotrophic energy metabolism in global deep-sea hydrothermal systems, (2) aerobic Fe^{2+} and Mn^{2+} oxidation are less abundant but widespread metabolisms, and (3) aerobic methanotrophy and ammonia oxidizer are dominant in SED systems, respectively.

Diffusing hydrothermal fluids

Microbial communities in relatively low-temperature diffusing hydrothermal fluids have been investigated in a wide variety of geographically diverse hydrothermal systems (Huber and Holden 2008; Huber et al. 2009; Perner et al. 2011, 2013a, 2013b; Akerman et al. 2013; Campbell et al. 2013). These are mainly planktonic free-living communities under conditions in which the end-member fluids have already mixed with infiltrated seawater in relatively shallow subseafloor environments in the recharge and discharge regions of hydrothermal fluids (Bemis et al. 2012). In addition to planktonic free-living populations, these microbial communities also include detached epilithic and biofilm-forming components entrained by the fluid flows.

Hydrothermal fluids diffusing at the seafloor would be expected to combine all growing, living, and dead populations and components within the entire subseafloor fluid flow path in the discharge. This differs from microbial communities in hydrothermal plume habitats in which most microbial populations are indigenously growing from planktonic origins in deep-sea water (Lesniewski et al. 2012). Thus, it is challenging to clearly understand the relationships between the diversity and composition of the observed microbial populations and hydrothermal fluid chemistry based on existing biological and geochemical data for diffusing fluid habitats. However, several studies have recently suggested the dominance of sulfur oxidizers in typical MOR-B settings and the presence of hydrogen oxidizers in MOR-U settings (Perner et al. 2011, 2013a; Akerman et al. 2013). These results are generally consistent with our thermodynamic predictions of relationships among microbial metabolisms and functions, hydrothermal fluid chemistry, and geological background.

On the other hand, a very recent study producing function-focused quantitative estimates, based on H_2 and sulfide consumption and CO_2 assimilation activities as well as metagenomic 16S rRNA and functional gene analyses, has indicated that even in H_2-poor hydrothermal fluids in MOR-B settings, H_2 oxidation-based primary production is highly attractive despite the low H_2 content (Perner et al. 2013b). This leads us to consider that low-temperature diffusing flow is the product of complex subseafloor processes (including seawater-hydrothermal fluid mixing, conductive cooling, various redox reactions, and mineral precipitation), and thus, the simple batch-mixing model employed in this study may not be able to accurately reproduce all of these processes.

Geochemical data sets for low-temperature diffusing flows remain scarce. Thus, further quantitative estimates combined with in situ or detailed chemical measurements will be needed to more accurately describe interactions among the physical and chemical processes, bioavailable energy yields from chemolithotrophic metabolisms, and community development in diffusing hydrothermal fluids.

Chemosynthetic invertebrates

Chemolithotrophic symbionts hosted by deep-sea vent-endemic chemosynthetic invertebrates represent important primary producers in global deep-sea hydrothermal ecosystems. Their biomass production and functions may be constrained by the energy states of the habitats of both the symbionts and the host animals (Dubilier et al. 2008), as productivity and energy metabolism in a tight symbiotic association are strongly regulated by the interaction of the symbiont with the physiology and ecology of the host animal (Dubilier et al. 2008). There have been numerous studies of the basic relationships between the in situ physical and chemical conditions of habitats and the abundance and composition of energy metabolisms in deep-sea hydrothermal chemosynthetic symbioses (Beinart et al. 2012; Petersen et al. 2011; Suzuki et al. 2006; Watsuji et al. 2010, 2014). However, the energy conversion yield of the primary production of chemolithotrophic symbionts is likely one of the most important factors in the establishment of chemosynthetic symbioses, affecting the selection, acquisition, and breeding of specific bacterial counterparts. In particular, episymbioses and horizontally transmitted endosymbioses, in which symbionts are acquired from the proximal environments of habitats in their larval and adult stages in each generation, would be expected to be more strongly affected by the energy states of chemolithotrophic symbiont metabolisms than vertically transmitted endosymbioses.

Recently, several studies have been conducted of the relationship between the activity of chemolithotrophic symbionts and hydrothermal fluid chemistry, e.g., a series of studies on epibiotic microbial communities in the galatheid crab Shinkaia crosnieri in Okinawa Trough

(Watsuji et al. 2010, 2012, 2014). The epibiotic microbial communities of S. crosnieri abdominal setae consist primarily of the sulfur-oxidizing Epsilonproteobacteria and Gammaproteobacteria and methanotrophic Gammaproteobacteria (Watsuji et al. 2010, 2012, 2014). These results are generally consistent with the results of our thermodynamic modeling, indicating high bioavailable energy yields from both sulfur- and CH_4-oxidizing reactions at low temperatures (high seawater mixing ratios) in the Okinawa Trough (Figure 8). The abundance of epibiotic methanotrophic gammaproteobacterial populations and methanotrophic activity vary among the S. crosnieri populations recovered from different hydrothermal fields (Iheya North and Hatoma Knoll fields) (Watsuji et al. 2010). As suggested by our model calculations, these variations are likely correlated with the in situ CH_4 concentrations in the S. crosnieri colonies. This may, in turn, be associated with the CH_4 concentrations in the end-member hydrothermal fluids, although CH_4 concentrations in the Hatoma fluids have not yet been published.

Another notable example is the dual endosymbiotic (thiotrophic and methanotrophic) populations of vent mussels living in different geological settings of deep-sea hydrothermal vent sites in the MAR and the Eastern Pacific (Petersen et al. 2011). Based on thermodynamic modeling of preferred H_2-trophic energy metabolism, genetic characterization of a key enzyme gene for H_2-trophic energy metabolism, and functional analyses of H_2 consumption and CO_2 fixation of endosymbionts in Bathymodiolus puteoserpentis in a typical MOR-U hydrothermal system (Logatchev field), Peterson et al. (2011) clearly demonstrated that the thiotrophic population in dual endosymbiosis of the Logatchev B. puteoserpentis use H_2 and H_2S comparably as energy sources for primary production. This was the first reported example of chemosynthetic symbiosis sustained by an inorganic energy source (H_2) other than H_2S (reduced sulfur species) or CH_4. In other words, this study first revealed that compositional variations in H_2, H_2S, and CH_4 in the mixing zones of chemosynthetic macrofaunal habitats affect the patterns and functions of episymbioses and endosymbioses in chemosynthetic microbial-faunal associations. Peterson et al. (2011) also showed that the thio- and H_2-trophic endosymbionts hosted by different Bathymodiolus populations living in geographically and geologically distinct hydrothermal systems of the MAR and Pacific (all H_2-starved MOR-B hydrothermal systems) had much lower specific H_2 consumption activities than the endosymbiotic population from the H_2-abundant MOR-U hydrothermal system (Logatchev field). Although not fully quantitative, these results provide strong evidence that the metabolic activity of chemolithotrophic symbionts is controlled by hydrothermal fluid chemistry and is related to the geological background of the hydrothermal system, as suggested by our model calculations.

Recently, metagenomic and metatranscriptomic approaches have also been used to study possible biogeochemical associations with Alviniconcha (deep-sea hydrothermal-vent snail) holobionts (host/symbiont associations) that dwell in a broad region of the Eastern Lau Spreading Center (ELSC) (Beinart et al. 2012; Sanders et al. 2013). Different holobiont patterns have been found in the Alviniconcha populations obtained from the mixing zones of colonies in four different hydrothermal systems in the ELSC. The endosymbionts were classified into three groups: Sulfurimonas-type Epsilonproteobacteria, gamma-1-type Gammaproteobacteria, and gamma-Lau-type Gammaproteobacteria (Beinart et al. 2012), all of which are known to be typical sulfur (and H_2) oxidizers. The Alviniconcha holobiont and endosymbiont distributions varied from the northern to the southern region of the ELSC, which have somewhat different geological settings (Beinart et al. 2012). The ELSC has been well characterized from north to south in terms of geological properties (such as the spreading stage and rate, magmatic productivity, geochemistry, and host-rock geochemistry) (Tivey et al. 2012). Reflecting the change in geological setting, the chemical compositions of the end-member hydrothermal fluids also change from relatively reduced and H_2- and H_2S-enriched (approximately 500 μM and approximately 5 mM, respectively) in the north to relatively oxidized and H_2- and H_2S-depleted (approximately 30 μM and approximately 3 mM, respectively) in the south (Beinart et al. 2012). The endosymbiont species shift from predominantly Sulfurimonas-type Epsilonproteobacteria in the north to predominantly gamma-1-type Gammaproteobacteria and finally to both types of Gammaproteobacteria in the south (Beinart et al. 2012). Interestingly, the compositional abundance of both the endosymbiotic microbial components and the expressed key gene transcripts for various energy metabolisms vary with the regional geochemical gradient (Sanders et al. 2013). Under relatively reduced and H_2- and H_2S-enriched conditions, the compositional abundance of mRNAs related to H_2- and sulfur-oxidizing metabolisms throughout the entire endosymbiont transcriptome is highly elevated (Sanders et al. 2013). The compositional abundances of the mRNAs of these energy metabolisms throughout the entire endosymbiont transcriptome are likely associated with the activities of these energy metabolisms in the endosymbiotic populations of the Alviniconcha host individuals.

Our geochemical modeling suggests that changes in the H_2S concentration from north to south in the ELSC (approximately 5 mM to approximately 3 mM) should not significantly affect sulfur-oxidizing metabolisms, whereas H_2-oxidizing metabolisms could be affected by the change in H_2 concentration (approximately 500 μM to approximately 30 μM) (Figure 6). Indeed, based on differences in the expression of hydrogenases between Epsilonproteobacteria and Gammaproteobacteria individuals, Sanders et al. (2013) suggested that H_2 oxidation may play a larger role in the energy metabolism of holobionts with Epsilonproteobacteria. Thus, the change in H_2 concentration, rather than the H_2S concentration, may account for differences in the holobiont patterns found in the Alviniconcha populations in the ELSC. Although future studies are needed to elucidate the degree to which Epsilonproteobacteria endosymbionts rely on H_2 for energy production, this may be additional evidence supporting the relationship between geological constraints on hydrothermal fluid chemistry and the compositions and functions of chemosynthetic microbial symbionts in deep-sea hydrothermal environments.

Chimney structures and sediments

The compositional and functional diversity of microbial communities has been most extensively explored in deep-sea vent chimney structures that host high-temperature hydrothermal fluid emissions (Nakagawa and Takai 2008; Takai and Nakamura 2010, 2011; Takai et al. 2006b). These microbial communities consist of typical epilithic (mineral surface-attached) and/or biofilm-forming populations and are sustained by the redox disequilibria formed in the chimney wall via diffusive mixing between the internal hydrothermal fluid and external seawater. Thus, the abundance and composition of chemolithotrophic energy metabolisms in the chimney microbial communities are directly controlled by the energy states of the chimney habitats, resulting in good approximations by the thermodynamic calculations of the simple batch-mixing model.

Using data for both end-member hydrothermal fluids and chimney microbial communities obtained from six deep-sea hydrothermal systems with different geological settings, Takai and Nakamura (2010, 2011) constructed intra-field and inter-field comparisons between the abundance and composition of chemolithotrophic energy metabolisms predicted by thermodynamic modeling and the microbial community composition determined through quantitative cultivation methods. Based on both the thermodynamic modeling and microbiological characterization, statistically significant correlations between the H_2 concentration in the hydrothermal fluid or the estimated potential energy yield of H_2-trophic methanogenesis and the cultivatable population size of hydrogenotrophic methanogens were found. More recently, in addition to quantitative cultivation methods, statistical phylotype compositional analyses of 16S rRNA and other functional genes have also been used to demonstrate significant correlations between H_2-trophic chemolithotrophy in chimney microbial communities and hydrothermal fluid H_2 concentrations, particularly in MOR-U hydrothermal systems (Flores et al. 2011; Roussel et al. 2011). Our geochemical model provides consistent explanations for the results of these studies. Specifically, the energy yields of anaerobic (and aerobic, in most cases) H_2-trophic reactions were well correlated with the H_2 concentrations of hydrothermal fluids (Figure 6). In addition, thermophilic H_2-trophic methanogens and sulfate/sulfur reducers are predominantly found in H_2-enriched deep-sea hydrothermal systems in the MOR-U setting (Figure 8). It is thus very likely that serpentinization-driven H_2 enrichment in hydrothermal fluids is an important geochemical factor that directly controls the thermodynamic energy potential, biomass, and productivity of H_2-trophic chemolithotrophs.

Similar to the apparent relationship between H_2 concentrations in hydrothermal fluids and the abundance of H_2-trophic chemolithotroph populations, our thermodynamic calculations also identified a possible relationship between CH_4 enrichment and the abundance of anaerobic methanotroph populations (particularly thermophilic methanotrophic sulfate reducers) in SED hydrothermal systems (Figure 8). In low-temperature habitats (e.g., hydrothermal plumes, diffusing fluids, and chemosynthetic animal symbioses), many molecular signals and metabolic activities of aerobic methanotrophy have been observed. However, the existence of anaerobic thermophilic methanotrophic sulfate reducers has been inferred only from lipid biomarker signals (Schouten et al. 2003). Recently, thermophilic anoxic CH_4-oxidizing SO_4-reducing activities have been directly verified at up to 70°C in the sediments of the Guaymas Basin hydrothermal field (Holler et al. 2011). In addition, possible microbial components of thermophilic anaerobic methane oxidizers, consortia of ANME-1a Archaea and HotSeep-1 cluster Deltaproteobacteria, have been identified via FISH analysis (Holler et al. 2011). Similar phylotypes of potentially thermophilic ANME-1a Archaea have also been found in hydrothermal sediments and diffusing fluids of SED hydrothermal systems such as the Axial Seamount and Endeavour Segment fields in the Juan de Fuca Ridge and the Yonaguni Knoll IV field in the Okinawa Trough (Nunoura et al. 2010; Merkel et al. 2013). Because these phylotypes of thermophilic anaerobic methanotrophic Archaea remain resistant to cultivation, quantitative cultivation methods are not feasible. However, other quantitative techniques targeting 16S rRNA and functional gene transcripts together with direct metabolic activity measurements are now available (e.g., Beal et al. 2009; Watsuji et al. 2012, 2014). Such quantitative estimates will provide multiple lines of evidence for the relationship between CH_4 enrichment in SED hydrothermal systems and the abundance of thermophilic methanotrophic sulfate reducers in mixing zone habitats.

Takai and Nakamura (2010) have also noted that phase separation has a considerable impact on intra-field variations in hydrothermal fluid chemistry. In microbiological studies of the Iheya North field in the middle Okinawa Trough and the Mariner field in the Lau Basin, increased populations of anaerobic and aerobic H_2-trophic chemolithotrophs (e.g., Methanococcales and Aquificales) were observed, potentially due to phase separation of hydrothermal fluids (Nakagawa et al. 2005b; Takai et al. 2008). However, in studies of the Yonaguni Knoll IV field in the Southern Okinawa Trough, H_2-trophic (and thiotrophic) chemolithotrophs (e.g., Methanococcales, Desulfurococcales, Nautiliaceae, and Thioreductoraceae) were the dominant microbial components in both vapor- and brine-rich hydrothermal fluids (Nunoura and Takai 2009). Although H_2S concentration data for the Yonaguni Knoll IV fluid have not yet been reported, our geochemical model calculations indicate that if the H_2 concentrations of the end-member hydrothermal fluids were well below approximately 1 mmol/kg, the metabolic energy available from both aerobic and anaerobic H_2-trophic reactions would be directly correlated with the H_2 concentration of the fluids (Figure 6). The H_2 concentrations in the vapor- and brine-rich hydrothermal fluids venting at the Iheya North field (0.045 and 0.096 mmol/kg, respectively) and the Mariner field (0.012 and 0.13 mmol/kg, respectively) are well below approximately 1 mmol/kg and thus, phase separation-induced H_2 variations may directly affect the available energy yields from aerobic and anaerobic H_2-trophic reactions. In contrast, the H_2 concentrations in the fluids at the Yonaguni Knoll IV field (0.8 to 3.6 mmol/kg) are mostly above approximately 1 mmol/kg, resulting in saturation of the metabolic energy available from aerobic hydrogenotrophic reactions, particularly at higher temperatures (Figure 6). In addition, the bioavailable energy yield from hyperthermophilic anaerobic reactions is nearly comparable to that from aerobic thiotrophic reactions, even in brine-rich hydrothermal fluids (Figure 6). This may explain why there are no major differences in abundance or composition of microbial communities between the vapor- and brine-rich hydrothermal fluids in the Yonaguni Knoll IV field (Nunoura and Takai, 2009).

Conclusions

We have provided an overview of variations in the geology, geochemistry, microbial energy metabolisms, and community development associated with global deep-sea hydrothermal systems. Relationships among the geological backgrounds of hydrothermal activities (e.g., tectonic settings, basement rock geochemistry, abundance of sediments, magmatic volatile inputs, and phase separation related to subseafloor hydrothermal processes), physical and chemical variations in hydrothermal fluids, and compositional diversity of potentially bioavailable energy for various vent-endemic chemolithotrophic metabolisms have been elucidated through thermodynamic modeling of redox states in hydrothermal mixing zone habitats. In addition, these relationships have been empirically substantiated by recent multidisciplinary biogeochemical and microbiological studies of existing microbial communities and their metabolic functions in representative deep-sea hydrothermal systems in different geographic locations and geological settings.

However, thermodynamic estimates of potential energy for possible chemolithotrophic metabolisms are based on certain simplifications, particularly with respect to quantitative representations of mixing of source hydrothermal fluids and seawater and the kinetic effects of abiotic reactions, cellular uptake and excretion of energy sources, and enzymatic functions on energy metabolism. These shortcomings may result in differences between the theoretical predictions and actual configurations and patterns in some cases. To characterize the in situ physical and chemical characteristics of hydrothermal mixing zone habitats at high spatial and temporal resolution, several electrochemical sensors have been developed and applied to various habitats such as diffusing hydrothermal fluids and chemosynthetic animal colonies. However, these sensors have some technical challenges, including multiple chemical interferences, relatively narrow dynamic ranges, and the need for in situ calibration (Luther et al. 2012). A deep-sea in situ membrane inlet mass spectrometry (MIMS) method has also been developed and used for in situ measurement of volatile and time-sensitive chemical species (Wankel et al. 2010). These newly developed in situ measurement tools will enable more precise estimates of the energy states of habitats of chemolithotrophic microbial communities. Improvement and development of biogeochemical and microbiological characterization techniques are also necessary to understand the functional and metabolic abundance and composition of microbial communities in deep-sea hydrothermal environments. It is a significant challenge to quantify the biomass, productivity, and functions of dominant microbial components in deep-sea vents. Comprehensive characterizations of in situ mRNA assemblages using such tools as GeoChip technology (Wang et al. 2009) and the metatranscriptomic approach (Dahle et al. 2013) will likely prove more objective, quantitative, and effective than employing elaborate quantitative cultivations using numerous media for intra-field and inter-field comparisons of microbial communities and their functions. Successfully overcoming these challenges is essential for clarifying the relationships between geology, geochemistry, microbial energy metabolisms, and community development associated with global deep-sea hydrothermal activities.

Abbreviations
ABA: arc-backarc; ABA-F: felsic rock-hosted hydrothermal system in an arc-backarc setting; ABA-M: mafic rock-hosted hydrothermal system in an arc-backarc setting; MOR: mid-ocean ridge; MOR-B: basalt-hosted hydrothermal system in a mid-ocean ridge setting; MOR-U: ultramafic rock-hosted hydrothermal system in a mid-ocean ridge setting; SED: sediment-associated; SED-ABA: sediment-associated hydrothermal system in an arc-backarc setting; SED-MOR: sediment-associated hydrothermal system in a mid-ocean ridge setting.

Competing interests
The authors declare that they have no competing interests.

Authors' contributions
Both KN and KT proposed the topic, conceived of and designed the study, compiled and processed the data, and wrote the paper. Both authors read and approved the final manuscript.

Acknowledgements
We would like to thank the two anonymous reviewers for their constructive and helpful comments, which have significantly improved the manuscript. This research was financially supported by the Ministry of Education, Culture, Science, and Technology (MEXT) of Japan through the special coordination fund Project TAIGA: Trans-crustal Advection and In situ Biogeochemical Processes of Global Subseafloor Aquifer.

Author details
[1]Precambrian Ecosystem Laboratory (PEL), Japan Agency for Marine-Earth Science and Technology (JAMSTEC), 2-15 Natsushima-cho, Yokosuka, Kanagawa 237-0061, Japan. [2]Current address: Department of Systems Innovation, School of Engineering, The University of Tokyo, 7-3-1 Hongo, Bunkyo-ku, Tokyo 113-8656, Japan. [3]Subsurface Geobiology Advanced Research (SUGAR) Project, Japan Agency for Marine-Earth Science and Technology (JAMSTEC), 2-15 Natsushima-cho, Yokosuka, Kanagawa 237-0061, Japan.

References

Akerman NH, Butterfield DA, Huber JA (2013) Phylogenetic diversity and functional gene patterns of sulfur-oxidizing subseafloor Epsilonproteobacteria in diffuse hydrothermal vent fluids. Front Microbiol 4. doi:10.3389/fmicb.2013.00185

Allen DE, Seyfried WE, Jr (2003) Compositional controls on vent fluids from ultramafic-hosted hydrothermal systems at mid-ocean ridges: an experimental study at 400°C 500 bars. Gechim Cosmochim Acta 67:1531–1542

Alt JC, Shanks WC (1998) Sulfur in serpentinized oceanic peridotites: serpentinization processes and microbial sulfate reduction. J Geophys Res 103:9917–9929

Amend JP, McCollom TM, Hentscher M, Bach W (2011) Catabolic and anabolic energy for chemolithoautotrophs in deep-sea hydrothermal systems hosted in different rock types. Gechim Cosmochim Acta 75:5736–5748

Beal EJ, House CH, Orphan VJ (2009) Manganese- and iron-dependent marine methane oxidation. Science 325:184–187

Beinart RA, Sanders JG, Faure B, Sylva SP, Lee RW, Becker EL, Gartman A, Luther GW III, Seewald JS, Fisher CR, Girguis PR (2012) Evidence for the role of endosymbionts in regional-scale habitat partitioning by hydrothermal vent symbioses. Proc Natl Acad Sci U S A 109:E3241–E3250

Bemis K, Lowell RP, Farough A (2012) Diffuse flow on and around hydrothermal vents at mid-ocean ridges. Oceanography 25:182–191

Berndt ME, Allen DE, Seyfried WE, Jr (1996) Reduction of CO_2 during serpentinization of olivine at 300°C and 500 bar. Geology 24:351–354

Bischoff JL, Dickson FW (1975) Seawater-basalt interaction at 200°C and 500 bars implications for origin of seafloor heavy-metal deposits and regulations of seawater chemistry. Earth Planet Sci Lett 25:385–397

Bischoff JL, Seyfried WE, Jr (1978) Hydrothermal chemistry of seawater from 25° to 350°C. Am J Sci 278:838–860

Bischoff JL, Pitzer KS (1989) Liquid-vapor relations for the system NaCl-H_2O summary of the P-T-x surface from 300 to 500°C. Am J Sci 289:217–248

Bowers TS, Taylor HP, Jr (1985) An integrated chemical and stable isotope model of the origin of mid-ocean ridge hot spring systems. J Geophys Res 90:12583–12606

Bowers TS, Campbell AC, Measures CI, Spivack AJ, Khadem M, Edmond JM (1988) Chemical controls on the composition of vent fluids at 13°-11°N and 21°N East Pacific Rise. J Geophys Res 93:4522–4536

Butterfield DA, McDuff RE, Mottl MJ, Lilley MD, Lupton JE, Massoth GJ (1994) Gradients in the composition of hydrothermal fluids from the Endeavour Ridge vent field: phase separation and brine loss. J Geophys Res 99:9561–9583

Butterfield DA, Seyfried WE Jr, Lilley MD (2003) Composition and evolution of hydrothermal fluids. In: Halbach PE, Tunnicliffe V, Hein JR (ed) Energy and Mass Transfer in Marine Hydrothermal Systems. Dahlem University Press, Berlin

Campbell AC, Palmer MR, Klinkhammer GP, Bowers TS, Edmond JM, Lawrence JR, Casey JF, Thompson G, Humphris S, Rona P, Karson JA (1988) Chemistry of hot springs on the Mid-Atlantic Ridge. Nature 335:514–519

Campbell BJ, Polson SW, Allen LZ, Williamson SJ, Lee CK, Wommack KE, Cary SC (2013) Diffuse flow environments within basalt- and sediment-based hydrothermal vent ecosystems harbor specialized microbial communities. Front Microbiol 4. doi:10.3389/fmicb201300182

Charlou JL, Fouquet Y, Bougauit H, Donval JP, Etoubleau J, Jean-Baptiste P, Dapoigny A, Appriou P, Rona PA (1998) Intense CH_4 plumes generated by serpentinization of ultra-mafic rocks at the intersection of the 15°20'N fracture zone and the Mid-Atlantic Ridge. Geochim Cosmochim Acta 62:2323–2333

Charlou JL, Donval JP, Jean-Baptiste P, Radford-Knoery J, Fouquet Y, Dapoigny A, Stievenard M (2000) Compared geochemical signatures and the evolution of Menez Gwen (37°50'N) and Lucky Strike (37°17'N) hydrothermal fluids south of the Azores Triple Junction on the Mid-Atlantic Ridge. Chem Geol 171:49–75

Charlou JL, Donval JP, Fouquet Y, Jean-Baptiste P, Holm N (2002) Geochemistry of high H_2 and CH_4 vent fluids issuing from ultramafic rocks at the rainbow hydrothermal field (36°14'M MAR). Chem Geol 191:345–359

Cowen JP, Massoth GJ, Baker ET (1986) Bacterial scavenging of Mn and Fe in a mid- to far-field hydrothermal particle plume. Nature 322:169–171

Dahle H, Roalkvam I, Thorseth IH, Pedersen RB, Steen IH (2013) The versatile in situ gene expression of an Epsilonproteobacteria-dominated biofilm from a hydrothermal chimney. Environ Microbiol Repts 5:282–290

de Angelis MA, Lilley MD, Olson EJ, Baross JA (1993) Methane oxidation in deep-sea hydrothermal plumes of the endeavour segment of the Juan de Fuca Ridge. Deep-Sea Res I 40:1169–1186

Dick GJ, Tebo BM (2010) Microbial diversity and biogeochemistry of the Guaymas Basin deep-sea hydrothermal plume. Environ Microbiol 12:1334–1347

Dick GJ, Clement BG, Fodrie FJ, Webb SM, Bargar JR, Tebo BM (2009) Enzymatic microbial Mn(II) oxidation and Mn biooxide production in the Guaymas Basin hydrothermal plume. Geochim Cosmochim Acta 73:6517–6530

Ding K, Seyfried WE, Jr (1992) Determination of Fe-Cl complexing in the low pressure supercritical region (NaCl) Fe solubility constraints on the pH of subseafloor hydrothermal fluids. Geochim Cosmochim Acta 59:4769–4773

Ding K, Seyfried WE Jr, Tivey MK, Von Damm KL, Bradley AM, Zhang Z (2005) In situ pH measurement of hydrothermal fluids at mid-ocean ridges. Earth Planet Sci Lett 237:167–174

Dubilier N, Bergin C, Lott C (2008) Symbiotic diversity in marine animals: the art of harnessing chemosynthesis. Nat Rev Microbiol 6:725–740

Elderfield H, Schults A (1996) Mid-ocean ridge hydrothermal fluxes and the chemical composition of the ocean. Ann Rev Earth Planet Sci 24:191–224

Embley RW, Baker ET, Butterfield DA, Chadwick WW Jr, Lupton JE, Resing JA, de Ronde CEJ, Nakamura K, Tunnicliffe V, Dower JF, Merle SG (2007) Exploring the submarine ring of fire Mariana Arc-Western Pacific. Oceanography 20:68–79

Fisher RV, Schmincke H-U (1984) Pyroclastic Rocks. Springer, Berlin

Flores GE, Campbell JH, Kirshtein JD, Meneghin J, Podar M, Steinberg JI, Seewald JS, Tivey MK, Voytek MA, Yang ZK, Reysenbach AL (2011) Microbial community structure of hydrothermal deposits from geochemically different vent fields along the Mid-Atlantic Ridge. Environ Microbiol 13:2158–2171

Foustoukos DI, Seyfried WE, Jr (2007) Fluid phase separation processes in submarine hydrothermal systems. Rev Mineral Geochem 65:213–239

Frost RB (1985) On the stability of sulfides oxides and native metals in serpentinite. J Petrol 26:31–63

Früh-Green GL, Connolly JA, Plas A, Kelley DS, Grobety B (2004) Serpentinization of oceanic peridotites: implications for geochemical cycles and biological activity. In: Wilcock WS, DeLong EF, Kelley DS, Baross JA, Cary SC (ed) The Subseafloor Biosphere at mid-Ocean Ridges. Geophys Monogr Ser 144, AGU, Washington, DC

Gamo T, Okamura K, Charlou JL, Urabe T, Auzende JM, Ishibashi J, Shitashima K, Chiba H, Shipboard Scientific Party of the ManusFlux Cruise (1997) Acidic and sulfate-rich hydrothermal fluids from the Manus back-arc basin, Papua New Guinea. Geology 25:139–142

Gamo T, Ishibashi J, Tsunogai U, Okamura K, Chiba H (2006) Unique geochemistry of submarine hydrothermal fluids from arc-back-arc settings of the western Pacific. In: Christie DM, Fisher CR, Lee SM, Givens S (ed) Back-arc Spreading Systems—Geological, Biological, Chemical, and Physical Interactions. Geophys Monogr Ser 166, AGU, Washington, DC

German CR, Von Damm KL (2004) Hydrothermal processes. In: Turekian KK, Holland HD (ed) The Treatise on Geochemistry Chapter 6.07. Elsevier, New York

German CR, Bowen A, Coleman ML, Honig DL, Huber JA, Jakuba MV, Kinsey JC, Kurz MD, Leroy S, McDermott JM, de Lepinay BM, Nakamura K, Seewald JS, Smith JL, Sylva SP, Van Dover CL, Whitcomb LL, Yoerger DR (2010) Diverse styles of submarine venting on the ultraslow spreading Mid-Cayman Rise. Proc Natl Acad Sci U S A 107:14020–14025

Hajash A, Chandler GW (1981) An experimental investigation of high-temperature interactions between seawater and rhyolite andesite basalt and peridotite. Contrib Mineral Petrol 78:240–254

Hannington M, Jamieson J, Monecke T, Petersen S, Beaulieu S (2011) The abundance of seafloor massive sulfide deposits. Geology 39:1155–1158

Helgeson HC, Kirkham DH, Flowers GC (1981) Theoretical prediction of the thermodynamic behavior of aqueous electrolytes at high pressures and temperatures IV: calculation of activity coefficients osmotic coefficients and relative partial molal properties to 600°C and 5 kb. Am J Sci 281:1249–1516

Holler T, Widde F, Knittel K, Amann R, Kellermann MY, Hinrichs KU, Teske A, Boetius A, Wegener G (2011) Thermophilic anaerobic oxidation of methane by marine microbial consortia. ISME J 5:1946–1956

Huber JA, Holden JF (2008) Modeling the impact of diffuse vent microorganisms along mid-ocean ridges and flanks. In: Lowell RP, Seewald JS, Metaxas A, Perfit MR (ed) Magma to Microbe: Modeling Hydrothermal Processes at Oceanic Spreading Centers. Geophys Monogr Ser 178, AGU, Washington, DC

Huber JA, Morrison HG, Huse SM, Neal PR, Sogin ML, Mark Welch DB (2009) Effect of PCR amplicon size on assessments of clone library microbial diversity and community structure. Environ Microbiol 11:1292–1302

Humphris SE, Zierenberg RA, Mullineaux LS, Thomson RE (1995) Seafloor Hydrothermal Systems—Physical, Chemical, Biological, and Geological Interactions. Geophys Monogr Ser 91, AGU, Washington, DC

Inagaki F, Kuypers MMM, Tsunogai U, Ishibashi J, Nakamura K, Treude T, Ohkubo S, Nakaseama M, Gena K, Chiba H, Hirayama H, Nunoura T, Takai K, Jorgensen BB, Horikoshi K, Boetius A (2006) Microbial community in a sediment-hosted CO_2 lake of the southern Okinawa Trough hydrothermal. Proc Natl Acad Sci U S A 103:14164–14169

Ishibashi J, Urabe T (1995) Hydrothermal activity related to arc-backarc magmatism in the western Pacific. In: Taylor B (ed) Backarc Basins: Tectonics and Magmatism. Plenum, New York

Janecky DR, Seyfried WE, Jr (1986) Hydrothermal serpentinization of peridotite within the oceanic crust: experimental investigations of mineralogy and major element chemistry. Geochim Cosmochim Acta 50:1357–1378

Jannasch HW, Mottl MJ (1985) Geomicrobiology of deep-sea hydrothermal vents. Science 229:717–725

Karl DM (1995) Ecology of free-living hydrothermal vent microbial communities. In: Karl DM (ed) Microbiology of Deep-sea Hydrothermal Vents. CRC, Boca Raton

Kawagucci S, Chiba C, Ishibashi J, Yamanaka T, Toki T, Muramatsu Y, Ueno Y, Makabe A, Inoue K, Yoshida N, Nakagawa S, Nunoura T, Takai K, Takahata N, Sano Y, Narita T, Teranishi G, Obata H, Gamo T (2011) Hydrothermal fluid geochemistry at the Iheya North field in the mid-Okinawa trough: implication for origin of methane in subseafloor fluid circulation systems. Gechem J 45:109–124

Kawagucci S, Ueno Y, Takai K, Toki T, ito M, Inoue K, Makabe A, Yoshida N, Muramatsu Y, Takahata N, Sano Y, Narita T, Teranishi G, Obata H, Nakagawa S, Nunoura T, Gamo T (2013) Geochemical origin of hydrothermal fluid methane in sediment-associated fluids and its relevance to the geographical distribution of whole hydrothermal circulation. Chem Geol 339:213–225

Kawahata H, Nohara M, Ishizuka H, Hasebe S, Chiba H (2001) Sr isotope geochemistry and hydrothermal alteration of the Oman ophiolite. J Geophys Res 106:11083–11099

Kelley DS, Karson JA, Blackman DK, Fruh-Green GL, Butterfield DA, Lilley MD, Oison EJ, Schrenk MO, Roe KK, Lebon GT, Rivizzigno P, the AT3-60 Shipboard Party (2001) An off-axis hydrothermal vent field discovered near the Mid-Atlantic Ridge at 30°N. Nature 412:145–149

Kelley DS, Baross JA, Delaney JR (2002) Volcanoes fluids and life at mid-ocean ridge spreading centers. Annu Rev Earth Planet Sci 30:385–491

Kelley DS, Karson JA, Fruh-Green GL, Yoerger DR, Shank TM, Butterfield DA, Hayes JM, Schrenk MO, Oison EJ, Proskurowski G, Jakuba M, Bradley A, Larson B, Ludwig K, Glickson D, Buckman K, Bradley AS, Brazelton WJ, Roe K, Elend MJ, Delacour A, Bernasconi SM, Lilley MD, Baross JA, Summons RE, Sylva SP (2005) A serpentinite-hosted ecosystem: the lost city hydrothermal field. Science 307:1428–1434

Lam P, Cowen JP, Popp BN, Jones RD (2008) Microbial ammonia oxidation and enhanced nitrogen cycling in the Endeavour hydrothermal plume. Geochim Cosmochim Acta 72:2268–2286

Leat PT, Larter RD (2003) Intra-oceanic subduction systems introduction. In: Larter RD, Leat PT (ed) Intra-Oceanic Subduction Systems Tectonic and Magmatic Processes. Geol Soc London Spec Publ, London

Lesniewski RA, Jain S, Anantharaman K, Schloss PD, Dick GJ (2012) The metatranscriptome of a deep-sea hydrothermal plume is dominated by water column methanotrophs and lithotrophs. ISME J 6:2257–2268

Lilley MD, Butterfield DA, Olson EJ, Lupton JE, Macko SA, McDuff RE (1993) Anomalous CH_4 and NH_4-concentrations at an unsedimented mid-ocean-ridge hydrothermal system. Nature 364:45–47

Lilley MD, Butterfield DA, Lupton JE, Olson EJ (2003) Magmatic events can produce rapid changes in hydrothermal vent chemistry. Nature 422:878–881

Lupton J, Butterfield D, Lilley M, Evans L, Nakamura K, Chadwick W Jr, Resing J, Embley R, Olson E, Proskurowski G, Baker E, de Ronde C, Roe K, Greene R, Lebon G, Young C (2006) Submarine venting of liquid carbon dioxide on a Mariana Arc volcano. Geochem Geophys Geosys 7:Q08007. doi:10.1029/2005GC001152

Lupton J, Lilley M, Butterfield D, Evans L, Embley R, Massoth G, Christenson B, Nakamura K, Schmidt M (2008) Venting of a separate CO_2-rich gas phase from submarine arc volcanoes: examples from the Mariana and Tonga-Kermadec arcs. J Geophys Res 113:B08S12. doi:10.1029/2007JB005467

Luther GW, Gartman A, Yucel M, Madison AS, Moore RS, Nees HA, Nuzzio DB, Sen A, Lutz RA, Shank TM, Fisher CR (2012) Chemistry temperature and faunal distributions at diffuse flow hydrothermal vents comparisons of two geologically distinct ridge systems. Oceanography 25:234–245

Mandernack KW, Tebo BM (1993) Manganese scavenging and oxidation at hydrothermal vents and in vent plumes. Geochim Cosmochim Acta 57:3907–3923

McCollom TM (2007) Geochemical constraints on sources of metabolic energy for chemolithoautotrophy in ultramafic-hosted deep-sea hydrothermal systems. Astrobiology 7:933–950

McCollom TM, Shock EL (1997) Geochemical constraints on chemolithoautotrophic metabolism by microorganisms in deep-sea hydrothermal systems. Geochim Cosmochim Acta 61:4375–4391

McCollom TM, Shock EL (1998) Fluid-rock interactions in the lower oceanic crust: thermodynamic models of hydrothermal alteration. J Geophys Res 103:547–575

McCollom TM, Seewald JS (2001) A reassessment of the potential for reduction of dissolved CO_2 to hydrocarbons during serpentinization of olivine. Geochim Cosmochim Acta 65:3769–3778

Merkel AY, Huberb JA, Chernyha NA, Bonch-Osmolovskayaa EA, Lebedinskya AV (2013) Detection of putatively thermophilic anaerobic methanotrophs in diffuse hydrothermal vent fluids. Appl Environ Microbiol 79:915–923

Mottl MJ (1983) Hydrothermal processes at seafloor spreading centers: application of basalt-seawater experimental results. In: Rona PA, Bostrom K, Laubier L, Smith KL (ed) Hydrothermal Processes at Seafloor Spreading Centers. Plenum, New York

Mottl MJ, Holland HD (1978) Chemical exchange during hydrothermal alteration of basalt by seawater—I: experimental results for major and minor components of seawater. Geochim Cosmochim Acta 42:1103–1115

Nakagawa S, Takai K (2008) Deep-sea vent chemoautotrophs diversity biochemistry and ecological significance. FEMS Microbiol Ecol 65:1–14

Nakagawa S, Takai K, Inagaki F, Hirayama H, Nunoura T, Horikoshi K, Sako Y (2005a) Distribution phylogenetic diversity and physiological characteristics of epsilon-proteobacteria in a deep-sea hydrothermal field. Environ Microbiol 7:1619–1632

Nakagawa S, Takai K, Inagaki F, Chiba H, Ishibashi J, Kataoka S, Hirayama H, Nunoura T, Horikoshi K, Sako Y (2005b) Variability in microbial community and venting chemistry in a sediment-hosted backarc hydrothermal system: impacts of subseafloor phase-separation. FEMS Microbiol Ecol 54:141–155

Nakamura K, Kato Y, Tamaki K, Ishii T (2007) Geochemistry of hydrothermally altered basaltic rocks from the Southwest Indian Ridge near the Rodriguez Triple Junction. Mar Geol 239:125–141

Nakamura K, Morishita T, Bach W, Klein F, Hara K, Okino K, Takai K, Kumagai H (2009) Serpentinized troctolites exposed near the Kairei hydrothermal field Central Indian Ridge: insights into the origin of the Kairei hydrothermal fluid supporting a unique microbial ecosystem. Earth Planet Sci Lett 280:128–136

Nealson KH, Inagaki F, Takai K (2005) Hydrogen-driven subsurface lithoautotrophic microbial ecosystems (SLiMEs): do they exist and why should we care? TRENDS Microbiol 13:405–410

Nunoura T, Takai K (2009) Comparison of microbial communities associated with phase-separation-induced hydrothermal fluids at the Yonaguni Knoll IV hydrothermal field, the Southern Okinawa Trough. FEMS Microbiol Ecol 67:351–370

Nunoura T, Oida H, Nakaseama M, Kosaka A, Ohkubo SB, Kikuchi T, Kazama H, Hosoi-Tanabe S, Nakamura K, Kinoshita M, Hirayama H, Inagaki F, Tsunogai U, Ishibashi J, Takai K (2010) Archaeal diversity and distribution along thermal and geochemical gradients in hydrothermal sediments at the Yonaguni Knoll IV hydrothermal field in the Southern Okinawa Trough. Appl Environ Microbiol 76:1198–1211

Perner M, Kuever J, Seifert R, Pape T, Koschinsky A, Schmidt K, Strauss H, Imhoff JF (2007) The influence of ultramafic rocks on microbial communities at the Logatchev hydrothermal field, located 15°N on the Mid-Atlantic Ridge. FEMS Microbiol Ecol 61:97–109

Perner M, Petersen JM, Zielinski F, Gennerich H-H, Seifert R (2010) Geochemical constraints on the diversity and activity of H_2-oxidizing microorganisms in diffuse hydrothermal fluids from a basalt- and an ultramafic-hosted vent. FEMS Microbiol Ecol 74:55–71

Perner M, Hentscher M, Rychlik N, Seifert R, Strauss H, Bach W (2011) Driving forces behind the biotope structures in the low-temperature venting at 5°S and 9°S MAR. Environ Microbiol Repts 3:727–737

Perner M, Gonella G, Hourdez S, Böhnke S, Kurtz S, Girguis P (2013a) In-situ chemistry and microbial community compositions in five deep-sea hydrothermal fluid samples from Irina II in the Logatchev field. Environ Microbiol 15:1551–1560

Perner M, Hansen M, Seifert R, Strauss H, Koschinsky A, Petersen S (2013b) Linking geology fluid chemistry and microbial activity of basalt- and ultramafic-hosted deep-sea hydrothermal vent environments. Geobiology 11:340–355

Petersen JM, Zielinski FU, Pape T, Seifert R, Moraru C, Amann R, Hourdez S, Girguis PR, Wankel SD, Barbe V, Pellerier E, Fink D, Borowski C, Bach W, Dubiller N (2011) Hydrogen is an energy source for hydrothermal vent symbioses. Nature 476:176–180

Reysenbach A-L, Shock E (2002) Merging genomes with geochemistry in hydrothermal ecosystems. Science 296:1077–1082

Reed MH (1983) Seawater-basalt reaction and the origin of greenstones and related ore deposits. Econ Geol 78:466–485

Roussel EG, Konn C, Charlou JL, Donval JP, Fouquet Y, Querellou J, Prieur D, Bonavita MAC (2011) Comparison of microbial communities associated with three Atlantic ultramafic hydrothermal systems. FEMS Microbiol Ecol 77:647–665

Saccocia J, Seyfried WE, Jr (1990) Talc-quartz equilibria and the stability of magnesium chloride complexes in NaCl-$MgCl_2$ solutions at 300, 350, and 400°C 500 bars. Geochim Cosmochim Acta 54:3283–3295

Saccocia PJ, Ding K, Berndt ME, Seewald JS, Seyfried WE, Jr (1994) Experimental and theoretical perspectives on crustal alteration at mid-ocean ridges. In: Lentz D (ed) Alteration and Alteration Processes Associated With ore-Forming Systems. Geological Association of Canada Short Course Notes, Québec

Sakai H, Gamo T, Kim ES, Tsutsumi M, Tanaka T, Ishibashi J, Wakita H, Yamano M, Oomori T (1990) Venting of carbon dioxide-rich fluid and hydrate formation in mid-Okinawa Trough backarc basin. Science 248:1093–1096

Sanders JG, Beinart RA, Stewart FJ, Delong EF, Girguis PR (2013) Metatranscriptomics reveal differences in in situ energy and nitrogen metabolism among hydrothermal vent snail symbionts. ISME J 7:1556–1567

Schmincke H-U (2004) Volcanism. Springer, Berlin

Schouten S, Wakeham SG, Hopmans EC, Damsté JSS (2003) Biogeochemical evidence that thermophilic Archaea mediate the anaerobic oxidation of methane. Appl Environ Microbiol 69:1680–1686

Seyfried WE Jr (1987) Experimental and theoretical constraints on hydrothermal alteration processes at mid-ocean ridges. Annu Rev Earth Planet Sci 5:317–335

Seyfried WE Jr, Mottl MJ (1982) Hydrothermal alterations of basalt by seawater under seawater-dominated conditions. Geochim Cosmochim Acta 46:985–1002

Seyfried WE Jr, Ding K (1995) Phase equilibria in subseafloor hydrothermal systems: a review of the role of redox temperature pH and dissolved Cl on the chemistry of hot spring fluids at mid-ocean ridges. In: Humphris S, Mullineaux L, Zierenberg R, Thomson R (ed) Seafloor Hydrothermal Systems: Physical, Chemical, Biological, and Geological Interactions. Geophys Monogr Ser, vol. 91, AGU, Washington, DC, pp 248–273

Seyfried WE Jr, Berndt ME, Seewald JS (1988) Hydrothermal alteration processes at mid-ocean ridges—constraints from diabase alteration experiments, hot spring fluids and composition of the oceanic crust. Can Mineral 26:787–804

Seyfried WE Jr, Ding K, Berndt ME (1991) Phase equilibria constraints on the chemistry of hot spring fluids at mid-ocean ridges. Geochim Cosmochim Acta 55:3559–3580

Seyfried WE Jr, Ding K, Berndt ME, Chen X (1999) Experimental and theoretical controls on the composition of mid-ocean ridge hydrothermal fluids. Rev Econ Geol 8:181–200

Shock EL (1992) Chemical environments of submarine hydrothermal systems. Origin Life Evol Biosphere 22:67–107

Shock EL, Holland ME (2004) Geochemical energy sources that support the subsurface biosphere. In: Wilcock WSD, DeLong EF, Kelley DS, Baross JA, Cary SC (ed) The Subseafloor Biosphere at mid-Ocean Ridges. Geophys Monogr Ser vol. 144, AGU, Washington, DC, pp 153–165

Shock EL, Helgeson HC, Sverjensky DA (1989) Calculation of the thermodynamic and transport properties of aqueous species at high pressures and temperatures: standard partial molal properties of inorganic neutral species. Geochim Cosmochim Acta 53:2157–2183

Spiess FN, Macdonald KC, Atwater T, Ballard R, Carranza A, Cordoba D, Cox C, Diaz Garcia VM, Francheteau J, Guerrero J, Hawkins J, Haymon R, Hessler R, Juteau T, Kastner M, Larson R, Luyendyk B, Macdougall JD, Miller S, Normark W, Orcutt J, Rangin C (1980) East Pacific Rise: Hot springs and Geophysical experiments. Science 207:1421–1433

Sunamura M, Higashi Y, Miyako C, Ishibashi J, Maruyama A (2004) Two bacteria phylotypes are predominant in the Suiyo seamount hydrothermal plume. Appl Environ Microbiol 70:1190–1198

Suzuki Y, Kojima S, Sasaki T, Suzuki M, Utsumi T, Watanabe H, Urakawa H, Tsuchida S, Nunoura T, Hirayama H, Takai K, Nealson KH, Horikoshi K (2006) Host-symbiont relationships in hydrothermal vent gastropods of the genus *Alviniconcha* from the Southwest Pacific. Appl Environ Microbiol 72:1388–1393

Takai K, Horikoshi K (1999) Genetic diversity of Archaea in deep-sea hydrothermal vent environments. Genetics 152:1285–1297

Takai K, Nakamura K (2010) Compositional physiological and metabolic variability in microbial communities associated with geochemically diverse deep-sea hydrothermal vent fluids. In: Barton LL, Mandl M, Loy A (ed) Geomicrobiology: Molecular and Environmental Perspective. Springer, Berlin, pp 251–283

Takai K, Nakamura K (2011) Archaeal diversity and community development in deep-sea hydrothermal vents. Curr Opin Microbiol 14:282–291

Takai K, Komatsu T, Inagaki F, Horikoshi K (2001) Distribution of archaea in a black smoker chimney structure. Appl Environ Microbiol 67:3618–3629

Takai K, Gamo T, Tsunogai U, Nakayama N, Hirayama H, Nealson KH, Horikoshi K (2004a) Geochemical and microbiological evidence for a hydrogen-based, hyperthermophilic subsurface lithoautotrophic microbial ecosystem (HyperSLiME) beneath an active deep-sea hydrothermal field. Extremophiles 8:269–282

Takai K, Oida H, Suzuki Y, Hirayama H, Nakagawa S, Nunoura T, Inagaki F, Nealson KH, Horikoshi K (2004b) Spatial distribution of marine Crenarchaeota group I in the vicinity of deep-sea hydrothermal systems. Appl Environ Microbiol 70:2404–2413

Takai K, Nakamura K, Suzuki K, Inagaki F, Nealson KH, Kumagai H (2006a) Ultramafics-Hydrothermalism-Hydrogenesis-Hyperslime (UltraH^3) linkage: a key insight into early microbial ecosystem in the archean deep-sea hydrothermal systems. Paleontol Res 10:269–282

Takai K, Nakagawa S, Reysenbach AL, Hoek J (2006b) Microbial ecology of mid-ocean ridges and back-arc basins. In: Christie DM, Fisher CR, Lee SM, Givens S (ed) Back-arc Spreading Systems: Geological, Biological, Chemical, and Physical Interactions. Geophys Monogr Ser vol 166, AGU, Washington, DC, pp 185–213

Takai K, Nunoura T, Ishibashi J, Lupton J, Suzuki R, Hamasaki H, Ueno Y, Kawagucci S, Gamo T, Suzuki Y, Hirayama H, Horikoshi K (2008) Variability in the microbial communities and hydrothermal fluid chemistry at the newly discovered Mariner hydrothermal field southern Lau Basin. J Geophys Res 113:G02031. doi:0201001029/02007JG000636

Tivey MK (2004) Environmental conditions within active seafloor vent structures: sensitivity to vent fluid composition and fluid flow. In: Wilcock WS, DeLong EF, Kelley DS, Baross JA, Craig Cary S (ed) Subseafloor Biosphere at mid-Ocean Ridges. Geophys Monogr Ser vol. 144, AGU, Washington, DC, pp 137–152

Tivey MK (2007) Generation of seafloor hydrothermal vent fluids and associated mineral deposits. Oceanography 20:50–65

Tivey MK, Becker E, Beinart R, Fisher CR, Girguis PR, Langmuir CH, Michael PJ, Reysenbach AL (2012) Links from mantle to microbe at the Lau Integrated Study Site: insights from a back-arc spreading center. Oceanography 25:62–77

Van Dover CL (2000) The Ecology of Deep-sea Hydrothermal Vents. Princeton University Press, New Jersey

Von Damm KL (1988) Systematics of and postulated controls on submarine hydrothermal solution chemistry. J Geophys Res 93:4551–4561

Von Damm KL (1990) Seafloor hydrothermal activity: black smoker chemistry and chimneys. Ann Rev Earth Planet Sci 18:173–204

Von Damm KL (1995) Controls on the chemistry and temporal variability of seafloor hydrothermal fluids. In: Humphris SE, Zierenberg RA, Mullineaux LS, Thomson RE (ed) Seafloor Hydrothermal Systems: Physical, Chemical, Biological, and Geological Interactions. Geophys Monogr Ser vol 91, AGU, Washington, DC, pp 222–247

Von Damm KL (2000) Chemistry of hydrothermal vent fluids from 9°-10°N, East Pacific Rise: "Time zero" the immediate posteruptive period. J Geophys Res 105:11203–11222

Wallace PJ, Edmonds M (2011) The sulfur budget in magmas: evidence from melt inclusions submarine glasses and volcanic gas emissions. Rev Mineral Geochem 73:215–246

Wang F, Zhou H, Meng J, Peng X, Jiang L, Sun P, Zhang C, Van Nostrand JDV, Deng Y, He Z, Wu L, Zhou J, Xiao X (2009) GeoChip-based analysis of metabolic diversity of microbial communities at the Juan de Fuca Ridge hydrothermal vent. Proc Natl Acad Sci U S A 106:4840–4845

Wankel SD, Joye SB, Samarkin VA, Shah SR, Friederich G, Melas-Kyriazi J, Girguis PR (2010) New constraints on methane fluxes and rates of anaerobic methane oxidation in a Gulf of Mexico brine pool via in situ mass spectrometry. Deep-Sea Res II 57:2022–2029

Watsuji T, Nakagawa S, Tsuchida S, Toki T, Hirota A, Tsunogai U, Takai K (2010) Diversity and function of epibiotic microbial communities on the galatheid crab *Shinkaia crosnieri*. Microbes Environ 25:288–294

Watsuji T, Nishizawa M, Morono Y, Hirayama H, Kawagucci S, Takahata N, Sano Y, Takai K (2012) Cell-specific thioautotrophic productivity of epsilon-proteobacterial epibionts associated with *Shinkaia crosnieri*. PLoS One 7:e46282. doi:10.1371/journal.pone.0046282

Watsuji T, Yamamoto A, Takaki Y, Ueda K, Kawagucci S, Takai K (2014) Diversity and methane oxidation of active epibiotic methanotrophs on live *Shinkaia crosnieri*. ISME J. doi:10.1038/ismej.2013.226

Wetzel LR, Shock EL (2000) Distinguishing ultra-mafic from basalt-hosted submarine hydrothermal systems by comparing calculated vent fluid compositions. J Geophys Res 105:8319–8340

Wilcock WW, DeLong E, Kelley DS, Baross JA, Cary SC (2004) The Subseafloor Biosphere at mid-Ocean Ridges. Geophys Monogr Ser vol. 144, AGU, Washington, DC

2

Impact of cyclonic eddies and typhoons on biogeochemistry in the oligotrophic ocean based on biogeochemical/physical/meteorological time-series at station KEO

Makio C. Honda[1*], Yoshikazu Sasai[1], Eko Siswanto[1], Akira Kuwano-Yoshida[2], Hidenori Aiki[3] and Meghan F. Cronin[4]

Abstract

In 2014, a sediment trap mooring was deployed adjacent to station Kuroshio Extension Observatory (KEO)'s National Ocean and Atmosphere Administration (NOAA) surface mooring. These data, from July 2014 to July 2016, are used here to investigate nutrient supply mechanisms that support ocean productivity in this oligotrophic region of the subtropical western North Pacific. Both years of sediment trap data show that biogenic material fluxes at ~ 5000 m increased between late winter (March) and late spring (June). Based on sea surface temperature and upper ocean water temperature profiles, from the NOAA surface buoy, and satellite-based surface chlorophyll-*a* , this increase was likely due to an increase of ocean productivity in early spring (March) that was supported by nutrients supplied by winter mixing. On the other hand, biogenic material fluxes also increased in October 2014, and between late December 2014 and January 2015 when concentrations of nutrients near the surface typically are low. Sea surface height anomalies and vertical profiles of water temperature in the upper 500 m showed cyclonic eddies passing station KEO and causing upwelling in late July–early August 2014 and November 2014. It appears that these events supplied nutrients to the upper layer, which then caused ocean productivity in the subsurface layer to temporally increase, resulting in increased deep biogenic material fluxes in autumn and winter. This interpretation of the data is consistent with a simple 3D physical-biological model simulation that shows meso-scale cyclonic eddies can supply nutrient to support new production at KEO. During the 2-year-long sediment trap deployments, several typhoons also passed near station KEO and near-inertial internal waves were observed near the nitracline depth after the typhoons passed. Although turbulent mixing caused by near-inertial internal wave could have possibly supplied nutrient to upper oligotrophic euphotic layer, numerical simulations of the turbulent nutrient supply indicate that enhanced turbulent diffusion across the nutrient concentration gradient did not supply enough nitrate to support the increase in biogenic material flux in autumn.

Keywords: KEO, Oligotrophic, Carbon hot spot, Sediment trap, Cyclonic eddy, Nutrient, Typhoon, OceanSITES

Introduction

Recent increases of atmospheric CO_2 and associated global warming have caused warming, stratification, deoxygenation, and acidification of the ocean, collectively called "multiple stressors" (e.g., Bopp et al. 2013 and references in their paper). There is now great concern that these changes of the ocean environment will affect the ecosystem, carbon cycle, and the ocean's ability to mitigate future increases of atmospheric CO_2.

To collect essential baseline data about this ecosystem and its role in the biological pump, and to facilitate predictions of how these material cycles via biological activity may change, a comparative study of the ecosystem and its biogeochemistry was conducted at time-series stations in the western Pacific eutrophic subarctic gyre (station K2: Fig. 1) and oligotrophic subtropical gyre (station S1: Fig. 1). (For details and highlights on the K2S1 project, see special issue of Journal of

* Correspondence: hondam@jamstec.go.jp
[1]Japan Agency for Marine-Earth Science and Technology (JAMSTEC), Yokosuka, Kanagawa, Japan
Full list of author information is available at the end of the article

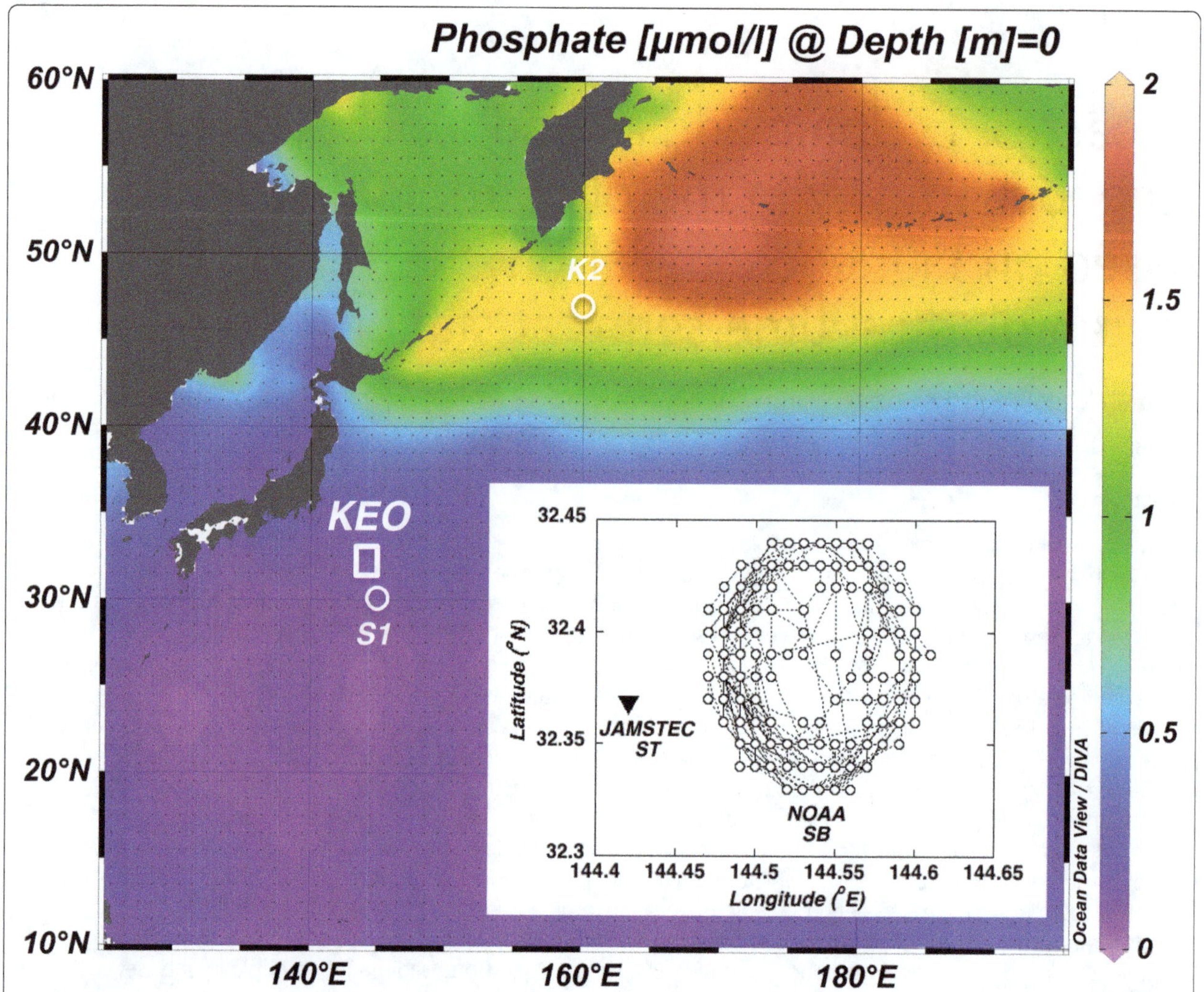

Fig. 1 Location of station KEO. Open circles in the insert (enlarged view of KEO) denote track of catenary-moored NOAA-KEO surface buoy (SB) between July 2014 and July 2016, and solid inverted triangle denotes JAMSTEC-sediment trap (ST) station. Background is horizontal distribution of annual mean of phosphate based on World Ocean Atlas data (https://www.nodc.noaa.gov/OC5/WOA09/pr_woa09.html) drawn by Ocean Data View (Schlitzer 2003). Small open circles in main figure denote former JAMSTEC subarctic and subtropical time-series stations (K2 47°N/160°E and S1 30°N/145°E, respectively)

Oceanography vol. 72, no. 3, 2016; Honda et al. 2017). An important finding of this project was that annual mean of phytoplankton biomass and primary productivity at the oligotrophic station S1 were comparable to that at the eutrophic station K2 (Matsumoto et al. 2014, 2016; Honda et al. 2017). Consistent with the K2S1 results, recently, using data from the National Oceanic and Atmospheric Administration (NOAA) Kuroshio Extension Observatory (KEO) surface mooring, Fassbender et al. (2017) estimated the annual net community productivity (aNCP) in the surface mixed layer of oligotrophic North Pacific subtropical region to be 7 ± 3 mol-C m^{-2} $year^{-1}$. This estimate was higher than that in the eutrophic North Pacific subarctic region reported previously (0.5–5 mol-C m^{-2} $year^{-1}$, references in Fassbender et al. 2017) and, thus, they called the KEO area "the biological hot spot for carbon cycle." Based on chemical/physical observations and numerical simulations, several likely "missing nutrient sources" at the oligotrophic station were suggested, including regeneration, meso-scale eddy driven upwelling, meteorological events, and eolian inputs in addition to winter mixing (Sasai et al. 2016).

Understanding vertical motion associated with mesoscale and submesoscale eddies is an area of active research. Cyclonic eddies generally have a shallow thermocline at their center and as these eddies move past a point (e.g., KEO), the subsurface nutricline can temporarily rise into the euphotic zone. As discussed by Martin and Pondaven (2003), eddies can be classified as either "linear" or "non-linear." If the eddy is "linear," the nutricline locally rises and falls as the eddy passes, and the local nutricline generally is in the euphotic zone for a short duration. In contrast, if the cyclonic eddy is "non-linear," the uplifted water tends to be trapped in the center of the eddy as it is advected past the local point. In the eddy-rich region of the KEO station, both types of eddies are found: Rossby waves generated by wind anomalies in the central Pacific, as well as non-linear eddies pinched off by large amplitude Kuroshio extension meandering (Qiu 2002). Wind acting on the eddies and eddies forming through a hydrodynamic instability can also have enhanced vertical velocity patterns (McGillicuddy Jr et al. 2007). In addition, Mahadevan et al. (2008) showed that submesoscale dynamics can cause very large vertical velocities along the periphery of the eddies. Effects of meso-scale cyclonic eddies on the supply of nutrients to the surface have been observed with in situ and/or remotely sensed observations at the Bermuda Atlantic Time-series Study (BATS) (McGillicuddy Jr et al. 1998, 2007; Sweeney et al. 2003) and in the Pacific (Sukigara et al. 2014; Inoue and Kouketsu 2016: Kouketsu et al. 2016; Inoue et al. 2016). Sasai et al. (2010) also studied these effects through numerical simulations. In previous studies, the contribution of eddy-induced nutrient to new production were estimated to be from less than 10% to ~ 100% (Falkowski et al. 1991: < 20%; McGillicuddy Jr and Robinson 1997: ~ 70%: McGillicuddy Jr et al. 1998: ~ 40%: Oschlies and Garcon 1999: ~ 100%; Siegel et al. 1999: ~ 50%; Oschlies 2002: < 10%).

Another mechanism of nutrient supply to the surface sun-lit but oligotrophic layer is meteorologically generated turbulence. Typhoons frequently pass through the NW Pacific subtropical area and, sometimes, near the KEO surface mooring. During the force stage, as the typhoon passes a site, Ekman pumping followed by strong wind- and surface cooling-mixing can lead to a rapid rise of the mixed layer depth followed by mixed layer deepening. Inertial oscillations are also generated preferentially on the right side of the storm, although can occur on the left side if it is transitioning to an extratropical storm (Bond et al. 2011). Convergences and divergences associated with the inertial oscillatory upwelling and downwelling (inertial pumping) after the typhoon has passed. As the oscillations propagate downward into the seasonal thermoclines as near-inertial internal waves, vertical shears can form and cause shear instabilities that result in enhanced turbulent mixing and vertical diapycnal fluxes of nutrients and heat (Inoue et al. 2016). Typhoon-enhanced surface water mixing resulting in ocean productivity has been seen in satellite data (Lin 2012; Lin et al. 2003; Babin et al. 2004; Siswanto et al. 2007, 2008) and numerical simulations (Shibano et al. 2011; Menkes et al. 2016; Pan et al. 2017). Inoue et al. (2016) showed that typhoon-generated near-inertial internal waves, propagating and breaking in the nutricline led to high export fluxes and high chlorophyll patches in the deep chlorophyll maximum layer. Subsurface nutrient supply and increased productivity have been observed following the passage of typhoon by Zhang et al. (2014) and Rumyantseva et al. (2015). Moreover, another possible nutrient source to the open ocean is eolian dust (Additional file 1). It has been reported that natural and anthropogenic eolian dust includes major/micro-nutrients such as nitrogen and Fe, which can affect ocean biogeochemistry (e.g., Jickells et al. 2005; Duce et al. 2008).

In order to study nutrient supply mechanisms that support ocean productivity in the western North Pacific oligotrophic subtropical region, since July 2014, a deep-sea (~ 5000 m) sediment trap mooring has been maintained at station KEO, adjacent to the NOAA surface mooring. The KEO surface mooring has made meteorological and physical oceanographic observations since 2004 and surface carbon measurements since 2006. The meteorological and physical KEO data were used by Cronin et al. (2013, 2015) to evaluate the surface heat budget and processes affecting the formation and erosion of the seasonal thermocline. Fassbender et al. (2017) then combined these physical and meteorological time-series with the carbon time-series to describe the mixed layer carbon budgets at KEO, and in particular, the net community production and calcium carbonate production there. Here, we focus on the sediment trap data to relate these physical, meteorological, and biological processes to the biogenic material flux data at 5000 m.

Methods/Experimental

Station KEO

The KEO time-series station in the western Pacific subtropical gyre was established in 2004 by Pacific Marine Environmental Laboratory (PMEL) of NOAA with the deployment of a surface mooring (Fig. 1). One of the primary purposes of the surface mooring has been to measure air-sea heat fluxes. Due to the cold, dry air of continental origin blowing over the warm Kuroshio Extension current, surface heat fluxes out of the ocean in this region are some of the largest found in the basin, a feature that is typical of western boundary current extension (WBCE) regions (Cronin et al. 2008, 2013). As is also characteristic of WBCE regions, meso-scale eddies are very energetic near KEO (e.g., Inoue and Kouketsu 2016). During the warm season, tropical cyclones routinely pass near KEO as they transition to extra-tropical storms (Tomita et al. 2010). In addition, there is large inter-annual and decadal variability in the KEO region that can affect the eddy energy, recirculation gyre size and shape, and Subtropical Mode Water ventilation, all of which can affect the biogeochemistry at station KEO (Qiu et al. 2014; Oka et al. 2015).

Based on climatological data from National Institute for Environmental Studies (NIES)-SOOP database (soop.jp), World Ocean Atlas (https://www.nodc.noaa.gov/OC5/WOA09/pr_woa09.html) and observed data (Japan Agency for Marine-Earth Science and Technology (JAMSTEC) K2S1 database (https://ebcrpa.jamstec.go.jp/k2s1/en/), concentrations of nutrients are quite low in the KEO region compared to the subarctic (Fig. 1), and thus this area can be categorized as oligotrophic.

KEO buoy data

The KEO NOAA surface mooring (sinker position for 2014–2016: 32°23′N/144°32′E), has been equipped with various physical and meteorology sensors since 2004 and a suite of carbon sensors since 2006. Multiple CTD (above 525 m) and 1–3 current meters (above 36 m) on the mooring provided data during experimental period. Data and details of the NOAA surface buoy mooring are available at https://www.pmel.noaa.gov/ocs/data/disdel/. Although drifting buoy, Argo float, and satellite are often used for time-series analyses, drifter and float data are not Eulerian observations and satellite can observe only sea surface condition. In contrast, this NOAA moored surface buoy provides high-resolution time-series of surface ocean data, and underwater physical oceanographic data, such as upper ocean profile time-series of temperature and density.

Sediment trap experiment and sample chemical analysis

The sediment trap experiment at KEO began in July 2014 and is ongoing. This study reports on results from the first 2 years, between July 2014 and June 2016, while drawing upon the longer meteorological and physical oceanographic time-series from the adjacent NOAA surface buoy in order to interpret seasonal variability in sediment trap data. A time-series sediment trap (McLane Mark VII-21) with 21 collecting cups was deployed at 32°22′N/144°25′E and about 5000 m (800 m above seafloor), ~ 10 km westward (6 nautical mile) from the NOAA-PMEL KEO buoy. The sediment trap mooring system was initially deployed by R/V Kaiyo in June 2014 and was then turned around by M/V Bluefin in September 2015 and by R/V Shinsei-maru in July 2016. Sampling interval (sampling period per one collecting cup) was 18 days. Before deployment, collecting cups were filled with seawater-based buffered 10% formalin solution. Non-filtered seawater for this preservative was collected from 2000 m. Its salinity was adjusted to approximately 39 by addition of NaCl (addition of 50 g NaCl to 10 L of seawater). Initial pH of this solution was ~ 8.1. After recovery of the sediment traps, samples were stored in a refrigerator at 4 °C until the chemical and biological analysis could be performed at a shore-based laboratory.

After pre-treatment (swimmer elimination, splitting, filtration, drying, weighing for total mass flux, and pulverization) at a shore-based laboratory, the major components of the sinking particles were measured. Concentrations of organic carbon, inorganic carbon and nitrogen were measured with an elemental analyzer (Perkin-Elmer 2400 CHN/O, USA). Concentrations of Al, Si, Ca, and other trace elements such as Fe and Ti were measured with an inductively coupled plasma emission spectrometer (Perkin-Elmer Optima 3300DV, USA). Concentrations of biogenic opal (opal, $SiO_2{\cdot}0.4H_2O$) and $CaCO_3$ were estimated with following equations:

$$\text{Opal}\ (\%) = (\text{Si} - 3.42 \times \text{Al}) \times 67.2/28 \tag{1}$$

$$\text{CaCO}_3\ (\%) = (\text{Ca} - 0.5 \times \text{Al}) \times 100/40 \tag{2}$$

where coefficients are based on crustal ratios (Taylor 1964). Honda et al. (2002, 2013) provide further details about pre-treatment and chemical analysis.

Numerical simulation of eddy-induced nutrient supply

In order to evaluate the effect of cyclonic eddies on biogeochemistry semi-quantitatively, eddy-induced increase of nitrate concentration in the upper 100 m were estimated between 2000 and 2012 using an eddy-resolving coupled physical-biological model of the North Pacific Ocean. In particular, we used the Ocean general circulation model For the Earth Simulator (OFES) with sea-ice process (Masumoto et al. 2004; Komori et al. 2005) coupled to a simple nitrogen-based Nitrate-Phytoplankton-Zooplankton-Detritus (NPZD) pelagic model (Oschilies 2001). The OFES domain extends from 20°S in the south Pacific to 66°N in the North Pacific Ocean and from 100°E to 70°W. The eddy-resolving model is 1/10° and has 54 vertical levels from 5 m thickness at the surface to 330 m thickness at the maximum depth of 6065 m. The OFES was integrated for 30 years under climatological forcing using the Japanese 25-year Reanalysis (JRA25) (Onogi et al. 2007), from the observed climatological fields of temperature and salinity fields (WOA09: https://www.nodc.noaa.gov/OC5/WOA09/pr_woa09.html) without motion. After 30 years of spin-up integration, the OFES was integrated from 1979 to 2012 using 6-hourly JRA25. Detail of the biological model is described elsewhere (Sasai et al. 2007, 2010). To reach equilibrated biological fields, the biological model was incorporated after 30 years spin-up of the OFES with JRA25 climatological forcing and was integrated over a 5-year period with the climatological forcing. The biological fields at the end of the 5 years were then used as the initial condition for the physical-biological model, driven by the JRA25 from 1995 to 2012.

Results and discussion

Seasonal variability in materials' flux

Over the experimental period, averages of biogenic material fluxes at KEO and their standard deviations (organic carbon 2.2 ± 1.3; opal 12.8 ± 12.2; $CaCO_3$ 14.6 ± 8.5 mg m^{-2} day^{-1}; Fig. 2) are comparable to those at S1 (organic carbon 2.3 ± 0.7; opal 7.0 ± 2.9; $CaCO_3$ 20.4 ± 6.1 mg m^{-2} day^{-1}; K2S1 database). On the other hand, fluxes at KEO are significantly smaller than that at K2 (organic carbon 5.5 ± 1.6; opal 80.7 ± 36.2; $CaCO_3$ 38.5 ± 8.6 mg m^{-2} day^{-1}; K2S1 database), while primary productivity at stations K2 and S1 were, interestingly, comparable (Matsumoto et al. 2014, 2016; Honda et al. 2017). The cause of larger vertical attenuation in biogenic settling particle in the subtropical region (such as station KEO or S1) than the subarctic region (such as station K2) is still under debate (e.g., Honda and Watanabe 2010).

During the first year (July 2014–July 2015), remarkably large increases in total mass flux (TMF), which were more than 40% larger than TMF in the preceding and subsequent samples, were observed three times: early October 2014, late December 2014–early January 2015, and late April 2015 (Fig. 2a). During the second year (September 2015–June 2016), although its intensity was

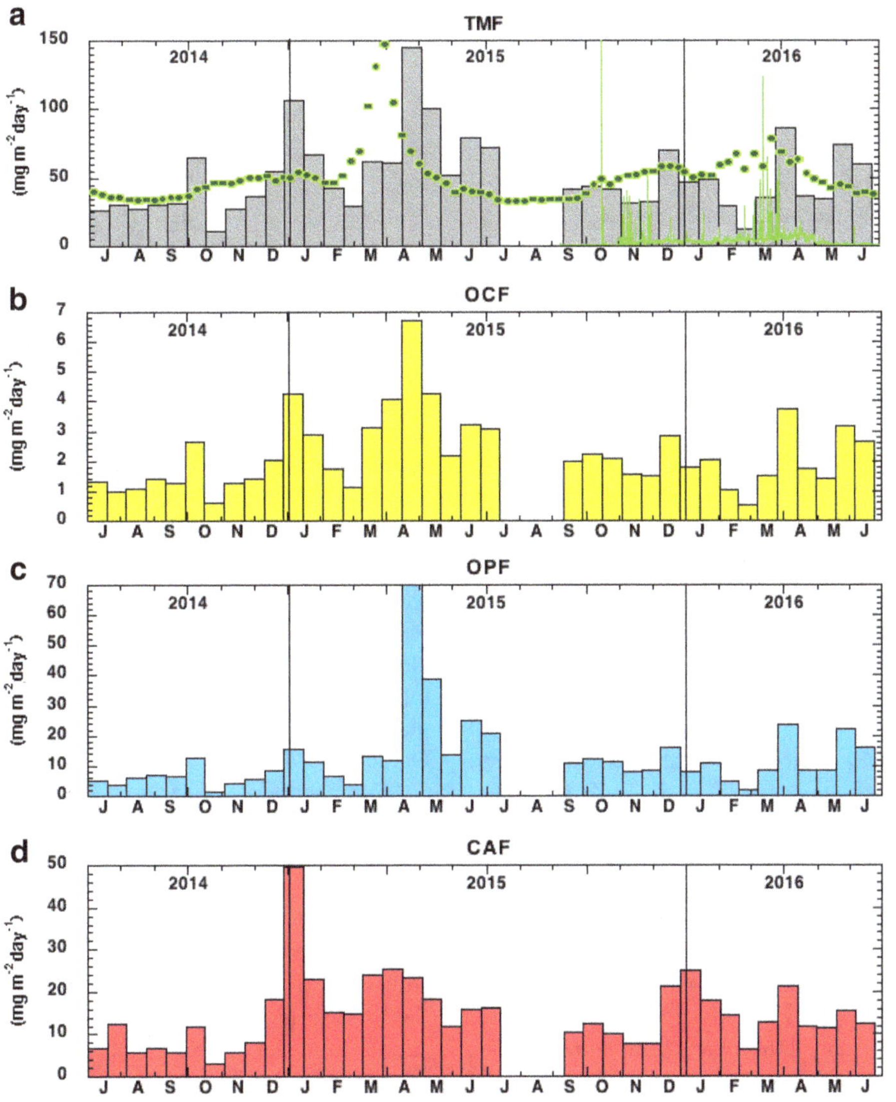

Fig. 2 Seasonal variability in fluxes of total mass (**a** TMF), organic carbon (**b** OCF), biogenic opal (**c** OPF), and $CaCO_3$ (**d** CAF) at 5000 m of KEO (bar graphs). Solid circles and yellow-green line graph in **a** TMF denote index of satellite-based surface Chl-*a* (8 days composite) and subsurface Chl-*a* (at about 35 m) observed between September 2015 and June 2016, respectively. Surface Chl-*a* with 4 km and 8-day spatiotemporal resolutions derived from the moderate resolution imaging spectroradiometer were processed by the NASA's Ocean Biology Processing Group at Goddard Space Flight Center

not as large as in the first year, TMF was increased two times during almost the same seasons as the first year: middle December 2015 and late March–the first half of April 2016. Average and seasonal variability (standard deviation of seasonal flux) of TMF between September and June in the first year (60.7 ± 34.2 mg m^{-2} day^{-1}) were slightly but significantly higher than that in the same period of the second year (45.3 ± 18.9 mg m^{-2} day^{-1}) ($p < 0.05$, F test).

Satellite-based surface chlorophyll-*a* concentration (Chl-*a*) started to increase substantially in late February and peaked in late March in both years (Fig. 2a). If it can be assumed that the sinking velocity of particulate materials in the deep sea (> 1000 m) is 100–200 m day^{-1} (Honda et al. 2009 and references therein), then peaks of TMF at about 5000 m in late April 2014 (the first half of April 2015) may reflect this Chl-*a* increase, i.e., phytoplankton increase in the surface sun-lit layer. On the other hand, not all TMF peaks synchronize with increase of surface Chl-*a*. However, subsurface Chl-*a* and backscatter at about 35 m observed in the second year by a fluorometer/backscatter meter package (Satlantic Eco-Triplet, Canada) installed on the KEO buoy mooring line increased in not only March 2016 but also November 2015 (Fig. 2a). Backscatter observed with this sensor also increased simultaneously with Chl-*a* (data not shown). The observed time-series and numerical simulation of vertical profile of Chl-*a* at the subtropical time-series station S1 (K2S1 project, Honda et al. 2017) showed occurrence of subsurface Chl-*a* increase or subsurface Chl-*a* maximum (SCM) in summer and autumn (Fujiki personal communication 2017; Inoue et al. 2016; Sasai et al. 2016). Thus, it can be presumed that the TMF increase observed in December 2015 reflected this subsurface phytoplankton increase in November 2015.

Seasonal variability in organic carbon flux (OCF) was similar to that of TMF (Fig. 2b) and OCF correlated well with TMF ($r^2 = 0.97$). Biogenic opal flux (OPF) also showed similar seasonal variability to TMF ($r^2 = 0.81$) and OCF ($r^2 = 0.81$) (Fig. 2c). Increases of OCF in April 2015 and March–April 2016 were likely attributed to increase of diatom and/or plankton with biogenic opal test. On the other hand, correlation coefficients between $CaCO_3$ flux (CAF) and TMF, and between CAF and OCF were slightly lower ($r^2 = 0.76$ and 0.71, respectively) (Fig. 2d). Unlike OPF, peaks of CAF appeared in December 2014–January 2015 and December 2015–January 2016. Thus, winter peak in OCF and CAF was likely attributed to the autumn increase of subsurface phytoplankton with $CaCO_3$ such as coccolithophorids. This seasonal variability in phytoplankton species in the western Pacific subtropical region (station S1) is supported by previous studies (Fujiki et al. 2016).

Cyclonic eddy and its effect on increases of phytoplankton and carbon flux

Based on NOAA surface buoy data, sea surface temperature (SST) at local noontime (12:00) varied from about 15 to 28 °C (Fig. 3a). Between January 2015 and April 2015, and between February 2016 and April 2016 winter cooling and mixing caused February subsurface subtropical mode water with low temperature (< 19 °C) to outcrop at the surface. As a result of mixing within this very deep surface layer, nutrients were supplied to the oligotrophic surface. When the layer began to re-stratify in spring, phytoplankton increased (spring bloom) in the nutrient-rich shallow surface layer as can be seen by the late March increase of surface Chl-*a* in Fig. 2a. Sequentially, deep sea biogenic material fluxes (OCF, CAF, and OPF) increased as observed at KEO in April–early May (Fig. 2b–d).

In addition to the deep winter mixing periods described above, shoaling of intermediate water with temperature lower than 15 °C and density higher than 25.5 (sigma theta) to depth of ~ 100 m (pressure of ~ 100 dbar) was observed in late July or early August and November in 2014 (Fig. 3a, b). During each of these periods, satellite-based sea surface height anomalies (SSHA) at KEO were remarkably low, with values of about − 50 cm (Fig. 3a) due to a nearby meso-scale cyclonic eddy (Additional file 2: Figure S1). We hypothesize that these isopycnal shoaling events are caused by cyclonic eddy-induced upwelling, which supplied nutrients to depths of near 100 m below the surface. Although nutrients were not supplied to the surface, 90–100 m is still within or at the bottom of the euphotic zone during these periods. As described above, the SCM during summer and winter was observed and numerically simulated for this region. Thus, passage of cyclonic eddies in late July and in November 2014 (hereinafter CE1 and CE2, respectively) might have introduced an increase of subsurface phytoplankton. If so, the TMF increase observed in early October 2014 and in late December 2014–early January 2015 might reflect a subsurface phytoplankton increase not observed by satellite. Based on SSHA and vertical profile of seawater temperature, the cyclonic eddy also passed KEO in late April 2015 and in February 2016. However, these periods were when the seasonal thermocline was very weak or absent, and, thus, it is not clear how these eddies could affect phytoplankton. We further note that the average biogenic materials' flux (sum of OCF, CAF, and OPF) in the first year (35.5 ± 24.5 mg m^{-2} year^{-1} for September 2014–June 2015) was significantly higher than that in the second year (27.0 ± 10.3 mg m^{-2} year^{-1} for September 2015–June 2016) ($p < 0.005$, F test). This might also be attributed to not only stronger winter mixing but also more cyclonic eddy passing near KEO in the first year during periods when the seasonal thermocline is present.

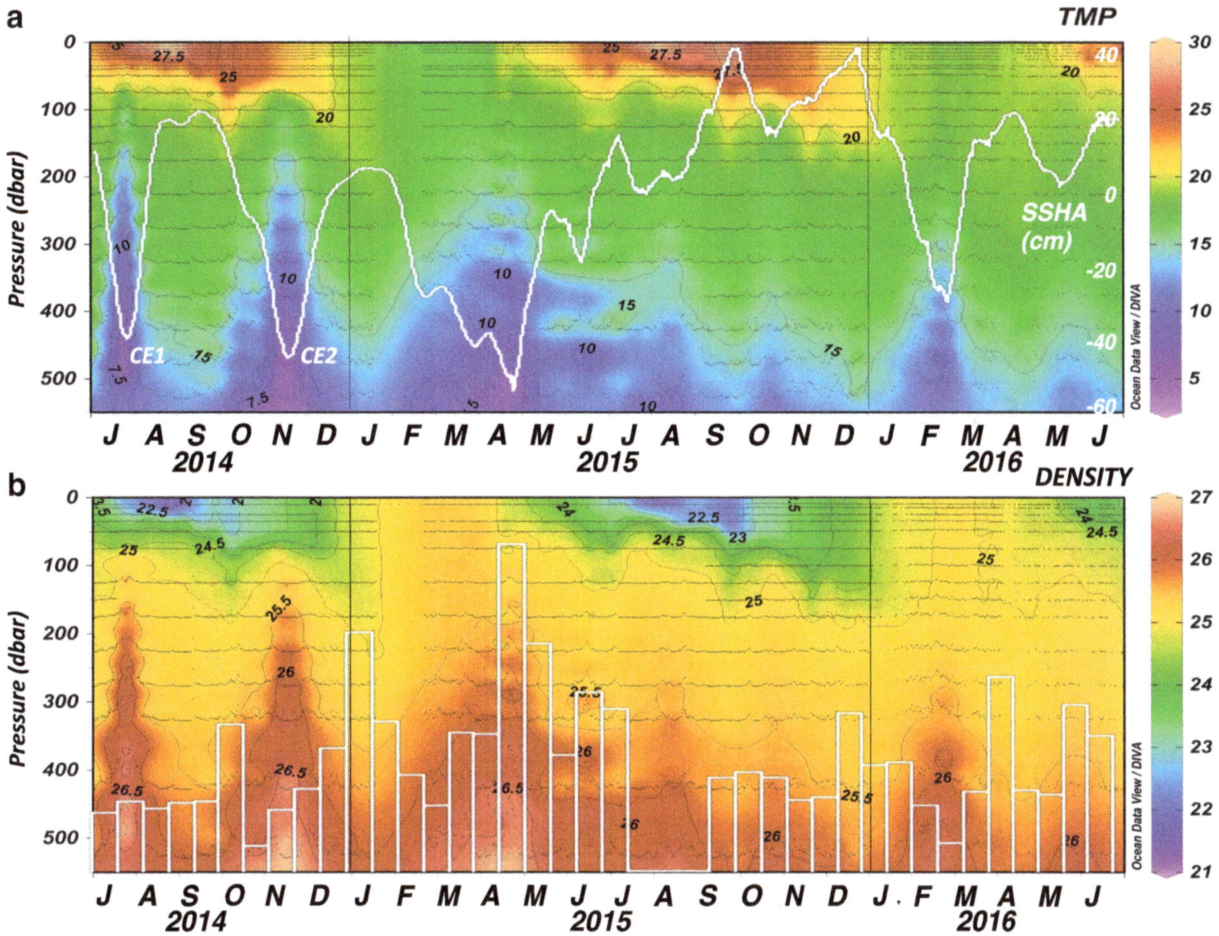

Fig. 3 Time-series contour plots of daily-averaged (**a**) seawater temperature (°C) and (**b**) density (sigma-theta, kg m^{-3}), for the upper ~ 550 m (less than 550 dbar) at station KEO between July 2014 and June 2016, using data available from https://www.pmel.noaa.gov/ocs/data/display-delivery/disdel/index.html. Solid line is isoline and dots are sampling points. White line in **a** denotes temporal variability in satellite-based sea surface height anomaly (SSHA) daily data. Daily SSHA data, which were acquired from Copernicus Marine Environment Monitoring Service, were specifically defined as departure from 20-year long-term mean SSH (1993–2012). White open bars in **b** denote TMF (same as Fig. 2a)

Quantitative evaluation of individual cyclonic eddy as a nutrient supplier

According to satellite-based SSHA and NOAA buoy data after 2008, cyclonic eddies with SSHA < − 50 cm typically passed near station KEO a few times (0–3 times) per year (Additional file 3: Figure S2 or Fig. 9 in Fassbender et al. 2017) and, at those times, subsurface water ascended to shallower than 100 m. In order to evaluate the effect of cyclonic eddy on biogeochemistry in more detail, the eddy-induced increase of nitrate concentration in the upper 100 m (from 100 to 0 m) was estimated between 2000 and 2012 using an eddy-resolving coupled physical-biological ocean model (see the "Numerical simulation of eddy-induced nutrient supply" section). The increased integrated nitrate concentration in the upper 100 m caused by cyclonic eddy with SSHA < − 50 cm ($\Delta_{int}N_{(eddy)}$) were estimated by taking the difference of the maximum integrated nitrate concentration in the upper 100 m when the cyclonic eddy passes KEO from the integrated nitrate concentration in the upper 100 m just before the cyclonic eddy passes. As shown in Fig. 4, the numerical simulation successfully reproduced the cooling and nutrient enrichment from wintertime mixing, as well as the upwelling of subsurface cold and nutrient rich water to shallower than 100 m that occurred a few (0–3) times a year when cyclonic eddies with SSHA < − 50 cm pass near station KEO. Each $\Delta_{int}N_{(eddy)}$ and total net of $\Delta_{int}N_{(eddy)}$ for respective year (annual $\Delta_{int}N_{(eddy)}$ or $\Delta_{int}AN_{(eddy)}$) were estimated to be 0.02–1.07 mol-N m^{-2} and 0.10–1.07 mol-N m^{-2} $year^{-1}$, respectively (Table 1). Between 2000 and 2012, average of $\Delta_{int}AN_{(eddy)}$ (with standard deviation) were estimated to be 0.40 (± 0.36) mol-N m^{-2} $year^{-1}$. Some eddies passed near KEO during the deep winter mixing (DWM) period. In this case, upwelling of subsurface nutrient-rich water is introduced by

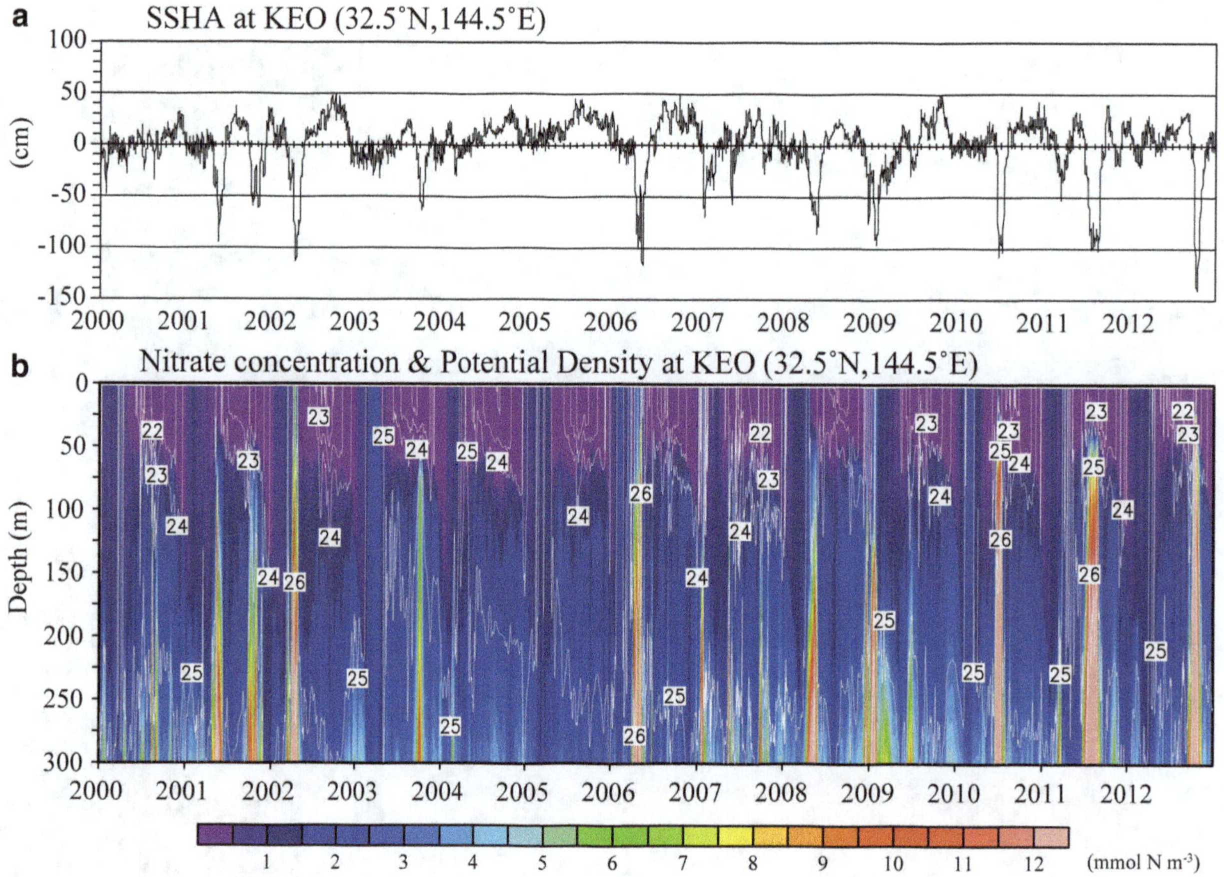

Fig. 4 a Simulated SSHA and (**b**) time-series contour plots of simulated nitrate concentration, at KEO between 2000 and 2012. White line and numbers in **b** denote potential density

DWM and $\Delta_{int}N_{(eddy)}$ might be overestimated. We also calculated $\Delta_{int}AN_{(eddy)}$ without $\Delta_{int}N_{(eddy)}$ during the DWM period (0.31 ± 0.38 mol-N m^{-2} year^{-1}, Table 1). This estimate was comparable or slightly higher than that estimated for the Sargasso Sea (0.35 mol-N m^{-2} year^{-1} by McGillicuddy Jr and Robinson 1997, 0.19 mol-N m^{-2} year^{-1} by McGillicuddy Jr et al. 1998, and 0.18 mol-N m^{-2} year^{-1} by Siegel et al. 1999).

Next, we attempt to evaluate semi-quantitatively the impact of the CE1 and CE2 eddies on 5000 m particle flux, recognizing that the variables have large uncertainties, particularly the amount of nutrient (N) supplied to the upper 100 m by CE1 and CE2, fraction of supplied N utilized by phytoplankton, and fraction of primary productivity (PP) transported vertically to 5000 m. Based on our simulation, a cyclonic eddy with SSHA of ~ − 50 cm, which is comparable in size to CE1 and CE2 (cyclonic eddies 2001–2, 2001–3, 2003–1, 2007–2, 2011–1 in Table 1), can potentially supply nutrient (N) to the upper 100 m ($\Delta_{int}N_{(eddy)}$) at a rate of ~ 0.08 mol-N m^{-2} on average. If supplied N is completely (or 100%) utilized by phytoplankton, primary productivity can increase by ~ 0.50 mol-C m^{-2} ($C/N_{(mole)} = 6.6$, Redfield et al. 1963) or ~ 6.02 g-C m^{-2}. Honda et al. (2017) reported that ~ 0.6% of PP was transported vertically to 5000 m at western Pacific subtropical area (S1) as particulate organic carbon at 5000 m (OCF) on annual average (PP 329 mg-C m^{-2} day^{-1}; OCF 2.0 mg-C m^{-2} day^{-1}). If this relation can be applied to PP and OCF at KEO, increase of OCF following increase of PP can be estimated to be 36.6 mg-C m^{-2}. This potential increase is comparable to increases in OCF observed in early October 2014 ((2.64–1.29) mg-C m^{-2} day^{-1} × 18 days = 24.3 mg-C m^{-2}) and in December 2014–January 2015 ((4.27–2.05) mg-C m^{-2} day^{-1} × 18 days = 40.0 mg-C m^{-2}, Fig. 2b). Thus $\Delta_{int}N_{(eddy)}$ can potentially support increase of OCF.

It is noted that if TMF or OCF increase observed in early October 2014 was due to increase of subsurface phytoplankton following an increase in nutrient supplied by CE1 in late July 2014, sinking velocity of the particulate biogenic materials is estimated to be less than 100 m day^{-1} (4900 m/60 days). This sinking velocity is slightly lower than that reported generally (> 100 m day^{-1}, e.g., Honda et al. 2009; Honda et al. 2013). However, it might be conceivable because

Table 1 Simulated increase of integrated nitrate concentration upper 100 m by pass of cyclonic eddy with SSHA < – 50 cm ($\Delta_{int}N_{(eddy)}$) and annual $\Delta_{int}N_{(eddy)}$ ($\Delta_{int}AN_{(eddy)}$) between 2000 and 2012. $\Delta_{int}AN_{(eddy)}$ without winter cooling mixing period (WCM) data are also shown. "0" of S/N denotes that no eddy passed over KEO in respective years

Year	S/N	Period with SSHA < – 50 cm	$\Delta_{int}N_{(eddy)}$ (mol-N m^{-2})	$\Delta_{int}AN_{(eddy)}$ (mol-N m^{-2} $year^{-1}$)	$\Delta_{int}AN_{(eddy)}$ (mol-N m^{-2} $year^{-1}$) without WCM data
2000	0		0.00	0.00	0.00
2001	1	11 May–30 May	0.22	0.32	0.32
	2[a]	3 October–11 October	0.08		
	3[a]	10 October–2 November	0.02		
2002	1	31 March–26 April[b]	0.68	0.68	0.00
2003	1[a]	1 October–12 October	0.10	0.10	0.10
2004	0		0.00	0.00	0.00
2005	0		0.00	0.00	0.00
2006	1	7 April–25 April	0.36	0.99	0.99
	2	28 April–11 May	0.63		
2007	1	18 January–26 January[b]	0.16	0.27	0.12
	2[a]	16 May–16 May	0.12		
2008	1	16 April–23 May	0.10	0.34	0.34
	2	16 December–21 December	0.24		
2009	1	8 January–30 January[a]	0.33	0.33	0.00
2010	1	20 June–16 July	0.65	0.65	0.65
2011	1[a]	13 March–13 March[b]	0.06	0.44	0.38
	2	9 July–26 August			
2012	1	4 October–31 October	1.07	1.07	1.07
			Average of ΔintAN(eddy)	0.40	0.31
			standard deviation	0.36	0.38

[a]Cyclonic eddy with SSHA of about – 50 cm
[b]This period is in the deep winter mixing (DWM) period

(1) Slow sinking particle with sinking velocity of about 50 m day^{-1} was reported based on measurement of particulate radiocesium emitted from the Fukushima Daiichi Nuclear Power Plant accident in March 2011 (Honda and Kawakami 2014), and
(2) When the surface ocean is well stratified and transparent, exopolymer particles (TEPs) are produced in this stratified layer by phytoplankton such as diatom (Mari 1999), coccolithophorids (Engel et al. 2004), and picophytoplankton, such as *Prochlorococcus* (Iuculano et al. 2017), TEPs, which is composed mainly of acidic polysaccharides and is sticky mucous, promoting the formation of marine aggregates (Passow et al. 2001). These aggregates are usually large, but loose and fluffy with low density and/or large drag, and therefore, might sink slowly or hardly at all (Bach et al. 2016). In late July when CE1 passed, KEO area was well stratified (surface mixed layer was about 20 m). It is also suspected that *Prochlorococcus* is predominant in autumn in this area (Fujiki et al. 2014). In addition, slow sinking particles might not be always transported downward: their motion might sometimes be upward and sometimes lateral. If this is the case, the apparent sinking velocity would become very small. Likewise,
(3) Slow sinking velocity of settling particles observed in early October 2014 might also be supported by its lower density: concentration of "biogenic ballast" (sum of opal and $CaCO_3$ concentrations), with density more than double that of organic matter ($CaCO_3$ 2.71; opal 2.10; organic matter 1.06 g cm^{-3}, e.g., Klaas and Archer 2002), bringing particle to the ocean interior quickly (e.g., Klaas and Archer 2002; Francois et al. 2002; Honda and Watanabe 2010), was about 30% while that of settling particle observed in late December 2014–early January 2015 was about 60%.

Contribution of annual eddy-induced nutrient to annual carbon flux

In addition, $\Delta_{int}AN_{(eddy)}$ (annual $\Delta_{int}N_{(eddy)}$) reported here (Table 1) can be compared to previous annual carbon (C) export studies at the western Pacific subtropical station S1 (30°N/145°E, e.g., Honda et al. 2016, 2017), about 140 miles southeast of station KEO. Assuming 100% utilization of $\Delta_{int}AN_{(eddy)}$ and application of the Redfield ratio, $\Delta_{int}AN_{(eddy)}$-supported annual PP can be estimated to be 2.0–2.6 mol-C m^{-2} $year^{-1}$ (0.31–0.40 mol-N m^{-2} $year^{-1}$ × 6.6) (Table 2). At station S1, net primary production was observed to be 10.6 mol-C m^{-2} $year^{-1}$ (Matsumoto et al. 2016; Honda et al. 2017). The *F* ratio, which is the uptake ratio of nitrate (an externally-supplied nutrient) to the sum of this nitrate and ammonium (regenerated nutrient) and is comparable to the ratio of annual C export flux to net primary production, was estimated to be ~ 0.3 by the numerical simulation (Sasai et al. 2016). Thus, annual C export flux or new production can be estimated to be 3.2 mol-C m^{-2} $year^{-1}$ (10.6 × 0.3). Annual C export fluxes at 100 and 200 m at S1 were estimated to be 1.7 and 1.3 mol-C m^{-2} $year^{-1}$, respectively, based on particulate organic carbon (POC) flux at respective depths observed by a drifting sediment trap (Honda et al. 2016). Sum of POC flux and dissolved organic carbon (DOC) flux at 200 m was also estimated from the carbon budget (2.7 mol-C m^{-2} $year^{-1}$: Wakita et al. 2016). Thus, $\Delta_{int}AN_{(eddy)}$-supported annual PP (2.0–2.6 mol-C m^{-2} $year^{-1}$) estimated in this study is comparable to the annual C export flux estimated from previous reports. As described earlier, aNCP at KEO was estimated to be 7 ± 3 mol-C m^{-2} $year^{-1}$ using KEO buoy pCO_2 and meteorological/physical oceanographic data (Fassbender et al. 2017). This value is slightly larger than our estimate and previous estimates. However, this might be reasonable because their aNCP was estimated for the surface mixed layer and, especially in summer, a portion of aNCP is likely to be regenerated in the upper layer (< 100 m) and not exported to the ocean interior.

Thus $\Delta_{int}AN_{(eddy)}$ estimated in this study was reasonable and it can be said that cyclonic eddies potentially play an important role in supplying nutrients to the oligotrophic upper layer and supporting much of the new production. However, our estimate is based on a simple numerical simulation without consideration of horizontal advection (i.e., efflux of nutrient from the eddy) and response rate of phytoplankton to temporal increase of nutrient (i.e., incomplete uptake). If efflux of nutrient and incomplete uptake had occurred, our estimate would have been overestimated. Inter-annual variability, which provides a measure of uncertainty to our interpretation of the annual average as a long-term mean, was also estimated to be about 100% (Table 1). On the other hand, if cyclonic eddies originate from the subarctic eutrophic region, they may already possess enough nutrient without vertical mixing (e.g., Sasai et al. 2010). For a more complete understanding of how cyclonic eddy affect productivity, we would need to determine whether the eddy was a non-linear cyclonic eddy (Martin and Pondaven 2003), and if so, if it was a coherent Lagrangian vortex (Haller and Beron-Vera 2013). We would also need to consider various factors such as its birth place, its path, and its overall time at the fixed point (e.g., KEO).

Effect of typhoon on physical oceanography and biogeochemistry

During the experimental period (July 2014–June 2016), abrupt barometric pressure drops observed by the KEO

Table 2 Comparison of simulated $\Delta_{int}AN_{(eddy)}$-supported C production with annual C export flux based on new production, particulate organic carbon (POC), and dissolved organic carbon (DOC) fluxes at station S1 and annual net community productivity (aNCP) in surface mixed layer at station KEO. POC flux at 100 and 200 m of station S1 were observed by seasonal drifting sediment trap deployment (Honda et al. 2016). Due to the many challenges of making sediment trap time-series from a mooring that extends into shallow water, for this KEO study, a deep sediment trap time-series mooring was used

	mol-C m^{-2} $year^{-1}$	References
$\Delta_{int}AN_{(eddy)}$-supported production[a]	2.0–2.6	This study
Annual N export flux based on:		
New production at S1[b]	3.2	Matsumoto et al. (2016); Honda et al. (2017); Sasai et al. (2016)
POC flux at 100 m of S1	1.7	Honda et al. (2016)
POC flux at 200 m of S1	1.3	Honda et al. (2016)
POC + DOC flux at 200 m of S1[c]	2.7	Wakita et al. (2016)
aNCP in surface mixed layer at KEO	7 ± 3	Fassbender et al. (2017)

[a]$\Delta_{int}AN_{(eddy)}$ (0.31–0.40 mol-N m^{-2} $year^{-1}$) multiplied by the Redfield ratio (6.6)
[b]Net primary productivity (10.6 mol-C m^{-2} $year^{-1}$) multiplied by *F* ratio (0.3)
[c]POC flux 2.49 mol-C m^{-2} $year^{-1}$; DOC flux 0.25 mol-C m^{-2} $year^{-1}$

surface mooring indicate that three typhoons in 2014 (Typhoon S/N: T1414, T1417, T1420) and two typhoons in 2015 (T1514, T1516) passed near KEO (Additional file 4: Figure S3). Here, we consider the possibility of whether one of these typhoons triggered the enhanced biogenic materials' flux. As described in the "Introduction" section, previous studies have reported that typhoons can play a significant role in subsurface nutrient supply resulting in increase of PP. However, in contrast to previous reports (Lin 2012; Lin et al. 2003; Babin et al. 2004; Siswanto et al. 2007, 2008), no remarkable increase of surface Chl-*a* during or after passage of these typhoons was observed (Fig. 2a). In addition, as can be seen in the profile time-series of seawater temperature (Fig. 5), large Ekman upwelling associated with the cyclonic winds, which could have supplied subsurface nutrient to the upper ocean, was also not evident.

On the other hand, it is notable that after typhoon T1414 passed, near-inertial internal waves occurred for about 10 days, with an inertial period of about 22 h and an amplitude of about 10 m in the upper seasonal pycnocline, i.e., around 50 m (Fig. 5a). Strong near-inertial oscillations could also be seen in the current meter at 36 m during T1414 (Additional file 5: Figure S4). Near-inertial internal waves are typically associated with vertical shears in horizontal velocity that can cause shear-instability mixing and enhanced diapycnal fluxes (Johnston et al. 2016, Inoue et al. 2017). Because this source of turbulent mixing within the euphotic zone extended for about 10 days, much longer than the duration of the passing typhoon, it is conceivable that this event could have provided a significant source of nutrients at a subsurface depth within the euphotic zone. If so, the turbulent mixing induced by typhoon passage might be responsible for

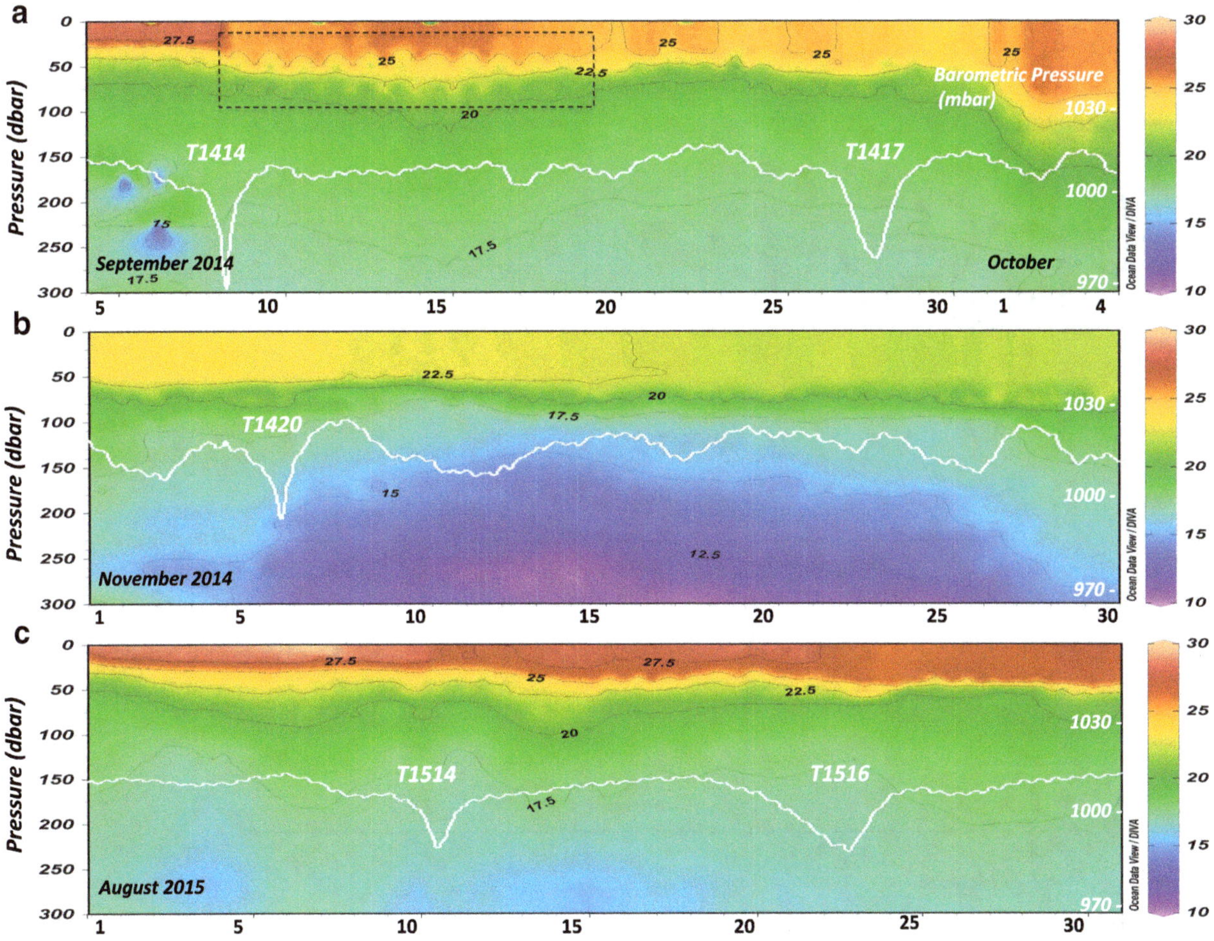

Fig. 5 Profile time-series of seawater temperature (°C) for the upper ocean (pressure less than 300 dbar, i.e., ~ 300 m) before and after typhoons (shown as T****, where **** is typhoon ID number) passed KEO. **a** Between 5 September and 4 October 2014. **b** In November 2014. (**c**) In August 2015. Hourly temperature data are obtained from NOAA KEO buoy database (https://www.pmel.noaa.gov/ocs/data/display-delivery/disdel/index.html). White line in each figure denotes hourly barometric pressure data from NOAA KEO buoy database. The near-inertial internal wave can be seen in broken square on **a**

the increase of OCF observed by the sediment trap mooring in October.

To test this hypothesis, the turbulent flux of NO_3 (F) at thermocline or nitracline by turbulent mixing following typhoon T1414 was calculated according to

$$F = \int_{t1}^{t2} \Delta Kp\,(C/Z)dt, \quad (3)$$

where ΔKp is the increase in turbulent diffusivity (diffusion coefficient, $m^2\ s^{-1}$) from an initial Kp ($10^{-5}\ m^2\ s^{-1}$ this study, see below) and t1 and t2 are the beginning and end of the mixing period. The C/Z is NO_3 concentration gradient (mmol-N m^{-4}). In this study, we suspected that C/Z in September between 50 and 100 m around KEO was 0.034 mmol m^{-4} based on climatological data (WOA2013: https://www.nodc.noaa.gov/cgi-bin/OC5/woa13/woa13ox-nu.pl?parameter=n, Additional file 6: Figure S5) and model simulation (Fig. 4b). This concentration is comparable to that at station S1 (0.042 mmol m^{-4}: Inoue et al. 2016).

Below the surface mixed layer in the KEO region, Kp has been reported to be ~ $10^{-5}\ m^2\ s^{-1}$ (Mori et al. 2008; Nagai et al. 2009; Kaneko et al. 2012; Inoue et al. 2016, 2017), but is expected to increase significantly during and following a typhoon. In order to investigate change in diffusion coefficient (ΔKp) at 50–70 m, where nitracline might exist (Additional file 6: Figure S5), a simple model simulation was conducted. Yablonsky and Ginis (2009) have shown that the cool SST wake following passage of typhoon is reproduced reasonably well by a one-dimensional model if the tropical cyclone translation speed is faster than 5 m s^{-1}. Because this is generally the case in the KEO region, a one-dimensional model could help in understanding the effect of typhoons on nutrient fluxes into the euphotic zone and resulting ocean productivity. Time evolution of the vertical profiles of the horizontal component of velocity, temperature and salinity and turbulent kinetic energy was solved using a one-dimensional model based on Mellor-Yamada-Nakanishi-Niino scheme (Furuichi et al. 2012). The turbulence closure model computed the development of turbulent kinetic energy (TKE) as well as a set of length scales. The eddy diffusivity ($m^2\ s^{-1}$) is set by a length scale times the square root of TKE. The model was initialized using the vertical profiles of the horizontal component of velocity, temperature, and salinity as observed by the KEO buoy on 00Z 1 September

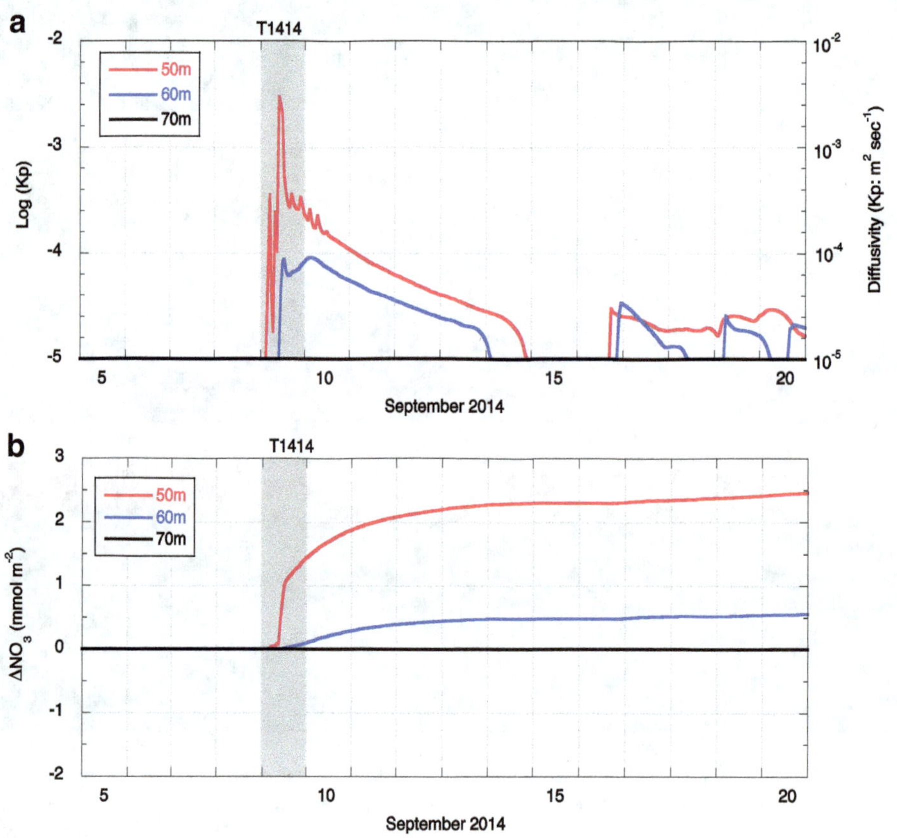

Fig. 6 **a** Time-series variability in simulated diffusion coefficient (Kp) at 50, 60, and 70 m before and after typhoons T1414 pass near KEO in September 2014. **b** Integrated turbulent NO_3 flux at 50, 60, and 70 m estimated using Kp (**a**) and concentration gradient (Additional file 6: Figure S5)

2009 and integrated for 1 month using the atmospheric forcing dataset as provided by https://www.pmel.noaa.gov/ocs/KEO.

Before typhoon T1414 passed near KEO buoy (i.e., before 8 September 2014), initial *Kp* at 50–70 m was set to be 1×10^{-5} m^2 s^{-1} (Fig. 6a). After typhoon T1414 passed on 9 September, *Kp* at 50 and 60 m largely increased to $\sim 10^{-2.5}$ m^2 s^{-1} and $\sim 10^{-4}$ m^2 s^{-1}, respectively. Thereafter, *Kp* at both depths decreased gradually and came back to initial coefficients (10^{-5}) after 15 September. On the other hand, *Kp* at 70 m did not increase after T1414 pass. Thus, these estimated increases in *Kp* relative to the background values of 10^{-5} m^2 s^{-1} (ΔKp) caused by typhoon T1414, by Eq. (3) can be associated with integrated turbulent NO_3 flux "F" of ~ 2.4 mmol-N m^{-2} that could have supplied NO_3 to the upper 50 m between 9 and 19 September (Fig. 6b). Following the same procedure used in the previous section ("Quantitative evaluation of individual cyclonic eddy as a nutrient supplier") for estimating the carbon flux supported by the eddy-induced N supply, the carbon flux supported by the typhoon-induced N supply was estimated. Doing so, we found that this typhoon-induced N supply could potentially introduce increases of PP of ~ 158 mg-C m^{-2} ($2.4 \times 6.6 \times 12$) and of 5000 m organic carbon flux (OCF) of ~ 0.95 mg-C m^{-2} (158×0.006). It is indicative that typhoon-induced N supply can support only ~ 4% of observed increase of OCF in early October (24.3 mg m^{-2}, see the "Quantitative evaluation of individual cyclonic eddy as a nutrient supplier" section). Thus, typhoon T1414 in early September 2014 did not contribute to OCF increase observed at 5000 m in early October. While typhoon is an intriguing potential mechanism for entraining nutrients into the euphotic zone, typhoon's significant role in nutrient supply was not observed in this study.

However, the above estimation was based on a climatological NO_3 vertical profile with a large standard deviation because its NO_3 data were collected from a broad area in the western Pacific subtropical region (Additional file 6: Figure S5). Seasonal observations of NO_3 vertical profile, especially during the typhoon season, are needed at station KEO to more accurately determine typhoon effect on physical oceanography and biogeochemistry.

Conclusions

Time-series observation of settling particle from a sediment trap deployed adjacent to a surface buoy, instrumented for meteorology, physical oceanography, and biogeochemistry, revealed that cyclonic eddies are likely an important mechanism for nutrient supply. Unlike the spring bloom, which occurs after deep winter mixing brings nutrients to the surface, eddies tend to supply the nutrients to a level below the surface that is still in the euphotic zone. As a consequence, the resulting bloom occurs below the surface and is not apparent in satellite data. For this reason, we refer to this as a "missing nutrient source." Surprisingly, typhoons were not observed to play a significant role in the nutrient supply in this study. While we have focused on the supply of nutrients from deep winter mixing, eddies, and typhoons, other sources may also contribute. In particular, further work is needed to quantify the role of N_2 fixation and eolian input for supplying nutrients in this oligotrophic ocean region.

Additional files

Additional file 1: Near-inertial internal wave. (DOCX 16 kb)

Additional file 2: Figure S1. SSHA around station KEO before, when and after the cyclonic eddy #1 and #2 (CE1 and CE2, respectively) passed over station KEO.

Additional file 3: Figure S2. Profile time-series of seawater temperature (°C) in the upper ocean above ~ 550 m at station KEO between 2008 and 2015. Figure is drawn with daily averaged data on KEO database website (https://www.pmel.noaa.gov/ocs/data/display-delivery/disdel/index.html). White line denotes temporal variability in satellite-based SSHA daily data.

Additional file 4: Figure S3. Tracks of typhoons during experimental 2 years. Lines with open circle, square, triangle, plus, and cross are for typhoons T1414, T1417, T1420, T1514, and T1516, respectively. Numbers in figure are date and time of typhoon position. Data are obtained online from the website of the Japan Meteorology Agency (https://www.data.jma.go.jp/fcd/yoho/typhoon/position_table/index.html).

Additional file 5: Figure S4. Hourly time-series of surface pressure (hPa, red line) and ocean current vector at 36 m depth (cm s^{-1}, blue vector) at KEO for T1414, T1417, T1420, T1514, and T1516.

Additional file 6 Figure S5. Vertical profile of NO_3 in the subtropical region near KEO in September. Data of respective depths with standard deviation are obtained from WOA 2013 climatological database (red circles 30.5–34.5°N/136.5–164.5°E; $n = 11$) and from numerical model simulation (blue circles based on Fig. 4). The NO_3 concentration gradient between 50 and 100 m is estimated to be ~ 0.034 mmol m^{-4} using 0.3 μmol kg^{-1} at 50 m and 2.0 μmol kg^{-1} at 100 m.

Abbreviations

BATS: Bermuda Atlantic Time-series Study; CAF: $CaCO_3$ flux; CE: Cyclonic eddy; Chl-*a*: Chlorophyll-*a* concentration; DOC: Dissolved organic carbon; DWM: Deep winter mixing; JAMSTEC: Japan Agency for Marine-Earth Science and Technology; JRA25: Japanese 25-year reanalysis; KEO: Kuroshio Extension Observatory; LDR: Local deepening rate; NCP: Net community productivity; NIES: National Institute for Environmental Studies; NIW: Near-inertial internal waves; NOAA: National Ocean and Atmosphere Administration; NPZD: Nitrate-Phytoplankton-Zooplankton-Detritus; OCF: Organic carbon flux; OFES: Ocean general circulation model For the Earth Simulator; OPF: Biogenic opal flux; PMEL: Pacific Marine Environmental Laboratory; POC: Particulate organic carbon; PP: Primary productivity; SCM: Subsurface Chl-*a* maximum; SSHA: Sea surface height anomaly; SST: Sea surface temperature; TEPs: Transparent exopolymer particles; TKE: Turbulent kinetic energy; TMF: Total mass flux; WBCE: Western boundary current extension; WOA: World Ocean Atlas

Acknowledgements

We acknowledge Dr. Kazuhiko Matsumoto from JAMSTEC, Technicians from Marine Works Japan, and ship crews of R/V Kaiyo, T/S Shinsei-maru, and M/V Bluefin for the assistance of mooring work and data acquisition. Dr. Yutaka Yoshikawa from Kyoto University kindly provided opportunity to submit this paper to this journal (Progress in Earth and Planetary Science). The NOAA surface mooring is supported by the Ocean Observing and Monitoring Division, Climate Program Office, National Oceanic and Atmospheric

Administration, US Department of Commerce. This is PMEL contribution 4643. OFES simulations were conducted on the Earth Simulator under the support of JAMSTEC. A part of this work was carried out by the joint research program of the Institute for Space-Earth Environmental Research (ISEE), Nagoya University. We deeply appreciate anonymous reviewers and the journal editor who provided constructive comments that helped improve this paper.

Funding
This study is financially supported by MEXT KAKENHI: grant-in-aid for scientific research in innovative areas JP15H05822, "Ocean Mixing Process; Impact on Biogeochemistry, Climate and Ecosystem" (URL: http://omix.aori.u-tokyo.ac.jp).

Authors' contributions
MCH designed this study and conducted time-series sediment trap experiment. YS was in charge of mathematical simulation. ES conducted satellite data analysis. AKY was in charge of analysis of typhoon and its impact on physical oceanography. MFC provided KEO buoy-derived meteorological/physical oceanographic data and edits to the text. HA conducted mathematical simulation on time-series variability in turbulent diffusion. All authors read and approved the final manuscript.

Competing interests
The authors declare that they have no competing interests.

Author details
[1]Japan Agency for Marine-Earth Science and Technology (JAMSTEC), Yokosuka, Kanagawa, Japan. [2]Disaster Protection Research Institution, Kyoto University, Shirahama, Wakayama, Japan. [3]Institute for Space-Earth Environmental Research, Nagoya University, Nagoya, Aichi, Japan. [4]National Ocean and Atmosphere Administration (NOAA), Pacific Marine Environment Laboratory (PMEL), Seattle, WA, USA.

References
Babin SM, Carton JA, Dickey TD, Wiggert JD (2004) Satellite evidence of hurricane-induced phytoplankton blooms in an oceanic desert. J Geophys Res 109:1–21. https://doi.org/10.1029/2003JC001938

Bach LT, Boxhammer T, Larsen A, Hildebrandt N, Schulz KG, Riebesell U (2016) Influence of plankton community structure on the sinking velocity of marine aggregates. Glob Biogeochem Cycles 30:1145–1165. https://doi.org/10.1029/2016GB005372

Bond NA, Cronin MF, Sabine C, Kawai Y, Ichikawa H, Freitag P, Ronnholm K (2011) Upper-ocean response to tyhpoon Choi-Wan as measured by the Kuroshio Extension Observatory (KEO) mooring. J Geophys Res 116(C02031). https://doi.org/10.1029/2010JC006548

Bopp L, Resplandy L, Orr JC, Doney SC, Dunne JP, Gehlen M, Halloran P, Heinze C, Ilyina T, Seferian R, Tjiputra J, Vichi (2013) Multiple stressors of ocean ecosystems in the 21 century: projections with CMIP5 models. Biogeosciences 10:6225–6245. https://doi.org/10.5194/bg-10-6225-2013

Cronin MF, Bond NA, Farrar JT, Ichikawa H, Jayne SR, Kawai Y, Konda M, Qiu B, Rainville L, Tomita H (2013) Formation and erosion of the seasonal thermocline in the Kuroshio extension recirculation gyre. Deep-Sea Res II 85:62–74

Cronin MF, Meinig C, Sabine CL, Ichikawa H, Tomita H (2008) Surface mooring network in the Kuroshio extension. IEEE Syst Spec Issue GEOSS 2(3):424–430

Cronin MF, Pelland NA, Emerson SR, Crawford WR (2015) Estimating diffusivity from the mixed layer heat and salt balances in the North Pacific. J Geophys Res 120. https://doi.org/10.1002/2015JC011010

Duce RA, LaRoche J, Altieri K, Arrigo KR, Baker AR, Capone DG, Cornell S, Dentener F, Galloway J, Ganeshram RS, Geider RJ, Jickells T, Kuypers MM, Langlois R, Liss PS, Liu SM, Middelburg JJ, Moore CM, Nickovic S, Oschlies A, Pedersen T, Prospero J, Schlitzer R, Seitzinger S, Sorensen LL, Uematsu M, Ulloa O, Voss M, Ward B, Zamora L (2008) Impacts of atmospheric anthropogenic nitrogen on the open ocean. Science 320:893_897

Engel A, Delille B, Jacquet S, Riebesell U, Rochelle-Newall E, Terbrüggen A et al (2004) Transparent exopolymer particles and dissolved organic carbon production by Emiliania huxleyi exposed to different CO2 concentrations: a mesocosm experiment. Aquat Microb Ecol 34:93–104. https://doi.org/10.3354/ame034093

Falkowski P, Zieman D, Kolber Z, Bienfang P (1991) Role of eddy-pumping in enhancing primary production in the ocean. Nature 352(55):55–55

Fassbender AJ, Sabines CL, Sutton AJ CMF (2017) Mixed-layer carbon cycling at the Kuroshio Extension Observatory. Glob Biogeochem Cycles 31. https://doi.org/10.1002/2016GB005547

Francois R, Honjo S, Krishfield MS (2002) Factors controlling the flux of organic carbon to the bathypelagic zone of the ocean. Glob Biogeochem Cycles 16(4):1087. https://doi.org/10.1029/2001GB001722

Fujiki T, Matsumoto K, Mino Y, Sasaoka K, Wakita M, Kawakami H, Honda MC, Watanabe S, Saino T (2014) Seasonal cycle of phytoplankton community structure and photophysiological state in the western subarctic gyre of the North Pacific. Limnol Oceanogr 59(1):887–900. https://doi.org/10.4310/lo.2014.59.0887

Fujiki T, Sasaoka K, Matsumoto K, Wakita M, Mino Y (2016) Seasonal variability of phytoplankton community structure in the subtropical western North Pacific. J Oceanogr 72:343–358. https://doi.org/10.1007/s10872-015-0346-9

Furuichi N, Hibiya T, Niwa Y (2012) Assessment of turbulence closure models for resonant inertial response in the oceanic mixed layer using a large eddy simulation model. J Oceanogr 68:285–294. https://doi.org/10.1007/s10872-011-0095-3

Haller G, Beron-Vera FJ (2013) Coherent Lagrangian vortices: the black holes of turbulence. J Fluid Mech 731:R4. https://doi.org/10.1017/jfm.2013.391

Honda MC, Imai K, Nojiri Y, Hoshi F, Sugawara T, Kusakabe M (2002) The biological pump in the northwestern North Pacific based on fluxes and major components of particulate matter obtained by sediment trap experiments (1997-2000). Deep-Sea Res II 49:5595–5625

Honda MC, Kawakami H (2014) Sinking velocity of particulate radiocesium in the northwestern North Pacific. Geophys Res Lett 41:3959–3965. https://doi.org/10.1002/2014GL060126

Honda MC, Kawakami H, Matsumoto K, Wakita M, Fujiki T, Mino Y, Sukigara C, Kobari T, Uchimiya M, Kaneko R, Saino T (2016) Comparison of sinking particles in the upper 200 m between subarctic station K2 and subtropical station S1 based on drifting sediment trap experiments. J Oceanogr 72:373–386. https://doi.org/10.1007/s10872-015-0280-x

Honda MC, Kawakami H, Watanabe S, Saino T (2013) Concentration and vertical flux of Fukushima-derived radiocesium in sinking particles from two sites in the Northwestern Pacific Ocean. Biogeosciences 10:3525–3534

Honda MC, Matsumoto K, Fujiki T, Siswanto E, Sasaoka K, Kawakami H, Wakita M, Mino Y, Sukigara C, Kitamura M, Sasai Y, Smith SL, Hashioka T, Yoshikawa C, Kimoto K, Watanabe S, Kobari T, Nagata T, Hamasaki K, Kaneko R, Uchimiya M, Fukuda H, Abe O, Saino T (2017) Comparison of carbon cycle between the western Pacific subarctic and subtropical time-series stations: highlights of the K2S1 project. J Oceanogr 73:647–667. https://doi.org/10.1007/s10872-017-0423-3

Honda MC, Sasaoka K, Kawakami H, Matsumoto K, Watanabe S, Dickey TD (2009) Application of underwater optical data to estimation of primary productivity. Deep-Sea Res I 56:2281–2292

Honda MC, Watanabe S (2010) Importance of biogenic opal as ballast of particulate organic carbon (POC) transport and existence of mineral ballast-associated and residual POC in the Western Pacific Subarctic Gyre. Geophys Res Lett 37(L02605). https://doi.org/10.1029/2009GL041521

Inoue R, Honda MC, Fujiki T, Matsumoto K, Kouketsu S, Suga T, Saino T (2016) Western North Pacific integrated physical-biogeochemical ocean observation experiment (INBOX): part 2. Biogeochemical responses to eddies and typhoons revealed from the S1 mooring and shipboard measurements. J Mar Res 74:71–99

Inoue R, Kouketsu S (2016) Physical oceanographic conditions around the S1 mooring site. J Oceanogr 72:453–464. https://doi.org/10.1007/s10872-015-0342-0

Inoue R, Watanabe M, Osafune S (2017) Wind-induced mixing in the North Pacific. J Phys Oceanogr 47(7): 1587-1603. https://doi.org/10.1175/JPO-D-16-0218.1

Iuculano F, Mazuecos IP, Reche I, Agustí S (2017) Prochlorococcus as a possible source for transparent exopolymer particles (TEP). Front Microbiol 8:709. https://doi.org/10.3389/fmicb.2017.00709

Jickells TD, An ZS, Andersen KK, Baker AR, Bergametti G, Brooks N, Cao JJ, Boyd PW, Duce RA, Hunter KA, Kawahata H, Kubilay N, laRoche J, Liss PS, Mahowald N, Prospero JM, Ridgwell AJ, Tegen S, Torres R (2005) Global iron connections between desert dust, ocean biogeochemistry, and climate. Science 308:67–71

Johnston TMS, Chaudhuri D, Mathur M, Rudnick DL, Sengupta D, Simmons HL, Tandon A, Venkatesan R (2016) Decay mechanisms of near-inertial mixed layer oscillations in the Bay of Bengal. Oceanography 29(2):180–191 https://doi.org/10.5670/oceanog.2016.50

Kaneko H, Yasuda I, Komatsu K, Itoh S (2012) Observations of the structure of turbulent mixing across the Kuroshio. Geophys Res Lett 39:L15602. https://doi.org/10.1029/2012GL052419

Klaas C, Archer DE (2002) Association of sinking organic matter with various types of mineral ballast in the deep sea: implications for the rain ratio. Glob Biogeochem Cycles 16(4):1116. https://doi.org/10.1029/2001GB001765

Komori N, Takahashi K, Komine K, Motoi T, Zhang X, Segawa G (2005) Description of sea-ice compartment of coupled ocean-sea-ice model for the earth simulator (OIFES). J Earth Simul 4:31–45

Kouketsu S, Kaneko H, Okunishi T, Sasaoka K, Itoh S, Inoue R, Ueno H (2016) Mesoscale eddy effects on temporal variability of surface chlorophyll a in the Kuroshio extension. J Oceanogr 72:439–451. https://doi.org/10.1007/s10872-015-0286-4

Lin I-I (2012) Typhoon-induced phytoplankton blooms and primary productivity increase in the western North Pacific subtropical ocean. J Geophys Res 117(C03039). https://doi.org/10.1029/2011JC007626

Lin I-I, Liu WT, Wu CC, Wong GTF, Hu C, Chen Z, Liang WD, Yang Y, Liu KK (2003) New evidence for enhanced ocean primary production triggered by tropical cyclone. Geophys Res Lett 30:1718. https://doi.org/10.1029/2003GL017141

Mahadevan A, Thomas LN, Tandon A (2008) Comment on "eddy/wind interactions stimulate extraordinary midocean plankton bloom". Science 320(5875):448. http://science.sciencemag.org/content/320/5875/448.2

Mari X (1999) Carbon content and C:N ratio of transparent exopolymeric particles (TEP) produced by bubbling exudates of diatoms. Mar Ecol Prog Ser 183:59–71. https://doi.org/10.3354/meps183059

Martin AP, Pondaven P (2003) On estimates for the vertical nitrate flux due to eddy pumping. J Geophys Res 108(C11):3359. https://doi.org/10.1029/2003JC001841

Masumoto Y, Sasaki H, Kagimoto T, Komori N, Ishida A, Sasai Y, Miyama T, Motoi T, Mitsudera H, Takahashi K, Sakuma H (2004) A fifty-year-eddy-resolving simulation of the world ocean: preliminary outcomes of OFES (OGCM for the earth simulator). J Earth Simul 1:35–56

Matsumoto K, Abe O, Fujiki T, Sukigara C, Mino Y (2016) Primary productivity at the time-series statins in the northwestern Pacific Ocean. J Oceanogr 72:369–371. https://doi.org/10.1007/s10872-016-0354-4

Matsumoto K, Honda MC, Sasaoka K, Wakita M, Kawakami H, Watanabe S (2014) Seasonal variability of primary production and phytoplankton biomass in the western Pacific subarctic gyre: control by light availability within the mixed layer. J Geophys Res Oceans 119(9):6523–6534. https://doi.org/10.1002/2014JC009982

McGillicuddy DJ Jr, Anderson LA, Bates NR, Bibby T, Buesseler KO, Carlson CA, Davis CS, Ewart C, Falkowski PG, Goldthwait SA, Hansell DA, Jenkins WJ, Johnson R, Kosnyrev VK, Ledwell JR, Li QP, Siegel DA, Steinberg DK (2007) Eddy/wind interactions stimulate extraordinary mid-ocean plankton blooms. Science 316:1021–1026

McGillicuddy DJ Jr, Robinson AR (1997) Eddy-induced nutrient supply and new production in the Sargasso Sea. Deep-Sea Res I 44:1427–1450

McGillicuddy DJ Jr, Robinson AR, Siegel DA, Jannasch HW, Johnson R, Dickey TD, McNeil J, Michaels AF, Knap AH (1998) Influence of mesoscale eddies on new production in the Sargasso Sea. Nature 394:263–265

Menkes CE, Lengaigne M, Levy C, Ethe L, Bopp O, Aumont E, Vincent J, Vialard J, Jullien S (2016) Global impact of tropical cyclones on primary production. Glob Biogeochem Cycles 30:767–786. https://doi.org/10.1002/2015GB005214

Mori K, Uehara K, Kameda T, Kakehi S (2008) Direct measurements of dissipation rate of turbulent kinetic energy of North Pacific subtropical mode water. Geophys Res Lett 35:L05601. https://doi.org/10.1029/2007GL032867

Nagai T, Tandon A, Yamazaki H, Doubell M (2009) Evidence of enhanced turbulent dissipation in the frontgenetic Kuroshio front thermocline. Geophys Res Lett 36:L12609. https://doi.org/10.1029/2009GL038832

Oka E, Qiu B, Takatani Y, Enyo K, Sasano D, Kosugi N, Ishii M, Nakano T, Suga T (2015) Decadal variability of subtropical mode water subduction and its impact on biogeochemistry. J Oceanogr 71(4):389–400. https://doi.org/10.1007/s10872-015-0300-x

Onogi K, Tsutusi J, Koide H, Sakamoto M, Kobayashi S, Hatsushika H, Matsumoto T, Yamazaki N, Kamahori H, Takahashi K, Kadokura S, Wada K, Kato K, Oyama R, Ose T, Mannoji N, Taira R (2007) The JRA-25 reanalysis. J Met Soc Jap 85:369–432

Oschilies A (2001) Model-derived estimates of new production: new results point towards lower values. Deep-Sea Res II 48:2173–2197

Oschlies A (2002) Can eddies make ocean deserts bloom? Glob Biogeochem Cycles 16:1106. https://doi.org/10.1029/2001GB001830

Oschlies A, Garcon (1999) An eddy-permitting coupled physical-biological model of the North Atlantic 1. Sensitivity to advection numerics and mixed layer physics. Glob Biogeochem Cycles 13:135–160

Pan S, Shi J, Gao H, Guo X, Yao X, Gong X (2017) Contributions of physical and biogeochemical processes to phytoplankton biomass enhancement in the surface and subsurface layers during the passage of typhoon Damrey. J Geophys Res Biogeosci 122. https://doi.org/10.1002/2016JG003331

Passow U, Shipe RF, Murray A, Pak DK, Brzezinski MA, Alldredge AL (2001) The origin of transparent exopolymer particles (TEP) and their role in the sedimentation of particulate matter. Cont Shelf Res 21:327–346

Qiu B (2002) The Kuroshio extension system: its large-scale variability and role in mid latitude ocean-atmosphere interaction. J Oceanogr 58:57–75

Qiu B, Chen S, Schneider N, Taguchi B (2014) A coupled decadal prediction of the dynamic state of the Kuroshio extension system. J Clim 27:1751–1764. https://doi.org/10.1175/JCL-D-13-00318.1

Redfield AC, Ketchum BH, Richards FA (1963) The influence of organisms on the composition of seawater. In: Hill MN (ed) The sea, vol 2. Interscience, New York, pp 26–77

Rumyantseva A, Lucas N, Rippeth T, Martin A, Painter SC, Boyd TJ, Henson S (2015) Ocean nutrient pathways associated with the passage of a storm. Glob Biogeochem Cycles 29:1179–1189. https://doi.org/10.1002/2015GB005097

Sasai Y, Richards KJ, Ishida A, Sasaki H (2010) Effects of cyclonic mesoscale eddies on the marine ecosystem in the Kuroshio extension region using an eddy-resolving coupled physical-biological model. Ocean Dyn 60:693–704. https://doi.org/10.1007/s10236-010-0264-8

Sasai Y, Sasaki H, Sasaoka K, Ishida A, Yamanaka Y (2007) Marine ecosystem simulation in the eastern tropical Pacific with a global eddy resolving coupled physical-biological model. Geophys Res Lett 34(L23601). https://doi.org/10.1029/2007GL031507

Sasai Y, Yoshikawa C, Smith SL, Hashioka T, Matsumoto K, Wakita M, Sasaoka K, Honda MC (2016) Coupled 1-D physical-biological model study of phytoplankton production at two contrasting time-series stations in the western North Pacific. J Oceanogr 72:509–526. https://doi.org/10.1007/s10872-015-0341-1

Schlitzer R (2003) Ocean data view, https://odv.awi.de

Shibano R, Yamanaka Y, Okada N, Chuda T, Suzuki S, Niino H, Toratani M (2011) Responses of marine ecosystem to typhoon passages in the western subtropical North Pacific. Geophys Res Lett 38(L18608). https://doi.org/10.1029/2011GL048717

Siegel DA, McGillicuddy DJ, Fields EA (1999) Mesoscale eddies, satellite altimetry, and new production in the Sargasso Sea. J Geophys Res 104:13359–13379

Siswanto E, Ishizaka J, Morimoto A, Tanaka K, Okamura K, Kristijono A, Saino T (2008) Open physical and biogeochemical response to the passage of typhoon Meari in the East China Sea observed from Argo float and multiplatform satellites. Geophys Res Lett 35(L15604). https://doi.org/10.1029/2008GL035040

Siswanto E, Ishizaka J, Yokouchi K, Tanaka K, Tan CK (2007) Estimation of interannual and interdecadal variations of typhoon-induced primary production: a case study for the outer shelf of the East China Sea. Geophys Res Lett 34(L03604). https://doi.org/10.1029/2006GL028368

Sukigara C, Suga T, Toyama K, Oka E (2014) Biogeochemical responses associated with the passage of a cyclonic eddy based on shipboard observations in the western North Pacific. J Oceanogr 70:435–445. https://doi.org/10.1007/s10872-014-0244-6

Sweeney EN, McGillicuddy DJ Jr, Buesseler KO (2003) Biogeochemical impacts due to mesoscale eddy activity in the Sargasso Sea measured at the Bermuda Atlantic Time-Series Study (BATS). Deep-Sea Res II 50:3017–3039

Taylor SR (1964) Abundance of chemical elements in the continental crust: a new table. Geochim Cosmochim Acta 28:1273–1285

Tomita H, Kako S, Cronin MF, Kubota M (2010) Preconditioning of the wintertime mixed layer at the Kuroshio Extension Observatory. J Geophys Res 115(C12053). https://doi.org/10.1029/2010JC006373

Wakita M, Honda MC, Matsumoto K, Fujiki T, Kawakami H, Yasunaka S, Sasai Y, Sukigara C, Uchimiya M, Kitamura M, Kobari T, Mino Y, Nagano A, Watanabe S, Saino T (2016) Biological organic carbon export estimated from the annual carbon budget observed in the surface waters of the western subarctic and subtropical North Pacific Ocean from 2004 to 2013. J Oceanogr 72(5):665–685. https://doi.org/10.1007/s10872-016-0379-8

Yablonsky R, Ginis I (2009) Limitation of one-dimensional ocean models for coupled hurricane–ocean model forecasts. Mon Weather Rev 137:4410–4419. https://doi.org/10.1175/2009MWR2863.1

Zhang S, Xie L, Hou Y, Zhao H, Qi Y, Yi X (2014) Tropical storm-induced turbulent mixing and chlorophyll-*a* enhancement in the continental shelf southeast of Hainan Island. J Mar Syst 129:405–414. https://doi.org/10.1016/j.jmarsys.2013.09.002

3

Authigenic carbonate precipitation at the end-Guadalupian (Middle Permian) in China: Implications for the carbon cycle in ancient anoxic oceans

Masafumi Saitoh[1,2,3,8*], Yuichiro Ueno[1,2,3], Yukio Isozaki[4], Takazo Shibuya[1], Jianxin Yao[5], Zhansheng Ji[5], Katsumi Shozugawa[6], Motoyuki Matsuo[6] and Naohiro Yoshida[3,7]

Abstract

Carbonate precipitation is a major process in the global carbon cycle. It was recently proposed that authigenic carbonate (carbonate precipitated in situ at the sediment–water interface and/or within the sediment) played a major role in the carbon cycle throughout Earth's history. The carbon isotopic composition of authigenic carbonates in ancient oceans have been assumed to be significantly lower than that of dissolved inorganic carbon (DIC) in seawater, as is observed in the modern oceans. However, the $\delta^{13}C_{carb}$ values of authigenic carbonates in the past has not been analyzed in detail. Here, we report authigenic carbonates in the uppermost Guadalupian (Middle Permian) rocks at Chaotian, Sichuan, South China. Monocrystalline calcite crystals <20 mm long are common in the black mudstone/chert sequence that was deposited on a relatively deep anoxic slope/basin along the continental margin. Textures of the crystals indicate in situ precipitation on the seafloor and/or within the sediments. The calcite precipitation corresponds stratigraphically with denitrification and sulfate reduction in the anoxic deep-water mass, as indicated by previously reported nitrogen and sulfur isotope records, respectively. Relatively high $\delta^{13}C_{carb}$ values of the authigenic carbonates (largely −1 ‰) compared with those of organic matter in the rocks (ca. −26 ‰) suggest that the main carbon source of the carbonates was DIC in the water column. The calcite crystals precipitated in an open system with respect to carbonate, possibly near the sediment–water interface rather than deep within the sediments. The $\delta^{13}C_{carb}$ values of the carbonates were close to the $\delta^{13}C_{DIC}$ value of seawater due to mixing of ^{13}C-depleted remineralized organic carbon (that was released into the water column by the water-mass anaerobic respiration) with the large DIC pool in the oceans. Our results imply that $\delta^{13}C_{carb}$ values of authigenic carbonates in the anoxic oceans might have been systematically different from the values in the oxic oceans in Earth's history, controlled by the depth of the redoxcline in the water column and sediments. If our model is correct, authigenic carbonates with relatively high $\delta^{13}C_{carb}$ values in the ancient anoxic oceans may have had a less substantial influence on the bulk $\delta^{13}C_{carb}$ values in geologic records than has been previously suggested.

Keywords: Middle Permian, South China, Anoxic deep-water, Authigenic carbonate, Carbon isotopic composition, Oxic/anoxic oceans, Redoxcline depth, Water-mass anaerobic respiration

* Correspondence: saitoh.m.ab@jamstec.go.jp
[1]Laboratory of Ocean-Earth Life Evolution Research (OELE), Japan Agency for Marine-Earth Science and Technology (JAMSTEC), 2-15 Natsushima-cho, Yokosuka 237-0061, Japan
[2]Department of Earth and Planetary Sciences, Tokyo Institute of Technology, 2-12-1 Ookayama, Meguro, Tokyo 152-8551, Japan
Full list of author information is available at the end of the article

Background

Carbonate precipitation is a major process in the global carbon cycle and is thought to have been closely related to climate changes throughout Earth's history (e.g., Walker et al. 1981; Kasting 1993; Ridgwell and Zeebe 2005). The precipitation of carbonate has also been an important factor in the carbon isotope mass balance in both ancient and modern oceans. In a general model of the global carbon cycle, the input of carbon to the oceans is balanced by its output via two major pathways, normal marine carbonate precipitation and organic carbon burial (Shackleton and Hall 1984). Biotic and/or abiotic carbonate precipitation of dissolved inorganic carbon (DIC) in the oceans exhibits little carbon isotope fractionation, whereas organic matter generated by biological carbon fixation is depleted in ^{13}C relative to seawater DIC. The carbon isotopic composition of seawater DIC ($\delta^{13}C_{DIC}$), and consequently that of the carbonate mineral ($\delta^{13}C_{carb}$), is interpreted to represent the partitioning of carbon between the two major sinks. On the basis of a compilation of $\delta^{13}C_{carb}$ values in geologic records from numerous previous studies, carbonate precipitation is suggested to have accounted for up to 80 % of total carbon removal from the ocean during most of Earth's history (Hayes et al. 1999; Shields and Veizer 2002).

Recently, Schrag et al. (2013) introduced a new component into a model of the global carbon cycle in the past. They proposed that authigenic carbonate, inorganically precipitated in situ at the sediment–water interface and/or within the sediments associated with anaerobic respiration (sulfate and iron reduction), has played a major role in Earth's history. They also suggested that the global authigenic carbonate sink represented a major component of the carbon isotope mass balance in the past, particularly in the anoxic oceans in the Precambrian and Paleozoic, although this sink may also be significant in the modern oxic oceans (Sun and Turchyn 2014). According to the model of Schrag et al. (2013), secular changes in the size of the global sink of ^{13}C-depleted authigenic carbonate could change the $\delta^{13}C_{DIC}$ value in the oceans without large fluctuations in atmospheric O_2 and CO_2 levels. They assumed that the $\delta^{13}C_{carb}$ values of authigenic carbonate have been substantially lower than the $\delta^{13}C_{DIC}$ value of seawater in the past, as observed in the modern oceans. However, the $\delta^{13}C_{carb}$ values of authigenic carbonate throughout Earth's history have yet to be studied in detail.

We analyzed the Guadalupian–Lopingian (Middle–Upper Permian) rocks in the Chaotian section in northern Sichuan, South China, to examine the environmental changes related to the end-Guadalupian extinction (Fig. 1; Isozaki et al. 2004). In our previous studies, we identified evidence for anoxia and for water-mass denitrification and sulfate reduction on the relatively deep slope/basin along the continental margin, prior to the extinction (Fig. 1c; Isozaki et al. 2008; Saitoh et al. 2013a, 2013b, 2014a). Here, we report on the occurrence of anomalous carbonate crystals in a characteristic interval of deep-water facies in the uppermost Guadalupian at Chaotian. To clarify the origin of these crystals, we describe the textures and analyze the inorganic carbon and oxygen isotopic compositions of the carbonates. We classify the Chaotian carbonates as authigenic carbonate and discuss their generation mechanisms. Moreover, based on our results, we argue the influence of authigenic carbonate on the bulk $\delta^{13}C_{carb}$ values in geologic records, considering the systematic difference in $\delta^{13}C_{carb}$ values between authigenic carbonates in the oxic and anoxic oceans throughout Earth's history.

Methods

The Chaotian section is located nearly 20 km north of the city of Guangyuan in northern Sichuan, South China (32° 37′ N, 105° 51′ E; Fig. 1a; Isozaki et al. 2004). At Chaotian, Middle Permian to lowermost Triassic carbonates are continuously exposed along the bank of the Jialingjiang River in a narrow gorge called Mingyuexia (Fig. 1c). We mapped the eastern bank of the gorge, particularly focusing on the ~30 m thick Middle–Upper Permian rocks. Fresh rock samples were collected by field mapping and by deep drilling to a depth of >150 m. We described the textures of the rocks based on petrographic observations of polished slabs and thin sections. To determine the mineral compositions of the carbonate samples, X-ray diffraction (XRD) analyses were conducted on a new D8 Advance (Bruker) diffractometer equipped with a graphite monochromator at The University of Tokyo and operated at 40 kV and 40 mA using Cu–Kα radiation. Selected samples were powdered and mounted on plastic holders (ø = 25 mm, depth = 1 mm). The oriented samples were scanned over an interval of 5°–60° (2θ) at a scanning speed of 1.375°/min for every 0.03° (2θ) step. The divergence, scattering, and receiving slits were 0.6, 0.6, and 0.10 mm, respectively.

In addition to the lithofacies description and XRD analysis, the inorganic carbon and oxygen isotopic compositions of carbonate samples were analyzed at Tokyo Institute of Technology. Powdered samples were reacted with >100 % phosphoric acid at 80 °C for >12 h using a Thermoquest GasBench II. The extracted CO_2 was separated in a chromatography line with a helium flow, and the carbon and oxygen isotope ratios were measured with a DELTA Plus XL mass spectrometer, following a modified version of the method in Revesz and Landwehr (2002). The carbonate carbon and oxygen isotope compositions ($\delta^{13}C_{carb}$ and $\delta^{18}O_{carb}$) are reported in per mil notation relative to Vienna Peedee Belemnite (‰ VPDB). The analytical reproducibility of the $\delta^{13}C_{carb}$ and $\delta^{18}O_{carb}$ values, determined by replicate analyses of the

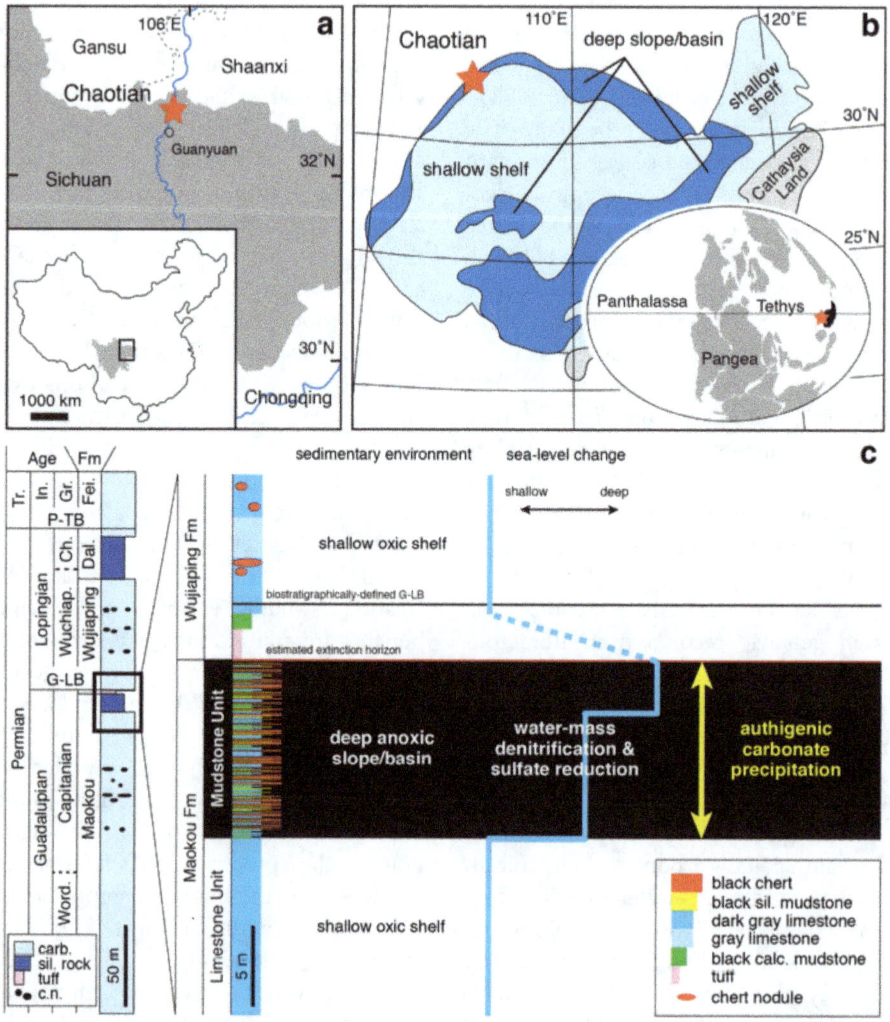

Fig. 1 Stratigraphy of the Chaotian section. **a** Location of the Chaotian section. **b** Paleogeography of South China in the Capitanian (Late Guadalupian) modified from Wang and Jin (2000). The global paleogeography is shown in the inset. South China is shown in *black*, and a *red star* represents the location of Chaotian. **c** Stratigraphy across the Guadalupian–Lopingian (Middle–Upper Permian) boundary (G–LB) at Chaotian. The sedimentary environments and sea-level changes are modified from Saitoh et al. (2013a). *carb.* carbonates, *calc.* calcareous, *sil.* siliceous, *c.n.* chert nodule, *Word.* Wordian, *Wuchiap.* Wuchiapingian, *Ch.* Changhsingian, *Dal.* Dalong, *Tr.* Triassic, *In.* Induan, *Gr.* Griesbachian, *Fei.* Feixianguan, *P–TB* Permian–Triassic boundary

laboratory standard, were better than 0.3 and 0.4 ‰, respectively.

We conducted simple thermodynamic calculations to examine changes in chemical compositions (e.g., the saturation state of carbonate) in hypothetical Permian seawater associated with anaerobic respiration. We used a one-box model with the chemical compositions of the initial seawater based on values taken from previous studies (Tables 1 and 2). We assumed that the total Na concentration ($\Sigma Na = Na^+ + NaCO_3^- + NaCl\ (aq) + NaHCO_3\ (aq) + NaOH\ (aq)$) and the total Cl concentration ($\Sigma Cl = Cl^- + CaCl^+ + CaCl_2\ (aq) + NaCl\ (aq) + HCl\ (aq)$) in the initial seawater were 500 mmol/L. We also assumed that the total S concentration ($\Sigma S = SO_4^{2-} + HSO_4^- + CaSO_4\ (aq) + H_2S + HS^- + S_2^-$) in the initial seawater was 20 mmol/L according to a recent reconstruction of the secular changes in seawater sulfate concentrations in the Phanerozoic based on sulfur isotope records (Algeo et al. 2015). The total Ca concentration ($\Sigma Ca = Ca^{2+} + CaCO_3\ (aq) + CaCl^+ + CaCl_2\ (aq) + CaHCO_3^+ + CaSO_4\ (aq)$) at the initial condition was assumed to be 15 mmol/L based on secular variations in the Phanerozoic reconstructed by Stanley and Hardie (1998). We assumed that the total DIC concentration ($\Sigma CO_2 = CO_2\ (aq) + HCO_3^- + CO_3^{2-} + CaCO_3\ (aq) + CaHCO_3^+ + NaCO_3^- + NaHCO_3\ (aq)$) in the initial seawater was 2 mmol/kg according to an estimation by Arvidson et al. (2014). Although nitrate concentrations in the Permian oceans are poorly constrained, we assumed that the total N concentration ($\Sigma N = NO_3^- + HNO_3\ (aq) +$

Table 1 Model input parameters

	Value	Units	References
ΣNa^a	500	mmol/L	
ΣCl	500	mmol/L	
ΣS	20	mmol/L	Algeo et al. (2015)
ΣCa	15	mmol/L	Stanley and Hardie (1998)
$\Sigma CO_2{}^b$	2	mmol/kg	Arvidson et al. (2014)
ΣN^c	15	μmol/L	Lam et al. (2009)
pH	7.8		Ridgwell (2005)
Temperature	15	°C	
$\delta^{13}C_{DIC}$	+2	‰	
$\delta^{13}C_{org}$	−26	‰	

[a] $\Sigma Na = Na^+ + NaCO3^- + NaCl\ (aq) + NaHCO3\ (aq) + NaOH\ (aq)$
The total elemental concentrations of the other elements are calculated in a similar way to ΣNa (see the "Methods" section and Table 3)
[b] Total dissolved inorganic carbon (DIC)
[c] The value at the estimated water depth of the sedimentary environment (>150 m deep)

N_2 (aq)) in the initial seawater was 15 μmol/L based on the vertical NO_3^- profile in the modern oceans (Lam et al. 2009) and on the estimated water depth of the sedimentary environment of the Mudstone Unit at Chaotian (>150 m deep) (Saitoh et al. 2013a). The pH of the initial seawater was assumed to be 7.8 on the basis of the model results by Ridgwell (2005). The temperature of the seawater was assumed to be 15 °C. The $\delta^{13}C_{DIC}$ value of the initial seawater was assumed to be +2 ‰ based on the previously reported $\delta^{13}C_{carb}$ records in the Mudstone Unit (Saitoh et al. 2013b). Model input parameters are listed in Table 1, and the calculated chemical composition of the initial seawater is listed in Table 2. We calculated the saturation index of calcite (Ω) as follows:

Table 2 Chemical compositions of the initial seawater (SW_i) for the thermodynamic calculations

		Value	Units
Aqueous species	Na^+	495.0	mmol/L
	Cl^-	483.1	mmol/L
	$SO_4{}^{2-}$	18.6	mmol/L
	Ca^{2+}	13.1	mmol/L
	DIC	2.0	mmol/L
Salinity		32.09	g/kg
Density		1.03	g/cm^3
Ionic strength		0.55	
Carbonate alkalinity		1.96	meq/L
Calcite saturation index (Ω)		1.97	
Temperature		15	°C
pH		7.8	
$\delta^{13}C_{DIC}$		2.0	‰

$$\Omega = \left(\gamma Ca^{2+} \cdot \gamma CO_3^{2-}\right)/K_{sp} \tag{1}$$

where γCa^{2+} and $\gamma CO_3{}^{2-}$ are the activities of Ca^{2+} and $CO_3{}^{2-}$ in seawater, respectively, and K_{sp} is the solubility constant for calcite. The net reaction formulas for denitrification and sulfate reduction can be written as

$$NO_3^- + 5/4(CH_2O) + H^+ \rightarrow 1/2N_2 + 5/4CO_2 + 7/4H_2O \tag{2}$$

$$SO_4^{2-} + 2(CH_2O) + 2H^+ \rightarrow H_2S + 2CO_2 + 2H_2O \tag{3}$$

The composition of the seawater is modified as these reactions proceed, which were calculated using the REACT module in the computer code Geochemist's Workbench (Bethke 2008). The reaction path mimics the incremental addition of organic matter (as the addition of C and H_2O because of the absence of thermodynamic data for CH_2O) into the seawater, in which the respiration reaction and the resulting compositional modification of the seawater is reevaluated at each step based on the reactions listed in Table 3. Minerals (e.g., carbonate and sulfide) were not allowed

Table 3 Reaction formulas for the thermodynamic calculations in association with sulfate reduction

Aqueous species
$CO_2\ (aq) + H_2O = HCO_3^- + H^+$
$CO_3{}^{2-} + H^+ = HCO_3^-$
$CaCO_3\ (aq) + H^+ = HCO_3^- + Ca^{2+}$
$CaCl^+ = Cl^- + Ca^{2+}$
$CaCl_2\ (aq) = 2Cl^- + Ca^{2+}$
$CaHCO_3{}^+ = HCO_3^- + Ca^{2+}$
$CaSO_4\ (aq) = Ca^{2+} + SO_4{}^{2-}$
$H_2\ (aq) + 1/2O_2\ (aq) = H_2O$
$H_2S\ (aq) + 2O_2\ (aq) = 2H^+ + SO_4{}^{2-}$
$HCl\ (aq) = H^+ + Cl^-$
$HO_2{}^- + H^+ = H_2O + 1/2O_2\ (aq)$
$HS^- + 2O_2\ (aq) = H^+ + SO_4{}^{2-}$
$HSO_4{}^- = H^+ + SO_4{}^{2-}$
$NaCO_3{}^- + H^+ = HCO_3^- + Na^+$
$NaCl\ (aq) = Cl^- + Na^+$
$NaHCO_3\ (aq) = HCO_3^- + Na^+$
$NaOH\ (aq) + H^+ = H_2O + Na^+$
$OH^- + H^+ = H_2O$
$S^{2-} + 2O_2\ (aq) = SO_4{}^{2-}$
Solids
$C\ (in\ CH_2O) + H_2O + O_2\ (aq) = HCO_3^- + H^+$
$H_2O\ (in\ CH_2O) = H_2O$
$CaCO_3 + H^+ = HCO_3^- + Ca^{2+}$

Table 4 Chemical compositions of the seawaters used in the mixing calculations

			Value	Units
SW_1	Aqueous species	Na^+	494.7	mmol/L
		Cl^-	483.1	mmol/L
		SO_4^{2-}	17.9	mmol/L
		Ca^{2+}	13.1	mmol/L
		DIC	3.5	mmol/L
	Salinity		32.13	g/kg
	Density		1.03	g/cm^3
	Ionic strength		0.55	
	Carbonate alkalinity		2.98	meq/L
	Calcite saturation index	(Ω)	0.49	
	Temperature		15	°C
	pH		7.0	
	$\delta^{13}C_{DIC}$		−10.1	‰
SW_2	Aqueous species	Na^+	492.8	mmol/L
		Cl^-	483.0	mmol/L
		SO_4^{2-}	14.0	mmol/L
		Ca^{2+}	13.2	mmol/L
		DIC	11.8	mmol/L
	Salinity		32.38	g/kg
	Density		1.03	g/cm^3
	Ionic strength		0.55	
	Carbonate alkalinity		9.15	meq/L
	Calcite saturation index	(Ω)	1.00	
	Temperature		15	°C
	pH		6.8	
	$\delta^{13}C_{DIC}$		−21.2	‰
SW_3	Aqueous species	Na^+	485.7	mmol/L
		Cl^-	482.6	mmol/L
		SO_4^{2-}	0.0	mmol/L
		Ca^{2+}	13.6	mmol/L
		DIC	42.0	mmol/L
	Salinity		33.28	g/kg
	Density		1.03	g/cm^3
	Ionic strength		0.53	
	Carbonate alkalinity		31.89	meq/L
	Calcite saturation index	(Ω)	3.32	
	Temperature		15	°C
	pH		6.8	
	$\delta^{13}C_{DIC}$		−24.7	‰

to precipitate in calculations of the calcite saturation state and alkalinity during the reaction process. In the calculations, we used a default thermodynamic dataset (thermo.com.V8.R6.full) (Bethke 2008) with the B-dot activity model (Helgeson 1969; Helgeson and Kirkham 1974), while Na^+ was used to compensate for unbalanced charges in the initial seawater. We also calculated the $\delta^{13}C_{DIC}$ value in the seawater at each step assuming that the $\delta^{13}C$ value of the remineralized organic carbon, added to the box by the anaerobic respiration, was −26 ‰ based on the $\delta^{13}C_{org}$ records in the Mudstone Unit (Saitoh et al. 2014a). In addition, we mixed the initial seawater (SW_i) with the seawaters generated in the box by anaerobic respiration (SW_1, SW_2, SW_3), assuming mixing of the deep-water mass with the ambient seawater in the water column. The chemical compositions of SW_1, SW_2, and SW_3 are shown in Table 4. Mixing ratios of the generated seawaters to SW_i are between 0 and 0.5. We then calculated the $\delta^{13}C_{DIC}$ value and calcite saturation index of the mixed waters.

Results

Geologic setting and stratigraphy

During the Permian, South China was located at low latitudes on the eastern side of Pangaea, where shallow-marine carbonates and terrigenous clastics with diverse fossils accumulated extensively on the continental shelves (Fig. 1b; Zhao et al. 1981; Scotese and Langford 1995; Jin et al. 1998). In northern Sichuan, along the northwestern edge of South China, carbonates and mudstones of relatively deep-water facies were thickly accumulated in a slope/basin environment (Fig. 1b; Wang and Jin 2000). At Chaotian, the Permo-Triassic rocks (>300 m thick in total) include the Guadalupian Maokou Formation, the Lopingian Wujiaping and Dalong formations, and the lowermost Triassic Feixianguan Formation, in ascending order (Fig. 1c; Isozaki et al. 2004, 2008; Saitoh et al. 2013a, 2013b, 2014b). The Maokou Formation, >150 m thick, consists mainly of massive dark gray bioclastic limestone with abundant shallow-marine fossils such as calcareous algae, brachiopods, and fusulines. The uppermost part (~11 m thick) of the Maokou Formation is composed of thinly bedded black mudstone and chert containing radiolarians and conodonts. The Wujiaping Formation, ~70 m thick, is composed mainly of massive dark gray bioclastic limestone containing shallow-marine fossils such as fusulines, calcareous algae, and brachiopods. The Dalong Formation, ~25 m thick, is composed mainly of thinly bedded black mudstone and siliceous mudstone with abundant radiolarians. The Feixianguan Formation, >30 m thick, is dominated by thinly bedded light gray micritic limestone containing few conodonts and ammonoids.

The Guadalupian–Lopingian boundary (G–LB) interval (~30 m thick) analyzed here consists of three stratigraphic units: (1) the early Capitanian (Late Guadalupian) Limestone Unit of the Maokou Formation, (2) the early–late Capitanian Mudstone Unit of the Maokou Formation, and (3) the lower part of the early Wuchiapingian

(Early Lopingian) Wujiaping Formation, in ascending order (Fig. 1c; Isozaki et al. 2008; Saitoh et al. 2013a). The Limestone Unit of the Maokou Formation, ~11 m thick, is composed of bioclastic limestone with abundant shallow-marine fossils (e.g., calcareous algae, fusulines, and corals) deposited on a euphotic and oxic shelf. In contrast, the overlying Mudstone Unit, ~11 m thick, is composed mainly of black calcareous mudstone, black chert/siliceous mudstone, and dark gray carbonates containing abundant radiolarians, conodonts, and ammonoids. This unit was deposited on a relatively deep disphotic slope/basin under anoxic conditions. The lower Wujiaping Formation, ~ 11.5 m thick, is mainly bioclastic limestone with shallow-marine fossils including calcareous algae deposited on a euphotic and oxic shelf. A ~2 m thick tuffaceous Wangpo bed occurs at the base of the Wujiaping Formation. The end-Guadalupian extinction horizon is assigned to the top of the Mudstone Unit, while the biostratigraphically defined G–LB is placed at the base of the lower Wujiaping bioclastic limestones based on index fossils such as conodonts, fusulines, radiolarians, and ammonoids.

Petrology of the carbonate

Characteristic carbonate crystals occur frequently in the Mudstone Unit of the Maokou Formation at Chaotian (Figs. 1c and 2). Their occurrence is limited to the unit of deep-water facies, and no similar crystals are observed in the underlying and overlying shallow-marine bioclastic limestones. The carbonate crystals form several-centimeter-thick beds (5 cm on average) in multiple horizons in the Mudstone Unit (Fig. 2a, b). There are two major types of calcite crystal textures: "randomly oriented" and "bundled." In the first texture, monocrystalline calcite crystals (~200–300 μm long) occur in a randomly oriented manner within a bed (Fig. 2c). The aspect ratios (the ratio of the lengths of the long and short axes of a crystal) of these "randomly oriented" calcite crystals are largely <5, with no remarkable size variation. Each calcite crystal is consistently elongate along the *c*-axis of the trigonal system (Fig. 2c–e), yet the *c*-axes themselves are randomly, and therefore isotropically, oriented in a bed. However, in certain beds, the long axes of the crystals are generally perpendicular to the bedding (Fig. 2f, g). The matrix of the crystal beds is composed of very fine clay-sized silicate particles, recognized as dark areas on photomicrographs. In the second crystal texture type, 0.2–20 mm long monocrystalline calcite crystals are bundled together within a bed (Fig. 2h–k). These "bundled" calcite crystals are consistently elongate along the *c*-axis, and their aspect ratios are largely >6. The bundled crystals mainly grow upward and are consistently perpendicular to the bedding. Certain crystals originate from brachiopod shells contained in the bed and grow both upward and downward. Inclusions of 1–2 μm in size are aligned along grain boundaries in the bundled monocrystalline crystals. The bundled calcite crystals generally do not coexist with the randomly oriented crystals, although in some beds, the randomly oriented crystals gradually transition upward to bundled crystals (Fig. 2a). A small number of a third type of calcite crystal also occurs in the Mudstone Unit (Fig. 2b, k). Calcite crystals <200 μm long are densely packed in certain beds. These "packed" calcite crystals are also elongate along the *c*-axis, though their aspect ratios are low (primarily <3). The bedding of the packed crystal beds is wavy, and few bioclasts are contained within the beds. The results of XRD analyses show that the Chaotian carbonates are composed solely of calcite (Fig. 3).

Isotopic composition of the carbonate

Table 5 lists all the measured $\delta^{13}C_{carb}$ and $\delta^{18}O_{carb}$ values of the carbonates. Figure 4 shows stratigraphic profiles of the $\delta^{13}C_{carb}$ values, and Fig. 5 shows a cross-plot of the $\delta^{13}C_{carb}$ and $\delta^{18}O_{carb}$ values of the analyzed samples. The $\delta^{13}C_{carb}$ values of the randomly oriented calcite crystals range from −6.6 to + 1.6 ‰, with an average value of approximately −0.8 ‰. The $\delta^{13}C_{carb}$ values of the bundled calcite crystals range from −6.7 to +2.6 ‰, with an average value of approximately −0.6 ‰. The $\delta^{13}C_{carb}$ values of the packed calcite crystals range from −11.7 to −5.1 ‰, with an average value of approximately −8.7 ‰. The $\delta^{18}O_{carb}$ values of the randomly oriented calcite crystals range from −13.8 to −3.5 ‰, with an average value of approximately −6.7 ‰. The $\delta^{18}O_{carb}$ values of the bundled calcite crystals range from −14.2 to − 3.3 ‰, with an average value of approximately −7.1 ‰. The $\delta^{18}O_{carb}$ values of the packed calcite crystals range from −10.4 to −4.9 ‰, with an average value of approximately −7.3 ‰. No linear correlation is observed between the $\delta^{13}C_{carb}$ and $\delta^{18}O_{carb}$ values of each calcite type, suggesting that any diagenetic overprinting is not significant (Fig. 5; Knauth and Kennedy 2009), although some of the $\delta^{18}O_{carb}$ values are considerably low and imply some diagenetic isotopic alteration. As described above, the analyzed carbonate crystals are monocrystalline calcite elongate along the *c*-axis. This mineralogical characteristic indicates that the carbonate crystals were not affected by dissolution/remineralization during diagenesis (as discussed in detail in the next section) and that secondary isotopic alteration of the carbonates by these processes was not significant. In each type of carbonate, most of the $\delta^{13}C_{carb}$ values are relatively high in the lower part and relatively low in the upper part of the Mudstone Unit.

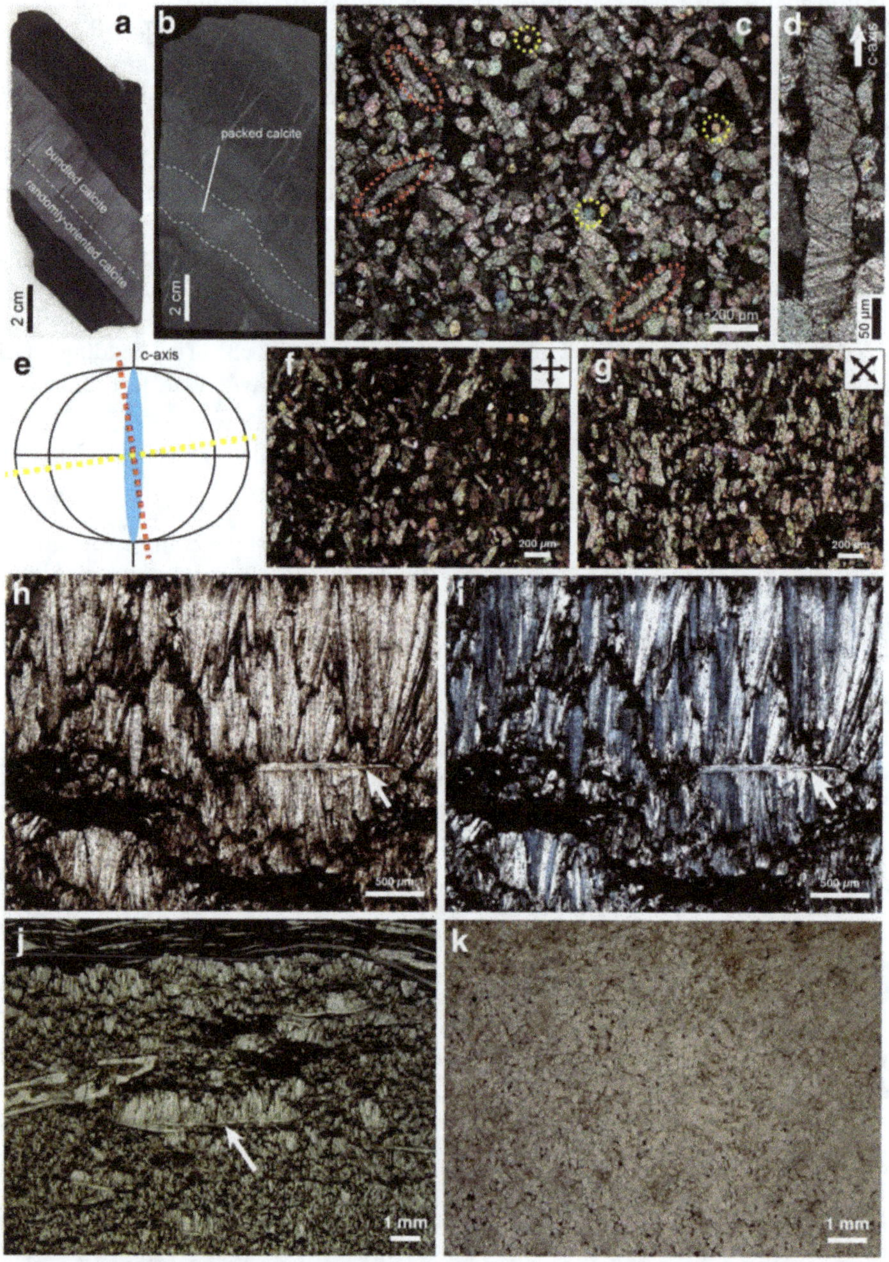

Fig. 2 Authigenic carbonates at Chaotian. **a**, **b** Drill core samples of a calcite crystal bed. In **a**, the lower and upper parts of the carbonate bed are composed of "randomly oriented" and "bundled" calcite crystals, respectively, and the bed is parallel to the overlying and underlying black mudstone beds. In **b**, "packed" calcite crystals form a wavy bed. **c** Photomicrograph of randomly oriented monocrystalline calcite crystals (crossed nicols). The horizontal direction corresponds to the bedding. Note that the interference colors of the elongate crystals (*circled in red*) are consistently high order (typical of calcite), whereas those of circular crystals (*circled in yellow*) are low order. **d** Enlarged photomicrograph of a crystal rotated 45° relative to the bedding (crossed nicols). **e** Normal velocity curve of an optically negative uniaxial calcite. The optical characteristics of monocrystalline crystals in C show that the crystals are elongate along the *c*-axis because the interference color of calcite is determined by the angle between the *c*-axis and the cross section. When the angle is low (e.g., the cross section represented by the *red dashed line* in **e**), the apparent shape of the crystals in thin section is elongate and the crystals' interference colors are high order (the crystals *circled in red* in **c**). The opposite is true for high-angle cross sections (the *yellow dashed line* in **e** and the circular crystals *circled in yellow* in **c**). **f**, **g** Photomicrographs of the same sample from a calcite bed. The horizontal direction corresponds to the bedding. The crossed polarizer and analyzer in **f** and **g** are parallel/perpendicular to and rotated 45° relative to the bedding, respectively. Note the contrast in brightness between the photos. This difference indicates that the crystals are largely perpendicular to the bedding because calcite exhibits straight extinction. **h**–**j** Photomicrographs of the bundled calcites. **h** (opened nicols) and **i** (crossed nicols) show the same field of view. The horizontal direction corresponds to the bedding in each photo. Some bundled crystals grow from brachiopod shells (*arrows*) upward and partly downward. **k** Photomicrograph of packed calcite crystals (opened nicols). Up to 200 µm long calcite crystals are densely packed in the bed

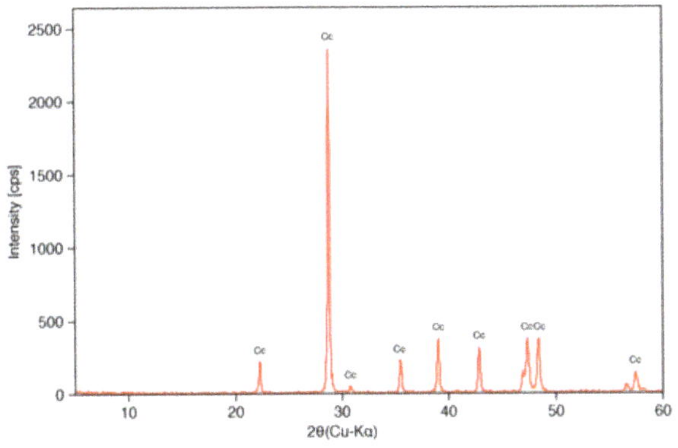

Fig. 3 Representative XRD pattern of the authigenic carbonates at Chaotian. The pattern shows that the carbonate is composed solely of calcite (Cc)

Thermodynamic and mixing calculations

Figure 6 shows calculated changes in the concentrations of inorganic carbon and sulfur chemical species, pH, carbonate alkalinity, saturation index of calcite (Ω), and the $\delta^{13}C_{DIC}$ value in response to sulfate reduction in the hypothetical Permian seawater. The pH value decreases abruptly from the initial value 7.8 to 6.9 in the early stage of the reaction, in association with dissociation of hydrogen sulfide generated by sulfate reduction and an addition of CO_2 species derived from decomposition of organic matter. It then remains nearly consistent, with values around 6.8, due to carbonate buffering in the middle to late stages. Carbonate alkalinity monotonously increases in response to sulfate reduction. Calcite is supersaturated under the initial conditions. In the early stage, calcite becomes undersaturated mainly due to the pH decrease, despite carbonate alkalinity increasing. Calcite is again supersaturated as alkalinity increases substantially in the late stage of the reaction. The $\delta^{13}C_{DIC}$ value in the seawater monotonously decreases to −25 ‰ in response to sulfate reduction. The results of this study are generally consistent with those of Meister (2013), who examined carbonate saturation states as a result of sulfate reduction. The pH, alkalinity, and Ω do not change significantly in response to denitrification, owing to the relatively low nitrate concentration (15 μmol/L) (results not shown).

We mixed the initial seawater (SW_i) with the seawaters generated by sulfate reduction (SW_1, SW_2, SW_3) (Fig. 6). Figure 7 shows the relationships between the $\delta^{13}C_{DIC}$ value and Ω of the mixed waters. When SW_1 is mixed with SW_i, the Ω value gradually and monotonously decreases from ~2.0 to 0.5 according to an increase of the mixing ratio of SW_1 to SW_i (from 0 to 0.5). When a small amount of SW_2 is mixed with SW_i, the mixed water becomes undersaturated with respect to calcite, possibly due to a pH decrease. The Ω value then gradually increases with increasing mixing ratios of SW_2 to SW_i but does not reach 1.0 even when the mixing ratio is 0.5. When SW_3 is mixed with SW_i, the Ω value also abruptly decreases to ~0.5 and then increases to >1.0 according to the increase in the mixing ratio. This is probably because of the high carbonate alkalinity of SW_3 (>30 meq/L) although the pH of the water is ~6.8. The Ω value reaches >3.0 when the mixing ratio is 0.5.

Discussion

Authigenic carbonate precipitation at Chaotian

The elongate morphology along the *c*-axis of the calcite crystals from Chaotian is an important feature that allows us to clarify their origin (Fig. 2c–k). This mineralogical characteristic suggests that the calcites were precipitated directly from oversaturated seawater, causing the calcite to grow along the *c*-axis. The Chaotian calcite crystals are neither bioclasts nor clastic grains, as it is unlikely that this remarkable mineralogical characteristic was generated by normal sedimentary processes. The possibility that the crystals were originally precipitated as other minerals, such as aragonite and gypsum, is also excluded because secondary calcite with a replacement origin usually exhibits a mosaic texture (e.g., Porter 2007), which is not the case for the monocrystalline calcites at Chaotian. Although apparently inconsistent with the general view that aragonite was preferentially precipitated (relative to calcite) in the Permian oceans (Sandberg 1983; Stanley and Hardie 1998), the calcite precipitation in the Chaotian section may have been caused by local "calcitic" conditions in the ocean of northwestern South China. No evidence for microbially induced precipitation of the crystals, such as "micropores" (Bosak et al. 2004), is recognized in the Chaotian carbonates.

Several lines of evidence suggest that the carbonate crystals at Chaotian were authigenically precipitated on the seafloor and/or within the soft sediments. The randomly oriented texture of the calcite crystals indicates that they were generated primarily on the seafloor and/or within the soft sediments below the seafloor in the early diagenetic stage (Fig. 2c). In either case, the crystals were generated in situ and are not allochthonous. If the crystals were originally precipitated in shallower environments and secondarily transported to the deeper slope/basin by gravity flow, the long axis of each crystal should be aligned parallel to the bedding, which clearly differs from the observed randomly oriented texture of the Chaotian crystals. Similarly, it is unlikely that the crystals were originally generated in the water column and subsequently sank to the slope/basin. The texture of the bundled crystals clearly indicates that they were generated in situ (Fig. 2h–j). Moreover, certain bundled crystals were generated from brachiopod shells in a bed and grew both upward and downward. This texture,

Table 5 Results of carbon and oxygen isotope analyses

Unit/Fm	Sample ID	Type	Thickness (m)	$\delta^{13}C_{VPDB}$	$\delta^{18}O_{VPDB}$
Mudstone Unit of the Maokou Fm	E83-41	bundled calcite	10.54	−2.5	−9.2
Mudstone Unit of the Maokou Fm	E83-40	bundled calcite	10.54	−2.4	−4.5
Mudstone Unit of the Maokou Fm	E83-39	bundled calcite	10.54	−3.0	−8.9
Mudstone Unit of the Maokou Fm	E83-38	bundled calcite	10.54	−2.1	−4.6
Mudstone Unit of the Maokou Fm	E83-37	bundled calcite	10.53	−1.7	−9.3
Mudstone Unit of the Maokou Fm	E83-36	bundled calcite	10.53	−2.4	−5.7
Mudstone Unit of the Maokou Fm	E83-35	bundled calcite	10.53	−1.6	−10.7
Mudstone Unit of the Maokou Fm	E83-34	bundled calcite	10.53	−2.1	−10.6
Mudstone Unit of the Maokou Fm	E83-33	bundled calcite	10.53	−2.0	−7.8
Mudstone Unit of the Maokou Fm	E83-32	bundled calcite	10.52	−2.1	−4.1
Mudstone Unit of the Maokou Fm	E83-31	bundled calcite	10.52	−1.8	−9.3
Mudstone Unit of the Maokou Fm	E83-30	bundled calcite	10.52	−1.8	−10.3
Mudstone Unit of the Maokou Fm	E83-29	bundled calcite	10.52	−0.8	−5.4
Mudstone Unit of the Maokou Fm	E83-28	bundled calcite	10.52	−0.4	−7.9
Mudstone Unit of the Maokou Fm	E83-27	bundled calcite	10.51	0.0	−5.5
Mudstone Unit of the Maokou Fm	E83-26	bundled calcite	10.51	−0.5	−7.2
Mudstone Unit of the Maokou Fm	E83-25	bundled calcite	10.51	−0.6	−8.1
Mudstone Unit of the Maokou Fm	E83-24	bundled calcite	10.51	−0.3	−4.4
Mudstone Unit of the Maokou Fm	E83-23	bundled calcite	10.51	−0.6	−5.9
Mudstone Unit of the Maokou Fm	E83-22	bundled calcite	10.50	−0.6	−6.2
Mudstone Unit of the Maokou Fm	E83-21	bundled calcite	10.50	−0.9	−5.3
Mudstone Unit of the Maokou Fm	E83-20	bundled calcite	10.50	−0.7	−10.0
Mudstone Unit of the Maokou Fm	E83-19	bundled calcite	10.50	−0.9	−10.3
Mudstone Unit of the Maokou Fm	E83-18	bundled calcite	10.50	−0.7	−5.9
Mudstone Unit of the Maokou Fm	E83-17	bundled calcite	10.49	−0.7	−5.4
Mudstone Unit of the Maokou Fm	E83-16	packed calcite	10.49	−8.7	−8.4
Mudstone Unit of the Maokou Fm	E83-15	packed calcite	10.49	−9.5	−7.9
Mudstone Unit of the Maokou Fm	E83-14	packed calcite	10.49	−9.1	−6.9
Mudstone Unit of the Maokou Fm	E83-13	packed calcite	10.49	−10.6	−8.3
Mudstone Unit of the Maokou Fm	E83-12	packed calcite	10.48	−7.5	−8.5
Mudstone Unit of the Maokou Fm	E83-11	packed calcite	10.48	−11.7	−10.4
Mudstone Unit of the Maokou Fm	E83-10	packed calcite	10.48	−10.2	−9.4
Mudstone Unit of the Maokou Fm	E83-9	bundled calcite	10.48	−0.1	−6.8
Mudstone Unit of the Maokou Fm	E83-8	bundled calcite	10.48	−0.7	−5.7
Mudstone Unit of the Maokou Fm	E83-7	bundled calcite	10.47	−0.4	−4.4
Mudstone Unit of the Maokou Fm	E83-6	bundled calcite	10.47	−0.1	−4.5
Mudstone Unit of the Maokou Fm	E83-5	bundled calcite	10.47	−0.1	−7.0
Mudstone Unit of the Maokou Fm	E83-4	bundled calcite	10.47	−0.4	−4.1
Mudstone Unit of the Maokou Fm	E83-3	bundled calcite	10.47	−0.3	−4.2
Mudstone Unit of the Maokou Fm	E83-2	bundled calcite	10.46	−0.2	−5.3
Mudstone Unit of the Maokou Fm	E83-1	bundled calcite	10.46	−1.4	−7.4
Mudstone Unit of the Maokou Fm	E65-7	bundled calcite	10.11	−0.9	−3.3
Mudstone Unit of the Maokou Fm	E65-6	bundled calcite	10.11	−1.0	−3.7
Mudstone Unit of the Maokou Fm	E65-5	bundled calcite	10.11	−1.1	−3.9

Table 5 Results of carbon and oxygen isotope analyses *(Continued)*

Mudstone Unit of the Maokou Fm	E65-4	packed calcite	10.10	−9.6	−6.1
Mudstone Unit of the Maokou Fm	E65-3	packed calcite	10.10	−8.7	−4.9
Mudstone Unit of the Maokou Fm	E65-2	packed calcite	10.10	−9.1	−5.6
Mudstone Unit of the Maokou Fm	E65-1	packed calcite	10.10	−9.6	−6.1
Mudstone Unit of the Maokou Fm	E54	bundled calcite	9.73	-6.7	−8.8
Mudstone Unit of the Maokou Fm	E31-7	randomly oriented calcite	9.02	−0.3	−7.7
Mudstone Unit of the Maokou Fm	E31-6	randomly oriented calcite	9.01	−0.5	−6.4
Mudstone Unit of the Maokou Fm	E31-5	randomly oriented calcite	9.00	−0.6	−4.7
Mudstone Unit of the Maokou Fm	E31-4	randomly oriented calcite	8.99	−0.7	−8.1
Mudstone Unit of the Maokou Fm	E31-3	randomly oriented calcite	8.98	−0.7	−4.7
Mudstone Unit of the Maokou Fm	E31-2	randomly oriented calcite	8.97	−0.7	−4.9
Mudstone Unit of the Maokou Fm	E31-1	randomly oriented calcite	8.96	−0.6	−5.4
Mudstone Unit of the Maokou Fm	E23R14	bundled calcite	8.75	−1.4	−8.8
Mudstone Unit of the Maokou Fm	E23R13	bundled calcite	8.75	−1.2	−8.4
Mudstone Unit of the Maokou Fm	E23R12	bundled calcite	8.75	−0.4	−7.7
Mudstone Unit of the Maokou Fm	E23R11	bundled calcite	8.74	−0.9	−10.5
Mudstone Unit of the Maokou Fm	E23R10	bundled calcite	8.74	−1.3	−11.4
Mudstone Unit of the Maokou Fm	E23R9	bundled calcite	8.74	−0.3	−14.2
Mudstone Unit of the Maokou Fm	E23R8	bundled calcite	8.74	−0.1	−10.1
Mudstone Unit of the Maokou Fm	E23R7	bundled calcite	8.74	−0.6	−12.3
Mudstone Unit of the Maokou Fm	E23R6	bundled calcite	8.73	−0.8	−7.3
Mudstone Unit of the Maokou Fm	E23R5	randomly oriented calcite	8.73	−1.4	−7.9
Mudstone Unit of the Maokou Fm	E23R4	randomly oriented calcite	8.73	−0.7	−5.6
Mudstone Unit of the Maokou Fm	E23R3	randomly oriented calcite	8.73	−1.0	−5.7
Mudstone Unit of the Maokou Fm	E23R2	randomly oriented calcite	8.73	−0.8	−8.8
Mudstone Unit of the Maokou Fm	E23R1	randomly oriented calcite	8.72	−0.7	−7.3
Mudstone Unit of the Maokou Fm	E1-10	randomly oriented calcite	8.32	−1.7	−4.5
Mudstone Unit of the Maokou Fm	E1-9	randomly oriented calcite	8.31	−1.3	−4.9
Mudstone Unit of the Maokou Fm	E1-8	randomly oriented calcite	8.31	−1.4	−9.3
Mudstone Unit of the Maokou Fm	E1-7	randomly oriented calcite	8.31	−1.3	−7.7
Mudstone Unit of the Maokou Fm	E1-6	randomly oriented calcite	8.31	−0.8	−4.9
Mudstone Unit of the Maokou Fm	E1-5	randomly oriented calcite	8.30	−0.6	−13.8
Mudstone Unit of the Maokou Fm	E1-4	randomly oriented calcite	8.30	−0.6	−6.1
Mudstone Unit of the Maokou Fm	E1-3	randomly oriented calcite	8.30	−0.9	−6.3
Mudstone Unit of the Maokou Fm	E1-2	randomly oriented calcite	8.29	−1.2	−8.3
Mudstone Unit of the Maokou Fm	E1-1	randomly oriented calcite	8.29	−1.4	−8.3
Mudstone Unit of the Maokou Fm	D120-4	packed calcite	8.17	−6.3	−6.4
Mudstone Unit of the Maokou Fm	D120-3	packed calcite	8.17	−5.9	−6.4
Mudstone Unit of the Maokou Fm	D120-2	packed calcite	8.16	−5.1	−7.1
Mudstone Unit of the Maokou Fm	D120-1	randomly oriented calcite	8.16	−0.6	−4.9
Mudstone Unit of the Maokou Fm	D98	randomly oriented calcite	7.41	0.7	−4.8
Mudstone Unit of the Maokou Fm	D84.5	randomly oriented calcite	6.95	0.3	−10.0
Mudstone Unit of the Maokou Fm	D70.6	bundled calcite	6.39	0.0	−6.6
Mudstone Unit of the Maokou Fm	D70.5	randomly oriented calcite	6.38	0.8	−3.5
Mudstone Unit of the Maokou Fm	D60-4	randomly oriented calcite	6.14	0.0	−6.0

Table 5 Results of carbon and oxygen isotope analyses *(Continued)*

Mudstone Unit of the Maokou Fm	D60-3	randomly oriented calcite	6.13	0.6	−9.9
Mudstone Unit of the Maokou Fm	D60-2	randomly oriented calcite	6.13	0.8	−7.7
Mudstone Unit of the Maokou Fm	D60-1	randomly oriented calcite	6.12	0.2	−8.0
Mudstone Unit of the Maokou Fm	D22-2	randomly oriented calcite	4.50	−6.6	−6.8
Mudstone Unit of the Maokou Fm	D22-1	randomly oriented calcite	4.49	−6.6	−5.7
Mudstone Unit of the Maokou Fm	C54-4	bundled calcite	2.95	1.7	−10.3
Mudstone Unit of the Maokou Fm	C54-3	bundled calcite	2.94	2.3	−4.2
Mudstone Unit of the Maokou Fm	C54-2	bundled calcite	2.94	2.4	−4.6
Mudstone Unit of the Maokou Fm	C54-1	bundled calcite	2.94	2.6	−3.5
Mudstone Unit of the Maokou Fm	C43	bundled calcite	2.62	1.3	−11.6
Mudstone Unit of the Maokou Fm	C28-5	bundled calcite	2.09	1.4	−10.5
Mudstone Unit of the Maokou Fm	C28-4	bundled calcite	2.09	1.6	−4.8
Mudstone Unit of the Maokou Fm	C28-3	bundled calcite	2.09	1.9	−4.8
Mudstone Unit of the Maokou Fm	C28-2	bundled calcite	2.09	2.3	−3.5
Mudstone Unit of the Maokou Fm	C28-1	bundled calcite	2.09	1.8	−4.2
Mudstone Unit of the Maokou Fm	C9-4	randomly oriented calcite	1.64	−5.8	−8.6
Mudstone Unit of the Maokou Fm	C9-3	randomly oriented calcite	1.63	−4.3	−6.4
Mudstone Unit of the Maokou Fm	C9-2	randomly oriented calcite	1.62	−6.2	−6.2
Mudstone Unit of the Maokou Fm	C9-1	randomly oriented calcite	1.61	−6.3	−6.2
Mudstone Unit of the Maokou Fm	B27	randomly oriented calcite	1.37	0.3	−4.3
Mudstone Unit of the Maokou Fm	B18-5	randomly oriented calcite	1.07	1.6	−4.4
Mudstone Unit of the Maokou Fm	B18-4	randomly oriented calcite	1.07	1.5	−9.9
Mudstone Unit of the Maokou Fm	B18-3	randomly oriented calcite	1.07	1.6	−9.1
Mudstone Unit of the Maokou Fm	B18-2	randomly oriented calcite	1.07	1.3	−6.2
Mudstone Unit of the Maokou Fm	B18-1	randomly oriented calcite	1.07	1.4	−6.5
Mudstone Unit of the Maokou Fm	B16-8	randomly oriented calcite	1.06	1.0	−4.5
Mudstone Unit of the Maokou Fm	B16-7	randomly oriented calcite	1.06	1.0	−4.7
Mudstone Unit of the Maokou Fm	B16-6	randomly oriented calcite	1.05	1.0	−4.5
Mudstone Unit of the Maokou Fm	B16-5	randomly oriented calcite	1.05	0.8	−4.6
Mudstone Unit of the Maokou Fm	B16-4	randomly oriented calcite	1.04	0.7	−4.7
Mudstone Unit of the Maokou Fm	B16-3	randomly oriented calcite	1.04	0.7	−8.0
Mudstone Unit of the Maokou Fm	B16-2	randomly oriented calcite	1.03	0.6	−9.0
Mudstone Unit of the Maokou Fm	B16-1	randomly oriented calcite	1.03	0.6	−8.2

particularly the downward growth, strongly suggests that the bundled crystals grew from the shells within the sediments. The packed calcite crystal beds contain few bioclasts, and it is unlikely that they are allochthonous carbonate debris flow deposits (Fig. 2k). The wavy bedding of the packed crystal beds may imply that they were generated below the seafloor, pushing the upper and lower sediments up and down, respectively, in an early diagenetic stage (Fig. 2b). The consistent *c*-axis elongation of all the Chaotian calcite crystals suggests that no solid obstacle prevented growth during crystallization; the crystals likely precipitated in a free space and/or soft sediments.

The occurrence of the carbonate crystals in the Mudstone Unit at Chaotian corresponds stratigraphically to anoxic conditions and water-mass anaerobic respiration in the unit (Fig. 1c). Saitoh et al. (2013a) suggested that the Mudstone Unit was deposited under anoxic conditions based on the absence of bioturbation, high total organic carbon contents (up to 16 %), and abundant pyrite framboids. An anoxic water mass likely developed in the deep part of the water column, although the surface ocean may have remained oxic. Moreover, Saitoh et al. (2014a) reported the nitrogen and sulfur isotope records of the Mudstone Unit. The substantially high $\delta^{15}N$ values (up to +14 ‰) and consistently low $\delta^{34}S$ values

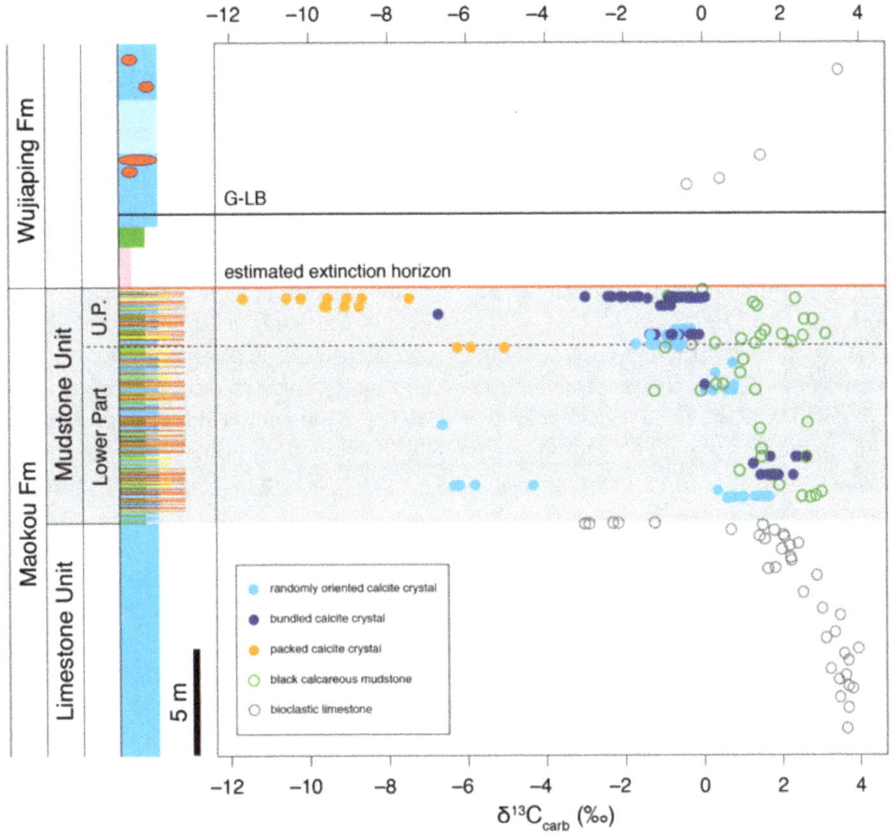

Fig. 4 Chemostratigraphic profile of the $\delta^{13}C_{carb}$ values of the authigenic carbonates at Chaotian. The $\delta^{13}C_{carb}$ values of the bioclastic limestone and black calcareous mudstone (*open circles*) were reported in Saitoh et al. (2013b). Except for some samples with lower values, the $\delta^{13}C_{carb}$ values of the carbonates are similar to or slightly lower than those of the black mudstone. *U.P.* upper part

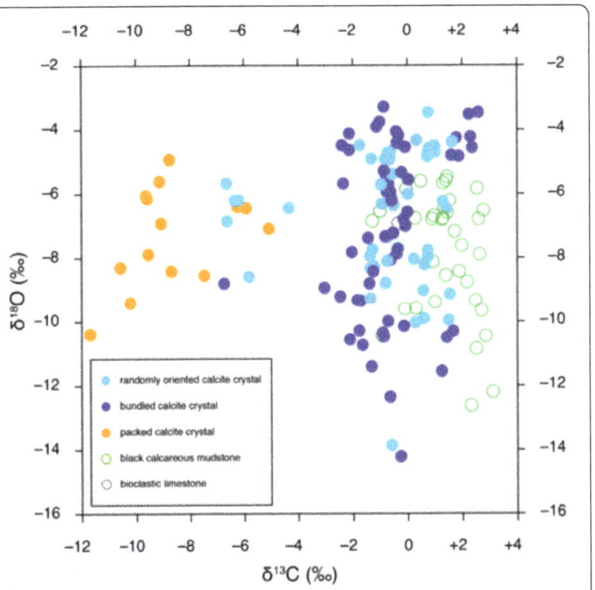

Fig. 5 Relationship between the $\delta^{18}O_{carb}$ and $\delta^{13}C_{carb}$ values of the authigenic carbonates at Chaotian

(approximately −37 ‰) in the unit indicate denitrification and sulfate reduction, respectively, in the anoxic deep waters. The clear stratigraphic correspondence implies that the authigenic carbonates at Chaotian may have precipitated in association with the water-mass anaerobic respiration.

Carbon isotopic composition of the carbonate

The inorganic carbon isotopic compositions of the authigenic carbonates at Chaotian are useful for identifying the carbon source of the carbonates. Generally, there are three potential carbon sources for authigenic carbonate: (1) seawater DIC, (2) remineralized organic carbon, and (3) remineralized methane carbon. The $\delta^{13}C_{DIC}$ value of the seawater is estimated to be approximately +2 ‰ based on the $\delta^{13}C_{carb}$ records of black mudstone in the Mudstone Unit in which carbonates are exclusively fragmented bioclasts of shallow-marine fossils (Fig. 4; Saitoh et al. 2013b). The $\delta^{13}C$ values of the organic carbon in the Mudstone Unit are approximately −26 ‰ (Saitoh et al. 2014a). The $\delta^{13}C$ value of biogenic methane is generally extremely low (less than −40 ‰; Whiticar 1999). The $\delta^{13}C_{carb}$ values of the authigenic carbonates at Chaotian were most likely determined by a degree of mixing of these three carbon sources. In the lower part of the Mudstone Unit, the $\delta^{13}C_{carb}$ values of the randomly oriented and bundled calcite crystals are largely +1 to +2 ‰ and are almost the same as or slightly lower than those of the black mudstone (Fig. 4). These observations indicate that almost all the carbon in these calcites was derived from the seawater DIC. In the upper part of the Mudstone Unit, the $\delta^{13}C_{carb}$ values of the randomly oriented and bundled calcites are largely about −1 ‰ (Figs. 4 and 5). These $\delta^{13}C_{carb}$ values suggest that >80 % of the carbon in these calcites was also derived from the seawater DIC. The main carbon source of the randomly oriented and bundled calcites in the Mudstone Unit is likely the seawater DIC. Although the textures of the randomly oriented and bundled calcite crystals suggest that the crystals may have grown within the sediments, these carbonates precipitated primarily in an open system with respect to carbonate, likely on the seafloor and/or immediately below the sediment–water interface.

The $\delta^{13}C_{carb}$ values of the packed calcites are clearly lower (ranging from −12 to −4 ‰) than those of the randomly oriented and bundled calcites (Figs. 4 and 5). The low $\delta^{13}C_{carb}$ values are attributed to greater contributions of ^{13}C-depleted remineralized carbon from decomposed organic matter and/or methane. The packed calcite crystals may have grown within the deep part of the sediments in a relatively closed system with respect to carbonate. The wavy bedding of the packed crystal beds suggests that the carbonates were generated within the sediments, which is consistent with their relatively low $\delta^{13}C_{carb}$ values (Fig. 2b). In ancient and modern oceans, the $\delta^{13}C_{carb}$ values of certain authigenic carbonates are relatively high due to methanogenesis in the sediments (e.g., Naehr et al. 2007; Heindel et al. 2015; Li et al. 2015). Microbial methanogenesis enriches the residual pore water within the sediments in ^{13}C, representing another potential mechanism for the relatively high $\delta^{13}C_{carb}$ values of the randomly oriented and bundled calcites at Chaotian. However, when authigenic carbonates are precipitated in sediments in which methanogenesis occurs, the $\delta^{13}C_{carb}$ values vary widely (up to 80 ‰) due to methane production and its subsequent anaerobic oxidation, as is observed in modern oceans (e.g., Meister et al. 2007; Naehr et al. 2007). In marked contrast, the $\delta^{13}C_{carb}$ values of the randomly oriented and bundled calcites at Chaotian largely fall within the relatively narrow range of −2 to +2 ‰ (Figs. 4 and 5). Only certain carbonates, including the packed calcites, exhibit relatively low $\delta^{13}C_{carb}$ values (less than −4 ‰). It is therefore unlikely that the observed carbon isotopic variations at Chaotian were caused by methanogenesis in the sediments.

The water-mass anaerobic respiration in the Mudstone Unit at Chaotian, as suggested by the nitrogen and sulfur isotope records, implies that remineralized carbon with a relatively low $\delta^{13}C$ value was released from suspended organic matter into the water column by the respiration. This ^{13}C-depleted carbon would have markedly decreased the $\delta^{13}C_{DIC}$ value in the deep-water mass in the water column (Fig. 6). However, the $\delta^{13}C_{carb}$ records of the authigenic carbonates imply that the $\delta^{13}C_{DIC}$ values in the deep-water mass at Chaotian were likely +1 to +2 ‰ in the lower part and approximately −1 ‰ in the upper part of the Mudstone Unit (Fig. 4). The vertical $\delta^{13}C_{DIC}$ gradients in the water column between the surface oceans and the deep-water mass were likely <1 and <3 ‰ in the lower and upper parts of the unit, respectively. The upper part of the Mudstone Unit was probably deposited on the deeper slope/basin than the lower part of the unit due to a relative deepening (Fig. 1c; Saitoh et al. 2013a). The $\delta^{13}C_{DIC}$ value of the anoxic deep water in the upper part of the Mudstone Unit was lower than the value in the lower part of the unit. The estimated vertical $\delta^{13}C_{DIC}$ gradients in the water column at Chaotian (<3 ‰) are similar to those in the modern and ancient oceans in the Phanerozoic (e.g., Kroopnick 1985; Kennett and Stott 1991; Huber et al. 1995; D'Hondt 2005). Although the $\delta^{13}C_{DIC}$ value in the deep water is lower than the value of the surface oceans as previously suggested at Chaotian (Saitoh et al. 2013b), our results demonstrate that the $\delta^{13}C_{DIC}$ value in the deep-water mass was not decreased remarkably by the addition of remineralized carbon released from the suspended organic matter by the water-mass anaerobic respiration. The relatively high $\delta^{13}C_{carb}$ values of the Chaotian authigenic carbonates suggest that the ^{13}C-depleted carbon released from the organic matter into the water column was quickly mixed with and diluted by a large DIC pool in the ocean. The addition of the remineralized organic carbon with low $\delta^{13}C$ values may have had only a small influence on the carbon isotopic composition of the large DIC pool.

The mixing of seawaters in the water column, suggested by the relatively high $\delta^{13}C_{carb}$ values of the Chaotian carbonates, is in concordance with our calculated results (Figs. 6 and 7). The thermodynamic calculations show that, although a deep-water mass would have become undersaturated locally with respect to carbonate in the early stage of the reaction, the successive sulfate reduction elevates alkalinity and Ω values, driving carbonate precipitation (Fig. 6). Moreover, the mixing calculations demonstrate that all of the $\delta^{13}C_{carb}$ values of the Chaotian carbonates can be explained by the mixing of the hypothetical Permian seawater (SW_i) with the deep-water masses where sulfate reduction had occurred (SW_1 and SW_3) (Fig. 7). The deep-water mass undersaturated with carbonate locally generated by the water-mass sulfate reduction could be diluted by the ambient seawater that is supersaturated with respect to carbonate,

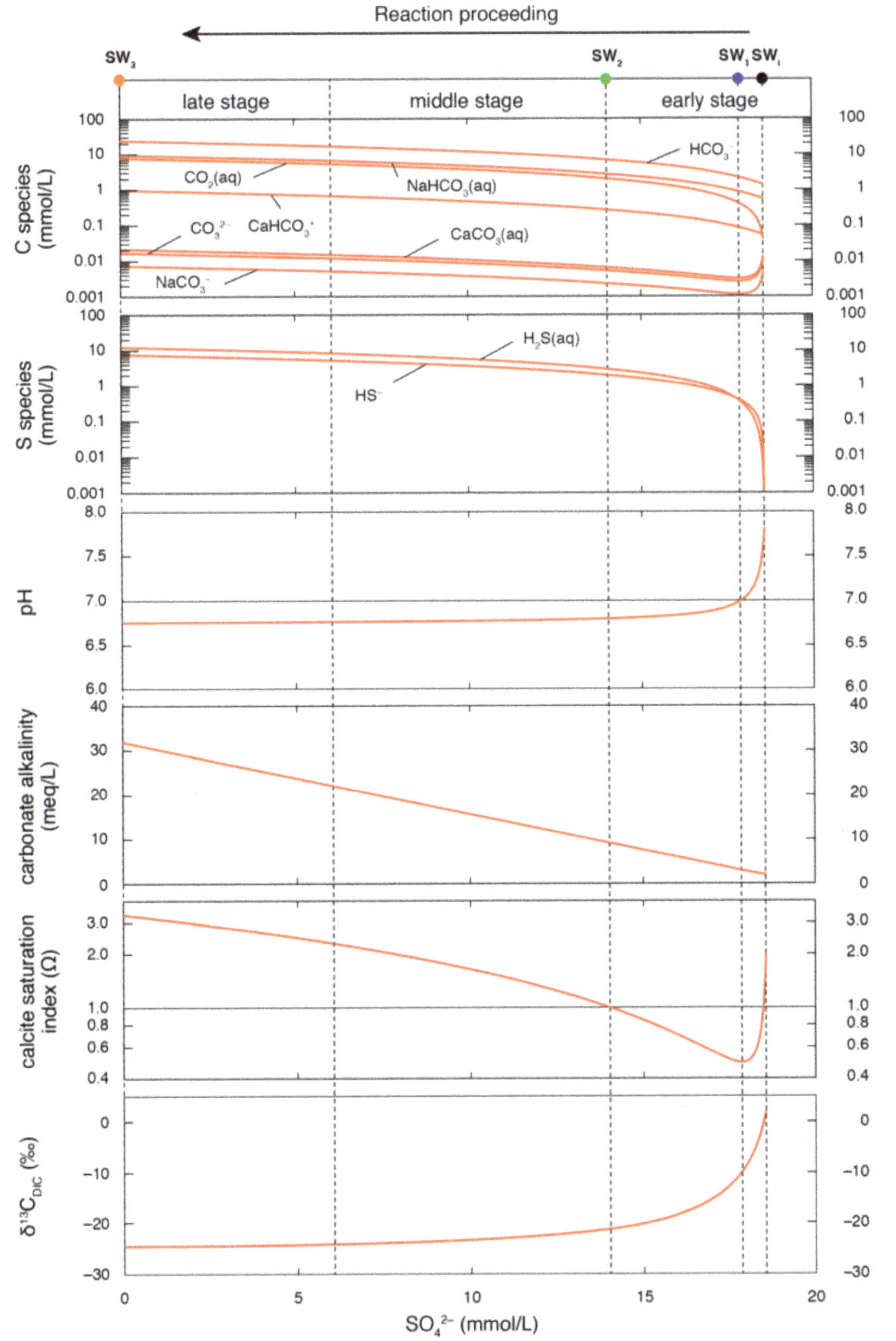

Fig. 6 Changes in the concentrations of inorganic carbon and sulfur chemical species, pH, carbonate alkalinity, calcite saturation index, and the $\delta^{13}C_{DIC}$ values in response to sulfate reduction in the hypothetical Permian seawater

stimulating the precipitation of the randomly oriented and bundled calcites at Chaotian. The mixing calculation results are also consistent with the greater contributions of ^{13}C-depleted remineralized organic carbon to certain carbonates with relatively low $\delta^{13}C_{carb}$ values, including the packed calcites, that were possibly generated within the sediments (Fig. 7). The relatively deep disphotic slope setting at Chaotian on the northwestern edge of South China at low latitudes might have been favorable to the seawater mixing by deep-water upwelling along the continental margin (Van der Kooij et al. 2010; Saitoh et al. 2013b). Moreover, Higgins et al. (2009) suggested that, in the ancient

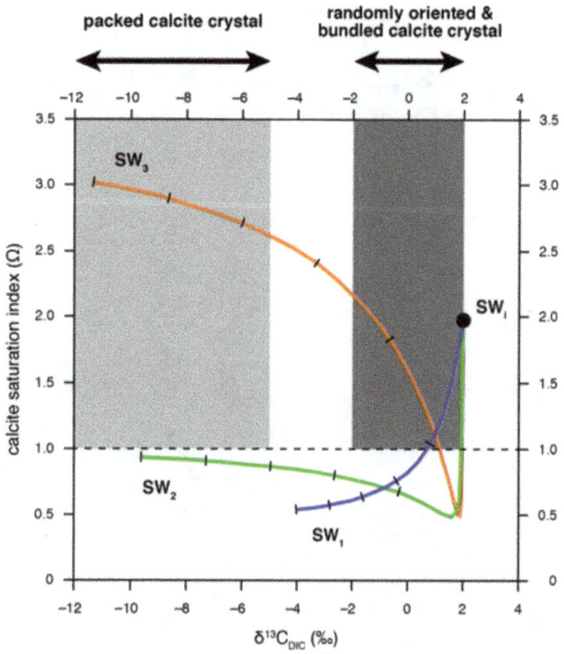

Fig. 7 Relationship between the $\delta^{13}C_{DIC}$ value and calcite saturation index (Ω) of the waters generated by mixing of the initial seawater (SW_i) with the seawaters in which sulfate reduction occurred (SW_1, SW_2, SW_3) (Fig. 6). The mixing ratios of SW_x to SW_i ($x = 1, 2, 3$) are 0 to 0.5, and the *tick marks* on each curve represent the ratios; 0.1, 0.2, 0.3, 0.4, and 0.5 from right to left on each curve. The *arrows* represent the main ranges of $\delta^{13}C_{carb}$ values of randomly oriented and bundled calcites (*right*) and of packed calcites (*left*)

anoxic oceans, a gradient in carbonate saturation in the water column and sediments might have been small, thus promoting seafloor carbonate precipitation. The small Ω gradient in the Permian anoxic water column may also have contributed to the Chaotian carbonate precipitation near the sediment–water interface.

Authigenic carbonate precipitation in the anoxic oceans

Schrag et al. (2013) proposed that authigenic carbonate played a major role in the ancient carbon cycle. According to their model, authigenic carbonates are generally generated within the sediments by the decomposition of sedimentary organic matter via anaerobic respiration (such as sulfate and iron reduction) (e.g., Irwin et al. 1977; Meister et al. 2007; Dela Pierre et al. 2012). The anaerobic respiration in the sediments increases the alkalinity in the pore waters and remineralized organic carbon with low $\delta^{13}C$ values is released, promoting the in situ carbonate precipitation. The waxing and waning of the global sink of ^{13}C-depleted authigenic carbonate may have changed the $\delta^{13}C_{DIC}$ value in the ancient oceans without large fluctuations in atmospheric O_2 and CO_2 levels. Schrag et al. (2013) suggested two mechanisms for $\delta^{13}C_{carb}$ excursion in geologic records associated with authigenic carbonate precipitation. The change in $\delta^{13}C_{DIC}$ values due to a change in the size of the global authigenic carbonate sink would be recorded as a $\delta^{13}C_{carb}$ excursion in geologic records ("type 1 excursion" according to Macdonald et al. 2013). Moreover, a local addition of ^{13}C-depleted authigenic carbonate to normal marine carbonate would have declined the bulk $\delta^{13}C_{carb}$ value in rock samples ("type 2 excursion" according to Macdonald et al. 2013).

Schrag et al. (2013) assumed that, in general, the $\delta^{13}C_{carb}$ values of authigenic carbonates have been substantially lower than the $\delta^{13}C_{DIC}$ value of seawater in Earth's history. However, our results from Chaotian imply that the $\delta^{13}C_{carb}$ values of the authigenic carbonates in the oxic and anoxic oceans in the past may have been systematically different (Fig. 8). We propose that the depth of the redoxcline in the oceans has controlled the generation depth and $\delta^{13}C_{carb}$ values of the authigenic carbonate. In the oxic oceans, anaerobic respiration is largely restricted within the sediments because the redoxcline is below the seafloor. The authigenic carbonate precipitation due to increased alkalinity caused by respiration is also restricted within the sediments, as is observed in the modern oceans. The upward diffusive transport of remineralized carbon released from sedimentary organic matter and the downward diffusive transport of seawater DIC within the sediments are low, and vertical mixing of these two inorganic carbon pools is suppressed. As a result, the $\delta^{13}C_{carb}$ values of authigenic carbonate generated within the sediments are relatively low compared with that of the DIC in the water column. In marked contrast, the redoxcline in the anoxic oceans is generally within the water column and water-mass anaerobic respiration is enhanced. When the water-mass anaerobic respiration occurs, it is likely that remineralized carbon from suspended organic matter is released into the water column and mixed with the large oceanic DIC pool. The low $\delta^{13}C$ values of remineralized organic carbon are likely diluted by the large DIC pool. In other words, the water-mass anaerobic respiration promotes the recycling of fixed organic carbon back to the large oceanic DIC pool. In the anoxic oceans, authigenic carbonate is mainly generated near the sediment–water interface rather than deep within the sediments (Higgins et al. 2009; Bergmann et al. 2013). As a result, the $\delta^{13}C_{carb}$ values of the authigenic carbonate precipitated near the sediment–water interface are close to the $\delta^{13}C_{DIC}$ value of seawater, although a minor amount of ^{13}C-depleted carbonate may also have been generated within the sediments (Fig. 8).

Carbonates precipitated abundantly on the seafloor in the Archean (e.g., Sumner and Grotzinger 2000), while the occurrence of carbonate seafloor precipitates and cements declined during the Proterozoic (e.g., Grotzinger and James 2000), except for a brief revival in the late

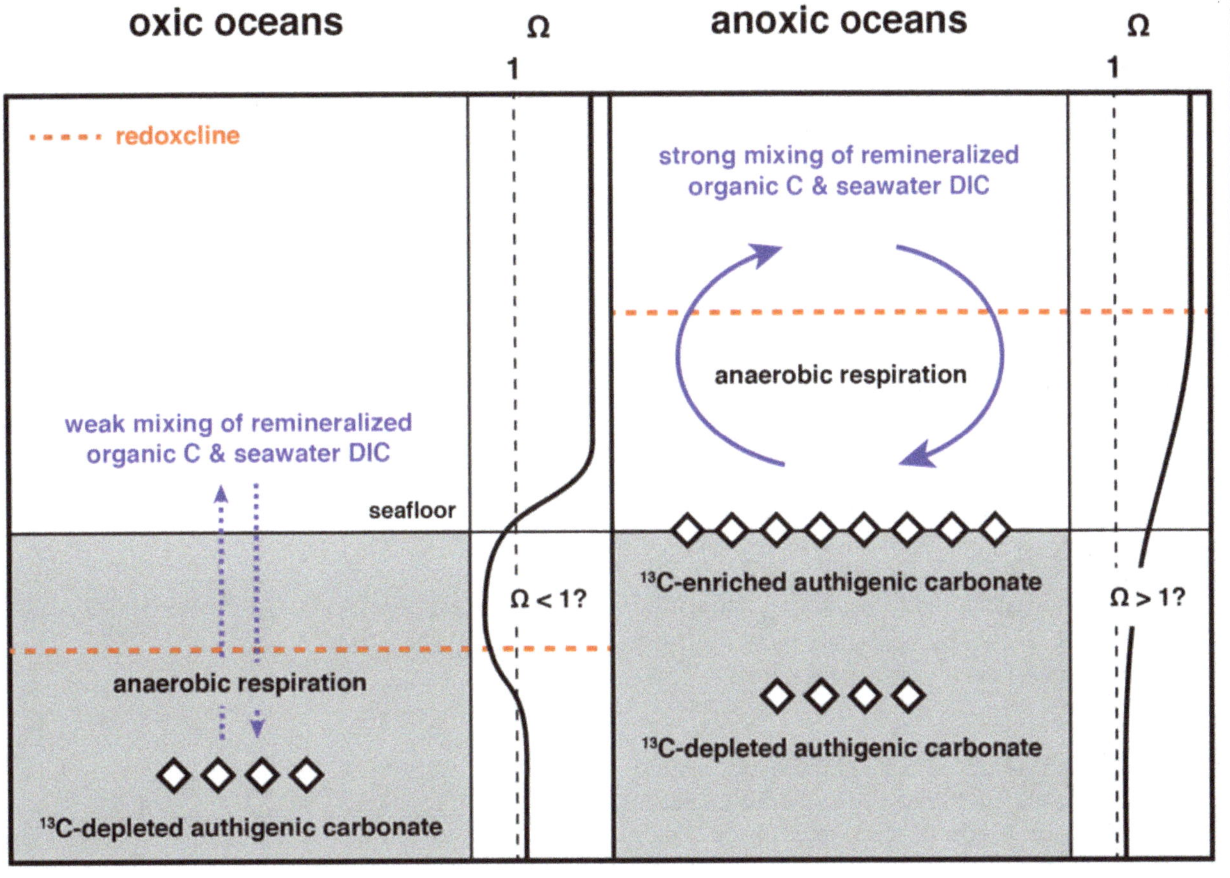

Fig. 8 Schematic diagram representing differences in the form of authigenic carbonate precipitation between the oxic and anoxic oceans, proposed in this study. In the oxic oceans, the redoxcline is generally below the sediment–water interface and a small amount of authigenic carbonate is generated within the sediments. The $\delta^{13}C_{carb}$ value of the carbonate is relatively low as a result of weak mixing between the remineralized carbon released from sedimentary organic matter and seawater DIC. In contrast, in the anoxic oceans, the redoxcline is above the sediment–water interface and large amounts of authigenic carbonate precipitate near the seafloor. The $\delta^{13}C_{carb}$ value of the carbonate is relatively high due to the strong mixing of remineralized organic carbon and the large DIC pool in the water column. Ω the carbonate saturation index

Neoproterozoic (e.g., Grotzinger and Knoll 1995; James et al. 2001; Pruss et al. 2008). In the Phanerozoic, carbonate seafloor precipitates are rare although they were widespread in the Paleozoic–Mesozoic transition interval where their precipitations are often associated with oceanic anoxia (e.g., Grotzinger and Knoll 1995; Woods et al. 1999, 2007, 2014; Heydari et al. 2003; Pruss et al. 2006; Baud et al. 2007). This secular trend in carbonate seafloor precipitation in Earth's history suggests that the carbonate precipitation has been generally related to oceanic anoxia (e.g., Grotzinger and Knoll 1995; Higgins et al. 2009). However, other factors may have also influenced the authigenic carbonate precipitation because oceanic anoxic events (OAEs) in the Phanerozoic were not always associated with the widespread occurrence of authigenic carbonates (e.g., Bergmann et al. 2013). Nonetheless, our results from Chaotian imply that when the authigenic carbonates precipitate near the sediment–water interface in anoxic oceans, their $\delta^{13}C_{carb}$ values are close to the $\delta^{13}C_{DIC}$ value of seawater. The systematic $\delta^{13}C_{carb}$ differences in authigenic carbonate due to its generation depth have also been recognized in some Phanerozoic geologic records; the $\delta^{13}C_{carb}$ values of Lower Triassic carbonate cements precipitated on the anoxic seafloor in California are relatively high (+2.8 to +4.2 ‰) (Woods et al. 1999), whereas those of early diagenetic subseafloor carbonates around the Triassic–Jurassic boundary in British Columbia are relatively low (−7.6 to −3.7 ‰) (Greene et al. 2012). Certain authigenic carbonates have been generated by anaerobic oxidation of methane in the ancient and modern oceans (e.g., Ritger et al. 1987; Michaelis et al. 2002; Allison et al. 2008; Bayon et al. 2013), although a detailed discussion of methane-derived carbonates is beyond the scope of this study.

Our model for authigenic carbonate precipitation in the oxic and anoxic oceans does not exclude the significance of the authigenic carbonate sink over the course of Earth's history (Schrag et al. 2013). However, when the $\delta^{13}C_{carb}$

value of authigenic carbonate was relatively high and close to the $\delta^{13}C_{DIC}$ value in the oceans, the influence of a local addition of authigenic carbonate to normal marine carbonate on the bulk $\delta^{13}C_{carb}$ value, causing a type 2 excursion in geologic records (Macdonald et al. 2013), would have been small. In the Precambrian anoxic oceans, diagenetic carbonates with relatively high $\delta^{13}C_{carb}$ values may have been extensively generated within the methane zone in the sediments (Li et al. 2015). Low sulfate and high DIC concentrations in the Precambrian oceans may also have suppressed ^{13}C depletion in authigenic carbonates (Bristow and Grotzinger 2013). Our model implies that, in the ancient anoxic oceans, the type 2 excursions may be less significant in geologic records than previously suggested (Schrag et al. 2013). In other words, the type 2 excursions might be limited to geologic records from oxic intervals in Earth's history because these excursions are caused by authigenic carbonates with substantially low $\delta^{13}C_{carb}$ values. The existence of type 2 excursions, such as the Shuram anomaly in the Ediacaran (e.g., Grotzinger et al. 2011; Macdonald et al. 2013; Schrag et al. 2013), may therefore suggest oxic conditions in the oceans.

Conclusions

Authigenic carbonates are common in the anoxic interval of the upper Guadalupian Maokou Formation at Chaotian, Sichuan, China. Calcite crystals precipitated in situ mainly near the sediment–water interface, possibly associated with the water-mass anaerobic respiration. The relatively high $\delta^{13}C_{carb}$ values of the carbonates (largely about −1 ‰) indicate that their main carbon source was seawater DIC. Our results suggest that the forms of authigenic carbonate precipitation in the oxic and anoxic oceans in the past may have been substantially different. In the anoxic oceans, authigenic carbonates likely precipitated near the seafloor with $\delta^{13}C_{carb}$ values close to that of seawater DIC. The influence of authigenic carbonate on the bulk $\delta^{13}C_{carb}$ values in geologic records from ancient anoxic oceans might have been overestimated in previous studies. A large negative $\delta^{13}C_{carb}$ excursion caused by the local addition of ^{13}C-depleted authigenic carbonate to normal marine carbonate (type 2 excursion in Macdonald et al. 2013) may have been limited to oxic intervals in Earth's history and thus might be a useful indicator of the oxic conditions in the ancient oceans.

Competing interests
The authors declare that they have no competing interests.

Authors' contributions
MS designed the study. MS, YI, JXY, and ZSJ conducted the fieldwork and sampling. MS and YI conducted the petrological analyses. KS and MM conducted the XRD analyses. MS, YU, and NY conducted the isotopic analyses. MS and TS conducted the calculation analyses. MS, YU, YI, and TS wrote the paper. All authors read and approved the final manuscript.

Authors' information
MS is a project researcher at the Research and Development (R&D) Center for Submarine Resources, Japan Agency for Marine-Earth Science and Technology (JAMSTEC). YU is an associate professor at the Department of Earth and Planetary Sciences and Earth-Life Science Institute, Tokyo Institute of Technology. YI is a professor at the Department of Earth Science and Astronomy, The University of Tokyo. TS is a scientist at the Laboratory of Ocean-Earth Life Evolution Research (OELE), Japan Agency for Marine-Earth Science and Technology (JAMSTEC). JXY is a professor at the Geology Institute, Chinese Academy of Geological Science. ZSJ is a research scientist at the Geology Institute, Chinese Academy of Geological Science. KS is an assistant professor at the Department of Chemistry, The University of Tokyo. MM is a professor, Department of Chemistry, The University of Tokyo. NY is a professor at the Department of Environmental Chemistry and Engineering and Earth-Life Science Institute, Tokyo Institute of Technology.

Acknowledgements
This study was supported by JSPS KAKENHI (16204040, 20224012, 26610159, 15H03740) and CGS (1212011120116, 1212011120143). YU is supported by the NEXT program of JSPS. Hodaka Kawahata and two anonymous reviewers provided constructive comments that improved the manuscript. Miyuki Tahata assisted with the isotopic analyses.

Author details
[1]Laboratory of Ocean-Earth Life Evolution Research (OELE), Japan Agency for Marine-Earth Science and Technology (JAMSTEC), 2-15 Natsushima-cho, Yokosuka 237-0061, Japan. [2]Department of Earth and Planetary Sciences, Tokyo Institute of Technology, 2-12-1 Ookayama, Meguro, Tokyo 152-8551, Japan. [3]Earth-Life Science Institute, Tokyo Institute of Technology, 2-12-1 Ookayama, Meguro, Tokyo 152-8551, Japan. [4]Department of Earth Science and Astronomy, The University of Tokyo, 3-8-1 Komaba, Meguro, Tokyo 153-8902, Japan. [5]Institute of Geology, Chinese Academy of Geological Science, Baiwanzhuang Road 26, Beijing 100037, China. [6]Department of Chemistry, The University of Tokyo, 3-8-1 Komaba, Meguro, Tokyo 153-8902, Japan. [7]Department of Environmental Chemistry and Engineering, Tokyo Institute of Technology, 4259 Nagatsuda, Midori, Yokohama 226-8502, Japan. [8]Present address: Research and Development (R&D) Center for Submarine Resources, Japan Agency for Marine-Earth Science and Technology (JAMSTEC), 2-15 Natsushima-cho, Yokosuka 237-0061, Japan.

References
Algeo TJ, Luo GM, Song HY, Lyons TW, Canfield DE. Reconstruction of secular variation in seawater sulfate concentrations. Biogeosciences. 2015;12:2131–51.
Allison PA, Hesselbo SP, Brett CE. Methane seeps on an early jurassic dysoxic seafloor. Palaeogeogr Palaeoclimatol Palaeoecol. 2008;270:230–8.
Arvidson RS, Mackenzie FT, Berner RA. The sensitivity of the Phanerozoic inorganic carbon system to the onset of pelagic sedimentation. Aquat Geochem. 2014;20:343–62.
Baud A, Richoz S, Pruss S. The lower Triassic anachronistic carbonate facies in space and time. Global Planet Change. 2007;55:81–9.
Bayon G, Dupré S, Ponzevera E, Etoubleau J, Chéron S, Pierre C, et al. Formation of carbonate chimneys in the Mediterranean Sea linked to deep-water oxygen depletion. Nat Geosci. 2013;6:755–60.
Bergmann KD, Grotzinger JP, Fischer WW. Biological influences on seafloor carbonate precipitation. Palaios. 2013;28:99–115.
Bethke CM. Geochemical and biogeochemical reaction modeling. 2nd ed. Cambridge: Cambridge University Press; 2008.
Bosak T, Corsetti FA, Newman DK. Micrometer-scale porosity as a biosignature in carbonate crusts. Geology. 2004;32:781–4.
Bristow TF, Grotzinger JP. Sulfate availability and the geological record of cold-seep deposits. Geology. 2013;41:811–4.
Dela Pierre F, Clari P, Bernardi E, Natalicchio M, Costa E, Cavagna S, et al. Messinian carbonate-rich beds of the Tertiary Piedmont Basin (NW Italy): microbially-mediated products straddling the onset of the salinity crisis. Palaeogeogr Palaeoclimatol Palaeoecol. 2012;344–345:78–93.
D'Hondt S. Consequences of the Cretaceous/Paleogene mass extinction for marine ecosystems. Annu Rev Ecol Evol Syst. 2005;36:295–317.
Greene SE, Bottjer DJ, Corsetti FA, Berelson WM, Zonneveld JP. A subseafloor carbonate factory across the Triassic-Jurassic transition. Geology. 2012;40:1043–6.

Grotzinger JP, Knoll AH. Anomalous carbonate precipitates: is the Precambrian the key to the Permian? Palaios. 1995;10:578–96.

Grotzinger JP, James NP. Precambrian carbonates: evolution of understanding. In: Grotzinger JP, James NP, editors. Carbonate Sedimentation and Diagenesis in the Evolving Precambrian World, SEPM Special Publication, vol. 67. Tulsa: Oklahoma; 2000. p. 3–20.

Grotzinger JP, Fike DA, Fischer WW. Enigmatic origin of the largest-known carbon isotope excursion in Earth's history. Nat Geosci. 2011;4:285–92.

Hayes JM, Strauss H, Kaufman AJ. The abundance of ^{13}C in marine organic matter and isotopic fractionation in the global biogeochemical cycle of carbon during the past 800 Ma. Chem Geol. 1999;161:103–25.

Heindel K, Richoz S, Birgel D, Brandner R, Klügel A, Krystyn L, et al. Biogeochemical formation of calyx-shaped carbonate crystal fans in the shallow subsurface of the Early Triassic seafloor. Gondwana Res. 2015;27:840–61.

Helgeson HC. Thermodynamics of hydrothermal systems at elevated temperatures and pressures. Am J Sci. 1969;267:729–804.

Helgeson HC, Kirkham DH. Theoretical prediction of the thermodynamic behavior of aqueous electrolytes at high pressures and temperatures; II, Debye-Huckel parameters for activity coefficients and relative partial molal properties. Am J Sci. 1974;274:1199–261.

Heydari E, Hassanzadeh J, Wade WJ, Ghazi AM. Permian-Triassic boundary interval in the Abadeh section of Iran with implications for mass extinction: part 1—sedimentology. Palaeogeogr Palaeoclimatol Palaeoecol. 2003;193:405–23.

Higgins JA, Fischer WW, Schrag DP. Oxygenation of the ocean and sediments: consequences for the seafloor carbonate factory. Earth Planet Sci Lett. 2009; 284:25–33.

Huber BT, Hodell DA, Hamilton CP. Middle-Late Cretaceous climate of the southern high latitudes: stable isotopic evidence for minimal equator-to-pole thermal gradients. GSA Bull. 1995;107:1164–91.

Irwin H, Curtis C, Coleman M. Isotopic evidence for source of diagenetic carbonates formed during burial of organic-rich sediments. Nature. 1977;269:209–13.

Isozaki Y, Yao JX, Matsuda T, Sakai H, Ji ZS, Shimizu N, et al. Stratigraphy of the Middle-Upper Permian and lowermost Triassic at Chaotian, Sichuan, China—record of Late Permian double mass extinction event. Proc Jpn Acad Ser B. 2004;80:10–6.

Isozaki Y, Yao JX, Ji ZS, Saitoh M, Kobayashi N, Sakai H. Rapid sea-level change in the Late Guadalupian (Permian) on the Tethyan side of South China: litho- and biostratigraphy of the Chaotian section in Sichuan. Proc Jpn Acad Ser B. 2008;80:344–53.

James NP, Narbonne GM, Kyser TK. Late Neoproterozoic cap carbonates: Mackenzie Mountains, northwestern Canada: precipitation and global glacial meltdown. Can J Earth Sci. 2001;38:1229–62.

Jin YG, Mei SL, Wang W, Wang XD, Shen SZ, Shang QH, et al. On the Lopingian series of the Permian System. Palaeoworld. 1998;9:1–18.

Kasting JF. Earth's early atmosphere. Science. 1993;259:920–6.

Kennett JP, Stott LD. Abrupt deep-sea warming, palaeoceanographic changes and benthic extinctions at the end of the Palaeocene. Nature. 1991;353:225–9.

Knauth LP, Kennedy MJ. The late Precambrian greening of the Earth. Nature. 2009;460:728–32.

Kroopnick PM. The distribution of 13C of ∑CO2 in the world oceans. Deep-Sea Res. 1985;32:57–84.

Lam P, Lavik G, Jensen MM, van de Vossenberg J, Schmid M, Woebken D, et al. Revising the nitrogen cycle in the Peruvian oxygen minimum zone. Proc Natl Acad Sci U S A. 2009;106:4752–7.

Li C, Planavsky NJ, Love GD, Reinhard CT, Hardisty D, Feng LJ, et al. Marine redox conditions in the middle Proterozoic ocean and isotopic constraints on authigenic carbonate formation: insights from the Chuanlinggou Formation, Yanshan Basin, North China. Geochim Cosmochim Acta. 2015;150:90–105.

Macdonald FA, Strauss JV, Sperling EA, Halverson GP, Narbonne GM, Johnston DT, et al. The stratigraphic relationship between the Shuram carbon isotope excursion, the oxygenation of Neoproterozoic oceans, and the first appearance of the Ediacara biota and bilaterian trace fossils in northwestern Canada. Chem Geol. 2013;362:250–72.

Meister P, McKenzie JA, Vasconcelos C, Bernasconi S, Frank M, Gutjahr M, et al. Dolomite formation in the dynamic deep biosphere: results from the Peru Margin. Sedimentology. 2007;54:1007–31.

Meister P. Two opposing effects of sulfate reduction on carbonate precipitation in normal marine, hypersaline, and alkaline environments. Geology. 2013;41:499–502.

Michaelis W, Seifert R, Nauhaus K, Treude T, Thiel V, Blumenberg M, et al. Microbial reefs in the Black Sea fueled by anaerobic oxidation of methane. Science. 2002;297:1013–5.

Naehr TH, Eichhubl P, Orphan VJ, Hovland M, Paull CK, Ussler W, et al. Authigenic carbonate formation at hydrocarbon seeps in continental margin sediments: a comparative study. Deep-Sea Res II. 2007;54:1268–91.

Porter SM. Seawater chemistry and early carbonate biomineralization. Science. 2007;316:1302.

Pruss SB, Bottjer DJ, Corsetti FA, Baud A. A global marine sedimentary response to the end-Permian mass extinction: examples from southern Turkey and the western United States. Earth-Sci Rev. 2006;78:193–206.

Pruss SB, Corsetti FA, Fischer WW. Seafloor-precipitated carbonate fans in the Neoproterozoic Rainstorm Member, Johnnie Formation, Death Valley Region, USA. Sed Geol. 2008;207:34–40.

Revesz KM, Landwehr JM. $\delta^{13}C$ and $\delta^{18}O$ isotopic composition of $CaCO_3$ measured by continuous flow isotope ratio mass spectrometry: statistical evaluation and verification by application to Devils Hole core DH-11 calcite. Rapid Commun Mass Spectrom. 2002;16:2102–14.

Ridgwell A. A Mid Mesozoic Revolution in the regulation of ocean chemistry. Mar Geol. 2005;217:339–57.

Ridgwell A, Zeebe RE. The role of the global carbonate cycle in the regulation and evolution of the Earth system. Earth Planet Sci Lett. 2005;234:299–315.

Ritger S, Carson B, Suess E. Methane-derived authigenic carbonates formed by subduction-induced pore-water expulsion along the Oregon/Washington margin. GSA Bull. 1987;98:147–56.

Saitoh M, Isozaki Y, Yao JX, Ji ZS, Ueno Y, Yoshida N. The appearance of an oxygen-depleted condition on the Capitanian disphotic slope/basin in South China: Middle-Upper Permian stratigraphy at Chaotian in northern Sichuan. Global Planet Change. 2013a;105:180–92.

Saitoh M, Isozaki Y, Ueno Y, Yoshida N, Yao JX, Ji ZS. Middle-Upper Permian carbon isotope stratigraphy at Chaotian, South China: pre-extinction multiple upwelling of oxygen-depleted water onto continental shelf. J Asian Earth Sci. 2013b;67–68:51–62.

Saitoh M, Ueno Y, Isozaki Y, Nishizawa M, Shozugawa K, Kawamura T, et al. Isotopic evidence for water-column denitrification and sulfate reduction at the end-Guadalupian (Middle Permian). Global Planet Change. 2014a;123:110–20.

Saitoh M, Ueno Y, Nishizawa M, Isozaki Y, Takai K, Yao JX, et al. Nitrogen isotope chemostratigraphy across the Permian-Triassic boundary at Chaotian, Sichuan, South China. J Asian Earth Sci. 2014b;93:113–28.

Sandberg PA. An oscillating trend in Phanerozoic non-skeletal carbonate mineralogy. Nature. 1983;305:19–22.

Schrag DP, Higgins JA, Macdonald FA, Johnston DT. Authigenic carbonate and the history of the global carbon cycle. Science. 2013;339:540–3.

Scotese CR, Langford RP. Pangea and the Paleogeography of the Permian. In: Scholle A, Peryt TM, Ulmer-Scholle DA, editors. The Permian of Northern Pangea, 1. Berlin: Springer; 1995.

Shackleton NJ, Hall MA. Carbon isotope data from Leg 74 sediments. Initial Reports Deep Sea Drill Project. 1984;74:613–9.

Shields G, Veizer J. Precambrian marine carbonate isotope database: Version 1.1. Geochem Geophys Geosyst 2002. doi:10.1029/2001GC000266

Stanley SM, Hardie LA. Secular oscillations in the carbonate mineralogy of reef-building and sediment-producing organisms driven by tectonically forced shifts in seawater chemistry. Palaeogeogr Palaeoclimatol Palaeoecol. 1998;144:3–19.

Sumner DY, Grotzinger JP. Late Archean aragonite precipitation: petrography, facies associations, and environmental significance. In: Grotzinger JP, James NP, editors. Carbonate Sedimentation and Diagenesis in the Evolving Precambrian World, SEPM Special Publication, vol. 67. Tulsa: Oklahoma; 2000. p. 123–44.

Sun XL, Turchyn AV. Significant contribution of authigenic carbonate to marine carbon burial. Nat Geosci. 2014;7:201–4.

Van der Kooij B, Immenhauser A, Steuber T, Bahamonde JR, Meríno-Tomé O. Controlling factors of volumetrically important marine carbonate cementation in deep slope settings. Sedimentology. 2010;57:1491–525.

Walker JCG, Hays PB, Kasting JF. A negative feedback mechanism for the long-term stabilization of Earth's surface temperature. J Geophys Res. 1981;86:9776–82.

Wang Y, Jin YG. Permian palaeogeographic evolution of the Jiangnan Basin, South China. Palaeogeogr Palaeoclimatol Palaeoecol. 2000;160:35–44.

Whiticar MJ. Carbon and hydrogen isotope systematics of bacterial formation and oxidation of methane. Chem Geol. 1999;161:291–314.

Woods AD, Bottjer DJ, Mutti M, Morrison J. Lower Triassic large sea-floor carbonate cements: their origin and a mechanism for the prolonged biotic recovery from the end-Permian mass extinction. Geology. 1999;27:645–8.

Woods AD, Bottjer DJ, Corsetti FA. Calcium carbonate precipitates from the outer shelf to slope facies of the Lower Triassic (Smithian-Spathian) Union Wash Formation, California, USA: sedimentary and palaeobiologic significance. Palaeogeogr Palaeoclimatol Palaeoecol. 2007;252:281–90.

Woods AD. Assessing Early Triassic paleoceanographic conditions via unusual sedimentary fabrics and features. Earth-Sci Rev. 2014;137:6–18.

Zhao JK, Sheng JZ, Yao ZQ, Liang XL, Chen CZ, Rui L, et al. The Changhsingian and Permian-Triassic boundary of South China. Bull Nanjing Inst Geol Paleontol Acad Sinica. 1981;2:1–112.

4

Structural context and variation of ocean plate stratigraphy, Franciscan Complex, California: insight into mélange origins and subduction-accretion processes

John Wakabayashi

Abstract

The transfer (accretion) of materials from a subducting oceanic plate to a subduction-accretionary complex has produced rock assemblages recording the history of the subducted oceanic plate from formation to arrival at the trench. These rock assemblages, comprising oceanic igneous rocks progressively overlain by pelagic sedimentary rocks (chert and/or limestone) and trench-fill clastic sedimentary rocks (mostly sandstone, shale/mudstone), have been called ocean plate stratigraphy (OPS). During accretion of OPS, megathrust slip is accommodated by imbricate faults and penetrative strain, shortening the unit and leading to tectonic repetition of the OPS sequence, whereas OPS accreted at different times are separated by non-accretionary megathrust horizons. The Franciscan subduction complex of California accreted episodically over a period of over 150 million years and incorporated OPS units with a variety of characteristics separated by non-accretionary megathrust horizons. Most Franciscan OPS comprises MORB (mid-ocean-ridge basalt) progressively overlain by chert and trench-fill clastic sedimentary rocks that are composed of variable proportions of turbidites and siliciclastic and serpentinite-matrix olistostromes (sedimentary mélanges). Volumetrically, the trench-fill component predominates in most Franciscan OPS, but some units have a significant component of igneous and pelagic rocks. Ocean island basalt (OIB) overlain by limestone is less common than MORB-chert assemblages, as are abyssal serpentinized peridotite slabs. The earliest accreted OPS comprises metabasite of supra-subduction zone affinity imbricated with smaller amounts of metaultramafic rocks and metachert, but lacking a clastic component. Most deformation of Franciscan OPS is localized along discrete faults rather than being distributed in the form of penetrative strain. This deformation locally results in block-in-matrix tectonic mélanges, in contrast to the sedimentary mélanges making up part of the clastic OPS component. Such tectonic mélanges may include blocks and matrix derived from the olistostromes. Franciscan subduction and OPS accretion initiated in island arc crust at about 165–170 Ma, after which MORB and OIB were subducted and accreted following a long (tens of mega-ampere) gap with little or no accretion. Following subduction initiation, a ridge crest approached the trench but probably went dormant prior to its subduction (120–125 Ma), after which the subducted oceanic crust became progressively older until about 95 Ma. From 95 Ma, the age of subducted oceanic crust decreased progressively until arrival of the Pacific-Farallon spreading center led to termination of subduction and conversion to a transform plate boundary.

Keywords: Ocean plate stratigraphy, Subduction complex evolution, Subduction megathrust slip accommodation, Tectonic and sedimentary mélanges

Correspondence: jwakabayashi@csufresno.edu
Department of Earth and Environmental Sciences, California State University Fresno, 2576 E. San Ramon Avenue, Mail Stop ST-24, Fresno, CA 93740, USA

Introduction

Ocean plate stratigraphy (OPS) comprises assemblages of trench-fill clastic sediments overlying pelagic sediments and oceanic igneous rocks that make up the world's subduction complexes (Wahrhafig 1984; Isozaki et al. 1990; Isozaki and Blake 1994; Wakita and Metcalfe 2005; Kusky et al. 2013; Wakita 2015). The specific OPS terminology was first applied by Isozaki et al. 1990, although the concept dates back at least as far as Wahrhaftig (1984). OPS is interpreted to record initial submarine volcanism at a mid-ocean ridge generating mid-ocean-ridge basalt (MORB) or at an off-axis location generating ocean island basalt (OIB), followed by open ocean pelagic sedimentation as the ocean plate drifts and by deposition of trench-fill clastic sediments as that part of the subducting oceanic plate reaches the trench (e.g., Chipping 1971; Wahrhaftig 1984; Isozaki et al. 1990; Osozawa 1994; Wakita and Metcalfe 2005; Kusky et al. 2013; Fig. 1).

OPS is transferred from the subducting plate to the subduction complex by the process of subduction-accretion during which the subduction megathrust cuts into the subducting plate. During accretion of OPS, subduction megathrust slip is accommodated by imbrication and penetrative deformation of the OPS. During much of the history of subduction zones, no accretion of any sort takes place (e.g., von Huene and Scholl 1991) so that accretion in any given subduction complex is expected to be sporadic. Accretion is associated with an accretionary mode of megathrust slip, whereas non-accretionary megathrust horizons separate OPS accreted at different times in a subduction complex (Wakabayashi 2016a). The incorporation of OPS into a subduction complex leads to intricate age relationships. The age of the igneous and the oldest pelagic components of a slice of OPS may be much older than the clastic sedimentary component that approximates the arrival time of the

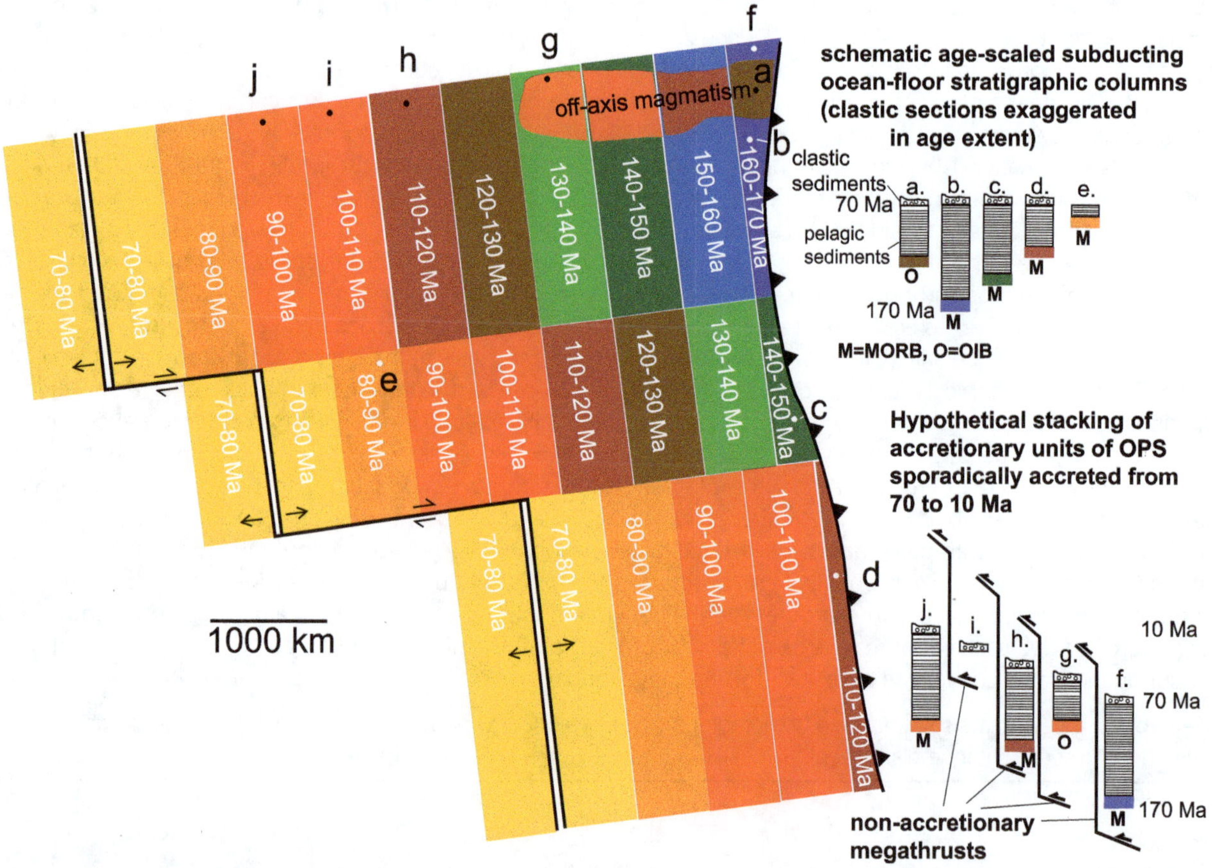

Fig. 1 Cartoon showing OPS evolution and age patterns versus accretion. The schematic stratigraphic columns are age-scaled rather than thickness-scaled. Columns **a-d** show how OPS arriving at the trench at the same time may vary along strike, in this case owing to subducting of fracture zones and crust formed off-axis. Obliquity of subduction relative to isochrons on the oceanic plate, which should be considered the general case, will introduce another level of complexity. Columns **a-e** show that pelagic sections formed far apart may share a common age range for part of their pelagic section. This suggests caution in correlating incomplete OPS sections where the basal pelagic rocks directly overlying sea floor volcanic rocks are not exposed. Columns **f-j** show the stacking of sporadically accreted OPS at one position along a trench. Compare to Franciscan Complex columns shown in Fig. 5b, c

piece of ocean floor at a trench shortly before subduction-accretion (e.g., Hamilton 1969; Wahrhaftig 1984; Osozawa 1994; Wakita and Metcalfe 2005; Kusky et al. 2013) (Fig. 1).

Successive episodic subduction-accretion of OPS can build a tectonic stack in a subduction complex wherein the clastic components young structurally downward across non-accretionary megathrust horizons, but the ocean crust age can have a less systematic age pattern that may trend both younger and older structurally downward (columns f to j in the lower right of Fig. 1) (Wakabayashi 2015). Nearly all of the igneous oceanic crust entering a subduction zone completely subducts and only a tiny fraction transfers to the upper plate as a part of OPS. In some cases, accretion of trench sediments may take place, but without incorporation of any of the underlying pelagic sedimentary and oceanic igneous rocks (Fig. 1, lower right "i" column).

Obliquity of plate convergence relative to isochrons of the ocean crust as well as the subduction of oceanic fracture zones, spreading ridges, and crust formed off-axis, such as seamounts, oceanic plateaus, can lead to complex along-strike patterns in the ages of oceanic crust accreted at the same time along the length of a subduction zone (Fig. 1; Osozawa 1994). Parts of the pelagic sections formed at different places and times on the ocean floor may be age correlative, but their actual sites of deposition may be far removed from one another (Fig. 1). Correlation of OPS sections lacking igneous rocks should not be made on the basis of similarity in age range of pelagic sections alone.

Whether OPS is or is not a synonym or subset of the term "ophiolite" depends on the varied definitions of ophiolite (e.g., Moores 1982; Dilek and Furnes 2011; Kusky et al. 2013). This has important implications for the connection between OPS packages and tectonic history. The largest ophiolite bodies, such as the Troodos, Semail, Bay of Islands, and Coast Range ophiolites, are assemblages that were part of the upper plate above a subduction zone, rather than the much smaller oceanic crust scraps transferred from the subducting plate to the upper plate as part of a subduction complex (e.g. Moores 1982). Large upper plate ophiolitic bodies are successively overlain by pelagic and clastic sedimentary rocks, so these have the same components as OPS slices in subduction complexes, but the tectonic-sedimentary history of such units differs markedly.

In the California Cordillera, the largest bodies of oceanic rocks (igneous rocks and overlying pelagic sedimentary rocks) are associated with the upper plate settings instead of subduction complexes. Figure 2 summarizes these relationships with various upper plate ophiolitic igneous rock bodies designated with acronyms vo, so, voc, soc, and bodies of island arc volcanic rocks designated with the acronym va. For example, the Coast Range ophiolite (voc, soc) is the upper plate ophiolite unit of the Coast Ranges, whereas "so" in the northwest Klamath Mountains is part of the Josephine ophiolite, "so" in the eastern Klamath Mountains is part of the Trinity ophiolite, and "va" along the northwestern margin of the Sierra Nevada is the Smartville arc. Large upper plate oceanic assemblages and subduction complex OPS are easily distinguished from one another in coastal California, owing to the lack of post-subduction disruption, whereas this distinction is much more difficult in many (most?) other orogenic belts (Wakabayashi 2015, 2016a). This paper will use the term OPS solely for the components of subduction complexes and will exclude the upper plate oceanic rocks, in contrast to Kusky et al. (2013; their Fig. 1) that included units interpreted by many as upper plate, but considered them to have been derived from the downgoing plate. The oceanic crustal components of Franciscan subduction complex crop out over much smaller areas than the upper plate oceanic fragments as shown by the size of outcrops labeled "vf, vfs, vrf, and va" on Fig. 2

This paper presents new information on the variety of OPS in the Franciscan Complex of the California Coast Ranges and its relationship to Franciscan Complex tectonic history and subduction processes. Kusky et al. (2013) reviewed OPS within the Franciscan Complex, but this paper presents more detail bearing on a greater variety of OPS types, and proposes different interpretations about the relationship between mélanges and the formation and deformation of OPS. Wakabayashi (2015) presented much of the field-based framework of this paper, but all of the outcrop photos and photomicrographs presented in this paper are new. In addition, the connection between OPS, subduction process, and tectonic versus sedimentary mélanges is more explicitly explained in this paper than in Wakabayashi (2015).

Review

California coast ranges regional relationships

The Coast Ranges preserve the relationship between paleoforearc basin deposits (Great Valley Group or GVG) that rest in depositional contact on a remnant of oceanic lithosphere (Coast Range Ophiolite or CRO) tectonically overlying an exhumed subduction complex (Franciscan Complex) (Hamilton 1969; Dickinson 1970: Ernst 1970; Bailey et al. 1970). These three megaunits crop out together for more than 700 km along and 100 km across strike (Fig. 2). This outcrop length is greater than 1000 km if strike-slip faulting is restored (e.g., Wakabayashi 2015). OPS, as defined herein is restricted to the Franciscan Complex. In contrast to the materials of the Franciscan Complex, that were transferred from the subducting plate to the upper plate, the

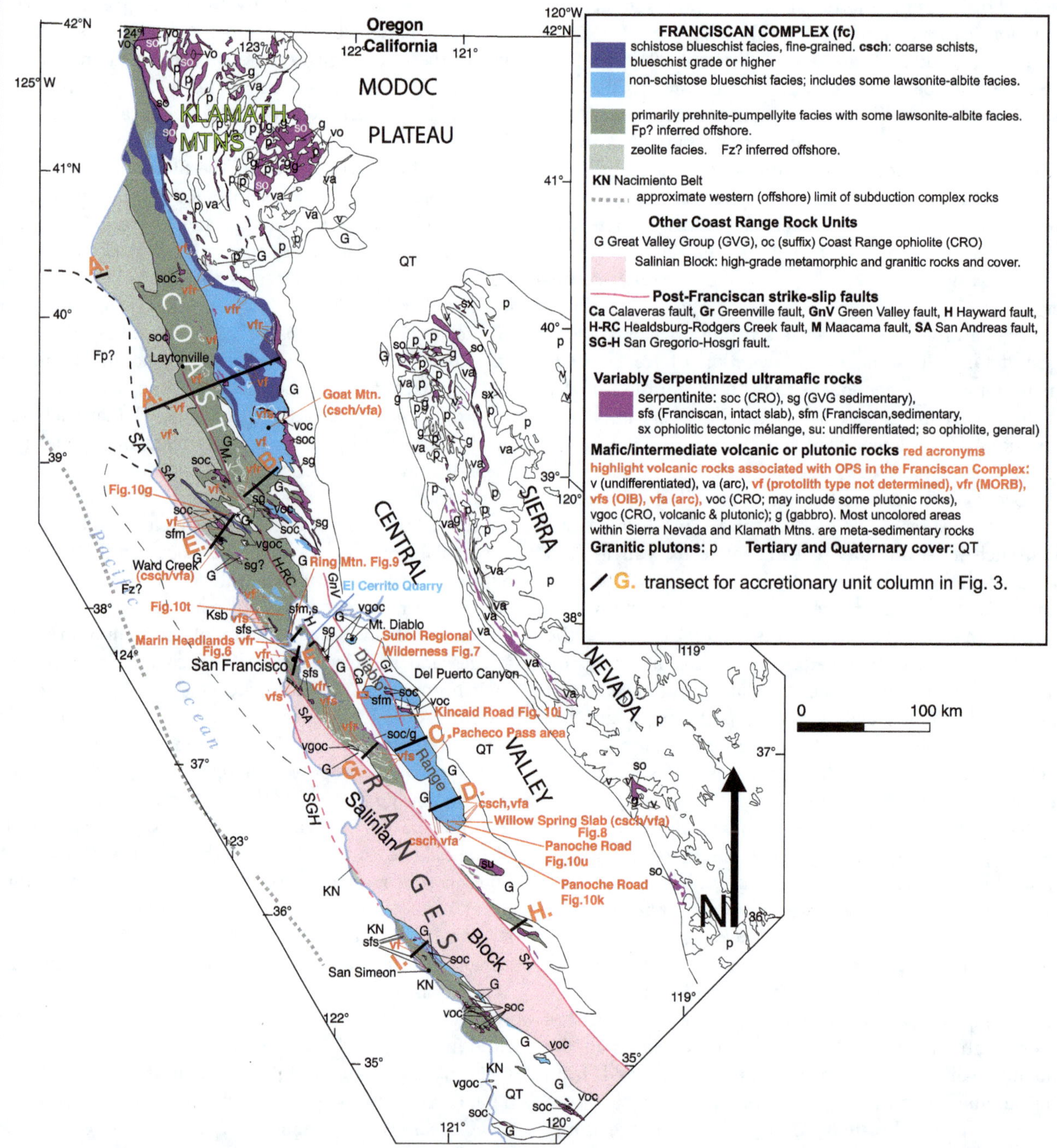

Fig. 2 Map showing general geology of north-central California and special emphasis on the California Coast Ranges. Oceanic rocks are emphasized on this map, and oceanic volcanic rocks associated with OPS are highlighted by red labels in the Franciscan subduction complex of the California Coast Ranges. In some cases the actual size of the volcanic rock exposure is exaggerated because most of the actual exposures consist of imbricated basalt, chert, and siliciclastic rocks. The largest exposures of mafic-intermediate oceanic volcanic rocks and overlying pelagic sediments are associated with accreted island arc terranes (labeled "va") or large ophiolitic sheets (some examples are labeled "voc"). Revised from Wakabayashi (2016b)

CRO and GVG were part of the upper plate of a convergent plate margin throughout their history (e.g., Hamilton 1969, Dickinson 1970; Ernst 1970; Bailey et al. 1970).

Part of the CRO, however, consists of mafic volcanic rocks, successively overlain by chert and then the siliclastic rocks of the GVG (Hopson et al. 1981; 2008). Locally, the upper part of the CRO may be imbricated along with the basal part of the GVG so that there are repeated slices of basalt-chert-siliclastic rocks (Wakabayashi 2016b) that resemble the types of field relationships seen in typical OPS in subduction complexes. There are widespread imbricate and block-in-matrix relationships within the Tehama-Colusa mélange whose affinity as CRO or Franciscan (or neither) is unclear (Hopson and Pessagno 2005; Shervais et al. 2011). In addition, large (up to hundredths of meters) blocks of basalt, chert, and basalt with overlying chert are present in olistostrome units of the basal GVG (Phipps 1984; 1992; Wakabayashi 2016a). These units can be distinguished from Franciscan rocks on the basis of the lack of burial metamorphism (excluding some blocks in mélanges/olistostromes) and their regional field relationships.

General spatial-temporal relationships of Franciscan Complex OPS

The Franciscan Complex accreted during a period of >150 Ma of continuous east-dipping subduction in coastal California from ca.165 to 12 Ma, with most accretion between ca. 120 and 32 Ma (e.g., McLaughlin et al. 1982; Dumitru et al. 2010; 2015; Wakabayashi 2015). Episodic accretion transferred material (OPS) from the downgoing plate to the subduction complex to form a structural stack of accretionary slices/units with different formational and accretion ages (Fig. 3); Dumitru et al. (2016); Chapman et al. (2016). These sheetlike accretionary units, bounded above and below by narrow (<100 m thick) non-accretionary paleomegathrust horizons, have low-angle regional dips, but dip steeply in many areas as a result of later folding (Wakabayashi 2015; 2016a) (Fig. 3). The accretionary units extend up to tens of kilometers along strike and attain structural thicknesses of up to ~3 km. In this paper, "accretionary units" and OPS units will be considered synonymous. The age of subduction-accretion of various accretionary units has been determined by ages of metamorphism where metamorphism was sufficient to generate datable metamorphic minerals and by the depositional age of clastic sedimentary rocks for rocks too low grade to yield metamorphic ages (e.g., Wakabayashi 2015) (Fig. 3).

About a fourth of the exposed accreted rocks underwent high-pressure/low-temperature (HP-LT) lawsonite-albite, blueschist or higher grade metamorphism; nearly all of the remainders are prehnite-pumpellyite or zeolite grade (Blake et al. 1984, 1988; Ernst 1993; Terabayashi and Maruyama 1998; Ernst and McLaughlin 2012) (Fig. 2). A small fraction (<<1%) of the Franciscan rocks are coarse-grained blueschist, amphibolite, garnet-amphibolite, and eclogite. Most of these rocks crop out as blocks-in-mélange termed "high-grade blocks" (Coleman and Lanphere 1971).These blocks are higher in metamorphic grade than surrounding matrix, whereas rare intact fault-bounded sheets of these rocks are found ("high-grade coherent sheets" of Wakabayashi et al. 2010; "csch/vfa" on Fig. 2).

Clastic sedimentary rocks make up most of the collective stack accreted OPS units, and the proportion of intact igneous and pelagic varies between various accreted units as well as in collective accreted stacks in different parts of the Franciscan (Wakabayashi 2015) (Figs. 3 and 4). This variability will be described further in "Examples of the variety of OPS in the franciscan complex" on specific types of OPS assemblages. The clastic sedimentary rocks of the subduction complex comprise turbidites and mélanges with non-native (exotic) blocks, such as those of higher metamorphic grade than the matrix (Wakabayashi 2015). Some have interpreted these types of mélanges having resulted from sedimentary (olistostromal) mixing of exotic blocks into matrix (e.g., Aalto 1981; Cowan 1978; Macpherson et al. 1990; Platt 2015; Wakabayashi 2015) whereas others have proposed that the introduction of blocks into matrix resulted from tectonic strain as tectonic mélanges (e.g., Cloos 1984; Cloos and Shreve 1988; Gerya et al. 2002; Hsü 1968).

The origin of mélanges with exotic blocks is important interpreting their relationship to accreted OPS units and the process of accretion and exhumation, so I will list the main arguments in favor of sedimentary origins of these bodies presented in Wakabayashi (2015), as well as presenting new evidence for the sedimentary origin of mélanges with exotic blocks in "Olistostromes as part of the clastic component of OPS." The sedimentary origin of Franciscan mélanges with exotic blocks has been proposed based on the basis of multiple criteria, including (1) sedimentary structures or features in the least deformed parts of the mélange from outcrop to submillimeter scale, (2) interbedding and interfingering of mélanges with ordinary bedded turbidites, and (3) blocks that originated at shallow levels of the upper plate. New data will be presented in support of points 1 and 2 in "Olistostromes as part of the clastic component of OPS."

In addition to the evidence for sedimentary origin of mélanges with exotic blocks, geochronologic evidence indicates episodic accretion rather than continual mixing in a progressively widening subduction channel spanning the full thickness of the subduction complex, as envisaged in some numerical models (e.g., Gerya et al. 2002). Collectively, the evidence for origin of blocks with exotic blocks, episodic accretion within a subduction complex, and narrow (<100 m thick) zones between accreted units suggests that the accommodation of subduction megathrust slip is more localized than the prediction of numerical models (Wakabayashi 2015; Wakabayashi and Rowe 2015). The analysis of the internal deformation of OPS gives additional insight into subduction slip

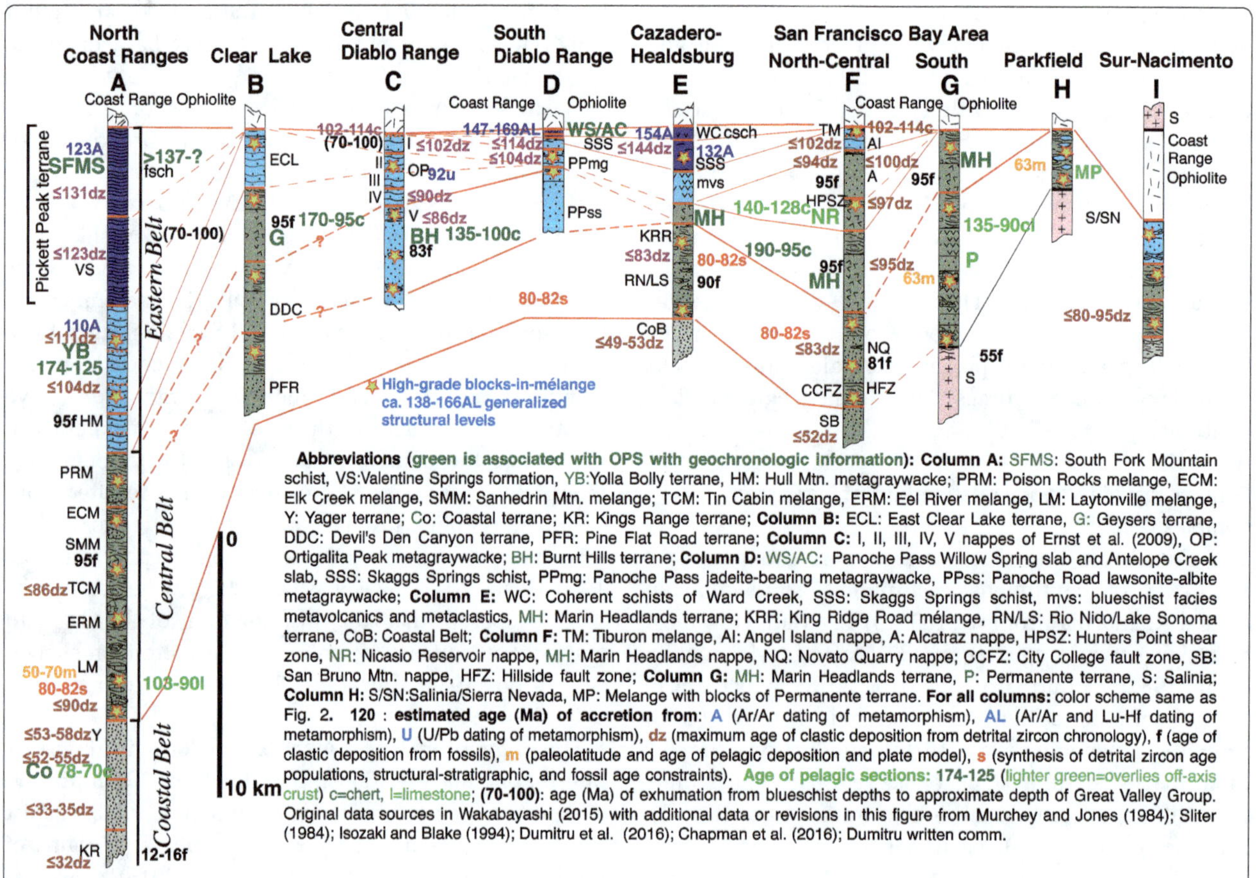

Fig. 3 Schematic diagram showing tectonic stacks of accretionary units in different parts of the Franciscan Complex. This shows the variability of the Franciscan accreted stack at different locations keyed to transects shown on Fig. 2. Units accreted at different times are separated by non-accretionary megathrust contacts shown in *bold red* (solid or dashed). Internal imbrication of each accreted unit is not depicted but the total structural thickness is estimated. The age of the pelagic sections of oceanic sections associated with accreted units is given in green but this information is not available for all units. The age of the oldest pelagic rocks in each unit approximates the age of the igneous oceanic crust. Accretion ages are also given based on different types of data, color coded with acronyms, updated from Wakabayashi (2015)

localization; new observations will be presented in "Imbrication and deformation of OPS and distinguishing 25 tectonic from sedimentary melanges."

The interpreted origin of the mélanges with exotic blocks as olistostromes suggests that they should be considered part of the trench-fill (clastic) component of larger scale OPS packages, whereas tectonic mélanges lack exotic blocks and formed by progressive deformation of OPS (Fig. 4) (Wakabayashi 2015). The proportion of mélange within the clastic units of the Franciscan varies from accreted unit to unit and spatially within each accreted unit.

Trace element and radiogenic isotopic data suggests that the high-grade coherent rocks and blocks-in-mélange have nascent island-arc protoliths (Wakabayashi et al. 2010), whereas lower grade coherent sheets of Franciscan metavolcanic rocks display ocean island basalt (OIB), or mid-ocean ridge (MORB), geochemical affinity (Shervais 1990; Ghatak et al. 2012). Metaigneous blocks-in-mélange, excluding the high-grade blocks, include those of OIB, MORB, and arc affinity (Macpherson et al. 1990).

Whereas the ages of clastic rocks in each accreted unit young structurally downward, the age of the oceanic crust subducted (and the age of this crust at time of subduction) shows a more complex relationship (Fig. 5a) (Wakabayashi 2015). Some of this complex age pattern resulted from periodic subduction of off-axis-generated oceanic crust, whereas the early (in subduction history) increase in the age of subducted oceanic crust has been proposed to have resulted from subduction of a dormant spreading ridge (Wakabayashi 2015). The subducted (dormant) ridge proposal conflicts with plate reconstructions that have not modeled a ridge crest near the trench at this time (Seton et al. 2012). The (dormant) ridge subduction proposal was based on geochronologic data as well as the temporal variation in peak metamorphism in the Franciscan (Wakabayashi 2015). Some of the data used to propose the (dormant) ridge subduction model was not available at the time Seton et al. (2012)) performed their analysis of paleoplate geometry. Data supporting the early increase in age of subducted oceanic

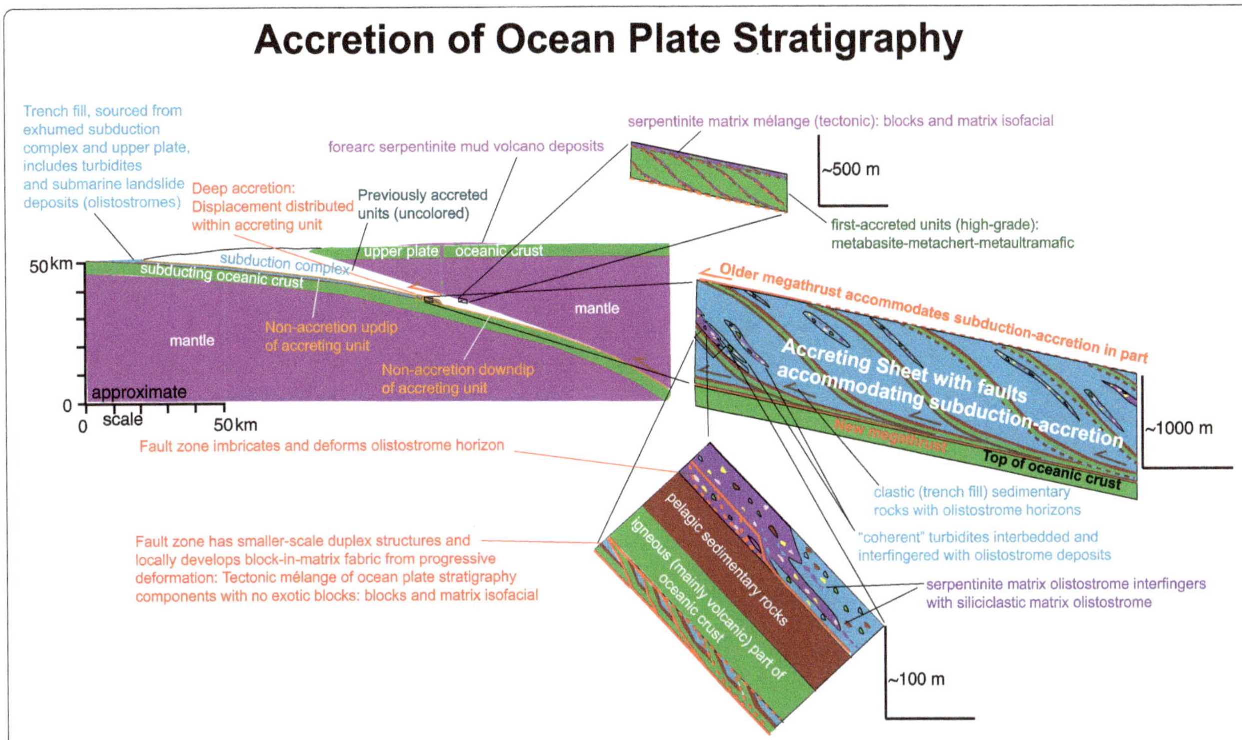

Fig. 4 Cross-sectional cartoon showing accretion of OPS and the connection between this process and the rock assemblages found in subduction complexes, including different types of mélanges. This schematic also illustrates that olistostromes of serpentinite and/or siliciclastic matrix comprise part of the clastic part of OPS

crust during Franciscan history was first presented Murchey and Blake (1993), who proposed that this age pattern resulted from subduction of an active spreading ridge at ca. 170, followed by conversion to a transform margin until initiation of new subduction at ca. 150.

Franciscan OPS sections have an apparent age gap of >10 Ma between the youngest pelagic rocks and the clastic cover, except for the OPS section of the Marin Headlands (Sliter and McGann 1992; Sliter 1984; Murchey and Jones 1984) (Fig. 5b). The age gap reflects the absence of the youngest part of the pelagic record rather than the oldest part of the clastic record because in several cases, the ages for the clastic section are maximum depositional ages. It is not clear whether the apparent age gaps are a product of the lack of original pelagic deposition during those times, excision of a deposited record by faulting and/or erosion, or incomplete biostratigraphic sampling.

Age-scaled depictions of OPS columns do not represent the relative proportions of the different OPS components in a subduction complex because the different components formed at different rates. For this reason, it is useful to consider the thickness-scaled columns of Fig. 5c for a graphical representation that better reflects the relative proportion of the various OPS components in the subduction complex.

The age relationships on Fig. 5 are a composite of the entire Franciscan Complex, and they do not necessarily reflect the subduction-accretion history at any specific reach of the trench. The accretionary unit stacks in Fig. 3 are arranged in approximate along-strike relative position after restoration of post-subduction dextral faulting. This figure illustrates the considerable along-strike variation in the Franciscan Complex, including along-strike variation in the age of oceanic crust subducted at approximately the same time. Age data is too sparse, however, to enable a more detailed analysis of the subducted plate history based on significant along-strike age variation as done by Osozawa (1994) for Japan.

For a number of the accretionary slices of the Franciscan shown in Fig. 3, ages of their assumed oceanic crustal base have not been determined because of the lack of identified pelagic-igneous imbricates, whereas the age of incorporation has been estimated from depositional ages of clastic sedimentary rocks. The schematic OPS accretionary stacks in Fig. 5b, c includes multiple accretionary units of Fig. 3 that lack pelagic-igneous imbricates and associated ocean crust age information.

In "Examples of the variety of **OPS** in the Franciscan Complex" that follows, Franciscan OPS assemblages will be categorized according to their lithologic assemblages. Following these descriptions, "Olistostromes as part of

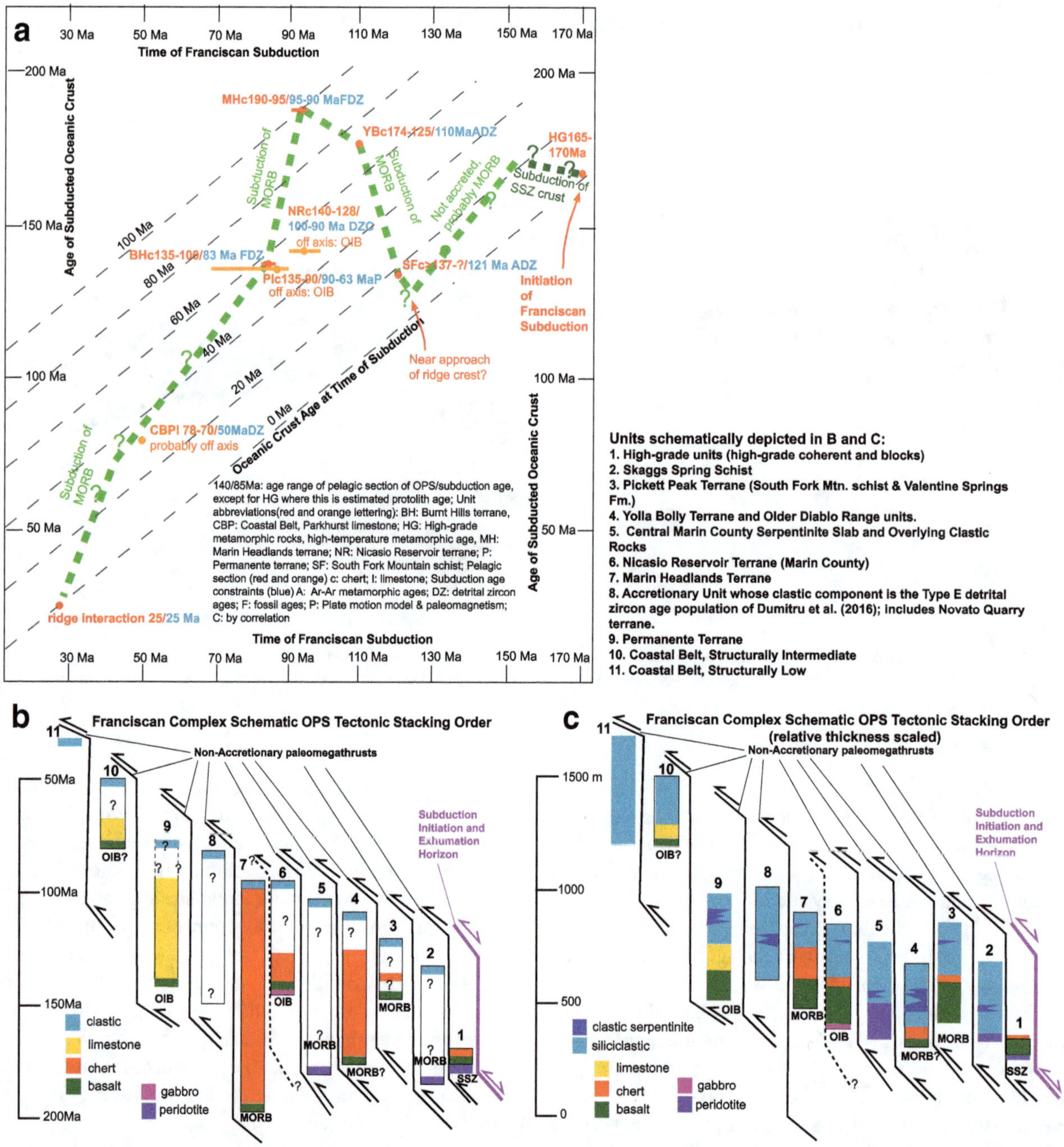

Fig. 5 **a** Subduction age versus ocean crust age for the Franciscan Complex. The *green dashed line* depicts the estimated age of subducted oceanic crust generated at an oceanic spreading center. For this reason, off-axis-generated oceanic crust plots younger than the *green dashed line* for any given age of subduction. Note that this diagram is a composite for the entire Franciscan Complex and does not address the potential for along-strike variation in the age of oceanic crust subducted. Existing age information (shown in Fig. 3) is insufficient at present to evaluate along-strike variation in age of subducted oceanic crust. Part a revised from Wakabayashi (2015). **b** Diagram showing the composite tectonic stacking of OPS that resulted from sporadic Franciscan subduction; the age ranges are from "**a**" and Fig. 3. As above, this is a composite of the Franciscan Complex and this does not represent a single reach along the subduction zone, and thus, it does not include all of the different accreted units. In addition, there are multiple accreted packages shown for which intact basalt and overlying pelagic sedimentary rocks have not been identified or dated. In this diagram, each accreted OPS package is separated by a non-accretionary paleomegathrust and the column height and vertical position is related to age. Accordingly, each column corresponds to an accreted unit between non-accretion megathrusts *within* a given accreted column stack on Fig. 3 rather than corresponding to an entire column on Fig. 3. The accreted OPS is imbricated by various faults that accommodated accretionary paleomegathrust zones (Fig. 4), but these are not shown on this figure. **c** Same OPS units as "**b**" but scaled to thickness within major imbricates rather than to age. This is not the total thickness of each accreted unit because each accreted unit is thickened as a result of imbricate fault repetition

the clastic component of OPS" will review published work and present new data suggesting that olistostromes make up a significant component of the clastic part of OPS packages. "Imbrication and deformation of OPS and distinguishing tectonic from sedimentary melanges" will summarize the imbrication and deformation of Franciscan OPS, and "Franciscan OPS and tectonic history" will summarize the history of OPS accretion in the context of Franciscan tectonic history.

Examples of the variety of OPS in the Franciscan Complex

Basalt-chert-clastic

The most common type of OPS in the Franciscan Complex consists of basalt overlain by chert and then clastic sedimentary rocks. The basalt of most such packages exhibits mid-ocean ridge (MORB) geochemical affinity, based on comparatively sparse sampling (Shervais 1990; Ghatak et al. 2012). An exception is the Nicasio Reservoir terrane (Murchey and Jones 1984; Blake et al. 1984) that has mafic rocks of OIB affinity (Ghatak et al. 2012).

Basalt-chert-clastic OPS packages range from prehnite-pumpellyite facies for units such as the Marin Headlands and Nicasio Reservoir terranes (Wahrhaftig 1984; Blake et al. 1984) to epidote blueschist facies for the highest grade parts of the South Fork Mountain schist (Worrall 1981). The proportion of basalt and chert to clastic sedimentary rocks varies from basalt and chert-rich packages of the Marin Headlands terrane (Fig. 6) to the more common types that are dominated by clastic sedimentary rocks, such as the Sunol Regional Wilderness (Fig. 7), or the Pacheco Pass areas of the northern Diablo Range (e.g., Ernst et al. 2009). The structurally higher parts of Tiburon Peninsula of the San Francisco Bay region include a blueschist facies OPS unit of the basalt-chert-clastic type with a proportion of clastic rocks greater than Marin Headlands but less than the Diablo Range examples noted above (Bero 2014). Part of this unit (primarily the clastic component) is shown in the geologic map on Fig. 8, and the significance of these exposures will be discussed further in "Olistostromes as part of the clastic component of OPS."

The thickest stratigraphic section of chert, about 80 m, is associated with the Marin Headlands terrane, but chert horizons are thickened by thrust faults, so that structural thicknesses locally reach 300 m (Murchey 1984; Wahrhaftig 1984) (Fig. 6). The thickest chert section also records the longest depositional history of any pelagic section in the OPS of the Franciscan Complex, from Pliensbachian to Cenomanian (Murchey 1984) (Fig. 5). This is an age range of ca. 191–94 Ma, based on the 2016–12 International Commission on Stratigraphy (ICS) Time Scale (Cohen et al. 2013, updated). In addition, the Marin Headlands terrane was one of the first global localities where the fault repetition of the pelagic biostratigraphy was recognized, and the basalt-pelagic-clastic succession and its

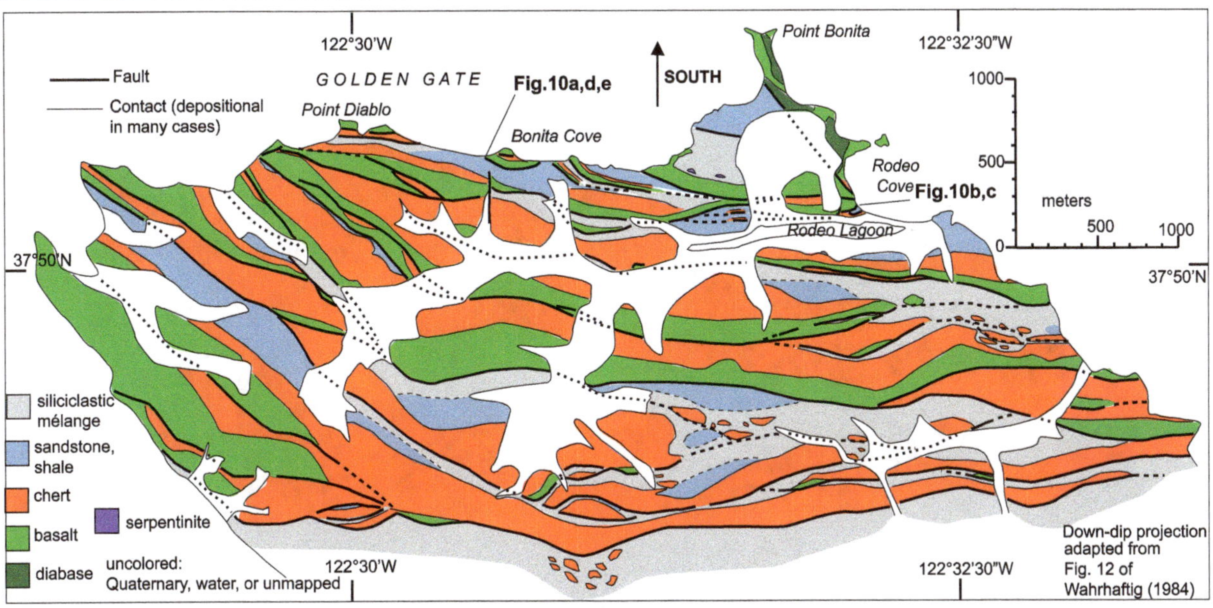

Fig. 6 Down-dip projection of the geology of the Marin Headlands region of the Franciscan Complex adapted from Wahrhafig (1984). Location on Fig. 2. This view illustrates the imbricate fault repetitions of prehnite-pumpellyite facies OPS and the true structural-stratigraphic thicknesses (from the down-dip projection) of the imbricates. This is the best example of OPS in the Franciscan Complex, but it has a higher proportion of volcanic rocks (basalt) and pelagic rocks (chert), than other OPS packages in the Franciscan

imbrication was linked to oceanic plate movement and subduction-accretion (Wahrhaftig 1984).

Chert horizons of tens of meters of structural thickness and less are typical. Volcanic sections attain structural thicknesses of up to about 500 m (Worrall 1981; Wahrhaftig 1984). Clastic sections reach thicknesses of up to about 1 km or more (e.g., Worrall 1981; Ernst et al. 2009; Prohoroff et al. 2012).

Basalt-limestone-clastic

A much less common type of OPS in the Franciscan Complex consists of basalt overlain by limestone overlain by clastic sedimentary rocks. The basalt of such packages exhibit ocean island basalt (OIB) geochemical affinity (Ghatak et al. 2012). Limestone is rare in the Franciscan Complex and much less abundant than chert (Sliter 1984). Limestone horizons are commonly tens of meters in thickness, and the thickest limestone horizon (130 m) is associated with the Permanente terrane (Sliter 1984). Owing to the faulting and deformation of the rocks, it is not clear whether this 130 m thickness was thickened relative to its original stratigraphic thickness by internal faults, but it probably was, by analogy to well-constrained chert exposures of the Marin Headlands as well as ubiquitous occurrence of imbricate faults in all OPS lithologies as will be reviewed in "Imbrication and deformation of OPS and distinguishing tectonic from sedimentary melanges." Much of the limestone section of the Permanente terrane is interbedded with chert (Sliter 1984; Murchey and Jones 1984). The longest duration of limestone deposition is recorded in the Permanente terrane ranges from early Albian to Turonian (Sliter 1984) (Fig. 5)

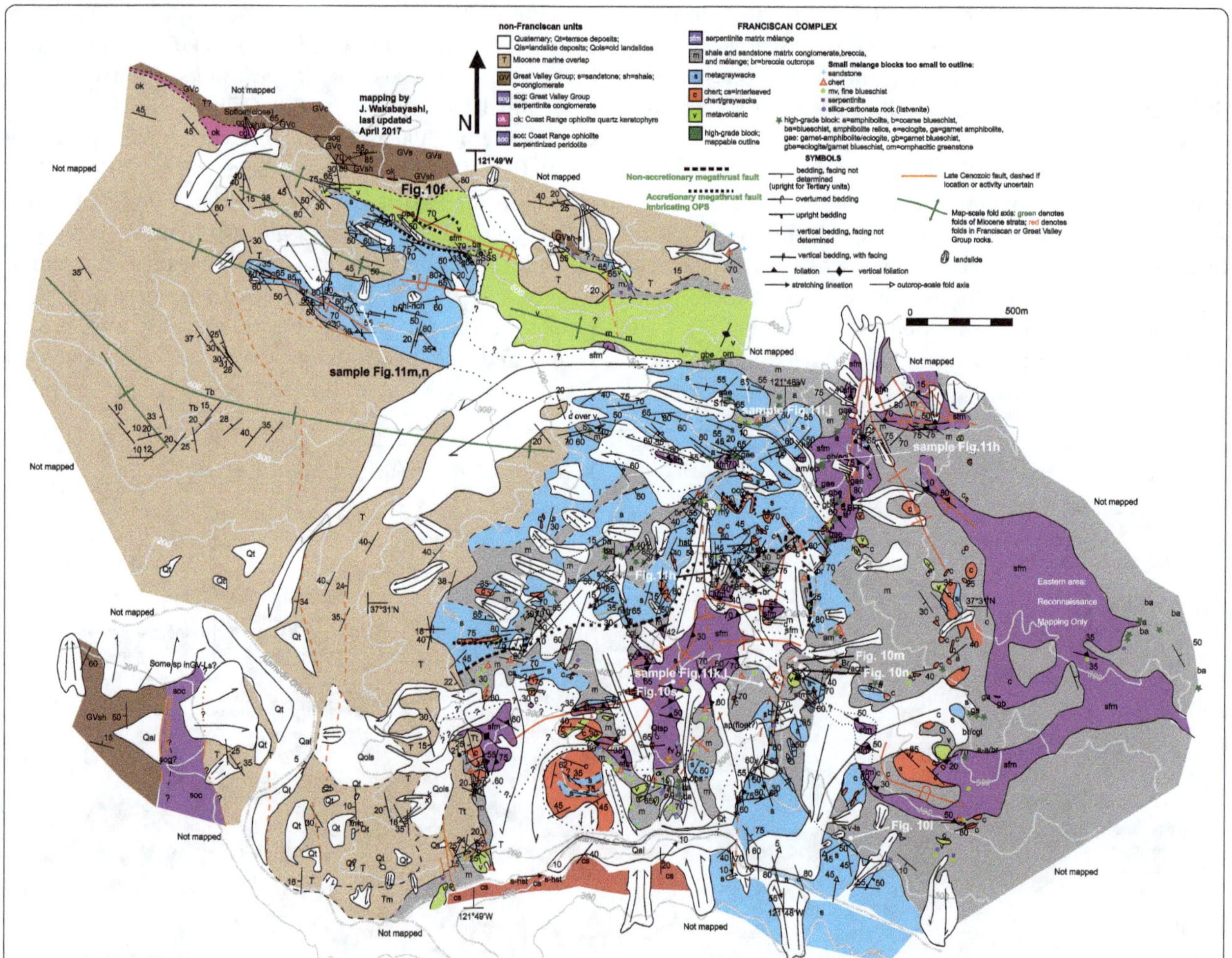

Fig. 7 Geologic map of part of Sunol Regional Wilderness in the northern Diablo Range. Location on Fig. 2. This is an example of OPS imbricates that are much richer in clastic rocks. Here, the OPS imbricates are metamorphosed to blueschist facies. The clastic rocks have a significant olistostromal component and include serpentinite matrix as well as siliciclastic matrix olistostromes, and these olistostromal units apparently interfinger with ordinary turbidite units. Repetitions of OPS in the most clastic rich parts of this region (southern part of map) are difficult to identify owing to the difficulty in distinguishing intact slices of basalt and chert from olistostrome blocks. In the more volcanic-rich area to the north, two of the imbricates are difficult to locate owing to the small amount of clastic and/or pelagic rocks associated with the fault repetition. Updated from Wakabayashi (2016b)

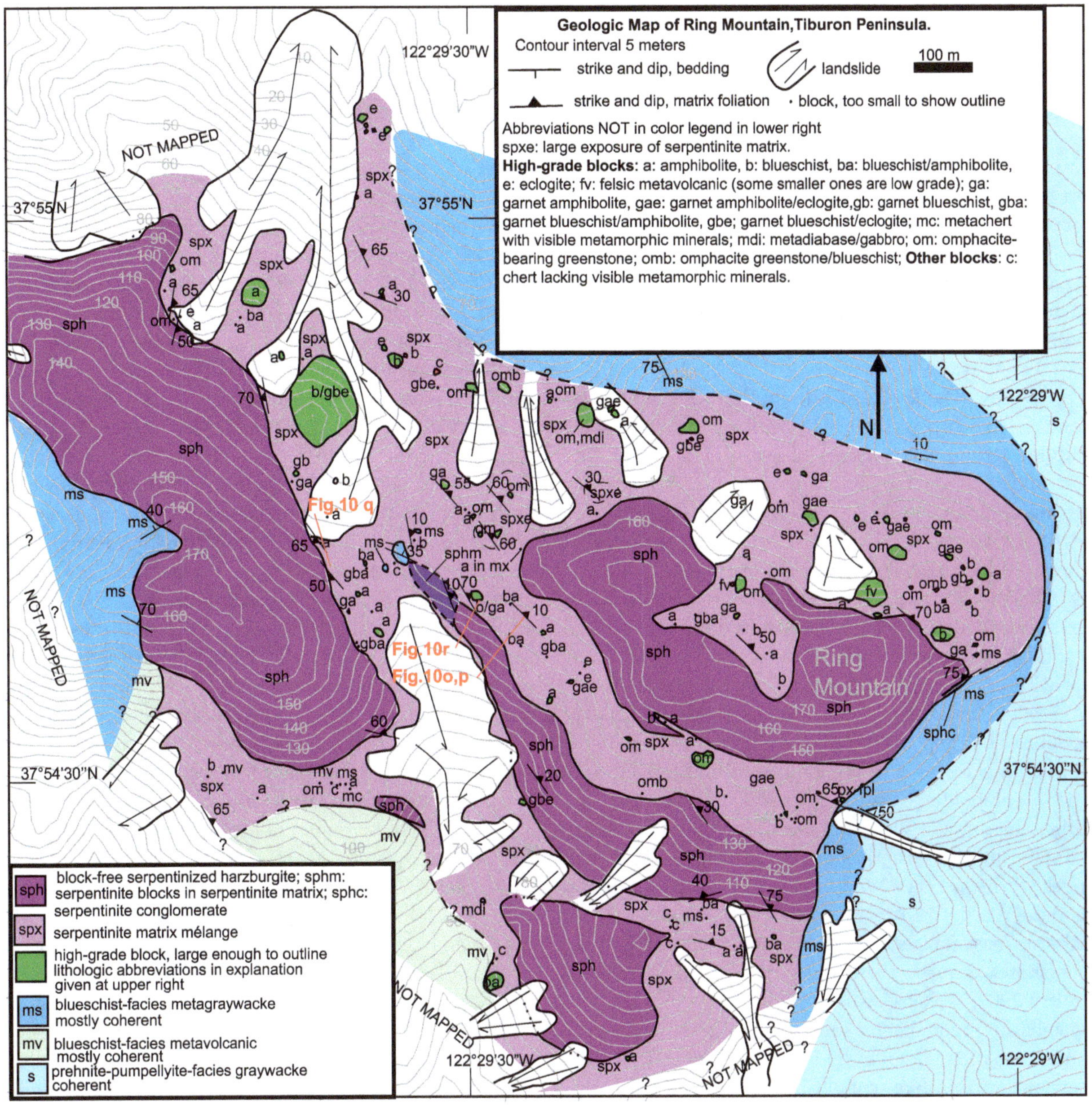

Fig. 8 Geologic map of the Ring Mountain area, San Francisco Bay area. Location on Fig. 2. This locality includes an example of sedimentary serpentinite mélange that is part of a clastic component of an accreted OPS unit, most likely of the basalt-chert-clastic type (Bero 2014). Some of the high-grade metamorphic blocks within the mélange may preserve metabasalt-metachert-metaultramafic OPS imbricates internally, preserving at a smaller scale the relationships seen in map view in the Willow Spring slab locality of Fig. 9 (see Fig. 10s). Updated from Wakabayashi (2015)

or ca. 110–90 Ma (Cohen et al. 2013, updated). The Permanente terrane has a significant clastic component, but it is less than most Franciscan units excluding the Marin Headlands. This clastic component is primarily siliciclastic, including siliciclastic-matrix mélange, but it may include sedimentary serpentinite mélange as well (Wakabayashi 2012).

Other limestone-bearing OPS units of the Franciscan include the well-known Laytonville limestone (Alvarez et al. 1980; Tarduno et al. 1990) that has a pelagic age range that overlaps with the Permanente terrane, as well as younger limestone units, such as the Parkhurst limestone (Sliter 1984). The far-traveled nature of limestone-bearing OPS has been shown by paleomagnetic studies, and the Laytonville limestone remains an enigma owing to the anomalously rapid plate motions indicated by the combination of paleomagnetic and biostratigraphic data (Alvarez et al. 1980; Tarduno et al. 1990). A new estimate for an older accretion time of the limestone based on new data from associated clastic rocks would require even higher apparent plate motions (Wakabayashi 2015).

Serpentinite-basalt-chert (with no siliciclastic rocks)

The structurally highest horizons of the Franciscan Complex include rare sheets of coarse-grained metamorphic rocks ranging up to several kilometers in long dimension and several hundred meters in thickness. These slabs have been called "high-grade coherent sheets" to distinguish them from high-grade blocks-in-mélange that have the same metamorphic mineral assemblages, lithologies, geochenmical affinity, and metamorphic ages (Wakabayashi et al. 2010).

The Willow Spring slab of the southern Diablo Range is one of the largest such slabs and serves as a good example of the characteristics of such units. It consists of an imbricate stack of primarily mafic schists ranging from medium to coarse-grained lawsonite blueschist at structurally lowest level to eclogite at the structurally highest level (Wakabayashi and Dumitru 2007) (Fig. 9). This stack of dominantly metabasite sheets structurally overlies a structural stack of metasandstones. The Willow Spring slab covers a map area of about 2 km long by 1 km wide with a structural thickness of several hundred meters. Several similar, but smaller, intact imbricated metamorphic slabs crop out in the southern Diablo Ranges, and these include slabs of predominantly amphibolite and garnet-amphibolite (Wakabayashi and Dumitru 2007) and another high-grade slab has been identified at Goat Mountain in the eastern part of the northern Coast Ranges (Ernst et al. 1970). In addition, part of the Cazadero-Ward Creek area in the western part of the northern Coast Ranges, several kilometers in long dimension, shows similar characteristics (Wakabayashi and Dumitru 2007; Wakabayashi et al. 2010).

Metabasite predominates in these high-grade rocks and has nascent island-arc affinity (Wakabayashi et al. 2010). Metachert horizons are spatially associated with antigorite schist bodies so that the inferred ocean floor stratigraphy was one where serpentinized mantle rocks were overlain by basalt and chert successively. Metaclastic rocks are absent in this package.

High-grade blocks in mélange have similar lithologic associations on a smaller (≤200 m) scale. They are composed predominantly of metabasite with small metachert layers and lack metaclastic components. Whereas they are commonly at least partly encased by actinolite-phengite-chlorite "rinds" derived from metasomatized ultramafic rock (e.g., Coleman and Lanphere 1971; Ghatak et al. 2012; Sorensen et al. 1997), some also include metaultramafic layers within them as described in "Imbrication and deformation of **OPS** and distinguishing tectonic from sedimentary melanges."

Serpentinite-clastic

Serpentinized peridotite lacking block-in-matrix character ("coherent") is also found within the Franciscan Complex (Wakabayashi 2004; Prohoroff et al. 2012; Wakabayashi 2015; Wakabayashi, in press). These serpentinite bodies range in structural thickness up to 1.5 km and extend for up to 30 km along strike, preserve primary peridotite textures in large domains, and lack non-serpentinite blocks other than rare gabbro (Wakabayashi 2004; 2012; 2015; in press). The serpentinized ultramafic rocks vary in texture from massive to areas with massive blocks surrounded by zones with brittle foliation, to pervasively foliated rocks. The dominant serpentine mineral in these rocks is lizardite (Wakabayashi 2004; in press). Prohoroff et al. (2012) proposed that clastic sedimentary rocks depositionally overlie the larger bodies in Marin County ("sfs" label directly above "Marin Headlands vfr" label on Fig. 2), but the contact relationships are disputed (Raymond and Bero 2015). Basalt and/or pelagic sedimentary rocks do not appear to depositionally overlie these serpentinite bodies. The parent peridotite has abyssal affinities, based on whole rock and mineral geochemistry (Barnes et al. 2013), which distinguishes it from the upper plate (Coast Range ophiolite) serpentinized peridotite, which has both arc and abyssal affinity (e.g., Choi et al. 2008; Jean et al. 2010).

Metaclastic rocks are imbricated with and may depositionally overlie serpentinized peridotite, exposed in the northwestern Diablo Range along Kincaid Road (Fig. 10i, j). The serpentinite makes up an exposure about 5.5 km long by up to 800 m wide (Wentworth et al. 1999). Most of this body is massive serpentinized harzburgite with local foliated zones. A schistose metasandstone that has been correlated to the Skaggs Springs schist of Sonoma County (Wakabayashi 1999) is locally imbricated with the serpentinite (Fig. 10k). At least one of the contacts of metasandstone on serpentinite may be depositional based on the presence of small blocks of silica-carbonate rock (Fig. 10j), a hydrothermal replacement of serpentinite also referred to as listvenite (e.g., Halls and Zhao 1995; Smith et al. 2008).

Serpentinized peridotite capping Tiburon Peninsula, including the Ring Mountain locality (Fig. 8), has been interpreted as being abyssal in origin and thus of intra-Franciscan affinity (Barnes et al. 2013), but clastic rocks have not been found in depositional contact atop this serpentinite unit (Bero 2014). This serpentininized harzburgite sheet tectonically overlies serpentinite matrix mélange interpreted to be sedimentary origin by Wakabayashi (2012; 2015; additional evidence for sedimentary origin presented in "**Sedimentary origins of mélanges with exotic blocks: interbedding with siliciclastic turbidites and internal sedimentary features**").

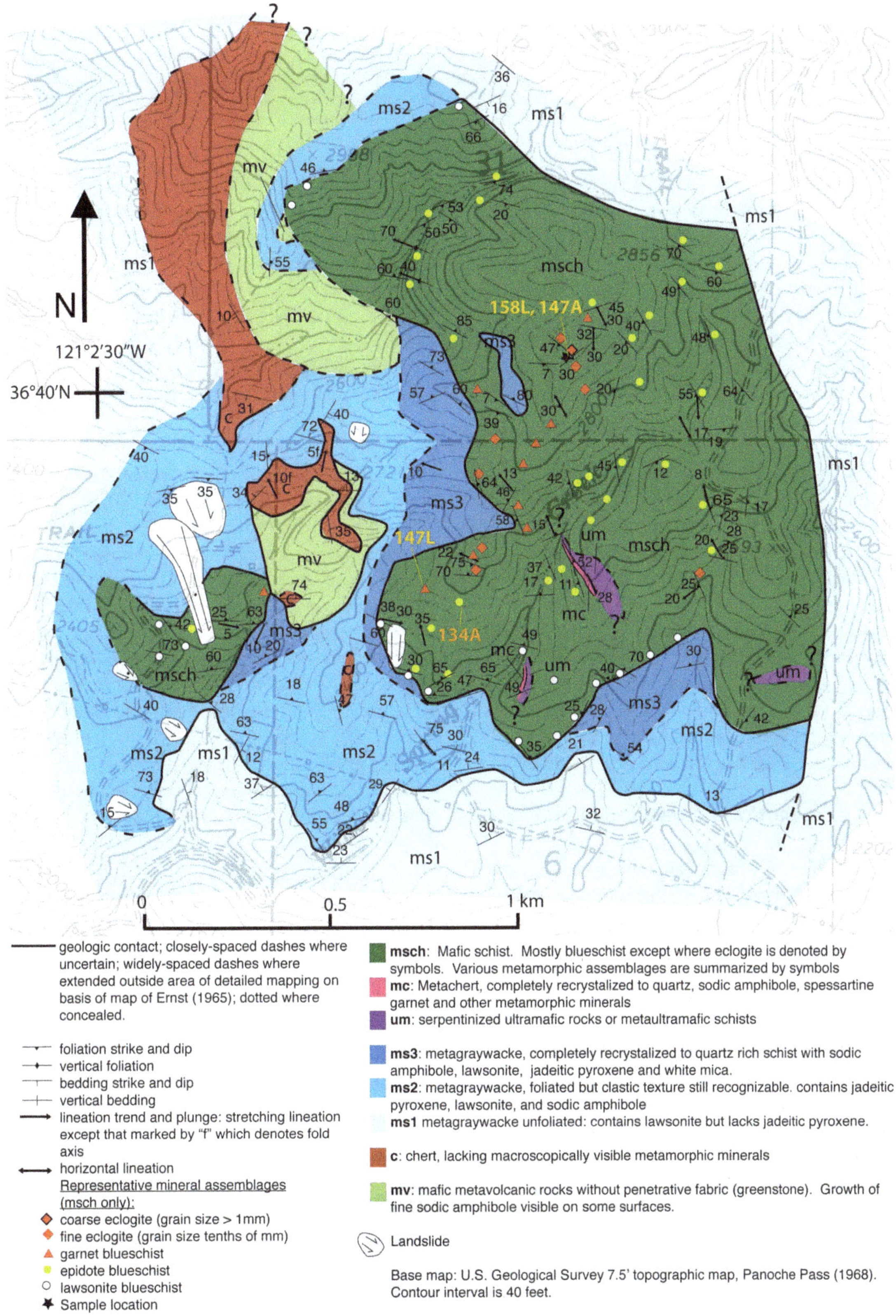

Fig. 9 (See legend on next page.)

(See figure on previous page.)
Fig. 9 Geologic map of the Willow Spring slab, southern Diablo Range. Location on Fig. 2. These (*dark green, purple,* and *magenta colors*) are among the highest grade coherent (non-mélange) rocks in the Franciscan Complex and are completely recrystallized and schistose. The Willow Springs slab is an example of OPS that is predominantly metabasite, with minor ultramafic schist and metachert, but no metaclastic rocks. These rocks tectonically overlie intact lower grade units including a *blueschist facies unit* with abundant metavolcanics (mov) and metachert (s) in addition to siliciclastic (mostly sandstone, ms2). The reconnaissance mapping done here did not completely define the extent of metachert and metaultramafic horizons in the high-grade rocks, but they certainly extend further along strike than shown. Adapted from Wakabayashi (2015)

Basalt-limestone-chert-clastic or basalt-chert-limestone-clastic

As noted earlier, limestone is rare in the Franciscan Complex and most of it is part of basalt-limestone-clastic sequences. A very small amount of limestone is present in sequences where basalt is successively overlain by chert and limestone. An example from an apparent coherent imbricated assemblage of blueschist facies rocks with this assemblage crops out along the Skaggs Road in Sonoma County (Fig. 10g). This unit may extend over a kilometer along strike based on my field reconnaissance; published data on this unit is otherwise lacking. The limestone appears limited to a thickness of 20 cm or less, whereas metchert, metabasalt, and metaclastic rocks may reach thicknesses of 10 m, several hundred meters, and about 1 km, respectively. Even rarer in the Franciscan is OPS composed of basalt overlain successively by limestone, chert, and clastic rocks. To my knowledge, these assemblages have not been previously reported in the literature on the Franciscan, but minor exposures are found as small (<20 m in size) blocks-in-mélange at Sunol Regional Wilderness (Fig. 10h).

Olistostromes as part of the clastic component of OPS

Clastic components of Franciscan OPS units are composed of variable proportions of turbidite and mélange with non-native (exotic) blocks, including those of higher metamorphic grade than the matrix. This type of mélange has been suggested to have formed by submarine sliding (as an olistostrome) prior to subsequent deformation as summarized previously (Macpherson et al. 1990; Wakabayashi 2015). The olistostromal horizons have siliciclastic or serpentinite matrix, and these matrix types may grade into and interfinger with each other as proposed by Wakabayashi (2015) and schematically shown in Fig. 4. New evidence for gradation between siliciclastic and clastic serpentinite rocks will be presented in "Interfingering, interbedding, and gradation between siliciclastic rocks and clastic serpentinite." Interbedding of siliciclastic olistostromal horizons and turbidites was proposed by Wakabayashi (2015), and additional information will be presented in "Sedimentary origins of mélanges with exotic blocks: interbedding with siliciclastic turbidites and internal sedimentary features," along with photographs showing sedimentary textures in siliciclastic and serpentinite matrix.

Sedimentary origins of mélanges with exotic blocks: interbedding with siliciclastic turbidites and internal sedimentary features

The interbedding and interfingering of sedimentary mélange and turbidites (bedded sandstone and shale) is suggested by 100 m to kilometer scale field relationships at Sunol Regional Wilderness, where the sedimentary mélange horizons include those of siliciclastic and serpentinite matrix (Fig. 7). Outcrop scale relationships from meter to 100 m scale show interbedding of siliciclastic turbidites and olistostromal horizons at many Franciscan localities (Wakabayashi 2015).

The clastic part of the Marin Headlands unit contains mélange horizons and beachcliff exposures locally show the matrix of these mélanges to be mud-rich conglomerate, pebbly sandstone, and clean sandstone (Fig. 10a, d, and e). These block-in-matrix horizons grade into or are interbedded with block-free sandstone, shale, and conglomerate (Fig. 10a). The blocks in the mélange are predominantly sandstone, chert, and basalt of prehnite-pumpellyite grade, but detailed studies have not been done to determine whether these rocks are the same as the intact Marin Headlands imbricates that have similar general character. Serpentinite blocks to tens of meters in size are present in some of the mélange zones of the Marin Headlands are (Wahrhaftig 1984). Some serpentinite blocks are found in the colluvial and landslide material draping part the outcrop shown in Fig. 10a, and the position of these blocks suggests derivation from the exposed block-in-matrix horizons. Serpentinite clasts are found in the pebbly sandstone matrix of Fig. 10d, and the tremolite-bearing metamorphic assemblage of some of these clasts (Fig. 11a) records a higher temperature metamorphism than the prehnite-pumpellyite metamorphic assemblages of the host rock and the other blocks observed.

Along Panoche Road in the southern Diablo Range, blocks of up to tens of meters in size, including serpentinite and high-grade (amphibolite, garnet-amphibolite, eclogite) rocks, as well as lower-grade (blueschist facies) blocks of a variety of lithologies, are embedded in blueschist facies siliciclastic conglomerate/sedimentary matrix (Wakabayashi 2015). The matrix contains centimeter- and millimeter-sized clasts that reflect the same lithologic range as the larger blocks. These mélange horizons are interbedded with block-free sandstone and pebbly

Fig. 10 (See legend on next page.)

(See figure on previous page.)
Fig. 10 Field photographs of OPS in the Franciscan Complex. In these photos, all contacts (*dashed lines*) of apparent "inverted" OPS stratigraphy (such as basalt over chert or sandstone, or chert over sandstone) and "omitted" OPS (clastics over basalt when pelagic rocks are present in the full sequence) are faulted. Some of the contacts in "normal" OPS stratigraphic order are depositional whereas others are faults. *Dashed lines* drawn within the same lithology represent faults. **a** Marin Headlands, prehnite-pumpellyite facies OPS imbricates at the western side of Black Sand Beach, Bonita Cove (Fig. 6), where basalt and chert (itself internally imbricated) is thrust over siliciclastic rocks. Whereas the siliciclastic section is primarily bedded with local broken formation, sedimentary mélange/conglomerate horizons are present. "**b**" and "**c**" show outcrops within the Rodeo Cove shear zone of the Marin Headlands (Fig. 6), a locality described in detail in Meneghini and Moore (2007). At the scale of Fig. 6, the Rodeo Cove shear zone is one of the faults repeating the OPS section and this imbrication is seen on a smaller scale within the shear zone itself as shown in the outcrop photos. The imbrication locally leads to block-in-matrix relationships such as the chert-in-sandstone seen in the "ch/ss" parts of "**b**", the basalt in basalt matrix part (labeled as "bs"), and the basalt and chert in shale matrix ("ch/sh/bs") part of "**c**". "**d**" and "**e**" show close ups of sedimentary block-matrix horizons in the outcrops of ("**a**"). **d** Pebbly mud-rich sandstone/conglomerate. This conglomerate grades into cleaner sandstone with blocks shown in ("**e**"), and this is interbedded with deformed shale with blocks of other lithologies. These sorts of textures and relationships are common in siliciclastic matrix mélange as seen in "**k-n**" as well as Figs. 16, 17, 18, and 20 of Wakabayashi 2015). **f** Siliciclastic-chert-basalt OPS from blueschist facies rocks, Sunol Regional Wilderness. The preserved parts of chert and siliciclastic sections in the imbricates of mainly metavolcanic exposures are very thin in this area (commonly <10 m and too small to show on the Fig. 7 map). **g** OPS stack of sandstone-limestone-chert-basalt in blueschist facies rocks along the Skaggs Road, Sonoma County (location on Fig. 2). This type of OPS stack is very rare in the Franciscan. **h** Block-in-sandstone matrix, Sunol Regional Wilderness showing OPS of sandstone overlying chert overlying limestone that overlies basalt. Coherent imbricates of this type of OPS have not been found to date in the Franciscan. **i** Skaggs Spring schist (completely recrystallized quartz-rich glaucophane-lawsonite-phengite ± jadeite schist) imbricated with intact serpentinized peridotite. Close up in "**j**" shows silica carbonate rock (hydrothermally replaced serpentinite, also called listvenite) blocks along one of the contacts that places Skaggs Springs Schist on serpentinite. This may represent a depositional contact. Note that there is a lot of surficial material draped over the outcrop in this photo. **k** Photo showing interbedded sandstone and pebbly sandstone and shale matrix sedimentary breccia/conglomerate, blueschist facies rocks along Panoche Road (location on Fig. 2). This shows the interbedding of ordinary turbidites and sedimentary mélange matrix. These horizons locally have blocks of tens of meters in size (compare with different photos in Fig. 18 of Wakabayashi 2015). **l** Interbedding of sedimentary breccia/conglomerate and sandstones and gradational depositional contact between sedimentary breccia/conglomerate (sedimentary mélange matrix), blueschist facies rocks, Sunol Regional Wilderness (location on Fig. 7). The elevation difference between the bottom of the photo and top is about 15 m. **m, n** Sunol Regional Wilderness exposures showing local deformational overprint on conglomerate/sedimentary breccia that makes up the siliciclastic mélange matrix in this area. In "**m,**" note the pencil on the center of the upper photo border for scale. In addition to the locally developed deformational fabric, note the late brittle fault in ("**m**"). The most deformed zones in these outcrops are typical of what has been interpreted as "tectonic" mélange matrix in the Franciscan Complex. **o–r** Sedimentary serpentinite mélange, Ring Mountain (locations on Fig. 8) **o** metamafic block in sedimentary serpentinite matrix. The block is an omphacite-chlorite schist on the left side of the photo (chlorite replaces most of the omphacite and there is late growth of lawsonite), and this grades to or is bordered by nearly monomineralic chlorite schist on the right. The matrix is serpentinite conglomerate and sandstone with many exotic clasts ranging from <0.1 mm to cm scale. **p** Close up of part of "**o**" showing texture of serpentinite conglomerate/pebbly sandstone. **q** Close of a separate outcrop of sandy/pebbly serpentinite matrix with many rounded grains, including a coarse blueschist clast. **r** Mafic block in serpentinite matrix, showing internal faulting and folding. The block is garnet amphibolite (grt-am) nearly completely overprinted by blueschist facies assemblages (bsch) in most areas. The block is internally imbricated and tightly folded resulting in repetitions of mafic (bsch/grt-am) and metaultramafic (UM) schist. The latter is composed of tremolite/actinolite + phengite + chlorite. Progressive deformation has created local block-in-matrix textures in which mafic blocks (bsch) are enclosed in metaultramafic schist (UM). Most of these block-in-matrix features are too small to be seen in this photo. Similar features are present in other blocks in this mélange. They apparently predate the incorporation into the clastic serpentinite matrix which is not part of the imbricates. The features preserved in these blocks may preserve an ocean plate stratigraphic relationship similar to that shown at map scale in Fig. 9 (Willow Spring Slab). **s** Pebbly serpentinite sandstone matrix from Sunol Regional Wilderness (location on Fig. 7) showing well-preserved sedimentary texture. This type of material lacks erosional resistance and tends to be poorly exposed, except on the steepest slopes or artificial excavations. **t, u** Examples of imbrication of siliciclastic (sandstone and shale) (locations on Fig. 2). Without good exposures, these structures are much more difficult to identify in these rocks in contrast to rock assemblages with volcanic and/or pelagic sedimentary rocks. "**t**" shows exposures of the prehnite-pumpellyite facies Novato Quarry terrane along Nave Drive in Marin County and "**u**" shows lawsonite-albite facies rocks that may be correlative to the Burnt Hills terrane along Panoche Road of the southern Diablo Range. For a higher resolution image, please see Additional file 1

sandstone at scales as small as centimeters, and the mélange matrix is commonly more mud rich than the interbedded sandstones as reflected by the darker gray color (Fig. 10k; compare with Fig. 18b, f of Wakabayashi 2015).

The blueschist facies rocks of Sunol Regional Wilderness of the northern Diablo Range include units of siliciclastic and serpentinite matrix mélange with high-grade blocks that include eclogite, garnet-amphibolite, and amphibolite to tens of meters in size, as well as a variety of lower grade (blueschist facies) lithologies ranging to block sizes of several hundred meters (Wakabayashi 2015) (Fig. 7). Several outcrops show interbedding of siliciclastic matrix, composed of conglomerate and sedimentary breccia, with block-free sandstone and shale (Fig. 10l). This matrix has centimeter and millimeter size clasts with the same lithologic range as the larger blocks (Wakabayashi 2015). Nearly undeformed conglomerate and sedimentary breccia is variably overprinted by later deformation, so that it grades into strongly deformed mélange matrix (Fig. 10m, n). The derivation of sheared matrix from conglomerate and sedimentary breccia has been shown for multiple Franciscan localities by Wakabayashi (2015, Figs. 16–18, 20). The inhomogeneous distribution of the later deformation results blocks of less-

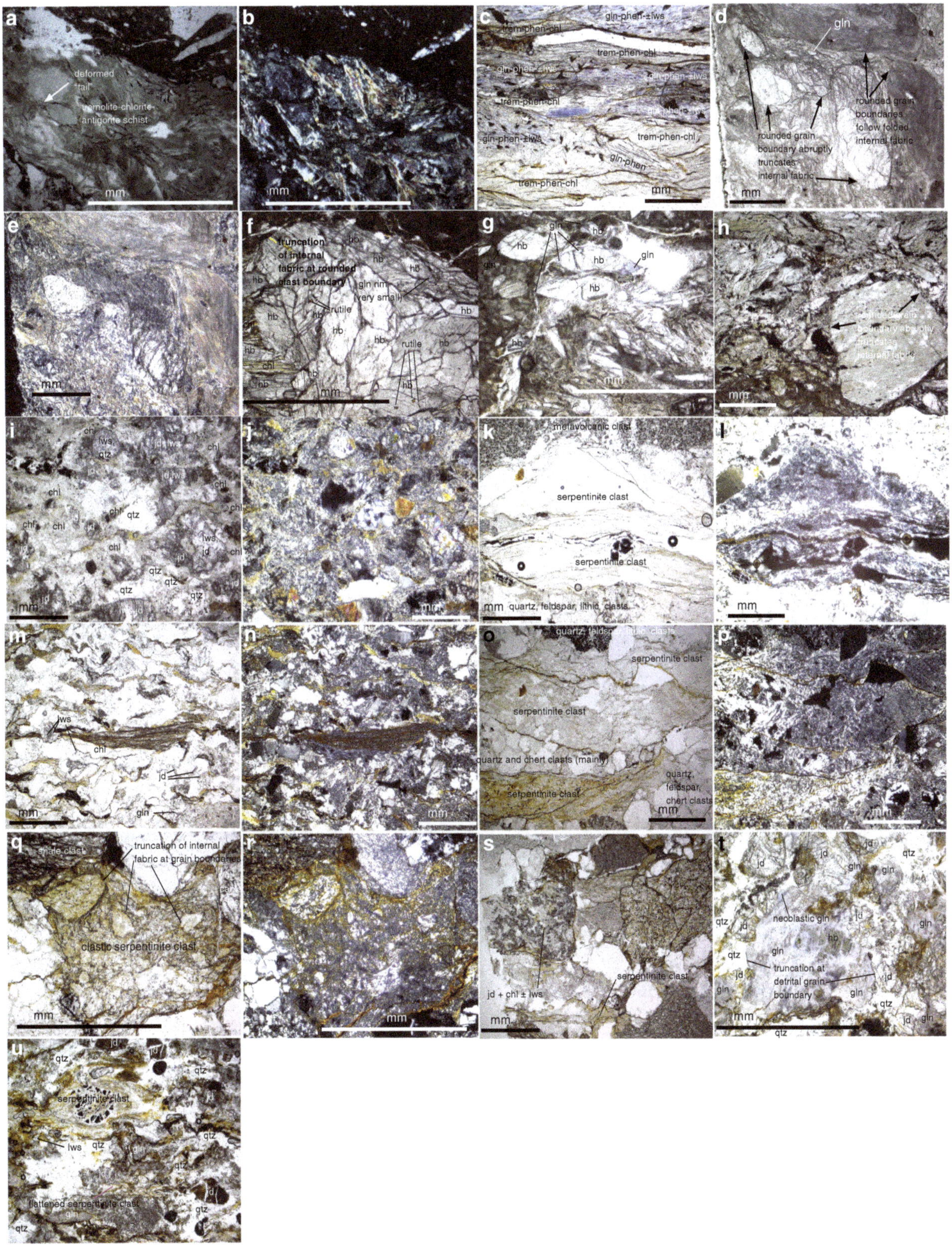

Fig. 11 (See legend on next page.)

(See figure on previous page.)
Fig. 11 Photomicrographs of clastic parts of OPS of the Franciscan Complex with emphasis on the clastic/olistostromal horizons. Mineral abbreviations are *chl* chlorite, *gln* glaucophane, *hb* hornblende, *jd* jadeitic clinopyroxene, *lws* lawsonite, *phen* phengite, *qtz* quartz, *trem* tremolite/actinolite. **a** Metaultramafic schist clast in the pebbly sandstone from Marin Headlands shown in Fig. 10d. The metamorphic assemblage in this clast is higher grade than the prehnite-pumpellyite facies host rock. Note that this clast exhibits an internal fabric that predates deposition, and this fabric appears to be the result of progressive deformation of serpentinized peridotite, rather than a deformational overprint on metaultramafic sandstone of the sort shown in "**e**" and ("**i**"). Plane light. **b** Cross-polarized light view of ("**a**"). **c** Imbricate and block-in-matrix zone in metaultramafic schist from the block in Fig. 10r, showing alternating blueschist layers and lenses with abundant sodic amphibole (*blue* in this plane light view) and pale tremolite/actinolite (trem)-phengite (phen)-chlorite (chl) schist. Note the contrast in textures between the metaultramafic schist and the serpentinite/metaultramafic sandstones shown in "**d, e, g,** and **h**," as well as the clast in ("**q**"). **d** Serpentinite sandstone matrix from the outcrop in Fig. 10o at Ring Mountain. The clasts in this view are primarily antigorite schist (*lighter colored*) or tremolite schist (*darker*). Although some of the rounding of the grains is a result of the folding of these grains (two examples shown with *arrows*) and development of locally curved shears, there are many rounded grain boundaries that abruptly truncate the internal grain fabric without drag; these are original detrital grain boundaries (examples shown by *arrows*). Plane-polarized light. **e** Cross-polarized light view of ("**d**"). **f** Rutile-bearing hornblendite clast (long dimension about 5 cm) in the same serpentinite matrix sampled about 1 m away from the sample in ("**d**"). The internal fabric in this clast is truncated along the rounded grain boundary with the enclosing serpentinite sandstone/conglomerate (*dark*) without drag features (most apparent in *the upper left hand part* of the view). Hornblende (*hb*) locally has small glaucophane (*gln*) rims, and there is some late growth of chlorite (*chl*). This is a cm-scale analog of the larger high-grade blocks found in this mélange. Plane-polarized light. **g** Serpentinite matrix from the same locality showing hornblende (*hb*) grains, some of which are rounded and rimmed with glaucophane (*gln*). Note the truncation of the hornblende and rim against the rounded grain boundary, best seen at the left-rounded termination of the *hb* grain in *the upper left corner* of the photo. Other grains in this view include tremolite/actinolite, and chlorite-tremolite-antiorite schist. Plane-polarized light. **h** Serpentinite matrix from serpentinite mélange, Sunol Regional Wilderness (sample location on Fig. 7) showing abrupt truncation of internal grain fabric along grain boundaries, similar to that seen at the Ring Mountain locality. Clasts in this view are primarily antigorite-chlorite schist with some grains of tremolite/actinolite, hornblende, and chlorite. Plane-polarized light. **i** Chlorite-rich sandstone from Sunol Regional Wilderness (Fig. 7). This sandstone is *medium green* in outcrop and resembles clastic serpentinite in outcrop appearance. Much of the chlorite rich zones (*chl*) have fine greenish phengite, and polycrystalline growths of chlorite form textures similar to antigorite schist. Chrome spinel is locally present in the chlorite-rich zones. The chlorite-rich zones probably originated as serpentinite clasts. Other clasts or minerals include jadeitic clinopyroxene (*jd*), lawsonite (*lws*), quartz (*qtz*), and glaucophane. Some of the metamorphic mineral grains are part of polymineralic clots (*jd, lws* or *lws, qtz* or *jd, lws, qtz*) and the polymetamorphic clots, and many of the metamorphic mineral grains have rounded grain boundaries that truncate their internal fabric, suggesting a detrital origin. Many of these grains also exhibit late neoblastic overgrowths with sharp terminations. Plane-polarized light. **j** Cross-polarized light view of ("**i**"). **k** Serpentinite clast with chrome spinel in mostly siliciclastic conglomerate, Sunol Regional Wilderness. Long dimension of the clast is about 10 mm. Such serpentinite clasts make up 5–10% of the clasts in some horizons in this conglomerate. Plane-polarized light. **l** Cross-polarized view of ("**k**"). **m** Sandstone that makes up most of non-mélange siliciclastic outcrops at Sunol Regional Wilderness. This sandstone has much less chlorite than "**i**," but some chlorite (*chl*) clasts have the appearance of serpentinite (best viewed in the cross-polarized light view of "n") such as the clast labeled that is directly below the shale chip in the center of the photo. This sample has neoblastic lawsonite (*lws*), jadeitic clinopyroxene (*jd*), and glaucophen (*gln*), but relict albite is still present as seen by the albite twinning seen in the cross-polarized light view of "**n**" (good example along *the upper left edge of photo*). **o–u** El Cerrito quarry area, eastern San Francisco Bay area (general location on Fig. 2). Two serpentinite clasts from prehnite-pumpellyite siliciclastic conglomerate, matrix of the Hillside mélange unit. This conglomerate/pebbly sandstone contains up to 15% serpentinite clasts. This photo shows a lizardite-rich and an antigorite-rich clast, both of which appear to have been derived from intact peridotite (they are not themselves clastic). Plane-polarized light. **p** Cross-polarized light view of ("**o**"). **q** Serpentinite sandstone clast from the same thin section showing the internal rounded grains that truncate fabric within them. Compare to the serpentinite matrix samples shown in "**d**" and ("**h**") The grains are primarily antigorite schist. Plane-polarized light. **r** Cross-polarized light view of ("**q**"). **s** Clast with jadeite-chlorite ± lawsonite, along with another serpentinite clast in the same thin section. **t** View of blueschist facies metasandstone unit of the upper El Cerrito quarry. This particular sample consists of more than 50% detrital blueschist clasts, most of which have hornblende relics. Much of the *blue* amphibole growth is truncated on the detrital grain boundaries indicating that these were blueschist-overprinted amphibolite clasts (equivalent of the high-grade blocks) at the time of deposition. There is also some late neoblastic glaucophane growth along one of the margins of this clast. **u** Same unit as "**t**" collected about 10 m away showing serpentinite clast with chrome spinel. For a higher resolution image, please see Additional file 2

deformed mélange surrounded by more strongly deformed zones (Wakabayashi 2015).

It is more difficult to demonstrate sedimentary features of serpentinite mélanges than siliciclastic matrix mélanges because apparent rounding of grains in the former may take place by two different tectonic mechanisms: (1) the formation of curved shear surfaces and (2) the tight folding early-formed fabrics that result in rounded apparent clast borders formed from detached fold hinges. For this reason I present outcrop and petrographic thin section photos showing what I interpret as sedimentary textures with goal of presenting more convincing relationships than those shown in my earlier papers (Wakabayashi, 2012; 2015). In addition, recent work allows a direct comparison between the textures observed in these Franciscan serpentinite mélanges to the GVG serpentinite deposits that are indisputably sedimentary in origin (e.g., Moiseyev 1970; Lockwood 1971; Phipps 1984; (Wakabayashi 2016b; Wakabayashi, in press).

Franciscan clastic serpentinite mélanges have comparatively fine-grained matrix that crops out poorly, and when it does, it forms outcrops that appear powdery from a distance owing to the lack of large pieces of more intact rock (Wakabayashi, in press). This contrasts with the blockier appearance of intact (not mélange) Franciscan serpentinized peridotite slabs or the ultramafic sections of the CRO, but is identical to the appearance of the GVG sedimentary serpentinites (Wakabayashi in press). Exposures of Franciscan serpentinite matrix range from serpentinite sandstone to serpentinite conglomerate/sedimentary breccia (Fig. 10o–s; see also Fig. 6a, b, Fig. 9 of Wakabayashi 2012, and Fig. 21c–f of Wakabayashi 2015). These textures are identical to those displayed by the sedimentary serpentinite of the Great Valley Group, but differ from the Franciscan serpentinized peridotite and Coast Range ophiolite serpentinitinized peridotite slabs (Fig. 6 of Wakabayashi, in press.

At thin section scale, certain grain boundary geometries are compatible with original detrital grain boundaries but not with formation by fracturing or shearing. Internal fabric within clasts is abruptly truncated without any drag features along rounded grain boundaries, including those grain boundaries oriented at high angles to the foliation or cleavage direction in the rock (Fig. 11d–h). There are no deformed tails associated with these clasts. The clasts include (non-ultramafic) metamorphic rock fragments and metamorphic mineral grains (Fig. 11f, g). Deformation has variably overprinted the sedimentary textures leaving lenses of relatively undeformed domains (millimeter to tens of meter scales) between strongly deformed zones (millimeter scale examples in Fig. 11d, e, g, and h). These textures differ markedly by textures that appear formed by progressive deformation, such as those shown in Fig. 11a–c that lack the grain boundary relationships, whether it is between serpentinite domains, or between non-ultramafic and ultramafic material.

The matrix at the Ring Mountain locality includes millimeter- to centimeter-scale clasts of minerals that have the same characteristics of the larger high-grade blocks in the mélange. For example, the clasts and mineral grains amphibolite grade metamorphism overprinted by blueschist facies metamorphism (Fig. 11f, g) as is common in the high-grade blocks of the Franciscan, including those of the Ring Mountain locality (e.g., Wakabayashi 1990). Abrupt truncation of sodic amphibole in these clasts and mineral grains by detrital grain boundaries suggests the blueschist overprint of the earlier higher temperature metamorphism took place prior to incorporation (sedimentation) (Fig. 11g).

The serpentinite mélange at Ring Mountain and at Sunol Regional Wilderness also contains lower grade metamorphic blocks, such as metachert lacking macroscopic metamorphic minerals, metavolcanic rocks with preserved igneous textures and incipient growth of fine-grained (mostly <0.1 mm) blueschist facies minerals such as lawsonite and glaucophane, and metasandstone that displays neoblastic growth of jadeitic clinopyroxene, lawsonite, and glaucophane (Wakabayashi 2015). At Sunol Regional Wilderness, the (non-serpentinite) block types in serpentinite matrix mélange are the same as those in the adjacent siliciclastic matrix mélange.

Interfingering, interbedding, and gradation between siliciclastic rocks and clastic serpentinite

Interbedding and interfingering of siliciclastic rocks an clastic serpentinite is suggested by map relationships at 100 m and larger scales, as well as the comparatively common occurrence of blocks of serpentinite and hydrothermally altered serpentinite (silica-carbonate rock or listvenite) in siliciclastic matrix mélange, as well as similar (non-serpentinite) block populations in clastic serpentinite and siliciclastic mélanges at Sunol Regional Wilderness (Fig. 7). However, outcrops demonstrating depositional relationships between clastic serpentinite and siliciclastic horizons have not been found in the Franciscan Complex. This contrasts to the Great Valley Group where depositional contacts of serpentinite over siliciclastic rocks and gradation between clastic serpentinite and siliciclastic rocks has been shown (Wakabayashi 2016b).

Petrographic inspection provides evidence for mixed siliciclastic and clastic serpentinite deposition in Franciscan rocks. Serpentinite clasts are more common in horizons of siliciclastic mélange matrix (Fig. 11a, b, k, l, and o–r) than "ordinary" sandstones (Fig. 11m, n). Serpentinite clasts may make up at least 15% of the rock (Fig. 11o), and the detrital serpentinite consists of intact serpentinite (no internal detrital grain boundaries, Fig. 11a, b, k, l, o, and p) and redeposited serpentinite sandstone (Fig. 11q, r). The host rocks have high-pressure metamorphic detrital clasts that are higher grade than the post-sedimentary burial metamorphism including clasts equivalent to Franciscan high-grade blocks (Fig. 11s, t). Whereas serpentinite or metaultramafic mineralogy is preserved in some detrital serpentinite clasts, in other cases, the original mineralogy has been altered, commonly to chlorite, or chromium-rich white mica (fuchsite), or a combination of the two (Fig. 11i, j). The serpentinite origin of such detrital grains may be ascertained in some cases by the presence of chrome spinel or the pseudomorphing of serpentinite textures by polycrystalline chlorite clots. Sandstones rich in chlorite clasts of the type noted above have an unusual green color in outcrop and resemble serpentinite.

Most Franciscan siliciclastic sandstones have small amounts of detrital serpentinite, whether inferred by textures in chlorite clasts (Fig. 11m, n) or directly verified by presence of preserved serpentinite. Larger amounts of detrital serpentinite are found in siliciclastic mélange matrix, and in limited horizons of pebbly sandstone within otherwise block-free sandstone, such as one pebbly sandstone from El Cerrito Quarry that is locally so rich in blueschist clasts (Fig. 11t) as to have a blue color in outcrop.

In summary, the common occurrence of detrital serpentinite clasts in siliciclastic rocks (both siliciclastic mélange matrix and bedded sandstones), a similar range of block types in siliciclastic and serpentinite mélanges in contact with one another, map-scale contact relationships, and serpentinite blocks in siliciclastic matrix mélange, suggest that interbedding, interfingering, and gradation between siliciclastic and clastic serpentinite horizons is common in the Franciscan.

Imbrication and deformation of OPS and distinguishing tectonic from sedimentary melanges

Internal imbrication of OPS is easily seen where the igneous and pelagic components of OPS are common, such in the Marin Headlands unit (Fig. 10a–c). Internal imbrication of the clastic and volcanic units is difficult to evaluate in most cases owing to the lack of obvious marker horizons. In the metavolcanic-rich domains of northern Sunol Regional Wilderness, imbrication is shown by siliciclastic and/or chert horizons, commonly less than 10 m in thickness (Fig. 10f). Owing to the thinness of these non-volcanic layers, they have not been found outside of one continuous set of roadcut exposures. In mélange rich exposures, it is difficult to distinguish throughgoing imbricate faults from a series of isolated blocks such as in the central part of the Sunol Regional Wilderness map of Fig. 7. With high-quality exposures, however, internal imbrication is observed to be ubiquitous in clastic Franciscan rocks (examples in Fig. 10t, u, as well as the spatially variable deformation in mélange matrix of Fig. 10m, n). The magnitude of penetrative strain in most rocks is small, so most of the deformation and accordingly accommodation of accretionary megathrust slip takes place along discrete faults or narrow (less than tens of meters thick) fault zones (Fig. 10a–c, m, n, t, and u).

Fault zones that repeat the OPS section, such as the Rodeo Cove shear zone of the Marin Headlands, locally exhibit block-in-matrix textures (Meneghini and Moore 2007) (Fig. 10a, b). In such cases, none of the blocks are exotic and all are of the same metamorphic grade, unless this late imbrication and deformation involves a clastic section that has olistostromal horizons that have exotic blocks (schematically shown in Fig. 4). Such deformation can result in blocks of sedimentary mélange within a tectonic mélange, as illustrated by the deformational overprint shown in Fig. 10m, n (see also Wakabayashi 2015, Fig. 18c, d).

Examples of Franciscan tectonic serpentinite mélange, developed by the progressive deformation of metaultramafic and metabasite/metachert may be preserved in some high-grade blocks in sedimentary mélange, as shown at outcrop (Fig. 10r) and thin section scale (Fig. 11c, d). In such examples, all of the blocks in the metaserpentinite matrix are of the same metamorphic grade and compatible with the metamorphic mineralogy of the matrix. Such relationships are similar to serpentinite mélanges found in the Alps and Iran that have been shown to have developed by progressive deformation of ultramafic-bearing OPS (e.g., Angiboust et al. 2012; 2013; Balestro et al. 2015; Vitale Brovarone et al. 2014), but differ markedly from the sedimentary serpentinite mélanges described in this paper and by Wakabayashi (2012; 2015; in press). Intact equivalents of such tectonic serpentinite mélanges have not been found in Franciscan rocks, but imbricated high-grade slabs such as the structurally high Willow Spring slab, whose imbricates include serpentinite, may locally have such serpentinite-matrix mélanges (Fig. 9).

Franciscan OPS and tectonic history

The Franciscan OPS exposures exhibit a large range in lithologic combinations, reflecting a correspondingly large range in character of oceanic crust subducted during the >150 Ma of the assembly of this accretionary complex. The lack of post-subduction disruption in the Franciscan results in good preservation of the variation in OPS packages and their relationship to the architecture of the subduction complex, and hence their relationship to the tectonic history. In addition, the lack of post-subduction disruption facilitates distinction between Franciscan subduction complex rocks (transferred from lower to upper plate of subduction system) and rocks that were always part of the upper plate (CRO and GVG).

The earliest accreted OPS units of the Franciscan are high-grade coherent sheets ("Basalt-limestone-chert-clastic or basalt-chert-limestone-clastic") and related blocks-in-mélange that underwent high P-high T metamorphism at about 150–170 Ma (Ross and Sharp 1988; Anczkiewicz et al. 2004; Wakabayashi and Dumitru 2007; Ukar et al. 2012; Mulcahy et al. 2014; Cooper et al. 2011). These rocks may have formed during or shortly after subduction that initiated at ca.165–170 Ma within young oceanic arc lithosphere, in an environment lacking clastic sediment input (Wakabayashi et al. 2010).

Following subduction initiation and early accretion of island arc crust, subsequent oceanic igneous rocks subducted and accreted were of MORB affinity with lesser amounts of OIB. A significant gap in accretion of tens of million years followed the initial accretion of the high-grade material (Dumitru et al. 2010; Wakabayashi 2015). Following subduction initiation, the age progression of accreted oceanic crust, as well as the progression of metamorphism, suggests that a ridge crest approached the trench (120–125 Ma) but probably shut off prior to subduction (Wakabayashi 2015) (Fig. 5a). After this near ridge-trench encounter, the age of subducted oceanic crust became progressively older until about 95 Ma (subduction accretion of the Marin Headlands terrane), after which the age of the subducted oceanic crust became progressively younger until termination of subduction and conversion to a transform plate boundary. In contrast to the age of the subducted oceanic crust, the age of clastic sedimentary rocks in each OPS package was progressively younger with each successively accreted unit.

These OPS packages had significant components of trench fill overlying the pelagic sedimentary rocks. This trench fill included both turbidites as well as siliciclastic and serpentinite matrix olistostromes. Some of the oceanic crust that subducted had serpentinized peridotite exposed on the sea floor (e.g., Coleman 2000; Wakabayashi 2004), resulting in OPS composed of abyssal serpentinized peridotite and overlying clastic trench fill.

In contrast to the Franciscan OPS, imbricates of the upper CRO and basal GVG were formed by shortening of the upper plate rather than by transfer from the subducting plate to the upper plate. Some of this shortening may have taken place synchronous with Franciscan subduction, whereas some may have taken place after subduction termination and conversion to a transform plate boundary (Phipps 1984; 1992; Wakabayashi and Unruh 1995). This imbrication did not result in burial metamorphism or penetrative deformation in these rocks, in contrast to Franciscan OPS.

Conclusions

OPS in the California Cordillera consists both of oceanic assemblages that were never part of a subducting plate (upper plate assemblages) and were not incorporated into subduction-accretion complexes as well as materials transferred from the subducting plate to the upper plate to form subduction complexes such as the Franciscan. Whereas the largest oceanic crustal remnants in the California Cordillera are of the upper plate type, the subduction complex materials reflect the OPS concept as commonly applied (e.g., Wahrhaftig 1984; Isozaki et al. 1990; Isozaki and Blake 1994; Wakita and Metcalfe 2005; Kusky et al. 2013).

Franciscan OPS varies considerably in characteristics. This variation reflects both the variety of ocean crust subducted during the ca. 150 million years of Franciscan subduction as well as a variety of processes of subduction-accretion and trench sedimentation. The variation in OPS and its relationship to tectonic process is more complex than generally considered in studies of these types of rocks.

The clastic component of OPS includes olistrostromal (sedimentary mélange) horizons that are interbedded with and interfinger with bedded turbidites. The olistostromal component includes units of both siliciclastic and serpentinite matrix, and the different types of matrix may be interbedded, interfinger, or grade into one another.

Non-accretionary megathrust horizons separate OPS accreted at different times, whereas megathrust slip during accretion was accommodated by internal deformation of OPS, chiefly imbrication of OPS along discrete faults. Such deformation results in local block-in-matrix horizons (tectonic mélanges) that differ from the olistostromal horizons noted above in the lack of exotic blocks. However, because the deformed OPS packages include olistostromal horizons, pieces of the olistostromes that include exotic blocks may be incorporated as blocks in tectonic mélanges.

Abbreviations

CRO: Coast range ophiolite; GVG: Great valley group; MORB: Mid-ocean-ridge basalt; OIB: Ocean island basalt; OPS: Ocean plate stratigraphy

Acknowledgements

Parts of this research were supported by the National Science Foundation grant EAR-0635767. I thank the detailed and constructive reviews that led to significant improvements in the paper.

Funding

Not applicable.

Authors' contributions

All works were conducted by the (single) author.

Authors' information

Wakabayashi has conducted research on subduction complexes for over 30 years with emphasis on the Franciscan Complex of California, starting with his Ph. D. dissertation research with Eldridge Moores that began in the early 1980s. His first exposure to ocean plate stratigraphy came from geologic mapping of the Marin Headlands as an undergraduate student (1980) under the guidance of Clyde Wahrhaftig who several years later authored a well-known paper on the Marin Headlands in the context of the ocean plate stratigraphy concept (Wahrhaftig 1984 cited in the manuscript). Recent work by Wakabayashi in the Franciscan Complex has included many studies based on geologic mapping, with much new information published in Wakabayashi (2015, cited in manuscript). Much of the primary field observations presented in this paper are derived from Wakabayashi (2015) but there are new observations presented in this paper as well, in addition to review of work and concepts not previously presented.

Competing interests

The author declares that he has no competing interest.

References

Aalto KR (1981) Multistage mélange formation in the Franciscan Complex, northernmost California. Geology 9:602–607

Alvarez W, Kent DV, Premoli Silva I, Schweickert RA, Larson RL (1980) Franciscan Complex limestones deposited at 17° south paleolatitude. Geol Soc Am Bull 91:476–484

Anczkiewicz R, Platt JP, Thirlwall MF, Wakabayashi J (2004) Franciscan subduction off to slow start: Evidence from high-precision Lu-Hf garnet ages on high-grade blocks. Earth Planet Sci Lett 225:147–161

Angiboust S, Langdon R, Agard P, Waters D, Chopin C (2012) Eclogitization of the Monviso ophiolite (W. Alps) and implications on subduction dynamics. J Metamorph Geol 20:37–61

Angiboust S, Agard P, De Hoog JCM, Omrani J, Plunder A (2013) Insight on deep accretionary subduction processes from the Sistan ophiolitic "mélange" (eastern Iran). Lithos 156–159:139–158

Bailey EH, Blake MCJ, Jones DL (1970) On-land Mesozoic ocean crust in California Coast Ranges. In: U.S. Geological Survey Professional Paper 700-C., pp 70–81

Balestro G, Festa A, Tartarotti P (2015) Tectonic significance of different block-in-matrix structures in exhumed convergent plate marings: examples from oceanic and continental HP rocks in the Inner Western Alps (northwest Italy). Int Geol Rev 57:581–605

Barnes JD, Eldam R, Lee C-TA, Errico JC, Loewy S, Cisneros M (2013) Petrogenesis of serpentinites from the Franciscan Complex, western California, USA. Lithos 178:143–157

Bero J (2014) Geology of Ring Mountain and Tiburon Peninsula, Marin County, California. California Geological Survey Map Sheet 62. Scale 1: 12,000.

Blake MCJ, Howell DG, Jayko AS (1984) Tectonostratigraphic terranes of the San Francisco Bay Region. In: Blake MCJ (ed) Franciscan Geology of Northern California, vol 43, Pacific Section, Society of Economic Paleontologists and Mineralogists., pp 5–22

Blake MC Jr, Jayko AS, McLaughlin RJ, Underwood MB (1988) Metamorphic and tectonic evolution of the Franciscan Complex, northern California. In: Ernst WG (ed) Metamorphism and Crustal Evolution of the Western United States, vol VII, Rubey. Prentice-Hall, Englewood Cliffs, New Jersey, pp 1035–1060

Chapman AD, Jacobson CE, Ernst WG, Grove M, Dumitru T, Hourigan J, Ducea M (2016) Assembling the world's type shallow subduction complex: detrital zircon geochronologic constraints on the origin of the Nacimiento block, central California Coast Ranges. Geosphere 12(2):1–25. doi:10.1130/GES01257.1

Chipping DH (1971) Paleoenvironmental significance of chert in the Franciscan Formation of western California. Geol Soc Am Bull 82:1707–1712

Choi SH, Shervais JW, Mukasa SB (2008) Supra-subduction and abyssal peridotites of the Coast Range ophiolite, California. Contrib Mineral Petrol 156:551–576

Cloos M (1984) Flow mélanges and the structural evolution of accretionary wedges. in Raymond L (Ed) Mélanges: Their Nature, Origin, and Significance. Boulder, Colorado. Geol Soc Am Spec Pap 198:71–80

Cloos M, Shreve RL (1988) Subduction channel model of prism accretion, melange formation, sediment subduction, and subduction erosion at convergent plate margins: 1. Background and description. Pure Appl Geophys 128:455–500

Cohen KM, Finney SC, Gibbard PL, Fan J-X (2013, updated October 2016) The ICS International Chronostratigraphic Chart. Episodes 36: 199–204. URL: http://www.stratigraphy.org/ICSchart/ChronostratChart2016-12.

Coleman RG (2000) Prospecting for ophiolites along the California continental margin. In: Dilek YD, Moores EM, Elthon D, Nicolas A (eds) Ophiolites and Oceanic Crust: New Insights From Field Studies and the Ocean Drilling Program, vol 349, Geological Society of America Special Paper. Boulder, Colorado, pp 351–364

Coleman RG, Lanphere MA (1971) Distribution and age of high-grade blueschists, associated eclogites, and amphibolites from Oregon and California. Geol Soc Am Bull 82:2397–2412

Cooper FJ, Platt JP, Anczkiewicz R (2011) Constraints on early Franciscan subduction rates from 2-D thermal modeling. Earth Planet Sci Lett 312:69–79

Cowan DS (1978) Origin of blueschist-bearing chaotic rocks in the Franciscan Complex, San Simeon, California. Geol Soc Am Bull 89:1415–1423

Dickinson WR (1970) Relations of andesites, granites and derivative sandstones to arc-trench tectonics. Rev Geophys Space Phys 8:813–860

Dilek Y, Furnes H (2011) Ophiolite genesis and global tectonics: geochemcial and tectonic fingerprinting of ancient oceanic lithosphere. Geol Soc Am Bull 123:387–411

Dumitru TA, Wakabayashi J, Wright JE, Wooden JL (2010) Early Cretaceous transition from nonaccretionary behavior to strongly accretionary behavior within the Franciscan subduction complex. Tectonics 29:TC5001. doi:10.1029/2009TC002542

Dumitru TA, Ernst WG, Hourigan JK, McLaughlin RJ (2015) Detrital zircon U-Pb reconnaissance of the Franciscan subduction complex in northwestern California. Int Geol Rev 57:767–800

Dumitru TA, Elder WP, Hourigan JK, Chapman AD, Graham SA, Wakabayashi J (2016) Four Cordilleran paleorivers that connected Sevier thrust zones in Idaho to depocenters in California, Washington, Wyoming, and indirectly, Alaska. Geology 44:75–78

Ernst WG (1970) Tectonic contact between the Franciscan mélange and the Great Valley Sequence, crustal expression of a Late Mesozoic Benioff Zone. J Geophys Res 75:886–902

Ernst WG (1993) Metamorphism of Franciscan tectonostratigraphic assemblage, Pacheco Pass area, east-central Diablo Range, California. Geol Soc Am Bull 105:618–636

Ernst WG, McLaughlin RJ (2012) Mineral parageneses, regional architecture, and tectonic evolution of Franciscan metagraywackes, Cape Mendocino-Garberville-Covelo 30' x 60' quadrangles, northwest California. Tectonics 31:TC1001. doi:10.1029/2011TC002987

Ernst WG, Seki Y, Onuki H, Gilbert MC (1970) Comparative study of low-grade metamorphism in the California Coast Ranges and the outer metamorphic belt of Japan, vol 124, Geological Society of America Memoir. Boulder, Colorado, USA, p 276

Ernst WG, Martens U, Valencia V (2009) U-Pb ages of detrital zircons in Pacheco Pass metagraywackes: Sierran-Klamath source of mid-Cretaceous and Late Cretaceous Franciscan deposition and underplating. Tectonics 28:TC6011. doi:10.2019/2008TC002352

Gerya TV, Stöckert B, Perchuk AL (2002) Exhumation of high-pressure metamorphic rocks in a subduction channel: a numerical simulation. Tectonics 21 (6), doi:10.1029/2002TC001406

Ghatak A, Basu AR, Wakabayashi J (2012) Element mobility in subduction metamorphism: Insight from metamorphic rocks of the Franciscan Complex and Feather River ultramafic belt, California. Int Geol Rev 54:654–685. doi:10.1080/00206814.2011.567087

Halls C, Zhao R (1995) Listvenite and related rocks: perspectives on terminology and mineralogy with reference to an occurrence at Cregganbaum, Co. Mayo, Republic of Ireland. Mineral Deposita 30:303–313

Hamilton WB (1969) Mesozoic California and underflow of the Pacific mantle. Geol Soc Am Bull 80:2409–2430

Hopson CA, Pessagno EA Jr (2005) Tehama-Colusa serpentinite mélange: A remnant of Franciscan Jurassic oceanic lithosphere, northern California. Int Geol Rev 47:65–100

Hopson CA, Mattinson JM, Pessagno EA Jr (1981) Coast Range Ophiolite, western California. In: Ernst WG (ed) Metamorphism and Crustal Evolution of the Western United States, vol I, Rubey. Prentice-Hall, Englewood Cliffs, New Jersey, pp 418–510

Hopson CA, Mattinson JM, Pessagno EA Jr, Luyendyk BP (2008) California Coast Range Ophiolite: Composite Middle and Late Jurassic oceanic lithosphere. In: Wright JE, Shervais JW (eds) Ophiolites, Arcs, and Batholiths: A Tribute to Cliff Hopson, Geological Society of America Special Paper 483., pp 1–101

Hsü KJ (1968) The principles of mélanges and their bearing on the Franciscan-Knoxville paradox. Geol Soc Am Bull 79:1063–1074

Isozaki Y, Blake MC Jr (1994) Biostratigraphic constraints on formation and timing of accretion in a subduction complex: An example from the Franciscan Complex of northern California. J Geol 102:283–296

Isozaki Y, Maruyama S, Furuoka F (1990) Accreted oceanic materials in Japan. Tectonophysics 181:179–205

Jean MM, Shervais JW, Choi S-H, Mukasa SB (2010) Melt extraction and melt refertilization in mantle peridotite of the Coast Range ophiolite: an LA-ICP-MS study. Contrib Mineral Petrol 159:113–136

Kusky TM, Windley BF, Safonova I, Wakita K, Wakabayashi J, Polat A, Santosh M (2013) Recognition of oceanic plate stratigraphy in accretionary orogens through Earth history: A record of 3.8 billion years of sea floor spreading, subduction, and accretion. Gondwana Res 24:501–547

Lockwood JP (1971) Sedimentary and gravity-slide emplacement of serpentinite. Geol Soc Am Bull 82:919–936

MacPherson GJ, Phipps SP, Grossman JN (1990) Diverse sources for igneous blocks in Franciscan mélanges, California Coast Ranges. J Geol 98:845–862

McLaughlin RJ, Kling SA, Poore RZ, McDougall K, Beutner EC (1982) Post-Miocene accretion of Franciscan rocks, northwestern California. Geol Soc Am Bull 93: 595–605.

Meneghini F, Moore JC (2007) Deformation and hydrofracture in a subduction thrust at seismogenic depths: The Rodeo Cove thrust zone, Marin Headlands, California. Geol Soc Am Bull 119:174–183

Moiseyev AN (1970) Late serpentinite movements in the California Coast Ranges: New evidence and its implications. Geol Soc Am Bull 81:1721–1732

Moores EM (1982) Origin and emplacement of ophiolites. Rev Geophys Space Phys 20:735–760

Mulcahy S, Vervoort JD, Renne PR (2014) Dating subduction zone metamorphism with combined garnet and lawsonite Lu-Hf geochronology. J Metamorph Geol 32:515–533

Murchey BM (1984) Biostratigraphy and lithostratigraphy of chert in the Franciscan Complex, Marin Headlands, California. In: Blake MC Jr (ed) Franciscan Geology of Northern California, vol 43, Pacific Section, Society of Economic Paleontologists and Mineralogists., pp 51–70

Murchey BM, Blake MC Jr (1993) Evidence for subduction of a major ocean plate along the California margin during the middle to early late Jurassic. In: Dunn G, McDougall K (eds) Mesozoic Paleogeography of the Western United States-II, vol 71, Pacific Section SEPM Book., pp 1–18

Murchey BM, Jones DL (1984) Age and significance of chert in the Franciscan Complex of the San Francisco Bay Region. In: Blake MC Jr (ed) Franciscan

Geology of Northern California, vol 43, Pacific Section, Society of Economic Paleontologists and Mineralogists., pp 23–30

Osozawa S (1994) Plate reconstruction based upon age data of Japanese accretionary complexes. Geology 22:1135–1138

Phipps SP (1984) Ophiolitic olistostromes in the basal Great Valley sequence, Napa County, northern California Coast Ranges. In: Raymond LA (ed) Mélanges: Their Nature, Origin, and Significance, vol 198, Boulder, Colorado, Geological Society of America Special Paper., pp 103–126

Phipps SP (1992) Late Cenozoic tectonic wedging and blind thrusting beneath the Sacramento Valley and eastern Coast Ranges, Day 2. In: Erskine MC, Unruh JR, Lettis WR, Bartow JA (eds) Field Guide to the Tectonics of the Boundary Between the California Coast Ranges and the Great Valley of California: Bakersfield, California, Pacific Section, American Association of Petroleum Geologists, Field trip guidebook GB-70., pp 63–83

Platt JP (2015) Origin of Franciscan blueschist-bearing mélange at San Simeon, central California Coast. Int Geol Rev 57:843–853.

Prohoroff RE, Wakabayashi J, Dumitru TA (2012) Sandstone-matrix olistostrome deposited on intra-subduction complex serpentinite, Franciscan Complex, western Marin County, California. Tectonophysics 568-569:296–305

Raymond LA, Bero DA (2015) Sandstone-matrix mélanges, architectural subdivision, and geologic history of accretionary complexes: A sedimentological and structural perspective from the Franciscan Complex of Sonoma and Marin counties, California, USA. Geosphere11 (4), doi: 10.1130/GES001137.1

Ross JA, Sharp WD (1988) The effects of sub-blocking temperature metamorphism on the K/Ar systematics of hornblendes: $^{40}Ar/^{39}Ar$ dating of polymetamorphic garnet amphibolite from the Franciscan Complex, California. Contrib Mineral Petrol 100:213–221

Seton M, Müller RD, Zahirovic S, Gaina C, Torsvik T, Shephard G, Talsma A, Gurnis M, Turner M, Chandler M (2012) Global continentaland ocean basin reconstructions since 200 Ma. Earth-Sci Rev 113:212–270. doi:10.1016/j.earscirev.2012.03.002

Shervais JW (1990) Island arc and ocean crust ophiolites; contrasts in the petrology, geochemistry and tectonic style of ophiolite assemblages in the California Coast Ranges. In: Malpas J, Moores E, Panayiotou A, Xenophontos C (eds) Ophiolites Oceanic Crustal Analogues: Proceedings of the Symposium 'Troodos 1987': Nicosia, Cyprus, Geological Survey Department, Ministry of Agriculture and Natural Resources., pp 507–520

Shervais JW, Choi SH, Sharp WD, Ross J, Zoglman-Schuman M, Mukasa SB (2011) Serpentinite matrix mélange: Implications for mixed provenance for mélange formation. In: Wakabayashi J, Dilek Y (eds) Mélanges: Processes of Formation and Societal Significance, vol 480, Geological Society of America Special Paper., pp 1–30

Sliter WV (1984) Foraminifers from Cretaceous limestone of the Franciscan Complex, northern California. In: Blake MC Jr (ed) Franciscan Geology of Northern California, vol 43, Pacific Section, Society of Economic Paleontologists and Mineralogists., pp 5–22

Sliter RV, McGann ML (1992) Age and correlation of the Calera Limestone in the Permanente Terrane of northern California. In: U.S. Geological Survey Open File Report OF 92–0306., p 27

Smith CN, Kesler SE, Blum JD, Rytuba JJ (2008) Isotope geochemistry of mercury in source rocks, mineral deposits and spring deposits of the California Coast Ranges, USA. Earth Planet Sci Lett 269:399–407

Sorensen SS, Grossman JN, Perfit MR (1997) Phengite-hosted LILE enrichment in eclogite and related rocks: Implications for fluid-mediated mass transfer in subduction zones and arc magma genesis. J Petrol 38:3–34. doi:10.1093/petroj/38.1.3

Tarduno JA, McWilliams M, Sleep N (1990) Fast instantaneous oceanic plate velocities recorded by the Cretaceous Laytonville limestone: Paleomagnetic analysis and kinematic interpretations. J Geophys Res 95:15503–15527

Terabayashi M, Maruyama S (1998) Large pressure gap between the Coastal and Central Belts, northern and central California. Tectonophysics 285:87–101

Ukar E, Cloos M, Vasconcelos P (2012) First 40Ar-39/ar ages from low-T mafic blueschist blocks in a Franciscan mélange near San Simeon: Implications for initiation of subduction. J Geol 120:543–556

Vitale Brovarone A, Picatto M, Beyssac O, Lagabrielle Y, Castelli D (2014) Blueschist-eclogite transition in the Alpine chain: P-T paths and the role of slow-spreading extensional structures in the evolution of HP-LT mountain belts. Tectonophysics 615–616:96–121

von Huene R, Scholl DW (1991) Observations at convergent margins concerning sediment subduction, subduction erosion, and the growth of continental crust. Rev Geophys 29:279–316.

Wahrhaftig CW (1984) Structure of the Marin Headlands Block, California: A progress report. In: Blake MC Jr (ed) Franciscan Geology of Northern California, vol 43, Pacific Section, Society of Economic Paleontologists and Mineralogists., pp 31–50

Wakabayashi J (1990) Counterclockwise P-T-t paths of amphibolites, Franciscan Complex, California: Metamorphism during the early stages of subduction. J Geol 98:657–680

Wakabayashi J (1999) Distribution of displacement on, and evolution of, a young transform fault system: The northern San Andreas fault system, California. Tectonics 18:1245–1274

Wakabayashi J (2004) Contrasting settings of serpentinite bodies, San Francisco Bay area, California: Derivation from the subducting plate vs. mantle hanging wall. Int Geol Rev 46:1103–1118

Wakabayashi J (2012) Subducted sedimentary serpentinite mélanges: Record of multiple burial-exhumation cycles and subduction erosion. Tectonophysics 568–569:230–247

Wakabayashi J (2015) Anatomy of a subduction complex: Architecture of the Franciscan Complex, California at multiple length and time scales. Int Geol Rev 57:669–746

Wakabayashi J (2016a) Strain localization and megathrust slip accommodation modes in the Franciscan Complex, California. Geological Society of America Abstracts with Programs 48, no. 7, doi: 10.1130/abs/2016AM-278951

Wakabayashi J (2016b) Sedimentary serpentinite and chaotic units of the lower Great Valley Group forearc basin deposits, California: updates on distribution and characteristics. International Geology Review.doi 10.10180/00206814.2016. 1219679.

Wakabayashi J., in press, Serpentinites and serpentinites: Variety of origins and emplacement mechanisms of serpentinite bodies in the California Cordillera. Island Arc. doi: 10.1111/iar. 12205

Wakabayashi J, Dumitru TA (2007) 40Ar/39Ar ages from coherent high-pressure metamorphic rocks of the Franciscan Complex, California: Revisiting the timing of metamorphism of the world's type subduction complex. Int Geol Rev 49:873–906

Wakabayashi J, Unruh JR (1995) Tectonic wedging, blueschist metamorphism, and exposure of blueschist: are they compatible? Geology 23:85–88

Wakabayashi J, Ghatak A, Basu AR (2010) Tectonic setting of supra subduction zone ophiolite generation and subduction initiation as revealed through geochemistry and regional field relationships. Geol Soc Am Bull 122:1548–1568

Wakabayashi J, Rowe C (2015) Whither the megathrust? Localization of large-scale subduction slip along a contact of a mélange. Int Geol Rev 57:854–870.

Wakita K (2015) OPS mélange: A new term for mélange of convergent margins of the world. Int Geol Rev 57:529–539

Wakita K, Metcalfe I (2005) Ocean plate stratigraphy in East and Southeast Asia. Journal of Asian Earth Sciences 24:679–702

Wentworth MC, Blake MC Jr, McLaughlin RJ, Graymer RW (1999) Preliminary geologic map of the San Jose 30 x 60 minute quadrangle, California. In: United States Geological Survey Open File Report., pp 98–795

Worrall DM (1981) Imbricate low-angle faulting in uppermost Franciscan rocks, south Yolla Bolly area, northern California. Geological Society of America Bulletin 92:703–729

5

Construction of perfectly continuous records of physical properties for dark-light sediment sequences collected from the Japan Sea during Integrated Ocean Drilling Program Expedition 346 and their potential utilities as paleoceanographic studies

Tomohisa Irino[1*], Ryuji Tada[2], Ken Ikehara[3], Takuya Sagawa[4], Akinori Karasuda[2], Shunsuke Kurokawa[2], Arisa Seki[2] and Song Lu[5]

Abstract

Establishment of sedimentary sequence is essential for the interpretation of the sedimentary records. Integrated Ocean Drilling Program (IODP) Expedition (Exp.) 346 drilled high-quality sediment archives with more than three holes each at nine sites that enable us to establish continuous sedimentary sequences through splicing technique. After extensive efforts to replace the disturbed, missed, or duplicated intervals found in spliced sediment sequences constructed onboard during IODP Exp. 346 to the corresponding undisturbed intervals, nearly perfectly continuous sediment columns and physical property records for the Pleistocene intervals of Sites U1422-U1427 and U1430 collected from the Japan Sea have been established. Same kind of efforts to correct the shipboard splices were also achieved for Plio-Miocene intervals at Sites U1425 and U1430. The continuous records established for these sites will allow a much more detailed understanding of the long-term variability in East Asian paleoclimate and North Pacific oceanography.

Keywords: IODP Expedition 346, The Japan Sea, Hemipelagic sediment, Stratigraphy, Splice, Composite depth, Visual core description, Physical property, Dark-light cycles

Introduction

Establishment of a continuous sedimentary sequence is an unavoidable starting point for the interpretation of the sedimentary records as events in earth history (Doyle and Bennett 1998). Multi-hole drilling has been widely used by scientific drilling operations such as Ocean Drilling Program (ODP) and Integrated Ocean Drilling Program (IODP) that allow us to establish continuous sedimentary records through splicing fragmented intervals from top to bottom of the sequence (Ruddiman et al. 1987). Splicing is usually performed onboard using physical properties such as magnetic susceptibility (MS), gamma-ray attenuation (GRA) density, natural gamma-ray radiation (NGR), and the lightness and color of sediments measured immediately after drilling. These physical properties are compared between the drilled holes, and inter-hole correlations are recognized according to the fluctuation patterns of these properties.

Revisions of spliced sequences after the cruise have been also common for the high-resolution paleoceanographic researches, which were required for a precise orbital tuning and inter-site correlations. The splice for ODP Site 964 (Emeis et al. 1996) was fully reassessed using core photograph and shipboard color data and revised by Sakamoto et al. (1998). This revision was further modified by choosing better intervals in different

* Correspondence: irino@ees.hokudai.ac.jp
[1]Faculty of Environmental Earth Science, Hokkaido University, N10-W5 Kita-ku, Sapporo 060-0810, Japan
Full list of author information is available at the end of the article

holes (Lourens 2004). The original splice for ODP Site 982 (Jansen et al. 1996) was also revised by Khélifi et al. (2012) through inspection of magnetostratigraphic reversals and the addition of epibenthic $\delta^{18}O$ records although Lawrence et al. (2013) recommended to return to the original splice based on the re-examination of a detailed shore-based magnetostratigraphy (Channell and Guyodo, 2004). Splices for IODP Expedition (Exp.) 320 (Lyle et al., 2009) were revised in the Expedition research results and summarized by Westerhold et al. (2012). Although these works gave their own revised splice interval tables that describe the stratigraphic positions of top and bottom of chosen splice intervals as well as offsets necessary for adjusting cores to the revised depth, it was not easy to find correspondence between the original depth scale and the revised one. In the case of ODP Site 1092 (Gersonde et al. 1999), the shipboard splice was revised using X-ray fluorescence (XRF) scanner data collected after the cruise to improve inter-hole correlation for intervals with too uniform physical properties (Evans et al. 2004). In this example, differences between the original composite depth and revised one were listed in the table that may be helpful for the calculation of the revised depth. In all the examples above, careful visual inspections of core images and/or higher resolution measurements of sediment properties were essential for greater improvement of splices.

IODP Exp. 346 also adopted multi-hole drilling where high-quality splicing of sedimentary sequences at each drilled site has been performed (Tada et al. 2015). The aims of this expedition were to test the hypothesis that Pliocene–Pleistocene uplift of the Himalaya and Tibetan Plateau and the consequent emergence of the two discrete modes of Westerly Jet circulation could have caused the amplification of millennial-scale variability of the East Asian summer monsoon and East Asian winter monsoon, which could also provide teleconnection mechanisms for Dansgaard–Oeschger cycles. Multiple time scales were targeted to reconstruct the onset and evolution of orbital- and millennial-scale variations of summer and winter monsoons, Westerly Jet position and intensity, desertification in East and Central Asia, and their interrelationships since the middle Miocene. In order to accomplish these initial purposes of this expedition, establishment of sediment record as continuous as possible at each site was strongly required. Such continuous sediment records were also necessary to conduct precise inter-site stratigraphic correlations in millennial to orbital time scales to reconstruct spatio-temporal variability of paleoceanographic and paleoclimatic conditions in this region (Tada et al. 2018).

Since the splicing of cores collected during Exp. 346 was conducted by the correlations of physical properties among the drilled holes, minor sediment disturbance that can be only recognized by a careful visual inspection would be missed. In order to establish a more complete and continuous spliced section and profiles of physical properties of sediment core, we tried to identify minor disturbed, missing, or overlapped intervals unconsciously involved in the spliced sequences and replace such intervals with the corresponding undisturbed intervals from another hole. We have also taken care of traceability of processes of stratigraphic revisions to keep compatibility between the newly revised depth scale and the original one. Accomplishment of these tasks enabled the research community to access, examine, and utilize our revised stratigraphic frameworks for the most of the sediment cores collected in the Japan Sea during this expedition.

Materials and data

Drilled sites

The Japan Sea is a marginal sea located at the eastern edge of the far-east Asia (Fig. 1). The sea is connected to the outside through the Tsushima, Tsugaru, Soya, and Mamiya Straits from south to north. The Tsushima Warm Current (TWC) is an only ocean current flowing into the Japan Sea from the outside in modern times. TWC water flows out of the sea mostly from the Tsugaru Strait, and the remaining water flows further north to the Soya Strait (Hase et al. 1999). Salinity and nutrient in the TWC is controlled by the circulation in the East China Sea and the Yangtze discharge which could be controlled by precipitation over South China during the East Asian summer monsoon season. The deep water circulation in the Japan Sea is characterized by intense ventilation due to severe cooling during winter monsoon season, which promotes the hyper-oxic deep water condition and shallow carbonate compensation depth of the sea (Gamo et al. 2014). The Japan Sea is also located at downwind of the Asian dust sources and receives a significant amount of dust transported by westerlies in spring or northwestern monsoon wind in winter (Irino and Tada 2000, 2002). Therefore, the reconstructions of the spatio-temporal variation for fluxes of organic and carbonate carbon, bottom water redox condition, and dust from sediment archives in the Japan Sea are essential to understand the past variability of the East Asian summer and winter monsoon as well as the strength and latitudinal position of westerlies since the middle Miocene (Tada et al. 2015).

We tried to revise the original spliced sequences constructed at seven sites drilled during Exp. 346 such as Sites U1422-U1427 and U1430 (Fig. 1). Sediments mainly consist of silty clay to clayey silt with various amounts of biogenic silica and minor carbonate (Tada et al. 2015). Sites U1422-U1426 and U1430 are located at bathyal depths and characterized by decimeter to meter scale dark-light bandings which correspond to Dansgaard–Oeschger cycles

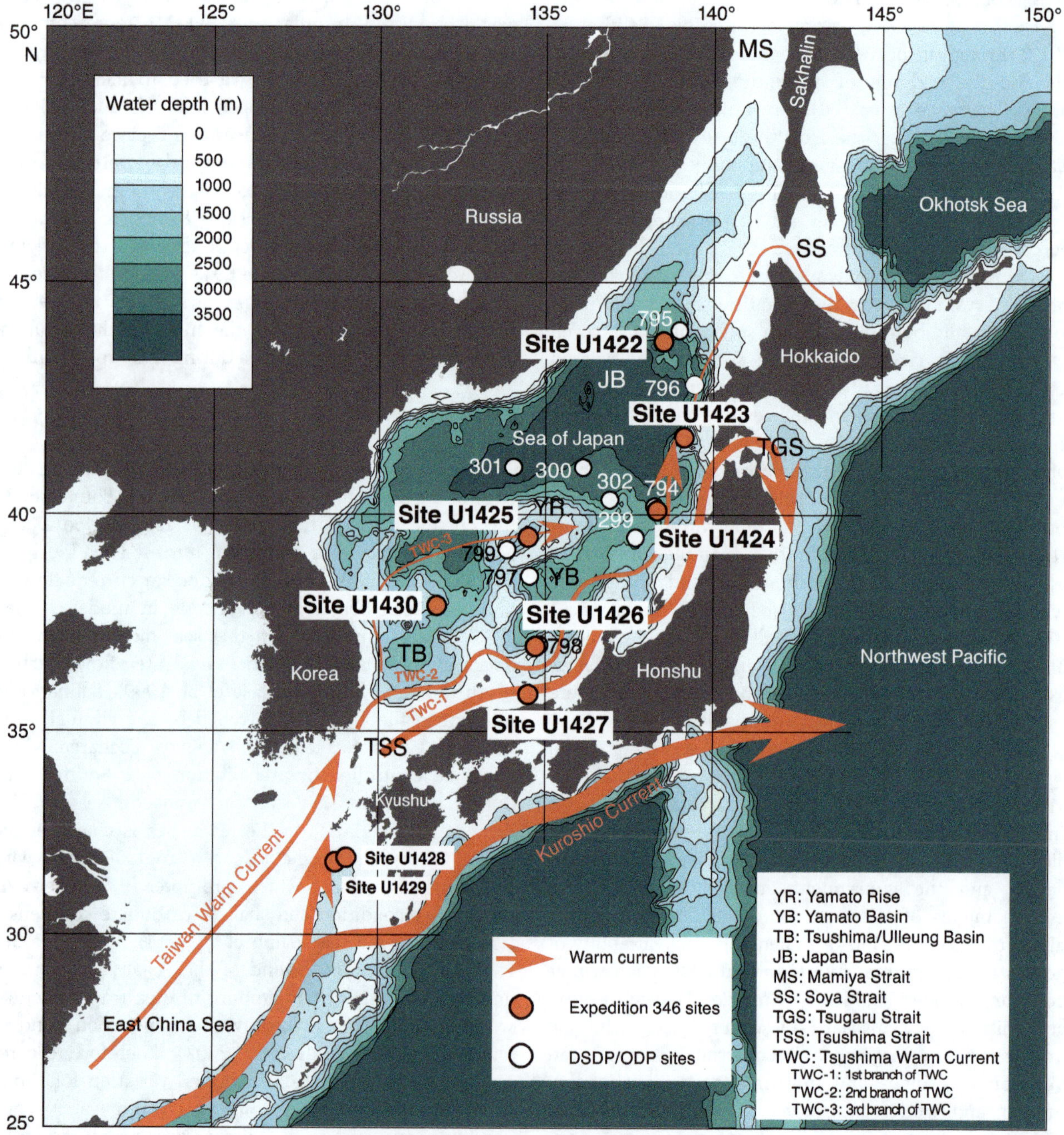

Fig. 1 Locality map of drilled sites during IODP Exp. 346 (modified from Tada et al. (2015))

(Tada et al. 1999), while Site U1427 is located at the upper slope and its sediments consist of homogenous or bioturbated clayey silt (Fig. 1). Water depths are, in ascending order, 330 m at Site U1427, 903 m at Site U1426, 1072 m at Site U1430, 1785 m at Site U1423, 1909 m at Site U1425, 2808 m at Site U1424, and 3429 m at Site U1422 (Tada et al. 2015). Therefore, an appropriate combination of these cores could provide a suitable set of depth transect to reconstruct the depth gradient of biogenic fluxes and deep water ventilation in the past.

Collected data

Depth information and physical property data collected during Exp. 346 and used in this study are all downloadable from the Laboratory Information Management System (LIMS) database (http://web.iodp.tamu.edu/LORE/). Depth scales used during the cruise were called CSF-A, CCSF-A, and CCSF-D. CSF-A is the depth (m) drilled from the sea floor regardless of potential gaps and overlaps between cores. Since more than one hole were drilled at one site during Exp. 346, the core gaps recognized at one hole could be fulfilled with the corresponding interval found at the other hole. In order to fill the core gap, offset values were added to the original drilled depth (CSF-A) of each core as to align the correlated horizons at the same depth, and then, a composite depth scale (CCSF-A) was defined for each core. The list of the offset values for adjustment of each core was called "affine table" which was also downloadable from LIMS database. A single continuous sediment sequence was constructed by splicing continuous intervals chosen from the multi-hole sequences, whose depth scale was called CCSF-D. The term CCSF-D is only used for splice and is same as CCSF-A for the intervals involved in the splice.

Physical properties utilized to establish composite depth scale at each site were MS, GRA density, NGR as well as red-green-blue (RGB) channel values extracted from digital images and reflectance spectroscopy and colorimetry (RSC). During Exp. 346, measurement intervals were routinely set as 2.5 cm for MS and GRA, 20 cm for NGR, and 1 cm for RSC. RSC was not measured for holes U1425E and U1427C due to time limitations. RGB values were routinely extracted onboard for every 5 mm interval and uploaded to LIMS database. These RGB values extracted onboard were different from those recorded in digital image of each section due to encoding procedure (Ebner 2007). Cores U1424C-1H through 3H and U1425C-1H through 3H which were recovered for optically stimulated luminescence (OSL) sampling in the dark further required collection of MS, GRA, photographs (RGB), and RSC later in the Kochi Core Center (KCC). NGR was not measured for these cores (Tada et al. 2015).

Methods/experimental

Strategy

The strategy here was to establish continuous sedimentary records for lithological subunits IA and IB intervals which were characterized by clear dark-light color banding at Sites U1422-U1426 and U1430 (Tada et al. 2015). We have also made same efforts for the intervals below 110 m CCSF-D of Site U1425, below 80 m CCSF-D of Site U1430, and 0 to 440 m CCSF-D of Site U1427 where banded stratifications were not necessarily clear, in order to establish precise age models for the Mio-Pliocene intervals at Sites U1425 and U1430 and for the whole spliced intervals at Site U1427 at a shallower water depth (Sagawa et al. 2018). As a result, we have worked out for revised stratigraphy down to 100 m CCSF-D for U1422, 140 m CCSF-D for U1423, 100 m CCSF-D for U1424, 350 m CCSF-D for U1425, 235 m CCSF-D for U1426, 440 m CCSF-D for U1427, and 260 m CCSF-D for U1430.

In this study, we aim to keep the correspondence between the original shipboard splice depth (m CCSF-D) and the newly revised splice depth scale compatible. The revised stratigraphic positions should be easily recognized, and the conversion from the revised depth scale to the shipboard original depth scale should be possible for everyone. In order to realize this policy, the workflow consisting of four steps is as follows:

1. Align all the core/section photographs in a same scale based on the affine table,
2. Identify disturbed or missing intervals involved in the spliced section,
3. Find the corresponding undisturbed intervals from another hole as the alternative "patch," and
4. Replace the disturbed or missing interval with this "patch,"

through careful inspections of core-section photographs available from IODP LIMS database. When aligning and connecting core-section photographs, the section widths were expanded and brightness and contrast were adjusted to easily recognize the sediment qualities such as lithology, stratification, and disturbance on the sediment split surface (Additional file 1: Figure S1; Additional file 2: Figure S2; Additional file 3: Figure S3; Additional file 4: Figure S4; Additional file 5: Figure S5; Additional file 6: Figure S6; Additional file 7: Figure S7).

On the inter-hole correlation photographs in Additional file 1: Figure S1, Additional file 2: Figure S2, Additional file 3: Figure S3, Additional file 4: Figure S4, Additional file 5: Figure S5; Additional file 6: Figure S6, and Additional file 7: Figure S7, originally spliced intervals are indicated by blue lines next to core photographs. Disturbed or missing intervals found at step 2 are indicated by blue broken lines. These disturbed or missing intervals were amended by the corresponding undisturbed "patch" intervals indicated by pink lines.

The top and bottom depth of disturbed intervals found at step 2 and "patch" intervals chosen at step 3 were defined in the precision of ± 5 mm. The correspondence between the disturbed intervals and the "patch" intervals are summarized as tables for patched intervals (Additional file 8). Patching may change the original composite depth (m CCSF-D) due to the difference of thickness between the disturbed and patch intervals. Therefore, the newly revised depth scale was defined as m CCSF-D_Patched_revYYYY MMDD where YYYYMMDD was the final date of the revision (Additional file 8). We also provide the revised splice interval table for each site (Additional file 9) defined as the patched sequence.

Patching physical properties data

Construction of the patched stratigraphic column resulted in the changes of depth values from the original splice "CCSF-D." The new "CCSF-D_Patched_revYYYYMMDD" should be also applied to the spliced records of physical properties such as MS, GRA, NGR, RGB, and RSC in order to obtain their nearly complete depth profiles. For this purpose, we conducted the following:

1. Translation of the splice (CCSF-D) depth-scale to the "CCSF-D_Pathced_revYYYYMMDD" depth-scale for each physical property data.
2. Replacement of the physical property data in disturbed intervals with those from the undisturbed patch intervals
3. Removal of outliers due to tephra, locally concentrated pyrites, or core disturbance such as core breaks, cracks, and styrofoam.

Then, we obtained a "patched and cleaned" profiles of physical properties that are provided in Additional file 10: Table S1, Additional file 11: Table S2, Additional file 12: Table S3, Additional file 13: Table S4; Additional file 14: Table S5, Additional file 15: Table S6, and Additional file 16: Table S7, and the examples using RGB profile are shown in Fig. 2a and b, Fig 2c and d, Fig 2e and f with "patched" sediment columns.

Compatibility of photographic RGB and RSC data

Color and brightness of the Japan Sea sediment has been treated as an essential property because the dark-light color cycles have been well-correlated basinwide (Tada 1994). Therefore, establishing completely continuous color profiles is essential for correlating between drill sites in the Japan Sea (Tada et al. 2018). When color data collected from core-sections such as photographic RGB values and RSC (L*, a*, b*, and tristimulus XYZ values) in disturbed intervals are patched by the data from another hole, it turned out that we need proper inter-calibrations among RGB values collected onboard, RSC data, and RGB values extracted from section images on LIMS. Sections subjected to OSL sampling in the dark further required collection of physical properties and color data later in the KCC (Tada et al. 2015). Therefore, additional calibrations were also necessary to use the data for RGB and RSC taken from Cores U1424C-1H, 2H, and 3H and U1425C-1H, 2H, and 3H. Inter-calibration between RGB and RSC is also necessary for holes U1425E and U1427C which were additionally drilled at the last moment of Exp. 346, and RSC data were not measured due to time limitations (Tada et al. 2015). Therefore, some special remarks related to color data handlings are described below.

Depth adjustment for U1424B-2H-2 RGB data

The assigned depth scale of RGB value for Section U1424B-2H-2 is 6.5 cm shallower in LIMS database because the top 6.5 cm interval of this section was not included in the cropped digital image due to an error during the operation. Therefore, we added 0.065 m to the depth of RGB values for Section U1424B-2H-2. The missing data could be fulfilled with RGB values extracted from the top 6.5 cm interval of the uncropped digital images stored in LIMS database.

Inter-calibration among RGB values, RSC data, and digital images collected onboard

RSC data on LIMS database include tristimulus X, Y, and Z that are used to calculate L*, a*, and b*. Inter-relationship between tristimulus XYZ and RGB (defined from 0 to 100) is defined under D65 illuminant and 2° observer:

$$R = 3.2404542\mathrm{X} - 1.5371385\mathrm{Y} - 0.4985314\mathrm{Z}$$

$$G = -0.9692660\mathrm{X} + 1.8760108\mathrm{Y} + 0.0415560\mathrm{Z}$$

$$B = 0.0556434\mathrm{X} - 0.2040259\mathrm{Y} + 1.0572252\mathrm{Z}$$

which means $R = G = B = 100$ when $X = 95.047$, $Y = 100$, and $Z = 108.883$ where D65 illuminant is reflected on a perfect white (International Electrotechnical Commission, 1999). These "true" RGB (linear for our eyes) are encoded in color digital photograph as its own RGB color space such as sRGB in the jpeg section images on LIMS.

Let the encoded RGB values calculated from raw digital images on D/V JOIDES Resolution as $\mathrm{R_{JR}}$, $\mathrm{G_{JR}}$, and $\mathrm{B_{JR}}$ and those extracted from the jpeg section images on LIMS as $\mathrm{R_{photo}}$, $\mathrm{G_{photo}}$, and $\mathrm{B_{photo}}$, where these encoded RGB are defined from 0 to 255 (8 bit). Transformations between "true" RGB and these encoded RGBs can be conducted with equations:

$$\frac{R}{100} = \left(\frac{R_{JR}}{255}\right)^{\gamma_R^{JR}}$$

$$\frac{G}{100} = \left(\frac{G_{JR}}{255}\right)^{\gamma_G^{JR}}$$

$$\frac{B}{100} = \left(\frac{B_{JR}}{255}\right)^{\gamma_B^{JR}}$$

and

$$\frac{R}{100} = \left(\frac{R_{\mathrm{photo}}}{255}\right)^{\gamma_R^{\mathrm{photo}}}$$

$$\frac{G}{100} = \left(\frac{G_{\mathrm{photo}}}{255}\right)^{\gamma_G^{\mathrm{photo}}}$$

$$\frac{B}{100} = \left(\frac{B_{\mathrm{photo}}}{255}\right)^{\gamma_B^{\mathrm{photo}}}$$

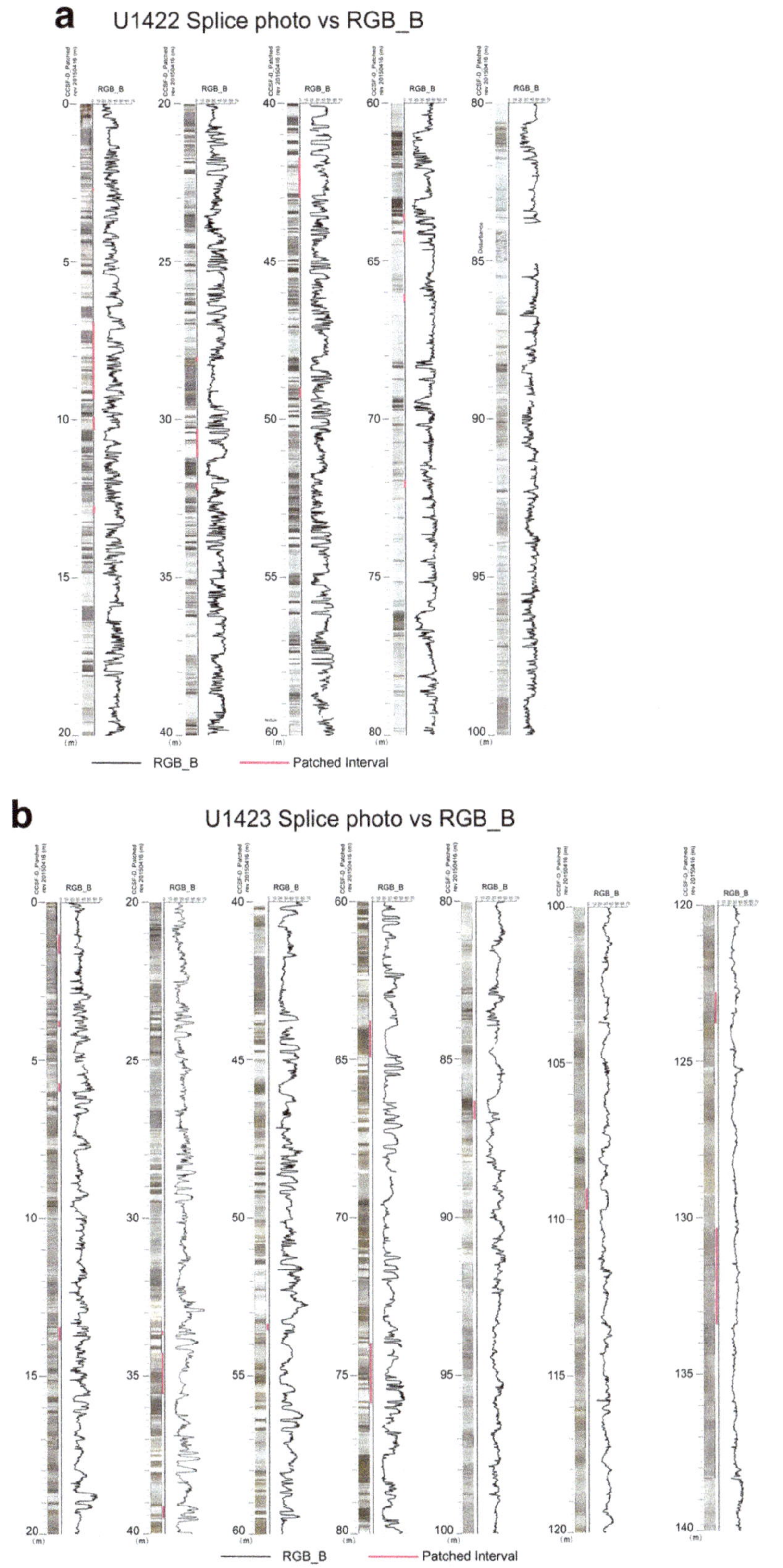

Fig. 2 Patched splices with RGB_B (brightness) profiles for IODP Exp. 346 Sites U1422 (**a**), U1423 (**b**), U1424 (**c**), U1425 (**d**), U1426 (**e**), and U1430 (**f**). Patched intervals are indicated by pink lines. For the full size figure 2 please see Additional file 24

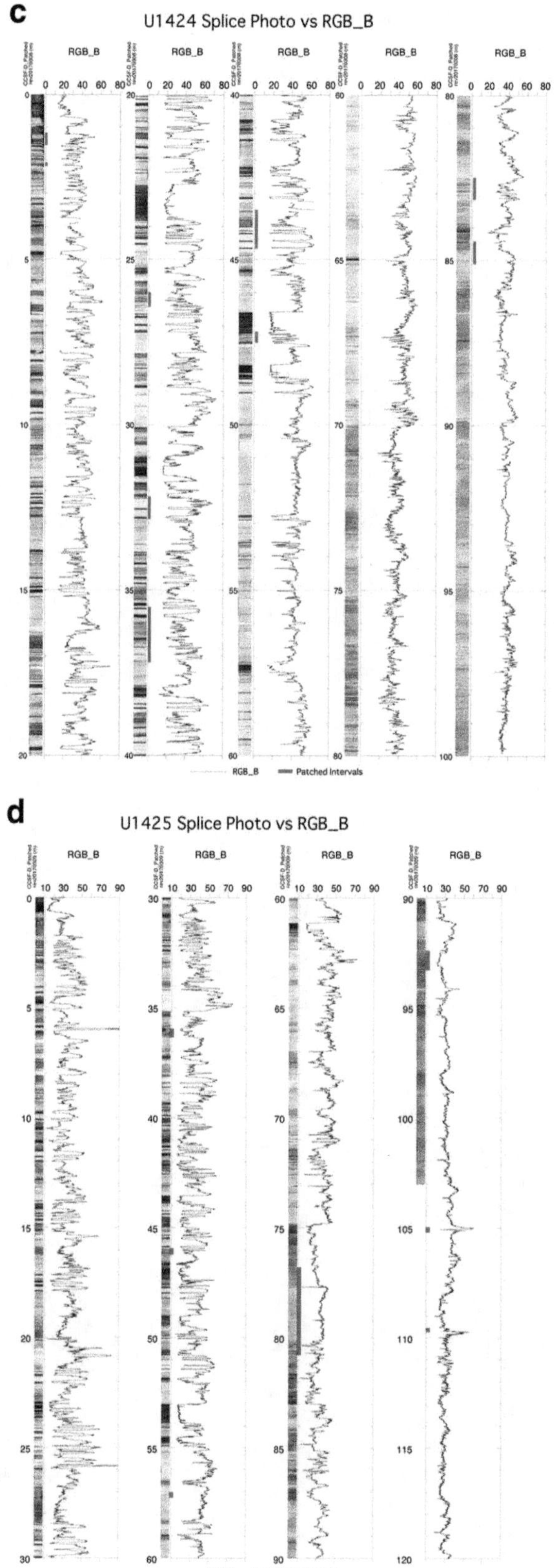

Fig. 2 Patched splices with RGB_B (brightness) profiles for IODP Exp. 346 Sites U1422 (**a**), U1423 (**b**), U1424 (**c**), U1425 (**d**), U1426 (**e**), and U1430 (**f**). Patched intervals are indicated by pink lines. For the full size figure 2 please see Additional file 24

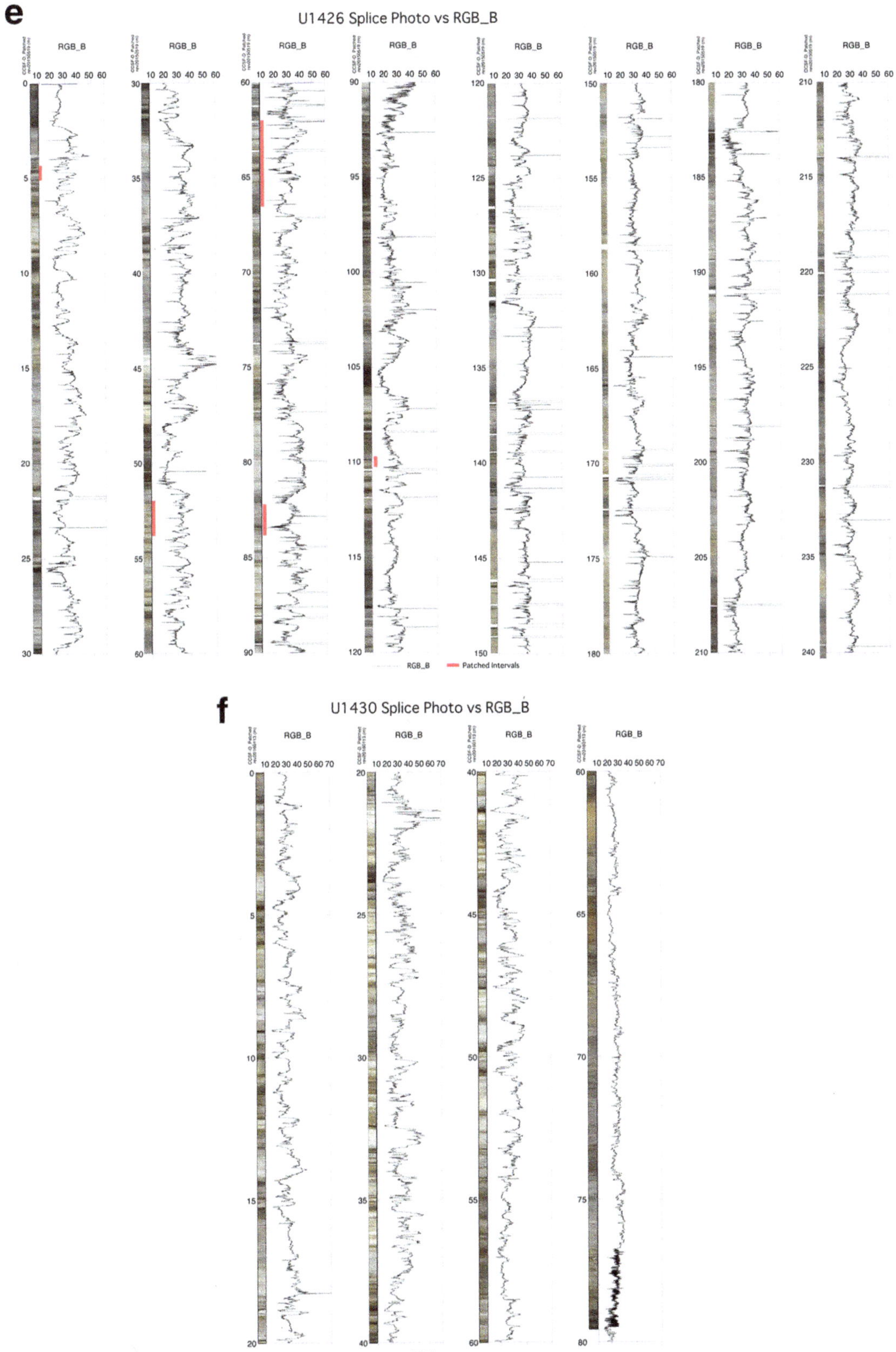

Fig. 2 Patched splices with RGB_B (brightness) profiles for IODP Exp. 346 Sites U1422 (**a**), U1423 (**b**), U1424 (**c**), U1425 (**d**), U1426 (**e**), and U1430 (**f**). Patched intervals are indicated by pink lines. For the full size figure 2 please see Additional file 24

where power terms γ are known as decoding gamma values (Ebner 2007). In the case that we could control the condition of data collection such as illumination and surface condition of sediments for digital image scanning and RSC measurement, these gamma values could be calibrated and defined well before data collection. However, actual scanning and measurement conditions were not consistent because of difference in illumination and use of GladWrap™ for RSC, and therefore, these decoding gamma values have to be estimated through direct comparisons between "true" RGB and the encoded RGB using actual data.

In order to determine practical gamma values, we chose the Core U1424A-1H and compared the RGB calculated from the tristimulus XYZ to the encoded RGB values taken onboard and extracted from the jpeg section images. Figure 3 shows the relationships between the RGB calculated from the tristimulus XYZ and the encoded RGB from digital images. Power function fitting could determine practical gamma values as

$$\gamma_R^{JR} = 1.2961, \gamma_G^{JR} = 1.3107, \gamma_B^{JR} = 1.3320$$

for the RGB collected onboard, and

$$\gamma_R^{\text{photo}} = 1.7748, \gamma_G^{\text{photo}} = 1.7815, \gamma_B^{\text{photo}} = 1.8146$$

for the RGB extracted from the jpeg section images.

Since the tristimulus *XYZ* can be calculated from "true" RGB using equations

$$X = 0.4124564\text{R} + 0.3575761\text{G} + 0.1804375\text{B}$$

$$Y = 0.2126729\text{R} + 0.7151522\text{G} + 0.0721750\text{B}$$

$$Z = 0.0193339\text{R} + 0.1191920\text{G} + 0.9503041\text{B}$$

we could transform the encoded RGB to XYZ using gamma values determined above. Under D65 illuminant and 2° observer conditions, L*, a*, and b* are defined as (CIE, 2004)

$$\text{L}* = 116\text{Y}' - 16, \text{Y}' = \begin{cases} \left(\frac{Y}{100}\right)^{\frac{1}{3}} \left(\frac{Y}{100} > 0.008856\right) \\ 7.787\left(\frac{Y}{100}\right) + \frac{16}{116} \left(\frac{Y}{100} \leq 0.008856\right) \end{cases}$$

$$\text{a}* = 500\left(\text{X}' - \text{Y}'\right), \text{X}' = \begin{cases} \left(\frac{X}{95.047}\right)^{\frac{1}{3}} \left(\frac{X}{95.047} > 0.008856\right) \\ 7.787\left(\frac{X}{95.047}\right) + \frac{16}{116} \left(\frac{X}{95.047} \leq 0.008856\right) \end{cases}$$

$$\text{b}* = 200\left(\text{Y}' - \text{Z}'\right), \text{Z}' = \begin{cases} \left(\frac{Z}{108.883}\right)^{\frac{1}{3}} \left(\frac{Z}{108.883} > 0.008856\right) \\ 7.787\left(\frac{Z}{108.883}\right) + \frac{16}{116} \left(\frac{Z}{108.883} \leq 0.008856\right) \end{cases}$$

Then, we can inter-calibrate among RGB values collected onboard, RSC, and digital images.

In Fig. 4, blue and orange lines indicate the calculated L*, a*, and b* from RGB taken onboard and extracted from the jpeg section images, respectively, and are compared with the measured RSC data (red dots). The three datasets agree well with one another. RGB and RSC data for lost or disturbed intervals could be amended or replaced using the RGB extracted from the jpeg section images for the corresponding intervals.

For the top 6.5 cm of Section U1424B-2H-2 where digital image was miscropped and RGB was missed, we calculated the RGB values equivalent to those collected onboard using the RGB extracted from the jpeg digital image of this section. For the top 8 cm of Section U1424B-2H-3, a light-colored sediment penetrated into a dark layer at the center of the core due to disturbance and the onboard RGB data were collected on the light color portion (not on the proper dark portion) because the data were routinely taken from the center of the core (shown as dotted lines in Fig. 5). Therefore, we extracted the RGB values from proper dark portion of this interval using the jpeg image, calculated the RGB values equivalent to those taken onboard by the gamma corrections described above (solid lines in Fig. 5), and replaced the data. L*, a*, and b* were replaced with those calculated from the jpeg extracted RGB at the proper dark portion as shown in Fig. 5 where L* was reasonably low for the dark layer in the top 8 cm interval.

Inter-calibration of RGB values and RSC data between JOIDES Resolution (JR) and KCC

For the patched interval from 1.22 to 1.53 m CCSF-D, the corresponding interval from the Core U1424C-1H was used to amend the data. Since physical properties for Cores U1424C-1H, 2H, and 3H were measured at KCC after Exp. 346, the RGB values were extracted from the digital images taken by the core scanner at KCC, and L*, a*, and b* (and tristimulus XYZ) values were transformed from those measured by the spectrophotometer used at KCC. In order to establish inter-calibration between data taken onboard at JR and those from KCC, we first construct peak by peak matching between B splice profile and B values extracted from digital images of Cores U1424C-1H, 2H, and 3H as shown in Fig. 6. Then, RGB and XYZ data for corresponding stratigraphic positions were compared between the splice data taken onboard and U1424C data taken at KCC (Fig. 7). Figure 7a–c shows that RGB_{JR} do not converge to zero when RGB_{KCC} get close to zero. This is probably due to the difference in sensitivity of scanners under very dark condition. Therefore, the transformation of RGB data from KCC values to JR ones requires the form of power function with bias factors *a* as

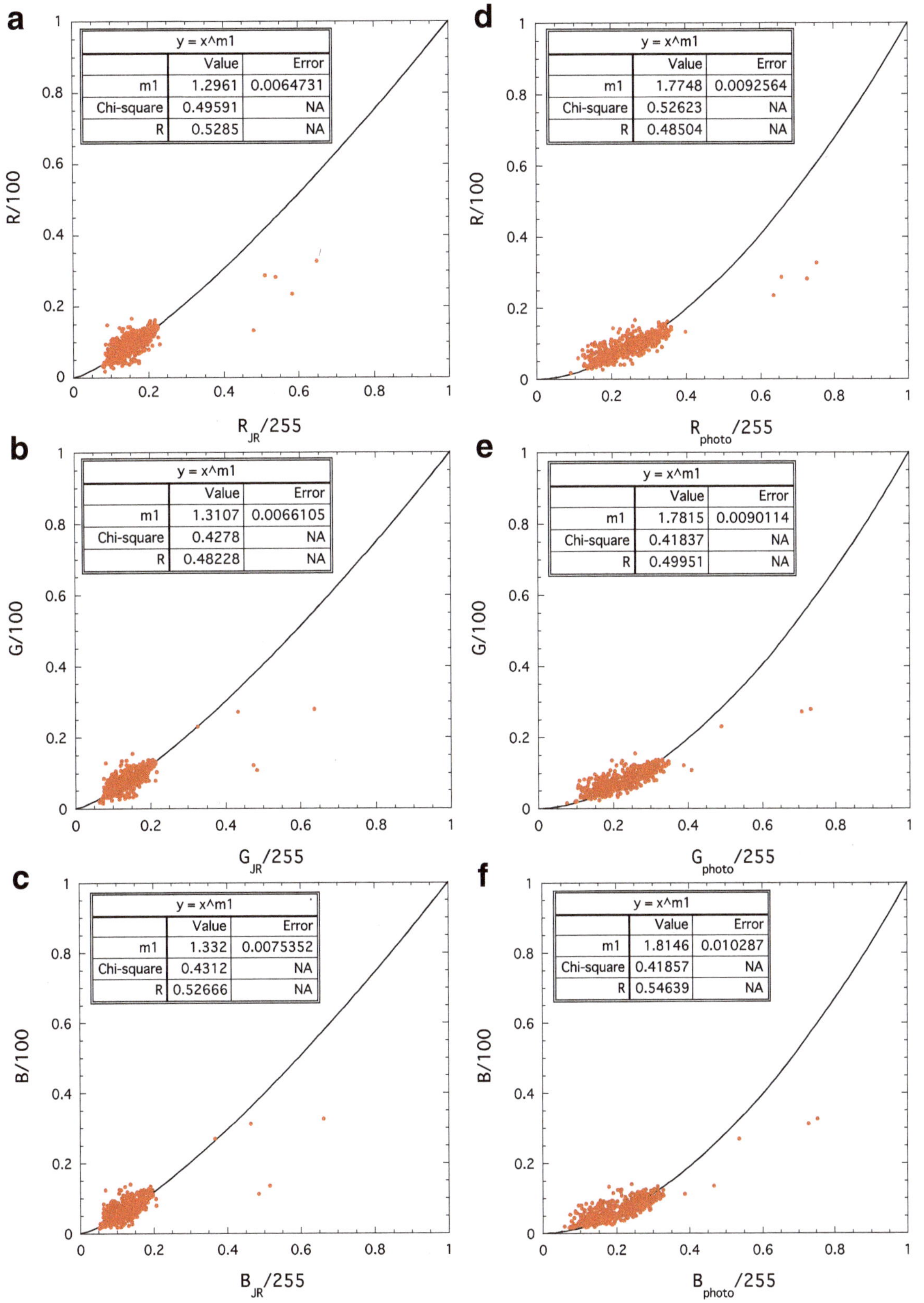

Fig. 3 Relationships between linear RGB estimated from RSC measurements and shipboard R (**a**), G (**b**), and B (**c**) or photo R (**d**), G (**e**), and B (**f**)

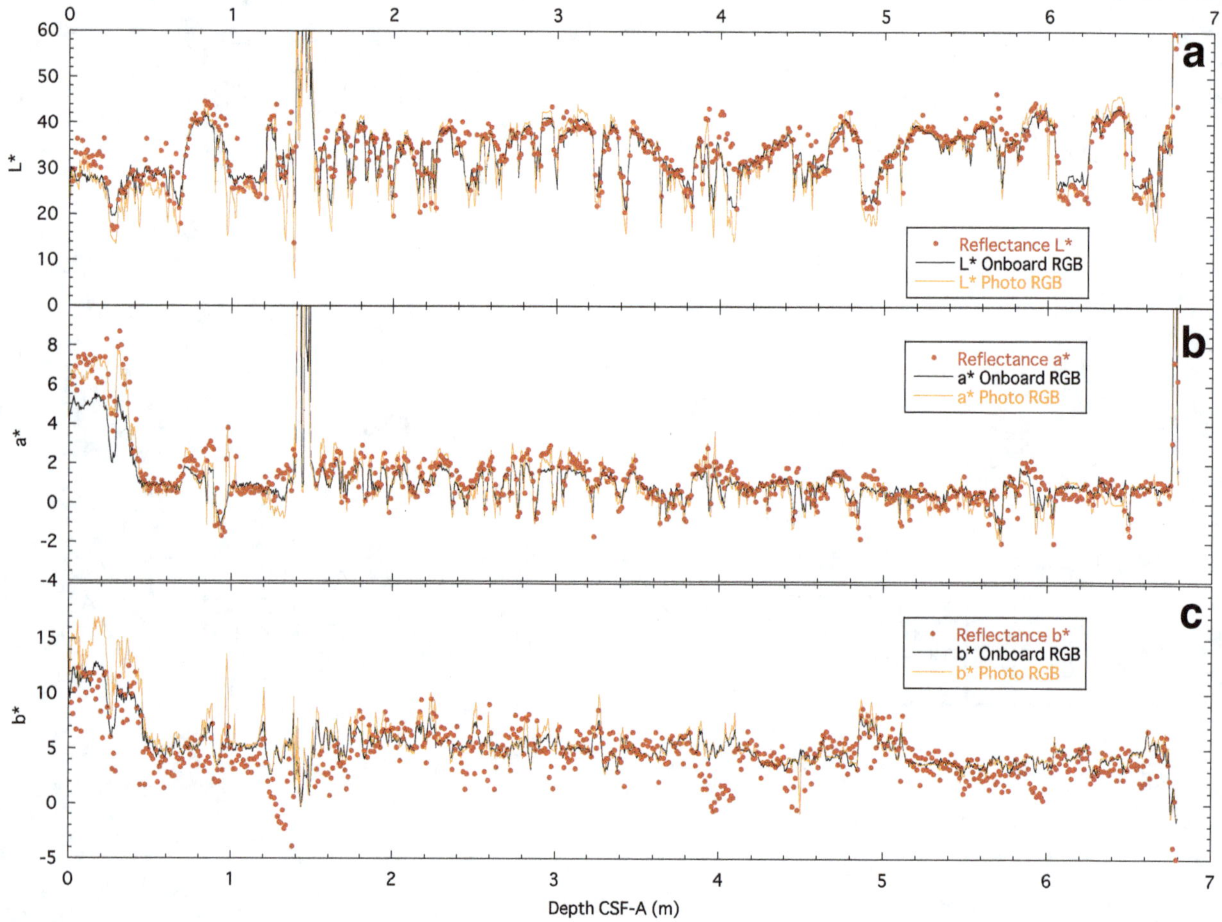

Fig. 4 Comparison between L* (**a**), a* (**b**), and b* (**c**) measured onboard (red dots) and those calculated from shipboard RGB (blue line) and photo RGB (orange line) in IODP Exp. 346 Site U1424A 1H

$$\frac{RGB_{JR}}{255} = a + (1-a)\left(\frac{RGB_{KCC}}{255}\right)^{\gamma}$$

where bias factors a are set as to RGB_{JR} be 255 when RGB_{KCC} is 255. The bias factors a are 0.0849, 0.080895, and 0.080709 for R, G, and B, respectively, while the power terms γ are 1.6572, 1.6212, and 1.5914 for R, G, and B, respectively (Fig. 7a–c).

Tristimulus XYZ are principally consistent between JR and KCC, but transparent film used in KCC was SaranWrap™ while GladWrap™ was used on JR. Therefore, we assumed that small bias factors have to be added to or subtracted from XYZ measured in KCC to match to those on JR such as

$$XYZ_{JR} = XYZ_{KCC} + b$$

The best estimates of this bias factors b are − 5.1763, − 5.6081, and − 5.7142 for X, Y, and Z, respectively (Fig. 7d–f). Negative bias factors suggest stronger surface reflection of SaranWrap™ than that of GladWrap™.

Results and discussions

Overview of patched splices

U1422

Fourteen disturbed intervals were patched (Additional file 8). The total length of the revised spliced sequence was 0.057 m shorter than the original shipboard splice at 72.21 m CCSF-D.

U1423

Fifteen disturbed intervals were patched (Additional file 8). The total length of the revised spliced sequence was 0.584 m longer than the original shipboard splice at 132.83 m CCSF-D.

U1424

Nine disturbed intervals were patched (Additional file 8). The total length of the revised spliced sequence was 1.310 m longer than the original shipboard splice at 83.76 m CCSF-D.

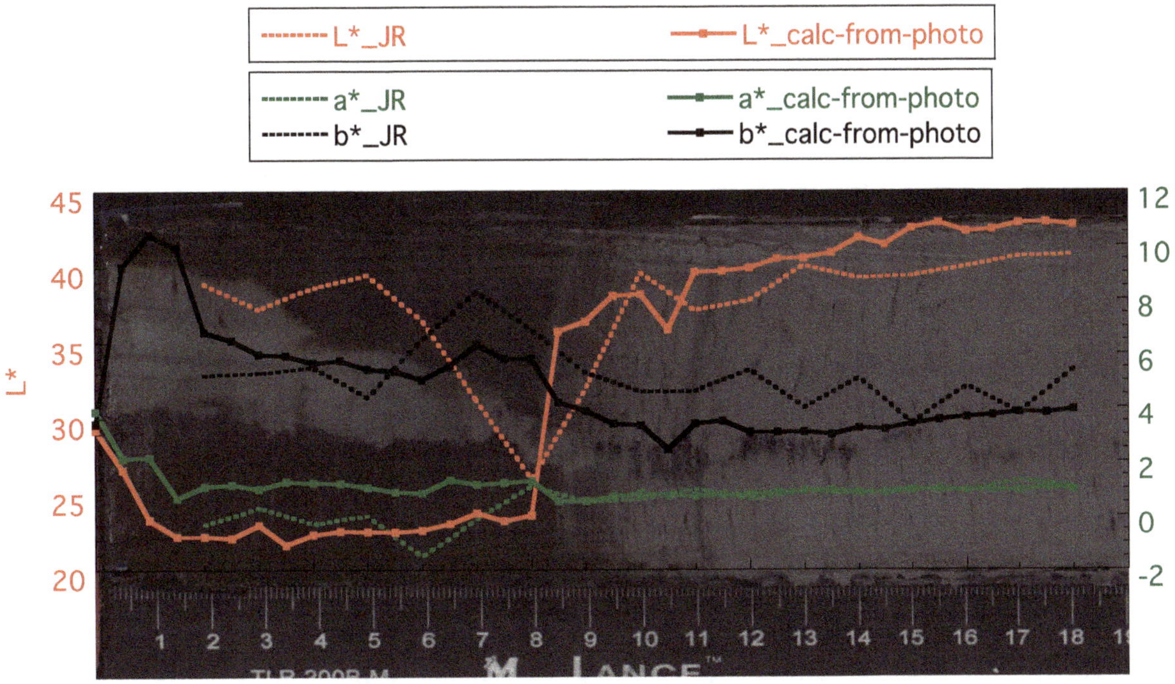

Fig. 5 Example of re-sampling of RGB from section photograph (U1424B-2H-3, top 8 cm) and the calculated L*a*b*

U1425

Twelve disturbed intervals were patched (Additional file 8). The total length of the revised spliced sequence was 0.827 m longer than the original shipboard splice at 320.86 m CCSF-D.

U1426

Five disturbed intervals were patched (Additional file 8). The total length of the revised spliced sequence was 4.212 m longer than the original shipboard splice at 105.95 m CCSF-D.

U1427

Thirteen disturbed intervals were patched (Additional file 8). The total length of the revised spliced sequence was 11.753 m longer than the original shipboard splice at 257.95 m CCSF-D.

U1430

Seven disturbed intervals recognized below 96.15 m CCSF-D were patched (Additional file 8). The total length of the revised spliced sequence was 1.071 m shorter than the original shipboard splice at 257.19 m CCSF-D.

Potential utilities of perfectly continuous stratigraphic records

Sediment sequences characterized by dark-light color cycles exhibit similar patterns of color banding (Fig. 2). After correction, each core-section is nearly continuous such that correlation among these sites using their color banding patterns allows reasonably isochronous horizons to be studied across wide regions of the Japan Sea over millennial time scales (Tada 1994; Tada et al. 1999; Tada et al. 2018). Combination of the reliably continuous stratigraphy at each site and its correlation to the reference site such as U1424 (Tada et al. 2018) enabled more precise tuning age construction at other drilled sites in the Japan Sea.

After inter-site correlations, differences in linear sedimentation rates, GRA densities, and other oceanographic and climatological variability and distribution can be examined for any combination of depth, latitudinal, and longitudinal transects. Further correlation between these dark-light banding sites and Site U1427 with no clear dark-light banding using other correlation tools such as tephrochronology would enable us to examine how the shallower sites such as U1427 respond to millennial-scale climate and oceanographic variability exhibited as dark-light cycles observed at sites with larger water depth (Sagawa et al. 2018).

Maintenance of revised splice records

Perfectly continuous physical properties' data such as GRA, MS, NGR, RGB, and RSC are summarized in Microsoft Excel books and provided here as Additional file 10: Table S1, Additional file 11: Table S2, Additional file 12: Table S3, Additional file 13: Table S4, Additional file 14: Table S5, Additional file 15: Table S6,

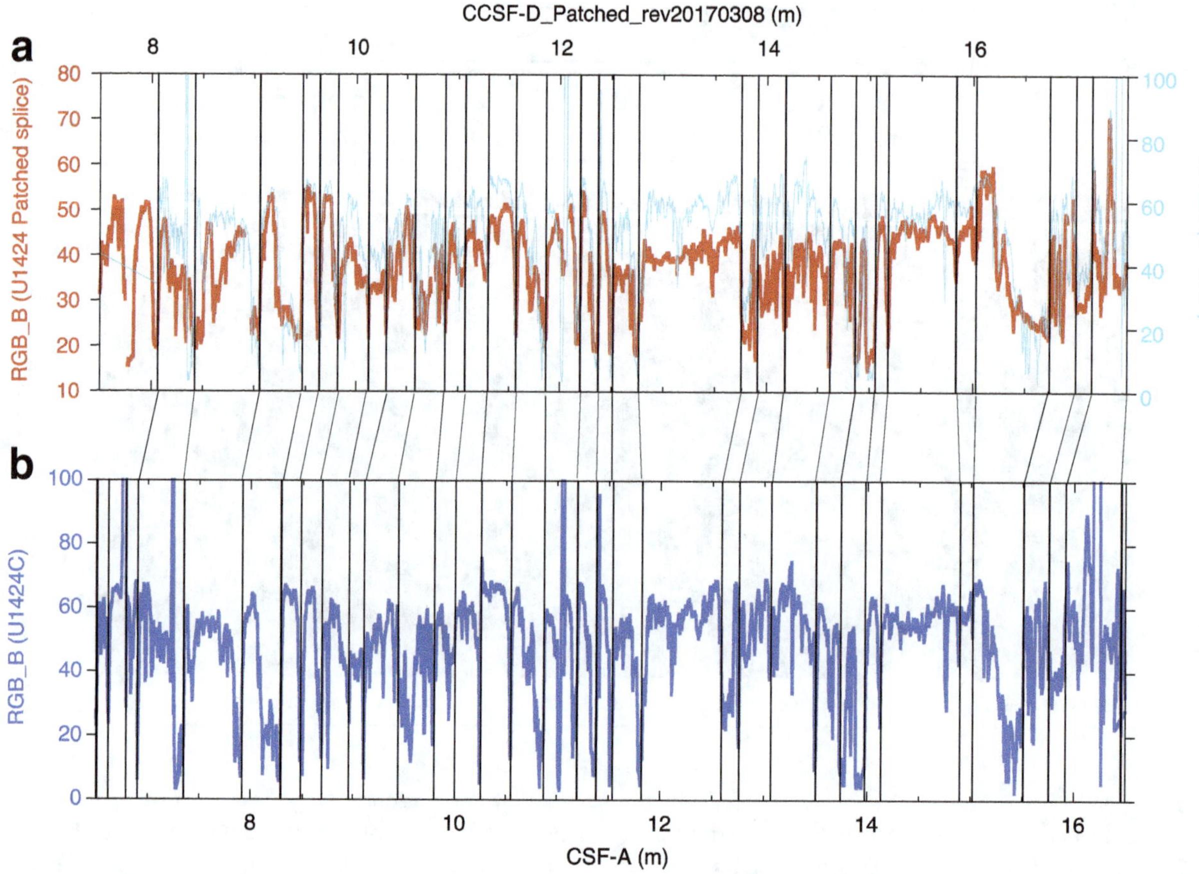

Fig. 6 Correlation of RGB_B profile between U1424 splice measured onboard (**a**) and Section U1424C-2H measured at Kochi Core Center (KCC) (**b**). RGB_B profile of KCC (blue line in **b**) was correlated to RGB_B of the splice (red line in **a**) at black vertical lines and was projected to splice depth scale (light blue line in **a**)

and Additional file 16: Table S7 for further research usage. Microsoft Excel workbooks which automate the calculation of the newly patched depth are also available as Additional file 17: Table S8, Additional file 18: Table S9, Additional file 19: Table S10, Additional file 20: Table S11, Additional file 21: Table S12, Additional file 22: Table S13, and Additional file 23: Table S14. These tables enabled the users to access, examine, and utilize our revised stratigraphic frameworks for the most of the sediment cores collected in the Japan Sea during this expedition. However, the users of these information should notice that the depth scale calculation for the interval out of splice is not necessarily consistent to the spliced section because perfect correlation among holes using dark-light color cycle has not been conducted. The users should check the patched depth on the spliced section corresponding to the horizon out of splice.

Since our policy for the revision of stratigraphic splice is traceability, we put the date of revision to our revised depth scale such as m CCSF-D_Patched_revYYYYMMDD. This means that we are always ready for (potential) further revision of stratigraphic sequences and provide revised "patch" tables to improve the stratigraphy of the drilled cores collected during Exp. 346.

Conclusions

After extensive efforts for improvement of sediment sequences drilled during IODP Exp. 346, nearly perfectly continuous sediment columns and physical property records for Sites U1422-U1427 and U1430 over the Pleistocene have been established. Stratigraphic revisions were also achieved for Plio-Miocene intervals at Sites U1425 and U1430. All the "patching" processes were clearly demonstrated in patch tables in this study to keep compatibility between depth scales of the original and revised splices. Revised depth scales and patched continuous physical property records are also available as additional file tables (Additional file 10: Table S1; Additional file 11: Table S2; Additional file 12: Table S3; Additional file 13: Table S4; Additional file 14: Table S5; Additional file 15: Table S6; Additional file 16: Table S7). Therefore, the users can access, examine, and utilize our

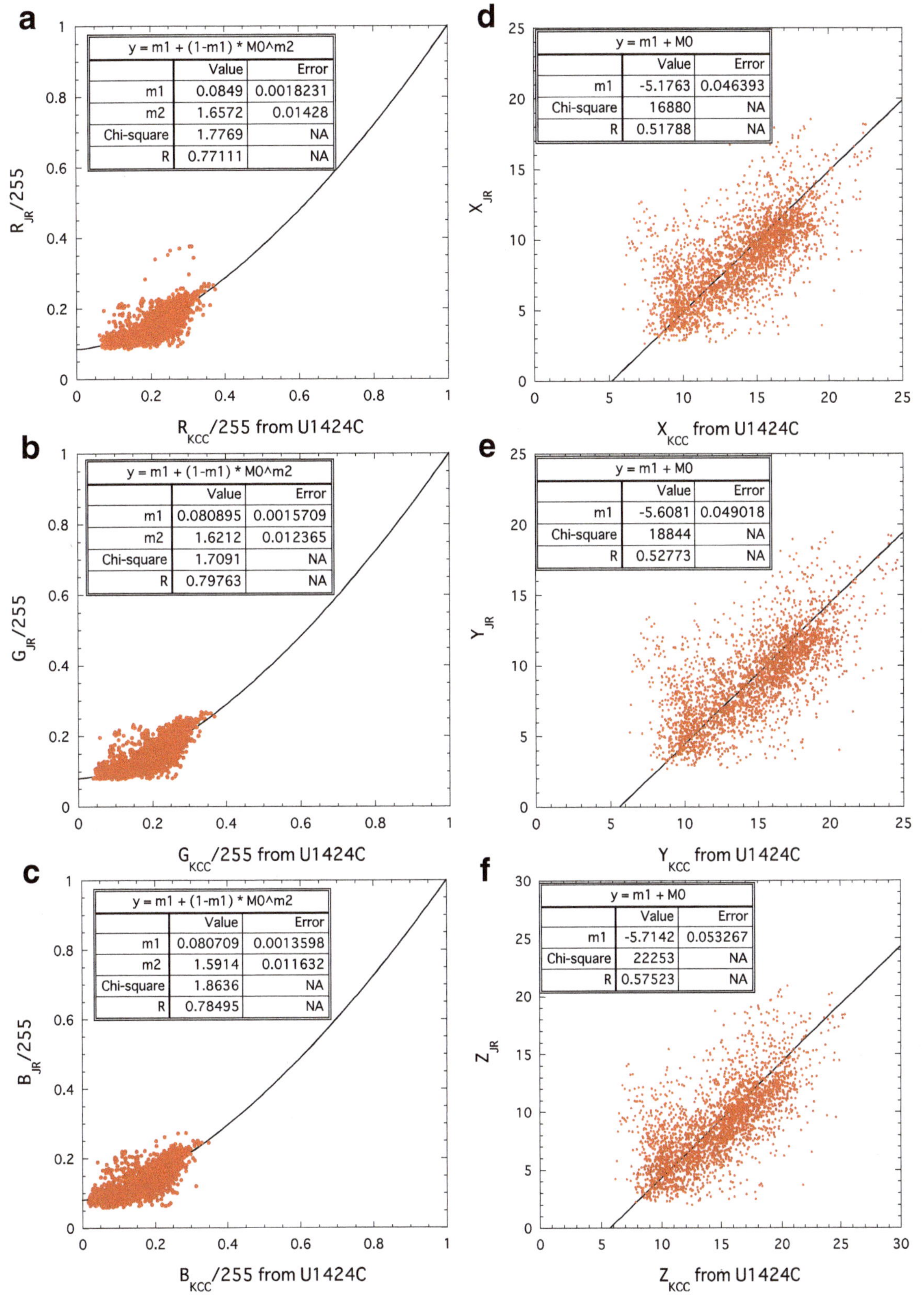

Fig. 7 Relationships between shipboard R (**a**), G (**b**), and B (**c**) and those measured at Kochi Core Center (KCC) and between shipboard X (**d**), Y (**e**), and Z (**f**) and those measured at KCC

revised stratigraphic frameworks. Continuous sedimentary records are exceedingly useful for inter-site correlation and enable the examination of the spatial and temporal distribution of material fluxes across wide areas of the Japan Sea for the entire Pleistocene. The continuous record established at these sites will allow a much more detailed understanding of the long-term variability in East Asian paleoclimate and North Pacific oceanography.

Additional files

Additional file 1: Figure S1. Graphic representations of inter-hole correlations for IODP Exp. 346 Sites U1422. Originally spliced intervals are indicated by blue lines next to core photographs. Disturbed or missing intervals found during visual inspection are indicated by blue broken lines. These disturbed or missing intervals were amended by the corresponding undisturbed "patch" intervals indicated by pink lines.

Additional file 2: Figure S2. Graphic representations of inter-hole correlations for IODP Exp. 346 Sites U1423. Originally spliced intervals are indicated by blue lines next to core photographs. Disturbed or missing intervals found during visual inspection are indicated by blue broken lines. These disturbed or missing intervals were amended by the corresponding undisturbed "patch" intervals indicated by pink lines.

Additional file 3: Figure S3. Graphic representations of inter-hole correlations for IODP Exp. 346 Sites U1424. Originally spliced intervals are indicated by blue lines next to core photographs. Disturbed or missing intervals found during visual inspection are indicated by blue broken lines. These disturbed or missing intervals were amended by the corresponding undisturbed "patch" intervals indicated by pink lines.

Additional file 4: Figure S4. Graphic representations of inter-hole correlations for IODP Exp. 346 Sites U1425. Originally spliced intervals are indicated by blue lines next to core photographs. Disturbed or missing intervals found during visual inspection are indicated by blue broken lines. These disturbed or missing intervals were amended by the corresponding undisturbed "patch" intervals indicated by pink lines.

Additional file 5: Figure S5. Graphic representations of inter-hole correlations for IODP Exp. 346 Sites U1426. Originally spliced intervals are indicated by blue lines next to core photographs. Disturbed or missing intervals found during visual inspection are indicated by blue broken lines. These disturbed or missing intervals were amended by the corresponding undisturbed "patch" intervals indicated by pink lines.

Additional file 6: Figure S6. Graphic representations of inter-hole correlations for IODP Exp. 346 Sites U1427. Originally spliced intervals are indicated by blue lines next to core photographs. Disturbed or missing intervals found during visual inspection are indicated by blue broken lines. These disturbed or missing intervals were amended by the corresponding undisturbed "patch" intervals indicated by pink lines.

Additional file 7: Figure S7. Graphic representations of inter-hole correlations for IODP Exp. 346 Sites U1430. Originally spliced intervals are indicated by blue lines next to core photographs. Disturbed or missing intervals found during visual inspection are indicated by blue broken lines. These disturbed or missing intervals were amended by the corresponding undisturbed "patch" intervals indicated by pink lines.

Additional file 8: Patch tables for IODP Exp. 346 Sites U1422-U1427 and U1430.

Additional file 9: Revised splice interval tables for IODP Exp. 346 Sites U1422-U1427 and U1430.

Additional file 10: Table S1. Physical properties such as MS, GRA, NGR, RGB, and RSC for U1422. Revised depth scale version is CCSF-D_Patched_rev20150416. L*a*b* values calculated from RGB are also given.

Additional file 11: Table S2. Physical properties such as MS, GRA, NGR, RGB, and RSC for U1423. Revised depth scale version is CCSF-D_Patched_rev20150416. L*a*b* values calculated from RGB are also given.

Additional file 12: Table S3. Physical properties such as MS, GRA, NGR, RGB, and RSC for U1424. Revised depth scale version is CCSF-D_Patched_rev20170308. L*a*b* values calculated from RGB are also given.

Additional file 13: Table S4. Physical properties such as MS, GRA, NGR, RGB, and RSC for U1425. Revised depth scale version is CCSF-D_Patched_rev20170309. L*a*b* values calculated from RGB are also given.

Additional file 14: Table S5. Physical properties such as MS, GRA, NGR, RGB, and RSC for U1426. Revised depth scale version is CCSF-D_Patched_rev20150519. L*a*b* values calculated from RGB are also given.

Additional file 15: Table S6. Physical properties such as MS, GRA, NGR, RGB, and RSC for U1427. Revised depth scale version is CCSF-D_Patched_rev20170310. L*a*b* values calculated from RGB are also given.

Additional file 16: Table S7. Physical properties such as MS, GRA, NGR, RGB, and RSC for U1430. Revised depth scale version is CCSF-D_Patched_rev20160113. L*a*b* values calculated from RGB are also given.

Additional file 17: Table S8. Calculation worksheet for U1422 revised depth scale (CCSF-D_Patched_rev20150416).

Additional file 18: Table S9. Calculation worksheet for U1423 revised depth scale (CCSF-D_Patched_rev20150416).

Additional file 19: Table S10. Calculation worksheet for U1424 revised depth scale (CCSF-D_Patched_rev20170308).

Additional file 20: Table S11. Calculation worksheet for U1425 revised depth scale (CCSF-D_Patched_rev20170309).

Additional file 21: Table S12. Calculation worksheet for U1426 revised depth scale (CCSF-D_Patched_rev20150519).

Additional file 22: Table S13. Calculation worksheet for U1427 revised depth scale (CCSF-D_Patched_rev20170310).

Additional file 23: Table S14. Calculation worksheet for U1430 revised depth scale (CCSF-D_Patched_rev20160113).

Additional file 24: Figure S8. Full size version of Figure 2.

Abbreviations

Exp.: Expedition; GRA: Gamma-ray attenuation density; IODP: Integrated Ocean Drilling Program; KCC: Kochi Core Center; LIMS: Laboratory Information Management System; MS: Magnetic susceptibility; NGR: Natural gamma-ray radiation; ODP: Ocean Drilling Program; OSL: Optically stimulated luminescence; RGB: Red-green-blue channel of digital image; RSC: Reflectance spectroscopy and colorimetry; TWC: Tsushima Warm Current

Acknowledgements

We would like to express our sincere thanks to all of the scientists, technicians, and crews of D/V JOIDES resolution during IODP Expedition 346. This work was financially supported by IODP Exp. 346 After Cruise Research Program, JAMSTEC.

Funding

Not applicable.

Authors' contributions

TI supervised each stratigraphic patching task, summarized the patch tables, calibrated and patched physical properties, and established the data share websites for all the drilled sites. RT and KI also supervised all the course of stratigraphic revisions and advised for improvement. TS mainly worked for the stratigraphic revision and figure drawing for U1427. AK and AS compiled core photographs and revised stratigraphy for U1424, U1425, U1426, and

U1430 mainly for Pleistocene intervals. SK worked for the Plio-Miocene intervals for U1425 and U1430. SL exclusively compiled the core photographs and revised the stratigraphy for U1422 and U1423. All authors read and approved the final manuscript.

Authors' information

RT was a co-chief scientist, TI was a stratigraphic correlator, and KI, TS, and AK were sedimentologists onboard for Exp. 346. RT is a supervisor for AK, AS, and SK in Graduate School of Science, University of Tokyo. TI is a supervisor for SL in Graduate School of Environmental Science, Hokkaido University.

Competing interests

The authors declare that they have no competing interest.

Author details

[1]Faculty of Environmental Earth Science, Hokkaido University, N10-W5 Kita-ku, Sapporo 060-0810, Japan. [2]Graduate School of Science, the University of Tokyo, 7-3-1 Hongo, Bunkyo-ku, Tokyo 113-0033, Japan. [3]Geological Survey of Japan, National Institute of Advanced Industrial Science and Technology, Tsukuba Central 7, 1-1-1 Higashi, Tsukuba, Ibaraki 305-8567, Japan. [4]Institute of Science and Engineering, Kanazawa University, Kakuma-machi, Kanazawa, Ishikawa 920-1192, Japan. [5]Graduate School of Environmental Science, Hokkaido University, N10-W5 Kita-ku, Sapporo 060-0810, Japan.

References

Channell JET, Guyodo Y (2004) The Matuyama Chronozone at ODP 982 (Rockall Bank): evidence for decimeter-scale magnetization lock-in depths. In: Channell JET, Kent DV, Lowrie W, Meert J (eds) Timescales of the internal geomagnetic field, vol 145. AGU Geophysical Monograph, Washington, D. C., pp 205–219.

CIE (2004) Colorimetry, vol 15, 3rd edn. Commission internationale de l'Eclairage, CIE Central Bureau, Vienna, p 72.

Doyle P, Bennett MR (eds) (1998) Unlocking the stratigraphic record—advances in modern stratigraphy. Wiley, Chichester.

Ebner M (2007) Color Constancy. Wiley, Chichester.

Emeis K-C et al (1996) Proceedings of the ocean drilling. Program, initial reports, vol 160. Ocean Drill. Program, College Station.

Evans HF, Westerhold T, Channell JET (2004) ODP Site 1092: revised composite depth section has implications for Upper Miocene 'cryptochrons'. Geophys J Int 156:195–99. https://doi.org/10.1111/j.1365-246X.2003.02189.x.

Gamo T, Nakayama N, Takahata N, Sano Y, Zhang J, Yamazaki E, Taniyasu S, Yamashita N (2014) The sea of Japan and its unique chemistry revealed by time-series observations over the last 30 years. Monogr Environ Earth Planets 2(1):1–22. https://doi.org/10.5047/meep.2014.00201.0001.

Gersonde R, Hodell DA, Blum P, Shipboard Scientific Party (1999) Proc. ODP, Init. Repts, vol 177. Ocean Drilling Program, Texas A & M University, USA, pp 1–82.

Hase H, Yoon JH, Koterayama W (1999) The current structure of the Tsushima Warm Current along the Japanese Coast. J Oceanogr 55(2):217–235. https://doi.org/10.1023/A:1007894030095.

International Electrotechnical Commission (1999) IEC 61966-2-1:1999.

Irino T, Tada R (2000) Quantification of aeolian dust (Kosa) contribution to the Japan Sea sediments and its variation during the last 200 ky. Geochem J 34(1):59–93.

Irino T, Tada R (2002) High resolution reconstruction of variation in aeolian dust (Kosa) deposition at ODP site 797, the Japan Sea, during the last 200 ky. Glob Planet Chang 35:107–120.

Jansen E, Raymo ME, Blum P et al (1996). Proc ODP, Init. Repts., 162: College Station, TX (Ocean Drilling Program). https://doi.org/10.2973/odp.proc.ir.162.1996.

Khélifi N, Sarnthein M, Naafs BDA (2012) Technical note: Late Pliocene age control and composite depths at ODP Site 982, revisited. Climate of the Past 8:79–87. https://doi.org/10.5194/cp-8-79-2012.

Lourens LJ (2004) Revised tuning of Ocean Drilling Program Site 964 and KC01B (Mediterranean) and implications for the d18O, tephra, calcareous nannofossil, and geomagnetic reversal chronologies of the past 1.1 Myr. Paleoceanography 19:PA3010. https://doi.org/10.1029/2003PA000997.

Lawrence KT, Bailey I, Raymo ME (2013) Re-evaluation of the age model for North Atlantic Ocean Site 982 – arguments for a return to the original chronology. Climate of the Past 9:2391–97. https://doi.org/10.5194/cp-9-2391-2013.

Lyle, M., Raffi, I., Pälike, H., Nishi, H., Gamage, K., Klaus, A., and the Expedition 320/321 Scientists (2009) Pacific equatorial transect. IODP Prel Rept, 321. https://doi.org/10.2204/ iodp.pr.321.2009.

Ruddiman WF, Cameron D, Clement BM (1987) Sediment disturbance and correlation of offset holes drilled with the hydraulic piston corer—Leg 94. Initial Rep Deep Sea Drill Proj 94:615–634.

Sagawa T, Nagahashi Y, Satoguchi Y, Holbourn A, Itaki T, Gallagher SJ, Saavedra-Pellitero M, Ikehara K, Irino T, Tada R (2018) Integrated tephrostratigraphy and stable isotope stratigraphy in the Japan Sea and East China Sea using IODP Sites U1426, U1427 and U1429, Expedition 346 Asian Monsoon. Prog Earth Planet Sci in press. https://doi.org/10.1186/s40645-018-0168-7.

Sakamoto T, Janecek T, Emeis K-C (1998) Continuous sedimentary sequences from the eastern Mediterranean Sea: composite depth sections. Proc. Ocean Drill. Program, Sci Results 160:37–59.

Tada R (1994) Paleoceanographic evolution of the Japan Sea. Palaeogeogr. Palaeoclimatol Palaeoecol 108(3–4):487–508. https://doi.org/10.1016/0031-0182(94)90248-8.

Tada R, Irino T, Ikehara K, Karasuda A, Sugisaki S, Xuan C, Sagawa T, Itaki T, Kubota Y, Lu S, Seki A, Murray RW, Alvarez-Zarikian C, Anderson, Jr WT, Bassetti M-A, Brace BJ, Clemens SC, da Costa Gurgel MH, Dickens GR, Dunlea AG, Gallagher SJ, Giosan L, Henderson ACG, Holbourn AE, Kinsley CW, Lee G-S, Lee K-E, Lofi J, Lopes CICD, Saavedra Pellitero M, Peterson LC, Singh RK, Toucanne S, Wan S, Zheng H, Ziegler M (2018) High-resolution and -precision correlation of dark and light layers in the quaternary hemipelagic sediments of the Japan Sea recovered during IODP Expedition 346. Prog Earth Planet Sci in press.

Tada R, Irino T, Koizumi I (1999) Land-ocean linkages over orbital and millennial timescales recorded in late quaternary sediments of the Japan Sea. Paleoceanography 14(2):236–247 https://doi.org/10.1029/1998PA900016.

Tada R, Murray RW, Alvarez Zarikian CA, and the Expedition 346 Scientists (2015) Proceedings of Integrated Ocean Drilling Program. IODP volume 346. College Station (Integrated Ocean Drilling Program). https://doi.org/10.2204/iodp.proc.346.101.

Westerhold, T., Röhl, U., Wilkens, R., Pälike, H., Lyle, M., Jones, T.D., Bown, P., Moore, T., Kamikuri, S., Acton, G., Ohneiser, C., Yamamoto, Y., Richter, C., Fitch, P., Scher, H., Liebrand, D., and the Expedition 320/321 Scientists (2012) Revised composite depth scales and integration of IODP Sites U1331–U1334 and ODP Sites 1218–1220. *In* Pälike, H., Lyle, M., Nishi, H., Raffi, I., Gamage, K., Klaus, A., and the Expedition 320/321 Scientists, Proc IODP, 320/321: https://doi.org/10.2204/iodp.proc.320321.201.2012 Tokyo (Integrated Ocean Drilling Program Management International, Inc.).

6

Geographical distribution of shear wave anisotropy within marine sediments in the northwestern Pacific

Takashi Tonegawa[1*], Yoshio Fukao[1], Gou Fujie[1], Shunsuke Takemura[2,3], Tsutomu Takahashi[1] and Shuichi Kodaira[1]

Abstract

In the northwestern Pacific, the elastic properties of marine sediments, including *P*-wave velocities (*Vp*) and *S* wave velocities (*Vs*), have recently been constrained by active seismic surveys. However, information on *S* anisotropy associated with the alignments of fractures and fabric remains elusive. To obtain such information, we used ambient noise records observed by ocean-bottom seismometers at 254 sites in the northwestern Pacific to calculate the auto-correlation functions for the *S* reflection retrieval from the top of the basement. For these *S* reflections, we measured differential travel times and polarized directions to reveal the potential geographical systematic distribution of *S* anisotropy. As a result, the observed differential times between fast and slow axes were at most 0.05 s. The fast polarization axes tend to align in the trench–parallel direction in the outer rise region. In particular, their directions changed systematically in accordance with the direction of the trench axis, which changes sharply across the junction of the Kuril and Japan trenches. We consider that a contributing factor for the obtained *S* anisotropy within marine sediments in the outer rise region is primarily aligned fractures due to the tensional stresses associated with the bending of the Pacific Plate. Moreover, numerical simulations conducted by using the three-dimensional (3D) finite difference method for isotropic and anisotropic media indicates that the successful extraction of *S* anisotropic information from the *S* reflection observed in this study is obtained from near-vertically propagating *S* waves due to extremely low *Vs* within marine sediments. In addition, we conducted an additional numerical simulation with a realistic velocity model to confirm whether *S* reflections below the basement can be extracted or not. The resultant auto-correlation function shows only *S* reflections from the top of the basement. It appears that such near-vertically propagating *S* waves obscure *S* reflections from interfaces below the basement.

Keywords: Ambient noise; Ocean-bottom seismometer; Shear wave reflection; Shear wave anisotropy; Northwestern Pacific

Background

Seismic information of marine sediments is important for understanding their historical development as well as for performing seismological techniques including travel time corrections for imaging underlying structure and for the determination of hypocenter locations. The Pacific Plate in the northwestern Pacific is one of the best seismically characterized oceanic plates (e.g., Kodaira et al., 2014). Constraints of the *P*-wave velocity (*Vp*) structure at the shallow part of the plate have been accomplished by seismic refraction and reflection surveys focusing on structures beneath the seawater (Cadet et al., 1987; Shimamura et al. 1983; Fujiwara et al., 2007; Shinohara et al. 2008). *S* wave velocity (*Vs*) structures have also been well confined, particularly at depths shallower than the topmost of the oceanic mantle by active seismic surveys (Fujie et al., 2013). At greater depths, surface wave (Isse et al., 2009) and receiver function (Kawakatsu et al., 2009; Kumar et al. 2011) analyses have revealed *Vs* structure beneath the ocean.

* Correspondence: tonegawa@jamstec.go.jp
[1]Research and Development Center for Earthquake and Tsunami, Japan Agency for Marine-Earth Science and Technology, 3173-25, Showa-machi, Kanazawa-ku, Yokohama 236-0001, Japan
Full list of author information is available at the end of the article

Despite the soft materials, it has been shown that the elastic properties of marine sediments have anisotropic structures by using signals from active seismic sources in the Nankai Subduction Zone, Japan (Tsuji et al. 2011). Such an anisotropic structure is potentially indicative of the existence of aligned cracks or fractures associated with the stress field at local or regional scales. When compressional stress is applied to a medium, cracks open under certain conditions with their major (long) axes aligned along the direction of compression. The fast polarization direction is taken to be the direction of the major axis in the case of ellipsoidal cracks (e.g., Hudson 1981). Thus, in addition to the fact that *Vs* is extremely sensitive to the presence of fractures, fluids, and hydrous minerals within a medium, because anisotropic parameters are key to understanding the alignments of cracks beneath the seafloor structure, more detailed material information within marine sediments could be obtained by combining these elastic parameters with *Vp* profiles and reflection images. It is therefore useful to construct a method to obtain *Vs* and its anisotropic information within marine sediments by using ocean-bottom seismometer (OBS) records.

Ambient noise correlation is capable of detecting surface (Campillo and Paul 2003; Shapiro and Campillo 2004), body (Roux et al. 2005; Miyazawa et al. 2008; Boué et al. 2013; Nakata et al. 2015), and reflected (Draganov et al. 2007, 2009; Zhan et al. 2010; Ryberg 2011; Poli et al. 2012a, 2012b; Ruigrok et al. 2012; Tibuleac and von Seggern 2012) waves propagating within a background wavefield. Tonegawa et al. (2013) calculated auto-correlation functions (ACFs) by using ambient noise at 1–3 Hz recorded by three broadband OBSs deployed on the outer rise region in the northwestern Pacific. The ACFs successfully showed *S* reflection retrieval from the basement and an *S* anisotropy of 2 % within marine sediments from differential travel times of the *S* reflections polarized in various directions. However, the result relevant to *S* anisotropy was determined from only one OBS site among the three; one of the remaining two sites represented relatively weak *S* anisotropy (<0.5 %) and the other one showed no *S* reflections. It is thus necessary to confirm whether the features reported by Tonegawa et al. (2013) are ubiquitous in the northwestern Pacific.

In this paper, we investigate *S* reflection retrieval by using continuous records observed at 254 OBS sites extending over an area of 500 × 500 km^2 in the northwestern Pacific (Fig. 1) and also investigate features related to *S* anisotropy within marine sediments, including its degree and fast polarization direction. Moreover, by conducting three-dimensional (3D) numerical simulations with finite difference methods, we verify what kind of environment causes bias for the estimation of anisotropic parameters. Further, we compare the retrieval of *S* reflection between our observation and synthetic waveforms obtained by using a realistic velocity model. Finally, we interpret the potential causes of observed anisotropy within marine sediments.

Methods

Deploying three-component, short-period (4.5 Hz) sensors with a station spacing of 6 km, we constructed six linear OBS arrays in the northwestern Pacific for seismic exploration surveys during 2009 and 2010 (Fujie et al. 2013). The observation period for each OBS was between 1 and

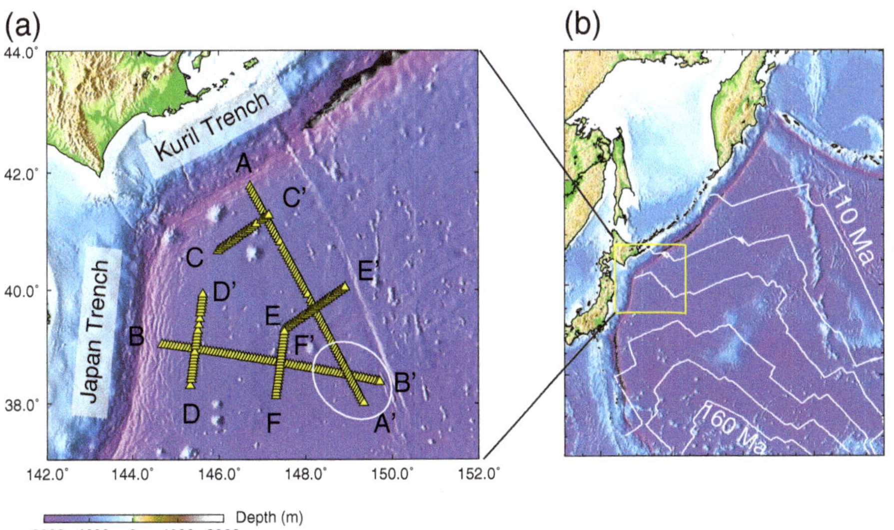

Fig. 1 Locations of the examined ocean bottom seismometer (OBS) sites. **a** The six linear arrays along A–A′, B–B′, C–C′, D–D′, E–E′, and F–F′ consisting of short-period (4.5 Hz) OBSs (*yellow triangles*). The *ellipse* shows an area of the abyssal plain in the region where OBSs were deployed. **b** Map showing the northwestern Pacific with contours (*white lines*) of the plate age. The *yellow rectangle* corresponds to the area shown in Fig. 1a

2 weeks. The continuous record for each OBS was dominated mainly by signals of air gun shots, but ambient noise without shot signals was also recorded with a time length of at least 1 day. This allowed us to use the record for ambient noise analysis for each OBS site. The six linear arrays thus totally provided a total of 254 sites with a 1-day ambient noise record. The horizontal direction of each OBS could be estimated by using air gun shot signals, because the locations of the shots and the OBSs were well constrained. In the case that the horizontal direction could not be estimated owing to the fewer available number of the shot signals with good signal-to-noise ratio (S/N), we did not use records observed by such OBSs, resulting in 201 available OBS sites.

The processing for *S* reflection retrieval from the OBS records was identical to that reported in previous research (Tonegawa et al. 2013) except for the frequency range used. We applied a band-pass filter of 2–5 Hz to the continuous records of two horizontal components. We then synthesized ambient noise records for each azimuth by rotating the two horizontal components at 5° increments, removed energetic seismic signals, calculated ACFs from continuous records with a time length of 600 s, and stacked the ACFs over the course of 1 day. Tonegawa et al. (2013) reported that the travel time of *S* reflection emerged in the ACF was stable when using a continuous record for 1 day. The rotation angle was measured clockwise from the north. The ACFs of ambient noise rotated for all directions allowed us to extract *S* reflections with various polarization directions.

To examine the differential travel times of *S* reflections polarized into various directions, we first calculated a reference waveform by stacking the ACFs over all of the rotated directions. We then cross-correlated the stacked *S* reflection in the waveform with those of every 1-day stacked ACF at every rotated direction. The same time window was adopted for both the reference waveform and each ACF trace: 1–2 s around the *S* reflection, as shown by the red dots in Fig. 2a. This time window was selected by roughly hand-picking the start and end times of the *S* reflection. ACF traces without red dots in Fig. 2b were not used in this study because of unclear *S* reflections. Similarly, we estimated differential travel times and fast polarization directions for other linear arrays. The degree of *S* anisotropy was estimated by dividing the differential travel times of the *S* reflection between fast and slow axes by the absolute travel time of the *S* reflection. We considered the lag time of the negative peak of the *S* reflection that emerged in the reference waveform as the absolute travel time.

Results

S reflection retrieval in the northwestern Pacific

Figure 2a shows the ACFs obtained by using the north–south (NS) component along line A–A′; we did not display ACFs at OBSs for which the horizontal direction could not be estimated. The ACFs show large amplitudes with negative polarities at a lag time of 2–6 s; the right (left) direction indicates positive (negative) polarity in the figure. Because such features are consistent with those obtained in the previous research (Tonegawa et al. 2013), these signals would correspond to the *S* reflection from the basement. The obtained lag time of the *S* reflection in the ACFs corresponds to the two-way travel time of an *S* wave propagating between the seafloor and basement. An additional phase with weak amplitude at a lag time of 4–12 s in the ACFs is the first-order multiple reflection. *S* reflections from other interfaces related to the crust and mantle cannot be observed in the ACFs.

Impulse-like reflections can be traced in the northern part of line A–A′ ($x < 350$ km), whereas *S* reflections in the southern part show multiple peaks, which is consistent with the seismic survey result. A post-stack time migration of multi-channel seismic (MCS) reflection survey data has been previously performed along A–A′ (Fig. 2c) by Fujie et al. (2013). Fujie et al. (2013) reported that the sub-seafloor structure in the southern part ($x > 350$ km) was formed near the spreading center and that the sub-seafloor structure around the region was deformed by magma-involved activity and normal faults during its formations. This explains the remarkably complex topography of the basement in the southern part, as shown in Fig. 2c, in addition to the multiple peaks of ACFs at the OBSs in the southern part (Fig. 2b).

Similar features can be identified along other lines; the ACFs along lines C–C′, D–D′, and E–E′ almost all show single, negative peak *S* reflections (Fig. 3b, c, and e). On the contrary, because the transect along line F–F′ is likely to show multiple peaks of *S* reflections, the structure around this region may be also affected by seafloor formation (Fig. 3d). Moreover, the features along line B–B′ appear to be somewhat complicated. The ACFs show unclear *S* reflections in the eastern part of line B–B′ (Fig. 3a) and multiple peaks of *S* reflections in the western part. To interpret the features of the emerged reflections in detail, it would be necessary to reveal the detailed structures from seismic surveys along line B–B′.

Vp/*Vs* within marine sediments

The time migration result (Fujie et al. 2013) provides us information relevant to the two-way travel time of *P*-wave between the seafloor and basement. Using the two-way travel times of *P* and *S* reflections, we can estimate *Vp*/*Vs* within the marine sediments. Given a *Vp* of 1.6 km/s for marine sediments (Shinohara et al. 2008) and an air gun shot frequency range of 5–20 Hz, as used by Fujie et al. (2013), the wavelengths of *P*-waves within the sediments are 80–320 m. In the case of *S* waves with an employed frequency range of 2–5 Hz and a *Vs* of 0.2 km/s

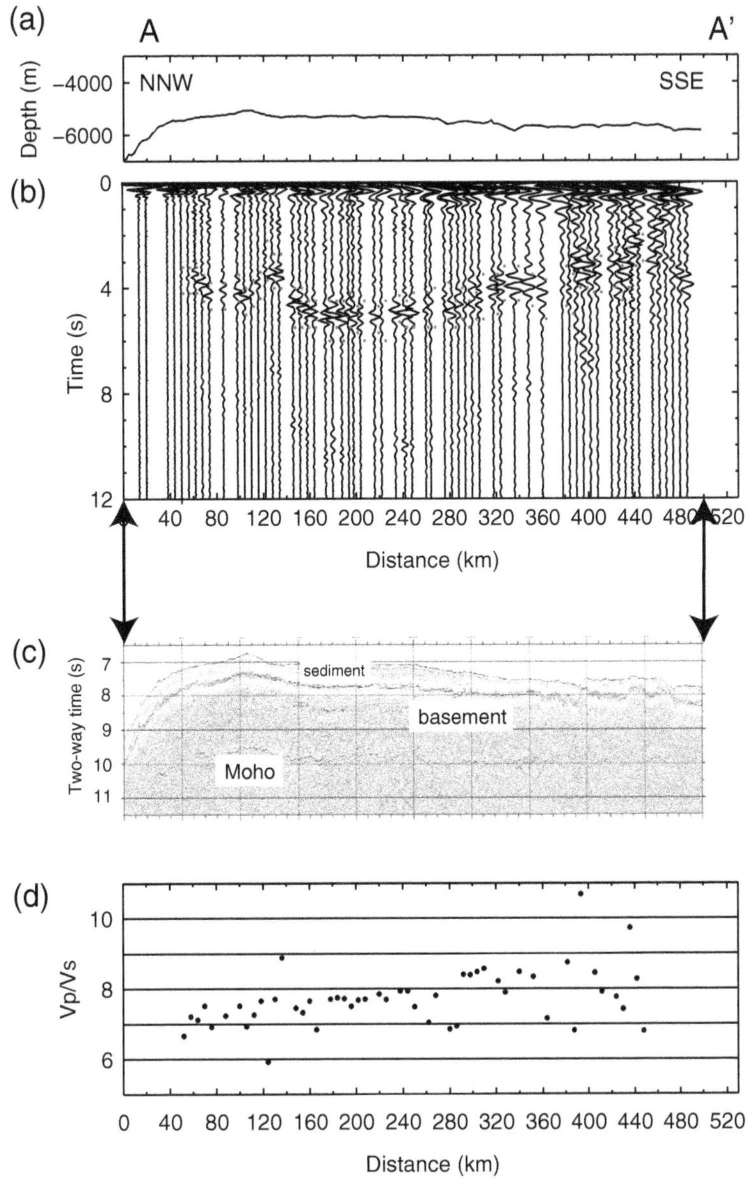

Fig. 2 *S* reflections that emerged in the auto correlation functions (ACFs) along A–A′. **a** The bathymetry along A–A′. **b** The ACF transect along A–A′. Left/right direction corresponds to negative/positive polarity for the ACF. An ACF is plotted for each ocean-bottom seismometer (OBS) site. *Red dots* indicate the time window for searching anisotropic parameters, differential travel time and fast polarization direction. **c** Post-stack time migration of multi channel seismic (MCS) data along A–A′ (after Fujie et al. 2013). **d** *Vp/Vs* along A–A′. The two-way travel time of *S* reflections obtained in this study is divided by that of *P* reflections from a seismic survey (after Fujie et al. 2013)

(Tonegawa et al. 2013), the wavelength becomes 40–100 m, indicating that estimation of *Vp*/*Vs* within marine sediments can be performed with mostly comparable wavelengths between *P* and *S* waves. Figure 2d shows the estimated value of *Vp*/*Vs* for each OBS site. Although the *Vp*/*Vs* is roughly concentrated around 8, it appears that the value decreases gradually from 9 to 7 northward from the location of $x = 350$ km. Because Fujie et al. (2013) found an increase of *Vp*/*Vs* at the basement around the outer rise region, this may correspond to water infiltration downward from the sediment to the basement. However, it is also possible to consider that water trapped inside the sediments escapes upward due to its buoyancy through fractures formed around the outer rise region. In the southern part, the estimated *Vp*/*Vs* may be unstable due to multiple peaks of *S* reflections.

S anisotropy within marine sediments

The estimation of a fast axis at each OBS site is summarized in Fig. 4a. The length and direction of the yellow

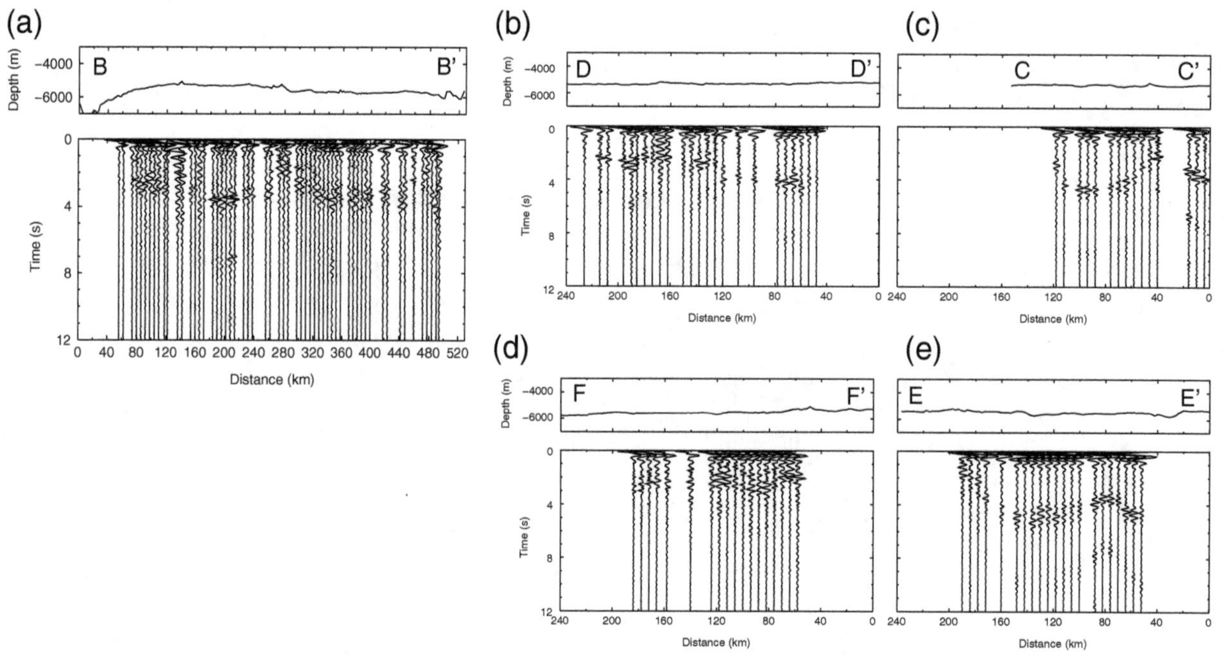

Fig. 3 *S* reflections that emerged in the auto correlation functions (ACFs) along five linear arrays. **a–e** Same as Fig. 2b but for lines B–B′, C–C′, D–D′, E–E′, and F–F′

bar in the figure indicate the degree of *S* anisotropy and its fast polarization direction. Herein, as shown in Fig. 4b, we define the following three regions: area A, the region from the top of the outer rise to the trench; area B, the region between the top of the outer rise and just short of the abyssal plain; and area C, the region of the abyssal plain where the bathymetry and basement are remarkably undulating. The most pronounced feature is that the fast axes are estimated to be in the trench–parallel direction in area A, except for the southern part of area A, as will be subsequently discussed. On the contrary, the degree of anisotropy decreases abruptly in area B (<0.5 %), compared with that in area A (1–2 %). These features are in good agreement with those reported previously (Fig. 4a; Tonegawa et al. 2013). Moreover, it appears that the anisotropic characteristics are complex in area C, probably reflecting large lateral heterogeneities within the sediments in the abyssal plain.

We plotted a rose diagram (Fig. 4c, d) by using the yellow bars shown in box 1 in Fig. 4a. The degree of the anisotropy was averaged at each azimuth bin with an interval of 10°. The fast polarization direction in box 1 is estimated to be ENE–WSW, which is in good agreement with the direction of the Kuril Trench (denoted by the red arrows in Fig. 4c). The degree of the anisotropy was approximately 1.5 %. The rose diagram in box 2 shows an NS-trending fast axis with an anisotropy of 1 % (denoted by the red arrows in Fig. 4d), the direction of which is parallel to the trench axis. Both diagrams show directions other than the trench–parallel. The rose diagram in box 1 shows a fast axis at N 30° E with an anisotropy of 1.5 %. The fast polarization direction in the rose diagram in box 2 shows several patterns, including fast axes at N 45° E and N 30° E with an anisotropy of 2–3 % (denoted by blue and pale-blue triangles in Fig. 4d) and a NW–SE-trending fast axis with an anisotropy of 1 % (denoted by the orange triangle in Fig. 4d). The features in box 2 can also be observed in the magnified map image in Fig. 4e.

Discussion

Numerical simulation for retrieving *S* reflection

An understanding of the mechanism inducing the differential travel times of *S* reflections polarized into various directions would be useful for interpretation of the obtained *S* anisotropy. Herein, we performed numerical simulations to investigate *S* reflection retrieval and the characteristics of their differential travel times with two-layered simple velocity models based on the following three cases: (1) a comparison of *S* reflections from horizontal and dipping interfaces in a two-layered isotropic medium with evenly distributed sources, (2) *S* reflection retrieval from a horizontal interface in anisotropic upper and isotropic lower layers with evenly distributed sources, and (3) *S* reflection retrieval from a horizontal interface in anisotropic upper and isotropic lower layers with unevenly distributed sources.

The retrieval of reflections with passive sources has been previously examined by using numerical simulations, in which the sources were randomly distributed and

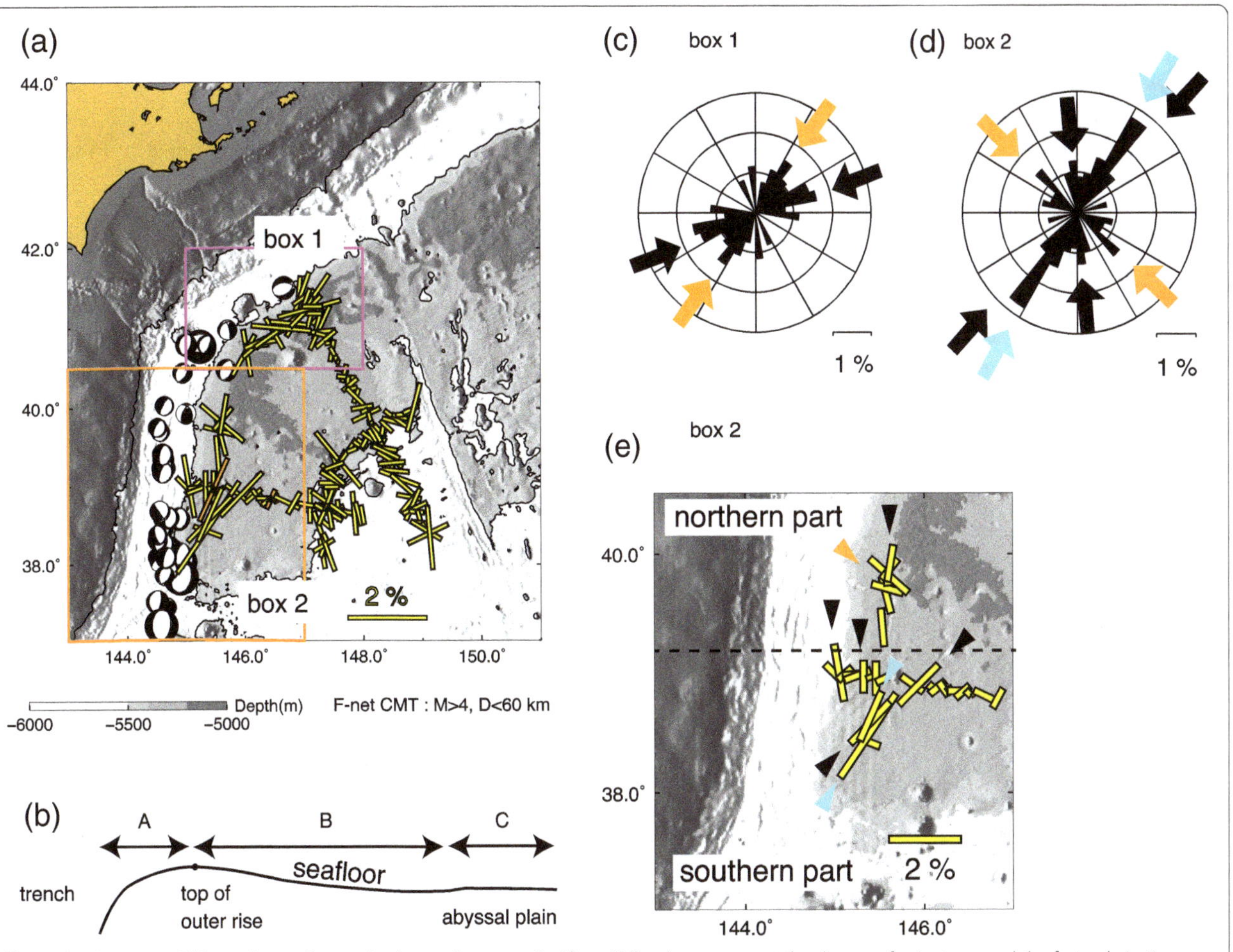

Fig. 4 *S* anisotropy within marine sediments in the northwestern Pacific. **a** *Yellow bars* represent the degree of anisotropy and the fast polarization direction at each ocean-bottom seismometer (OBS) site. Mechanism solutions, which were determined by F-net centroid moment tensor (CMT) with $M > 4$ and depths shallower than 60 km (Fukuyama et al. 1998; Okada et al. 2004), are displayed by *red beach balls. Red triangles* and *orange bars* indicate locations of three OBS sites and the results of *S* anisotropy at the sites in Tonegawa et al. (2013). **b** Definition of the regions for areas A, B, and C from the abyssal plain to the trench. **c** A rose diagram created using the anisotropic parameters inside *box 1* in Fig. 4a. The degree of anisotropy was averaged at each azimuth range of 10°. *Red arrows* represent the fast polarization direction parallel to the trench axis. *Orange arrows* indicate a N 30° E-trending fast axis. **d** Same as Fig. 4c but for *box 2. Red arrows* represent the fast polarization direction parallel to the trench axis. *Blue, pale blue,* and *orange arrows* indicate other directions than that parallel to the trench axis (see text in details). **e** A magnified inset map image in *box 2* in Fig. 4a. A *dashed line* is the boundary between the northern and southern parts in *box 2. Colored arrows* indicate OBS sites in which a relatively large degree of *S* anisotropy was observed

subsequently activated (e.g., Draganov et al. 2004; Thorbecke and Draganov 2011). Thorbecke and Draganov (2011) reported that the accuracy of retrieved reflections depends on several factors such as the duration and number of passive sources and the distribution and strength of the sources. Their findings indicate that careful considerations of the aforementioned source parameters should be required for the construction of accurate reflections. In addition, it is possible that the *S* reflections observed in this study were excited by fluid disturbances such as wave–wave interactions (Longuet-Higgins 1950) above the seafloor; this also requires careful parameterizations with regard to the source setting. In this study, we considered only ideal source parameterizations to mainly evaluate structural parameters affecting the differential travel times of *S* reflections polarized into various directions and to reduce computational costs. Sources with the same strengths were evenly distributed below the seafloor, as is subsequently discussed.

We employed the 3D finite difference method based on rotated staggered grids with second-order calculation accuracies in time and space (Saenger 2000). The calculation was performed in the displacement–stress scheme with a time step of 0.005 s. However, when more accurate information was required to evaluate travel time, we conducted the calculation with a time step of 0.001 s. For such cases, we inserted the term 0.001 s after the obtained time. We calculated wavefields for anisotropic media by incorporating a maximum of 21 elastic constants. The model space (x–y–z) was 16 × 16 × 12 km^3, which includes two layers with a grid spacing of 0.05 km (Fig. 5a). The physical parameters of the model are summarized in Table 1. Because this simulation explores only the characteristics of *S* reflections under various conditions, the model parameters represent a simplified two-layered submarine structure including the sediment and crust but do not correspond to a realistic structure. The location of the station was set at $(x, y, z) = (8, 8, 0)$. We simultaneously applied horizontal forces with a Ricker wavelet at a center frequency of 0.71 Hz with a maximum frequency of 1.02 Hz to 2337 locations, as indicated by the red dots in Fig. 5b. To make a wave incidence to the station with various azimuths and slownesses, sources should be located along a hemisphere beneath the station. These locations were randomly determined from grids, of which the distances to the station are between 7.95 and 8.05 km and at a vertically downward angle at the station of less than 45°. This resulted in a depth range of sources from 5.60 to 8.05 km (Fig. 5c). Herein, the direction of the horizontal source is the same as that of the horizontal component at the station, and the forcing direction is rotated from 0° to 180° with an interval of $\theta = 10°$ (Fig. 5d). Therefore, we conducted a total of 19 calculations for different directions of the source and horizontal component. In the case of the distribution and forcing direction of sources used in this study, the ACF calculated by using the wavefields from 2337 simultaneously excited sources is equivalent to a function stacking 2337 ACFs using the wavefield from one source, which enabled us to reduce the calculation times from 2337 × 19 to 1 × 19. Because the minimum velocity is 1.0 km/s, the simulation could be conducted under the condition of 20 grids per minimum wavelength at the used maximum frequency of 1.0 Hz, leading to sufficiently stable calculations (e.g., Bohlen and Saenger 2006). We finally plotted ACFs constructed by synthetic waveforms as a function of polarization direction and evaluated the differential

Table 1 Velocity model for numerical simulation

	H (km)	*Vp* (km/s)	*Vs* (km/s)	*Vp*/*Vs*	*ρ* (g/cm^3)
Layer 1	2.00	2.00	1.00	2.00	2.60
Layer 2	10.0	6.00	3.47	1.73	2.60

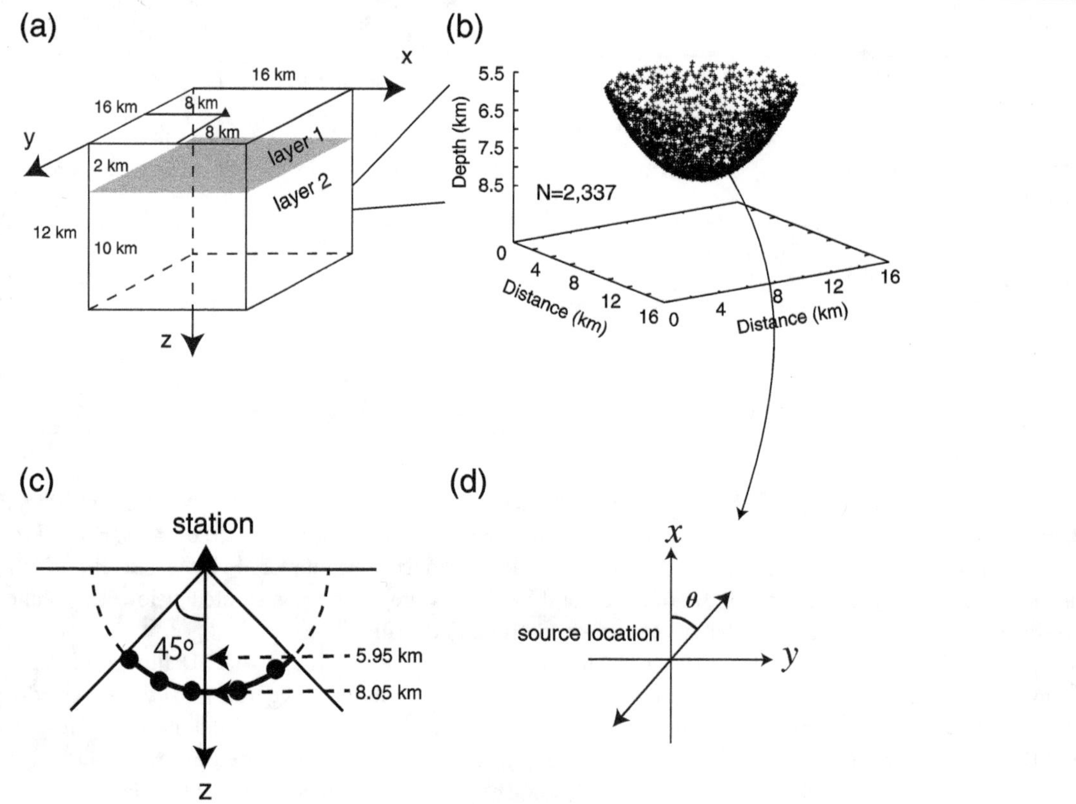

Fig. 5 Model space and source geometry employed in the numerical simulation. **a** Model space of 16 × 16 × 12 km^3 consisting of two layers for numerical simulation. Station is located at $(x, y, z) = (8, 8, 0)$. **b** Source distribution applying horizontal force at a depth of 5.60–8.05 km. **c** Sketch of source location. *Bold solid line* and *red dots* show allowable source locations and distributed sources. **d** The direction of applied horizontal force is rotated with an interval of $\theta = 10°$

travel times of S reflections for the aforementioned three cases.

Prior to examining the aforementioned source distribution, we first confirmed the travel time of S reflections from one source located at $(x, y, z) = (8, 8, 8)$ in the model space containing two isotropic layers with a horizontal interface at a depth of 2 km (Fig. 6a). This velocity model is also examined in the following case (case 1). As a

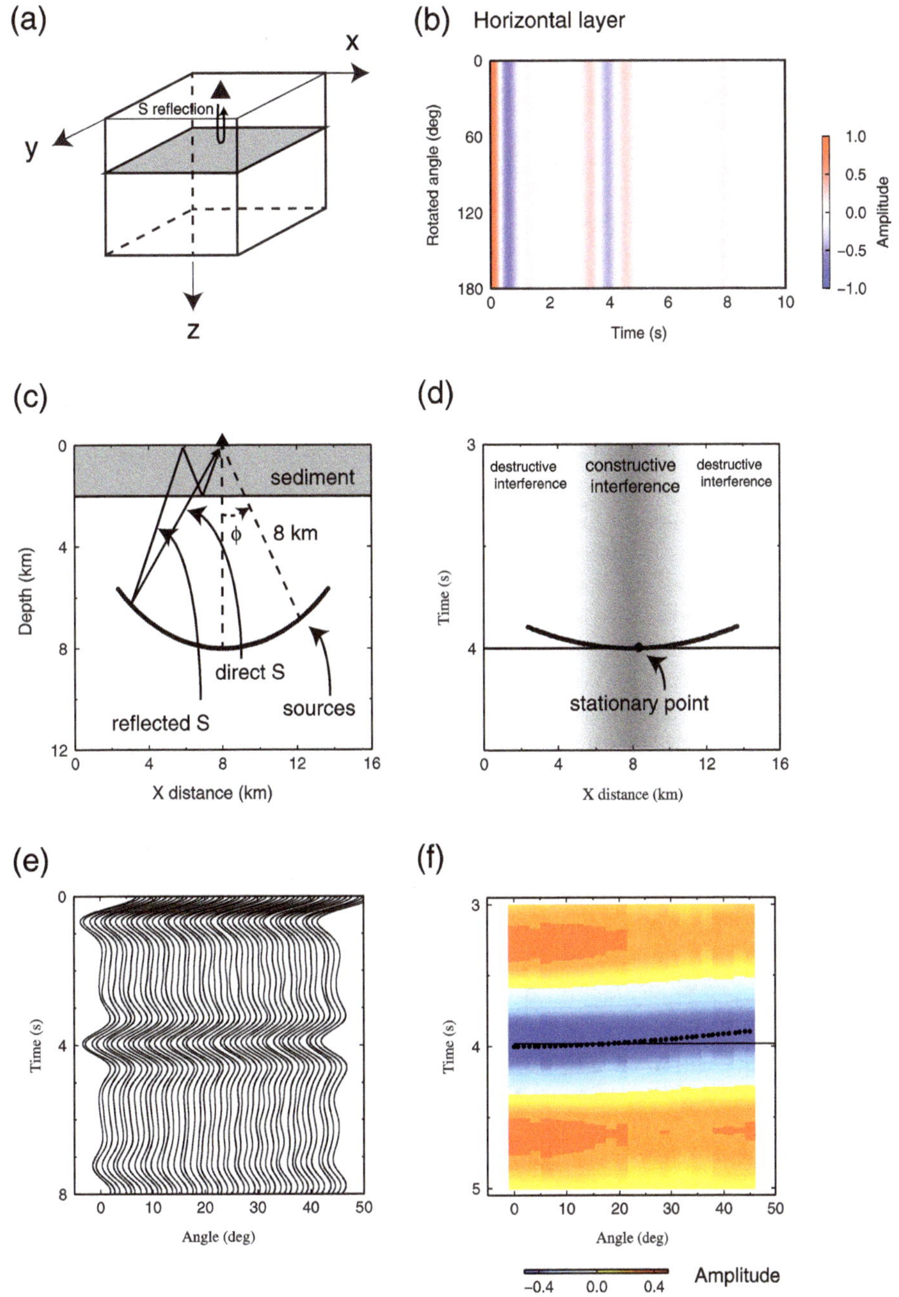

Fig. 6 Results of the numerical simulation for isotropic media. **a** Sketch for model space with two isotropic layers with a flat interface at a depth of 2 km. **b** Auto-correlation functions (ACFs) as a function of polarization direction, calculated with the model in Fig. 6a. **c** Model for the two-dimensional (2D) experiment for calculating differential travel time between direct S and reflected S waves. *Red dots* indicate locations of sources. The *triangle* shows the location of the station. *Gray shaded area* corresponds to low-velocity sediment. The parameters on the velocity model are the same as those in Fig. 6a and Table 1. **d** *Red dots* represent differential travel times between direct S and reflected S waves for each source (Fig. 7a). The *black solid line* corresponds to the travel time of two-way S reflection with zero offset for the one-shot case. The *gray shaded area* indicates the approximate region for constructive interference when stacking reflected S waves. **e** ACFs as a function of angle from vertically downward. Each ACF was calculated by the wavefield for each source shown in Fig. 6c. **f** Same as Fig. 6e but the range of lag time for 3–5 s and the amplitude is described by colors. *Red dots* are the same as those in Fig. 6d but for $0° \leqq \varphi \leqq 45°$ ($x \geqq 8$ km)

result, we obtained a lag time of 3.982 s (0.001 s) for the vertically propagating *S* reflection with zero offset.

Case 1: Fig. 6b shows the calculated ACFs as a function of polarization direction with the model shown in Fig. 6a. The *S* reflection is emerged in the ACFs at a lag time of 3.950 s irrespective of the polarization direction and arrived faster than that of the one-source case. Although we further examined five calculations by randomly changing the source distribution, which satisfies the aforementioned location condition, all results on the lag times of *S* reflection were 3.950 s (0.001 s), implying that we can obtain stable results on the two-way travel time. In the velocity model with low and high velocities for the top and bottom layers, respectively, the travel time of *S* reflections constructed from interferometry would depend mainly on the following three conditions: differential travel times between direct *S* and reflected *S* waves, amplitude of reflected *S* waves, and wave duration. Hereafter, we distinguish the term "S reflection," which means the primary *S* wave reflection between the seafloor and basement constructed by the interferometry (Fig. 6a), with the term "reflected *S* wave," as coming from one source (Fig. 6c). According to the idea of a stationary point (Snieder 2004), at which the gradient of the travel time curve of a phase with respect to the wave incidence angle is 0, a vertically propagating *S* reflection from interferometry is constructed by the enhancement of effects of reflected *S* waves coming with small angles from vertically downward by constructive interference and reduction of those with other incident angles by destructive interference. These interferences produce a vertically downgoing *S* wave from the surface and its upgoing reflection from the horizontal interface.

The range of the wave incidence angle to the constructive and destructive interferences was determined by the aforementioned three conditions. For instance, Fig. 6c shows a 2D case for evenly distributed sources within a two-layered structure. The physical parameters of the velocity model are the same as those in Fig. 6a and Table 1. The sources are located within an angle range of $-45° \leqq \phi \leqq 45°$ and within a distance from the station of 8 km (Fig. 6c). The measurement of differential travel times between direct *S* and reflected *S* waves is displayed in Fig. 6d. The gradient of the differential travel time curve with respect to wave incidence angle is small around the stationary point. This implies that vertically propagating *S* reverberations can be easily dominated at the top layer with a low velocity by various incident waves because upgoing waves with various slownesses at the bottom layer produce waves with near vertical incidence within the top layer. It is considered that such waves and their reflections, i.e., reverberations, are dominant within marine sediments. However, at the same time, this indicates that stacking reflected *S* waves around the stationary point produce a slightly faster vertically propagating *S* reflection than the theoretical travel time. The following two cases would construct an *S* reflection with a travel time closer to the theoretical time: (1) the case of smaller velocity contrast at the top of the basement, which results in a larger gradient of differential travel time curves; and (2) the case of lower velocity in the top layer, which results in a smaller gradient of differential travel time. For the seafloor observation, the latter case would be realistic. This 2D experiment considered only the effect of differential travel time to waveform stacking. However, the amplitude of the reflected *S* wave, or reflection coefficient, and its wave duration, or frequency component, would also contribute to the travel time of the *S* reflection. Figure 6e, f shows individual ACFs for each source ($0° \leqq \phi \leqq 45°$) shown in Fig. 6c. We applied a horizontal force at each source with a maximum frequency of 1.02 Hz, which is the same as that in the 3D case. As shown in Fig. 6f, the *S* reflection peak around a lag time of 4 s is slightly broad. When we stack these ACFs, the obtained travel time of S reflections would be deviated from the theoretical travel time. If we calculate the wavefield at higher frequency components, the *S* reflection peaks emerging in each ACF will be sharper, resulting in a more accurate travel time of *S* reflection in the stacked ACF.

In the case of the dipping interface (5°) within isotropic media (Fig. 7a), the two-way travel time of the *S* reflection became 3.940–3.969 s (0.001 s) with a travel time fluctuation of 0.029 s (Fig. 7b, d). This fluctuation is caused by the amplitude of the reflected *S* wave and its wave duration. If we consider only differential travel time to waveform stacking, an *S* reflection from a dipping interface should also be extracted in the ACFs at a lag time corresponding to the distance between the station and a point at which a perpendicular line from the station intersects to the dipping interface (d′ in Fig. 7c). In addition, its travel time has no fluctuations and becomes smaller than that in the case of the horizontal interface because of $d > d'$ in Fig. 7c. However, because such a dipping interface produces reflected *S* waves with various amplitudes (reflection coefficients depending on the incident angle to the dipping interface) as a function of propagation direction for an incident angle, these waves unevenly contribute to construct an *S* reflection around the stationary point, resulting in a slightly distorted waveform of the reflection. This is attributable to the travel time fluctuation in case of the dipping interface.

An important issue in the case of the dipping interface is whether this fluctuation affects *S* anisotropy measurement. We determined that such an effect was quite small in the seafloor observation and that the differential travel times of *S* reflections are caused only by *S* anisotropy because of the following reason. *S* waves coming from the bottom layer tend to propagate near normal to the dipping interface with decreasing *Vs* in the top layer.

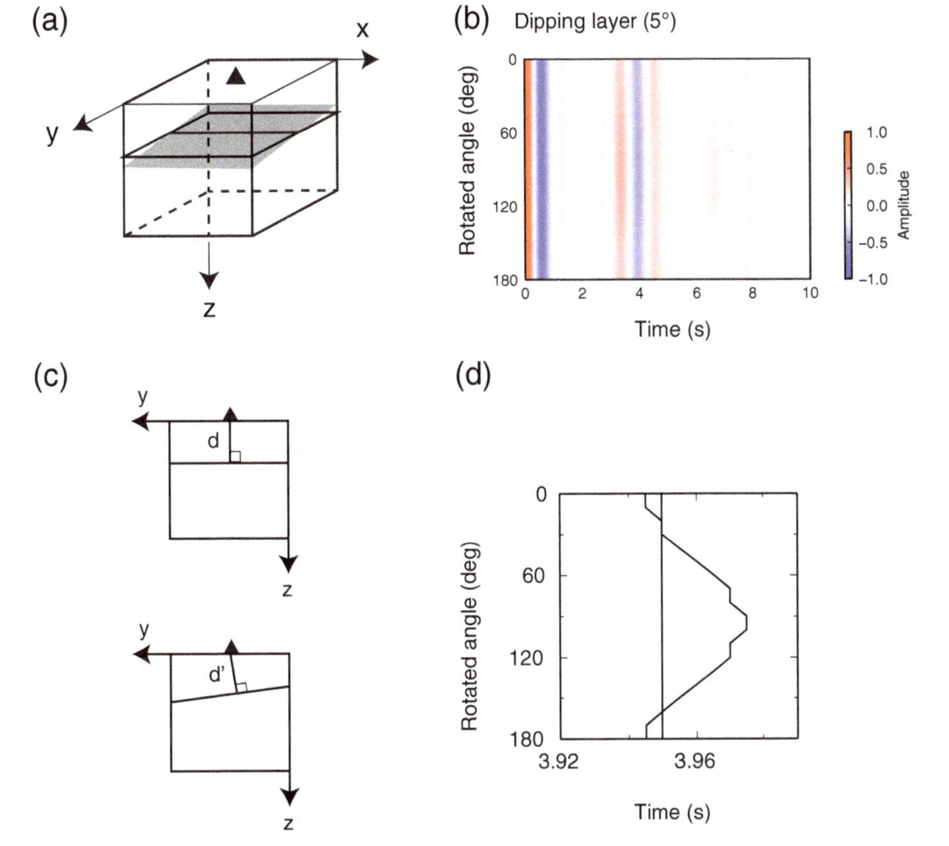

Fig. 7 **a** Same as Fig. 6a but for an interface with a dip angle of 5° toward the *y*-direction. The label *d'* indicates the distance from the station to the dipping interface. **b** Same as Fig. 6b but for the model in Fig. 7a. **c** The *y*–*z* cross-sections of Figs. 6a and 7a. Labels *d* and *d'* indicate the distance from station to interface. **d** *Red* and *black lines* indicate travel times of *S* reflection as a function of rotation angle for Figs. 6b and 7b, respectively

As a result, the reflection points of the *S* waves at the dipping interface are located around the stationary point, which produces reflected *S* waves with similar two-way travel time and their amplitudes. Stacking these waves constructs *S* reflections with significantly smaller travel time fluctuations with respect to the dipping interface. Indeed, we confirmed it by additional calculations. The velocity model was the same as the one in Fig. 7a but with a grid spacing of 0.04 km and a *Vs* of 0.9 km/s at the top layer, which ensured good calculation accuracy with 22 grids per minimum wavelength. As a result, the two-way travel time of the *S* reflection was 4.378–4.391 s (0.001 s) with a fluctuation of 0.013 s, which is much smaller than that in the case of *Vs* = 1.0 km/s at the top layer. Furthermore, another calculation for a grid spacing of 0.025 km and *Vs* of 0.5 km/s at the top layer with 20 grids per minimum wavelength resulted in 7.886–7.891 s (0.001 s) of two-way travel time of *S* reflection with a fluctuation of 0.005 s. Because the *Vs* of marine sediment is significantly slower (e.g., 0.2 km/s in Tonegawa et al. 2013), the dipping effect to the *S* anisotropy measurement would be plausibly negligible.

Case 2: We set anisotropies of $dVp = 0$ % and $dVs = -4$ %, which are respectively equivalent to $B = 0$ % and $E = -4$ % introduced by Park and Yu (1992) to the top layer of the model space with hexagonal symmetry (Fig. 8a). The symmetry axis was taken to be the *x*-direction, thereby producing a top layer with horizontal transverse isotropy (HTI) with a fast axis of the *y*-direction (Table 2); angles of η and ξ in Table 2 are measured from the *x*-axis and vertically downward, respectively (Fig. 8a). The degree of anisotropy in hexagonal symmetry can be converted to elastic constants (Park and Yu 1992; Nagaya et al. 2008). We assigned the same source distribution as that shown in Fig. 5b. The calculated ACFs show *S* reflections at a lag time of 3.880–4.035 s, which indicates that the travel times of the waves vary with the polarization direction of the *S* wave (Fig. 8b). The differential travel time between fast and slow axes is 0.155 s, which corresponds to a *Vs* anisotropy of 3.912 % (dividing it by the averaged travel time of 3.963 s of *S* reflection).

Case 3: The differential travel time of an *S* reflection polarized in various directions could be affected by inhomogeneous source distribution. Herein, we preserved

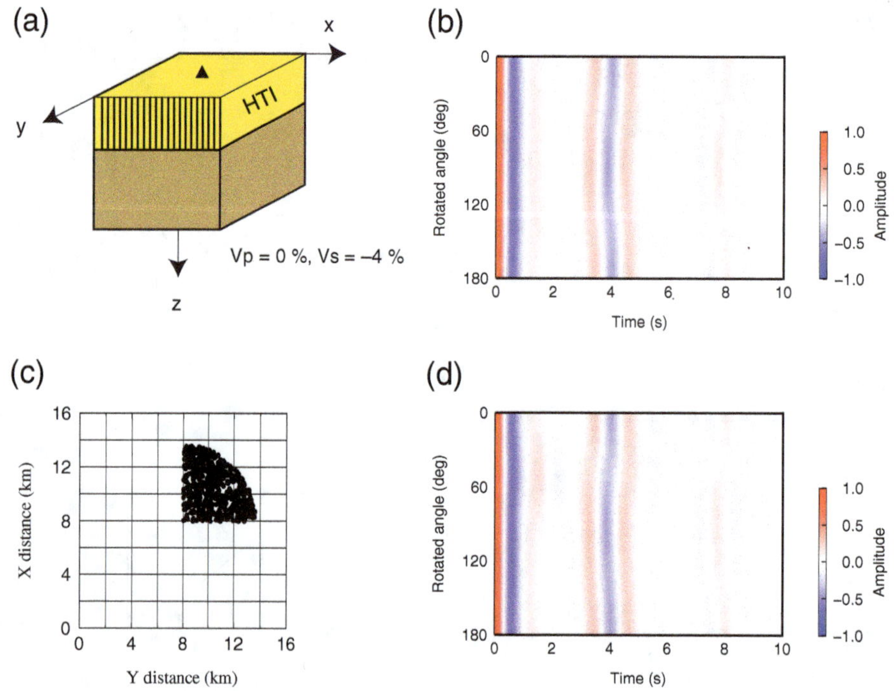

Fig. 8 Results of numerical simulation for anisotropic media. **a** Sketch for model space consisting of a horizontal transverse isotropy (HTI) top layer and an isotropic bottom layer with a flat interface at a depth of 2 km. The degree of anisotropy is $dVp = 0$ % and $dVs = -4$ %. The source distribution is homogeneous. **b** Auto-correlation functions (ACFs) as a function of polarization direction, calculated with the model in Fig. 8a. **c** Map view of the source distribution used in the calculation of the ACFs shown in Fig. 8d. **d** Same as Fig. 8b but for the sources in Fig. 8c

the sources at locations of $x \geqq 8$ km and $y \geqq 8$ km among those shown in Fig. 5b and discarded the other sources, which indicates that *S* waves excited at these locations are coming from a back-azimuth range of 0–90° (Fig. 8c). The synthetic ACF calculated in the model in Fig. 8a shows *S* reflections at a lag time of 3.875–4.035 s. A differential travel time of 0.160 s can be converted to an anisotropy of 4.040 %. Herein, although the obtained differential travel time was consistent with the case for evenly distributed sources, the behavior of the travel time fluctuation from unevenly distributed sources is slightly different relative to the case. The case of evenly distributed sources precisely represents the sinusoidal travel time curve of *S* reflection, whereas that of unevenly distributed sources shows a distorted travel time curve from the sinusoidal variation (Fig. 8d), which may provide inaccurate directions of fast and slow axes and a slight degree of anisotropy. Obtaining accurate fast and slow polarization directions depends strongly on whether ambient noise sufficiently contains either *S* waves with various slownesses for all azimuths or near-vertically propagating *S* waves. In marine sediments, it appears that the latter condition is satisfied as subsequently discussed.

Table 2 Model parameters for anisotropy

	dVp (%)	*dVs* (%)	*η* (°)	*ξ* (°)
Case 1	0	0	0	90
Case 2	0	−4	0	90
Case 3	0	−4	0	90

Summarizing the reflection extractions within a two-layered structure with isotropic media, the travel time of *S* reflection from a horizontal interface could be slightly faster than the theoretical time, and the one from a dipping interface is also faster and has a travel time fluctuation. However, in the case of seafloor observation, due to low *Vs* within marine sediments, it is expected that the travel time of constructed *S* reflection tends to be the accurate time and that the travel time fluctuation of the *S* reflection due to the dipping interface tends to be small with respect to that of *S* anisotropy. Moreover, the differential travel time of the *S* reflection constructed from both evenly and unevenly distributed sources precisely shows *S* anisotropy within the top layer.

S reflection from oceanic Moho

The reason as to why *S* reflections from the oceanic crust and mantle cannot be extracted from observed seismograms by the technique used in this study poses an interesting question. To investigate the reflection retrieval from interfaces below sediments, an effective method involves the investigation of the characteristics of the synthetic waveform/ACF calculated with a more realistic velocity model rather than the previous simulations.

We carried out another 3D finite difference method by using standard staggered grids with calculation accuracies of second order in time and fourth order in space (Graves 1996). However, the top boundary of the solid part of the model space, i.e., solid/air boundary, has a calculation accuracy of second order in space to take into account the free surface effect (e.g., Okamoto and Takenaka 2005; Nakamura et al. 2012; Takemura and Yoshimoto 2014; Takemura et al. 2015). This calculation was conducted under the condition of using higher maximum frequency and lower minimum velocity than those used in the previous simulations. Involving such parameters with finite difference calculations generally requires a large-scale computation with finer grids to obtain stable solutions (e.g., Levander 1988). The used velocity model was based on that of a previous study (Shinohara et al. 2008). We added a sediment layer with a *Vp* of 1.6 km/s, a *Vs* of 0.2 km/s, and a thickness of 0.4 km (e.g., Kanazawa et al. 2001; Shinohara et al. 2008) above the velocity model (Fig. 8a). By using an empirical relation between the *Vp* and the density (Brocher 2005), we created a sediment-to-mantle density profile (Fig. 9a). The calculation was performed in the velocity–stress scheme for isotropic media with a time step of 0.0002 s. The model space (x–y–z) was 12.288 × 12.288 × 15 km with a grid spacing of 0.008 km. The quality factors for *P* and *S* wave attenuation were set

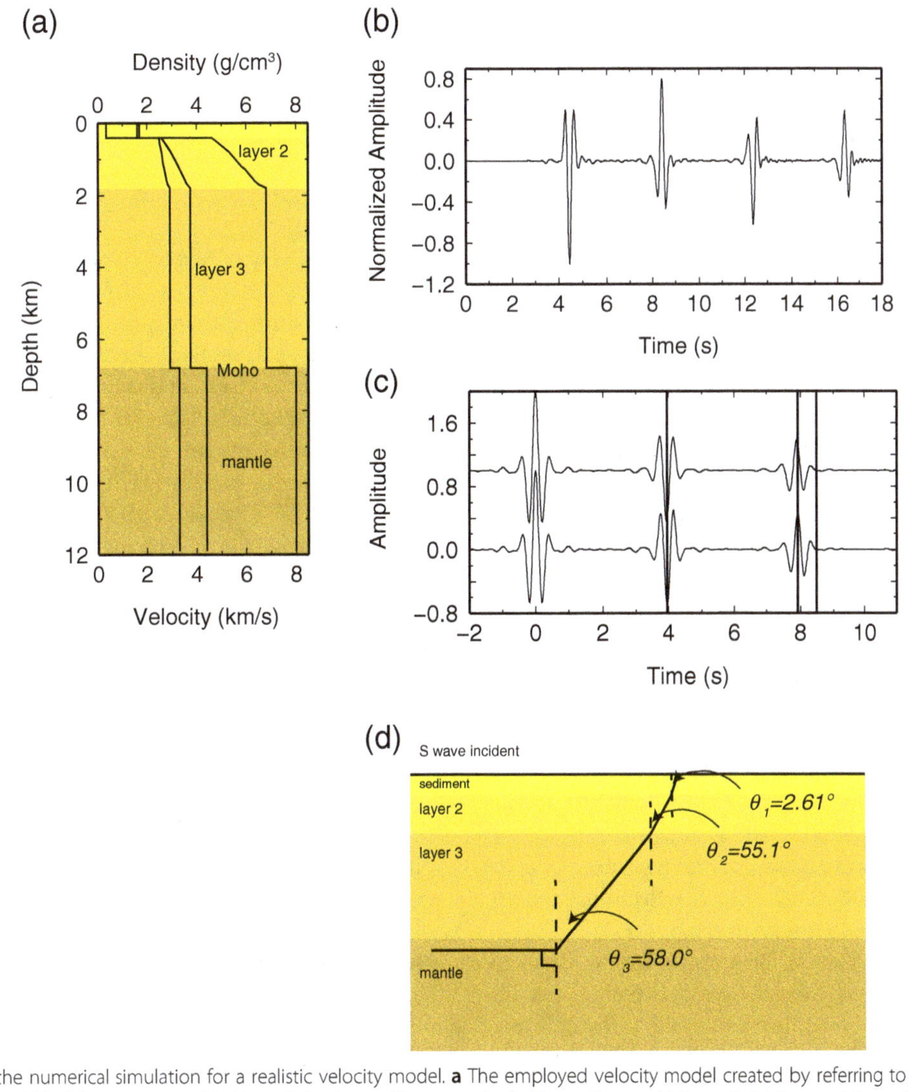

Fig. 9 Results of the numerical simulation for a realistic velocity model. **a** The employed velocity model created by referring to Shinohara et al. (2008). *Red*, *blue*, and *black lines* represent *Vp*, *Vs*, and density profile, respectively. The sediment thickness is 0.4 km, and *Vp* and *Vs* are 1.6 and 0.2 km/s, respectively. The density is estimated based on *Vp* of 1.6 km/s through an empirical relation (Brocher 2005). **b** The synthetic waveform calculated with the velocity model in Fig. 9a and for $Qs = 300$. **c** The auto-correlation functions (ACFs) constructed with the synthetic waveform shown in Fig. 9b for $Qs = 300$ (*top*) and $Qs = 110$ (*bottom*). *Red lines* correspond to the theoretical two-way travel times of the first and second *S* reflection from the top of the basement. The *blue line* indicates the theoretical two-way travel time of *S* reflection from the oceanic Moho. **d** The incident angles of *S* wave to layer 3, layer 2, and sediment, in the case of horizontally propagating *S* wave in the mantle

to be $Qs = 300$ and $Qp = 2Qs$, based on the method of Robertsson et al. (1994), within the entire model space. We adopted a single relaxation mechanism with a reference frequency of $f_0 = 2$ Hz, which provides peak Q^{-1} values at this frequency. To perform this large-scale simulation effectively, we employed a parallel 3D simulation code based on a domain-partitioning procedure that utilizes a large number of processors with a message-passing interface (based on the work of Furumura and Chen 2004).

The station was located at the center on the free surface: $(x, y, z) = (6.144, 6.144, 0.008)$. We simultaneously applied horizontal forces with a Ricker wavelet at a maximum frequency of 3.125 Hz to 821 locations. Source locations were randomly determined from grids whose distance to the station was between 7.95 and 8.05 km and angled from vertically downward at the station of less than 30°, resulting in a depth range of sources of 6.87–8.05 km. Herein, the direction of applied horizontal force and the observed horizontal component at the station are only one direction (the *y*-direction).

The simulation was conducted by using the Earth Simulator (ES2) supercomputer at the Japan Agency for Marine-Earth Science and Technology (JAMSTEC). A computer memory of 769 GB and a wall clock time of 2.1 h by parallel computing using 50 nodes (400 processors) of the supercomputer were required to evaluate seismic wave propagation of 18 s by 90,000 time-step calculations.

Figure 9b, c displays the calculated waveform and its ACF, respectively. The synthetic waveform shows a direct *S* wave and three multiple reflections, which were constructed from the summation of 821 signals with one calculation (Fig. 9b). The ACF shows the first and second *S* reflections with respective negative and positive polarities near the red lines that correspond to the theoretical zero-offset, two-way travel times of the reflections (Fig. 9c). The blue line indicates the theoretical arrival times of *S* reflections from the oceanic Moho. However, such a phase could not be recognized in the synthetic ACF. Furthermore, we calculated ACF by changing to $Qs = 110$ only at the marine sediment by referring to the result from Nolet and Dorman (1996). As a result, although the amplitude of the *S* reflection from the top of the basement became smaller than that for $Qs = 300$, the *S* reflection from the oceanic Moho also could not be found in the ACF.

It appears that the reason for an absence of Moho reflections in the ACF is caused by a relatively weak impedance contrast at the Moho compared with the top of the basement. However, in addition to that, we consider that the absence is also caused by the difference of the wavefield feature between the sediment and crust. For instance, given a horizontally propagating *S* wave in the mantle, i.e., an Sn wave, the incident angle to the marine sediment becomes 2.61° with the velocity model shown in Fig. 9a (Fig. 9d). This means that upgoing *S* waves with various slownesses in the mantle become near-vertically propagating *S* waves within the sediment layer. Because the seafloor plays the role of the free surface for near-vertically propagating *SV* and *SH* waves, when the wavelength of *S* wave is shorter than the thickness of the sediment layer, these waves are effectively trapped between the seafloor and the top of the basement that has a large impedance contrast. This would produce near-vertically propagating *S* waves within the sediment. However, in the case of oceanic crust, the various slownesses of *S* waves are presumably present within the layer due to relatively faster *Vs* than in sediment, indicating that the retrieval of vertically propagating *S* wave is not effective by the correlation technique with a 1-day continuous record compared with the case in the sediment. This means that it may be possible to extract *S* reflections from the oceanic Moho with more ambient noise record data, but whether the retrieval works well would also depend strongly on the impedance contrast at the boundary. In the case in which the amplitude of the reflected *S* wave is weak, it would be obscured by reverberations in the sediment layer.

Interpretation of sediment *S* anisotropy

As suggested by Tonegawa et al. (2013) based on anisotropic estimations in the northwestern Pacific region, *S* anisotropy within marine sediments in the outer rise region would be caused by aligned cracks and normal faulting formed by the bending of the Pacific Plate (Fig. 10a). The observed anisotropic structure within marine sediments has been explained by the presence of aligned cracks (e.g., Haacke and Westbrook 2006; Haacke et al. 2009). In the outer rise region, because the plate is bending along the Japan and Kuril trenches, the tensional stresses are induced within the uppermost of the Pacific Plate nearly in the trench–normal direction. As a result, cracks and normal faulting form due to stresses within the marine sediment, and these fractures are aligned in the trench–parallel direction (Fig. 10c), causing *S* anisotropy that has a trench–parallel fast axis. Such tensional stresses are also supported by earthquake mechanisms at shallower depths than 60 km (Fukuyama et al. 1998; Okada et al. 2004), which are mostly normal faulting with trench parallel strikes, around the outer rise region (Fig. 4a). The stress-induced anisotropy includes fractures within marine sediments due to the basement deformation associated with tensional stress, e.g., normal faulting within the basement or deeper (Fig. 10b). As discussed in the subsequent paragraphs, the expected trench–parallel fast axes were clearly observed along the Kuril Trench and along the northern part of the Japan Trench.

The fast polarization direction in box 1 in Fig. 4a, c shows the trench–parallel direction (red triangles in Fig. 4a, c), which reflects stress-induced cracks and fractures due to the basement deformation. In particular, the trench axis at this region coincides with the magnetic

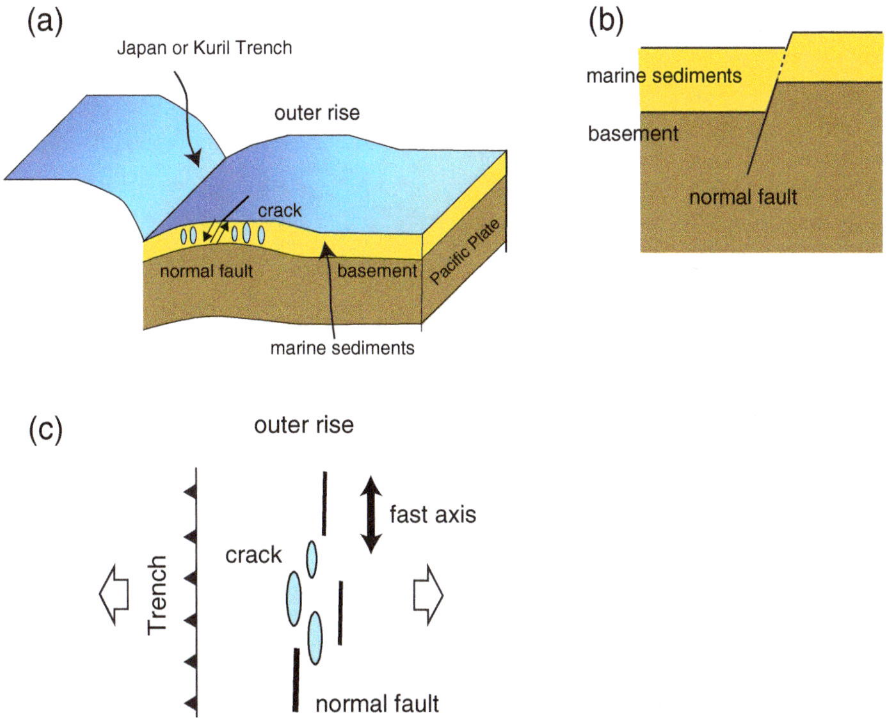

Fig. 10 Interpretation of the observed *S* anisotropy within the marine sediments. **a** Sketch showing cracks and normal faulting within sediments in the outer rise region. **b** Sketch showing fractures within sediments due to basement deformation. **c** Sketch showing cracks and faults in the outer rise region due to the tensional stress associated with the bending of the Pacific Plate

lineation direction (Nakanishi et al. 1989; Nakanishi 2011). The pre-existing fractures within the basement, which were formed at the creation of the Pacific Plate, gradually slide with the bending of the plate at the subduction zone, and the resulting displacement at the top of the basement produces the deformation of both sediments and the seafloor bathymetry above the basement. Because the region in this study has shifted southeasterly relative to the region where topographic features at the seafloor can often be observed, our study indicates that the development of these fractures begins landwards from the top of the outer rise region.

Compared with results in box 1, the patterns of the fast axis in box 2 appear to be scattered (Fig. 4d, e). Because the bathymetry and magnetic lineation in the oceanic crust have been investigated in detail (Nakanishi et al. 1989; Nakanishi 2011), we divided box 2 into the northern and southern parts (Fig. 4e) and interpreted the direction of the fast axis with a relatively large degree of anisotropy in box 2 with such information. In the northern part of box 2, the fast direction indicated by the red triangles in Fig. 4d, e corresponds to the trench axis at the west of the OBS sites, which is consistent with that in box 1. This reflects the trench–axis change from the Kuril Trench to Japan Trench. However, the fast axis oriented to the magnetic lineation could not be observed around this region, indicating fewer reactivations of the pre-existing fractures and high creations of newly formed fractures within the basement due to the bending of the plate.

In the southern part in box 2, we found that the fast axes at two OBS sites were oriented to N 30° E (denoted by the pale-blue triangles in Fig. 4e). It appears that this direction does not correspond to the direction of the trench axis at the west of the OBS sites. However, we consider that this direction reflects the trench axis south of 38° N (Fig. 4a). Nakanishi (2011) reported that landward- and oceanward-dipping escarpments with an orientation parallel to the trench axis south of 38° N have developed even at north of 38° N. Because the region of the OBS sites has shifted easterly relative to the region where such topographical features can often be observed, it appears that the fast axes observed at the two OBS sites corresponds to both stress-induced cracks and fractures due to small amounts of basement deformation developing landwards from the top of the outer rise region. This result may also indicate that the direction of the tensional stress is rotated clockwise in the southern part of box 2, according to the direction change of the trench axis occurring around 38° N.

The direction of the fast axis at an additional two OBS sites is N 45° E (denoted by the blue triangles in Fig. 4e).

Below the OBS sites, a NE–SW-trending lineation of the bathymetry can be recognized (Nakanishi 2011), and the direction almost coincides with the magnetic lineation pattern (Nakanishi et al. 1989; Nakanishi 2011). Such topography presumably reflects sediment deformation associated with fracture reactivations within the Pacific Plate formed at its creation, indicating that observed *S* anisotropy within the sediment is possibly caused by the basement deformation along the magnetic lineation.

In summary of our observations, the direction of the fast axis around the Kuril Trench can be explained by the fractures associated with tensional stress, including stress-induced cracks (and normal faulting) and fractures associated with basement deformation. The anisotropic structure within marine sediments in the northern part of the Japan Trench is also formed by similar mechanisms to those in box 1. In the southern part, although we found two OBS sites that show the fast axis corresponding to the magnetic lineation, the pattern of the fast axis can be presumably explained by the anisotropic structure associated with the tensional stress of the plate bending. In addition, we found fast polarization directions shown by the orange triangles in Fig. 4, which cannot be simply explained by mechanisms associated with the tensional stress. It would be necessary that further possible mechanisms for creating *S* anisotropy are explored through, e.g., MCS surveys and the bathymetry, in greater details.

Conclusions

We presented ACFs from ambient noise records observed by short-period OBSs that were deployed in the northwestern Pacific region. The ACFs clearly showed *S* reflection from the top of the basement at 201 OBS sites. The degree of *S* anisotropy was 1–2 % around the top part of the outer rise region, and the fast polarization axis was taken to be in the trench–parallel direction there. Based on our observations in the northwestern Pacific, we derived the conclusion that *S* anisotropy within marine sediments is produced by aligned fractures due to the tensional stresses that are associated with the bending of the Pacific Plate. The successful extractions of geographical patterns of *S* anisotropy indicate that our technique used in this study is capable of retrieving alignments of fractures and fabric within marine sediments through anisotropic information. We also extensively carried out numerical simulations with simple velocity models, taking into account anisotropy. Most importantly, these revealed that ambient noise should contain incident waves coming from either near-below or at various azimuths and slownesses to the station for the retrieval of *S* reflections containing accurate anisotropic information. It is predicted that the condition in which near-vertically propagating *S* waves are dominant is satisfied with low *Vs* of marine sediments. However, another numerical simulation with a realistic velocity model also elucidated that this condition possibly hinders other reflections from interfaces below the basement.

Abbreviations

Vp: *P*-wave velocity; *Vs*: *S* wave velocity; ACFs: auto correlation functions; OBSs: ocean-bottom seismometers; HTI: horizontal transverse isotropy; MCS: multi channel seismic.

Competing interests

The authors declare that they have no competing interests.

Authors' contributions

TTo performed data processing and numerical simulations, in addition to preparin the first version of the manuscript. TTo, YF, GF, and ST contributed to the interpretations. ST conducted a more sophisticated numerical simulation by using a realistic velocity model. GF, TTa, and KS designed seismic surveys and participated in data processing. All authors approved the final version of the manuscript.

Acknowledgements

We thank two anonymous reviewers for constructive comments that greatly improved the manuscript. The large-scale 3D simulation was conducted on the Earth Simulator at the Japan Agency for Marine-Earth Science and Technology (JAMSTEC) under the support of a joint research project between the Earthquake Research Institute, the University of Tokyo, and Center for Earth Information Science and Technology, titled "Seismic-wave propagation and strong ground motion in 3-D heterogeneous structure." Centroid moment tensor (CMT) solutions determined by using F-net records were provided by the National Research Institute for Earth Science and Disaster Prevention (NIED). TTo was funded by a Research Fellowship of the Japan Society for the Promotion of Science (JSPS) for Grants-in-Aid for Young Scientists (B: 15K17753).

Author details

[1]Research and Development Center for Earthquake and Tsunami, Japan Agency for Marine-Earth Science and Technology, 3173-25, Showa-machi, Kanazawa-ku, Yokohama 236-0001, Japan. [2]Department of Material System Science, Graduate School of Nanobioscience, Yokohama City University, 22-2, Seto, Kanazawa-ku, Yokohama 236-0027, Japan. [3]Present Address: Earthquake Research Institute, University of Tokyo, 1-1-1, Yayoi, 30 Bunkyo-ku, Tokyo 113-0032, Japan.

References

Bohlen T, Saenger EH (2006) Accuracy of heterogeneous staggered-grid finite-difference modeling of Rayleigh waves. Geophysics 71(4):T109–T115

Boué P, Poli P, Campillo M, Pedersen H, Briand X, Roux P (2013) Teleseismic correlations of ambient seismic noise for deep global imaging of the Earth. Geophys J Int 194:844–848

Brocher T (2005) Empirical relations between elastic wavespeeds and density in the Earth's crust. Bull Seismo Soc Am 95(6):2081–2092

Cadet JP, Kobayashi K, Aubouin J, Boulègue J, Deplus C, Dubois J, von Huene R, Jolivet L, Kanazawa T, Kasahara J, Koizumi K, Lallemand S, Nakamura Y, Pautot G, Suyehiro K, Tani S, Tokuyama H, Yamazaki T (1987) The Japan Trench and its juncture with the Kuril Trench cruise results of the Kaiko project, leg 3. Earth Planet Sci Lett 83:267–284

Campillo M, Paul A (2003) Long-range correlations in the diffuse seismic coda. Science 299:547–549

Draganov D, Wapenaar K, Thorbecke J (2004) Passive seismic imaging in the presence of white noise sources, The Leading Edge 23:889–892. doi:10.1029/2006GL028735

Draganov D, Wapenaar K, Mulder W, Singer J, Verdel A (2007) Retrieval of reflections from seismic background-noise measurements. Geophys Res Lett 34:L04305. doi:10.1029/2006GL028735

Draganov D, Campman X, Thorbecke J, Verdel A, Wapenaar K (2009) Reflection images from ambient seismic noise. Geophysics 74:A63–A67

Fujie G, Kodaira S, Yamashita M, Sato T, Takahashi T, Takahashi N (2013) Systematic changes in the incoming plate structure at the Kuril Trench. Geophys Res Lett 40:88–93. doi:10.1029/2012GL054340

Fujiwara T, Hirano N, Abe N, Takizawa K (2007) Subsurface structure of the "petit-spot" volcanoes on the northwestern Pacific Plate. Geophys Res Lett 34:L13305. doi:10.1029/2007GL030439

Fukuyama E, Ishida M, Dreger DS, Kawai H (1998) Automated seismic moment tensor determination by using online broadband seismic waveforms. Zisin 51:149–156

Furumura T, Chen L (2004) Large scale parallel simulation and visualization of 3-D seismic wavefield using Earth simulator. Comput Model Eng Sci 6:153–168. doi:10.3970/cmes/2004.006.153

Graves RW (1996) Simulating seismic wave propagation in 3D elastic media using staggered-grid finite differences. Bull Seis Soc Am 86(4):1091–1106

Haacke RR, Westbrook GK (2006) A fast, robust method for detecting and characterizing azimuthal anisotropy with marine PS converted waves, and its application to the west Svalbard continental slope. Geophys J Int 167:1402–1412. doi:10.1111/j.1365-246X.2006.03186.x

Haacke RR, Westbrook GK, Peacock S (2009) Layer stripping of shear-wave splitting in marine PS waves. Geophys J Int 176:782–804. doi:10.1111/j.1365-246X.2008.04060.x

Hudson JA (1981) Wave speeds and attenuation of elastic waves in material containing cracks. Geophys J Int 64:133–150

Isse T, Shiobara H, Tamura Y, Suetsugu D, Yoshizawa K, Sugioka H, Ito A, Kanazawa T, Shinohara M, Mochizuki K, Araki E, Nakahigashi K, Kawakatsu H, Shito A, Fukao Y, Ishizuka O, Gill JB (2009) Seismic structure of the upper mantle beneath the Philippine Sea from seafloor and land observation: implications for mantle convection and magma genesis in the Izu-Bonin-Mariana subduction zone. Earth Planet Sci Lett 278:107–119

Kanazawa T, Sager WW, Escutia C, Araki E, Arney JE, Carlson RL, Downey WS, Einaudi F, Haggas SL, Hayasaka Y, Hirata K, Horner-Johnson BC, Mandernack KW, McCarthy FMG, Moberly R, Mochizuki M, Pedersen RØ, Salimullah ARM, Shinohara M, Werner C (2001) Proc. ODP, Init. Repts. 191:College Station, TX (Ocean Drilling Program). doi:10.2973/odp.proc.ir.191.2001.

Kawakatsu H, Kumar P, Takei Y, Shinohara M, Kanazawa T, Araki E, Suyehiro K (2009) Seismic evidence for sharp lithosphere-asthenosphere boundaries of oceanic plates. Science 324:499–502. doi:10.1126/science.1169499

Kodaira S, Fujie G, Yamashita M, Sato T, Takahashi T, Takahashi N (2014) Seismological evidence of mantle flow driving plate motions at a palaeo-spreading centre. Nature Geosci 7:371–375. doi:10.1038/NGEO2121

Kumar P, Kawakatsu H, Shinohara M, Kanazawa T, Araki E, Suyehiro K (2011) *P* and *S* receiver function analysis of seafloor borehole broadband seismic data. J Geophys Res 116:B12308. doi:10.1029/2011JB008506

Levander AR (1988) Fourth-order finite-difference P-*SV* seismograms. Geophysics 53(11):1425–1436

Longuet-Higgins MS (1950) A theory of the origin of microseisms. Phil Trans R Soc A 243:1–35

Miyazawa M, Snieder R, Venkataraman A (2008) Application of seismic interferometry to extract *P*- and *S*- wave propagation and observation of shear-wave splitting from noise data at Cold Lake, Alberta, Canada. Geophysics 73(4):D35–D40

Nagaya M, Oda H, Akazawa H, Ishise M (2008) Receiver functions of seismic waves in layered anisotropic media: application to the estimate of seismic anisotropy. Bull Seis Soc Am 98(6):2990–3006. doi:10.1785/0120080130

Nakamura T, Takenaka H, Okamoto T, Kaneda Y (2012) FDM simulation of seismic-wave propagation for an aftershock of the 2009 Suruga Bay Earthquake: effects of ocean-bottom topography and seawater layer. Bull Seis Soc Am 102(6):2420–2435

Nakanishi M (2011) Bending-related topographic structures of the subducting plate in the northwestern Pacific Ocean. In: Ogawa Y, Anma R, Dilek Y (ed) Modern Approaches in Solid Earth Sciences, 8, Springer Science+ Business Media B.V, Netherlands, doi:10.1007/978-90-481-8885-7_1.

Nakanishi M, Tamaki K, Kobayashi K (1989) Mesozoic magnetic anomaly lineations and seafloor spreading history of the northwestern pacific. J Geophys Res 94:15437–15462

Nakata N, Chang JP, Lawrence JF, Boué P (2015) Body-wave extraction and tomography at Long Beach, California, with ambient-noise interferometry. J Geophys Res 120(2):1159–1173. doi:10.1002/2015JB011870

Nolet G, Dorman LM (1996) Waveform analysis of Scholte modes in ocean sediment layers. Geophys J Int 125:385–396

Okada Y, Kasahara K, Hori S, Obara K, Sekiguchi S, Fujiwara H, Yamamoto A (2004) Recent progress of seismic observation networks in Japan: Hi-net, F-net, K-NET and KiK-net. Earth Planets Space 56:xv–xxvii

Okamoto T, Takenaka H (2005) Fluid-solid boundary implementation in the velocity-stress finite-difference method, Zisin 57:355–364 (in Japanese with English abstract)

Park J, Yu Y (1992) Anisotropy and coupled free oscillations: simplified models and surface waves observations. Geophys J Int 110:401–420

Poli P, Pedersen HA, Campillo M, POLENET/LAPNET Working Group (2012a) Emergence of body waves from cross-correlation of short period seismic noise. Geophys J Int 188:549–558

Poli P, Campillo M, Pedersen H, LAPNET Working Group (2012b) Body-wave imaging of Earth's mantle discontinuities from ambient seismic noise. Science 338:1063–1065

Robertsson J, Blanch JO, Symes WW (1994) Viscoelastic finite-difference modeling. Geophysics 59:1444–1456. doi:10.1190/1.1443701

Roux P, Sabra KG, Gerstoft P, Kuperman WA, Fehler MC (2005) *P*-waves from cross-correlation of seismic noise. Geophys Res Lett 32:L19303. doi:10.1029/2005GL023803

Ruigrok E, Campman X, Wapenaar K (2012) Basin delineation with a 40-hour passive seismic record. Bull Seis Soc Am 102(5):2165–2176

Ryberg T (2011) Body wave observations from cross-correlations of ambient seismic noise: a case study from the Karoo, RSA. Geophys Res Lett 38:L13311. doi:10.1029/2011GL047665

Saenger EH, Gold N, Shapiro SA (2000) Modeling the propagation of elastic waves using a modified finite-difference grid. Wave Motion 32:77–92

Shapiro NM, Campillo M (2004) Emergence of broadband Rayleigh waves from correlations of the ambient seismic noise. Geophys Res Lett. doi:10.1029/2004GL019491

Shimamura H, Asada T, Suyehiro K, Yamada T, Inatani H (1983) Longshot experiments to study velocity anisotropy in the oceanic lithosphere of the northwestern Pacific. Phys Earth Planet Inter 31:348–362

Shinohara M, Fukano T, Kanazawa T, Araki E, Suyehiro K, Mochizuki M, Nakahigashi K, Yamada T, Mochizuki K (2008) Upper mantle and crustal seismic structure beneath the Northwestern Pacific Basin using a seafloor borehole broadband seismometer and ocean bottom seismometers. Phys Earth Planet Inter 170:95–106

Snieder R (2004) Extracting the Green's function from the correlation of coda waves: a derivation based on stationary phase. Phys Rev E 69:046610

Takemura S, Yoshimoto K (2014) Strong seismic wave scattering in the low-velocity anomaly associated with subduction of oceanic plate. Geophys J Int 197:1016–1032. doi:10.1093/gji/ggu031

Takemura S, Furumura T, Maeda T (2015) Scattering of high-frequency seismic waves caused by irregular surface topography and small-scale velocity inhomogeneity. Geophys J Int 201:459–474. doi:10.1093/gji/ggv038

Thorbecke J, Draganov D (2011) Finite-difference modeling experiments for seismic interferometry. Geophysics 76(6):H1–H18. doi:10.1190/GEO02010-0039.1

Tibuleac IM, von Seggern D (2012) Crust-mantle boundary reflectors in Nevada from ambient seismic noise autocorrelations. Geophys J Int 189:493–500. doi:10.1111/j.1365-246X.2011.05336.x

Tonegawa T, Fukao Y, Nishida K, Sugioka H, Ito A (2013) A temporal change of shear wave anisotropy within the marine sedimentary layer associated with the 2011 Tohoku-Oki earthquake. J Geophys Res 118:607–615. doi:10.1002/jgrb.50074

Tsuji T, Dvorkin J, Mavko G, Nakata N, Matsuoka T, Nakanishi A, Kodaira S, Nishizawa O (2011) *Vp*/*Vs* ratio and shear-wave splitting in the Nankai Trough seismogenic zone: insights into effective stress, pore pressure, and sediment consolidation. Geophysics 76(3):WA71–WA82

Zhan Z, Ni S, Helmberger DV, Clayton RW (2010) Retrieval of Moho-reflected shear wave arrivals from ambient seismic noise. Geophys J Int 182:408–420

7

Paleoceanographic history of the Japan Sea over the last 9.5 million years inferred from radiolarian assemblages (IODP Expedition 346 Sites U1425 and U1430)

Kenji M. Matsuzaki[1,2*], Takuya Itaki[3], Ryuji Tada[1] and Shin-ichi Kamikuri[4]

Abstract

Previous studies showed that the evolution of the Japan Sea paleoceanography since the Miocene has been influenced by the regional tectonism (e.g., opening/closing of the connecting seaways) and regional/global climate. In the Japan Sea, Expedition 346 of the Integrated Ocean Drilling Program (IODP) retrieved core sediments dating back to the Miocene at two sites (U1425 and U1430). In this study, we reconstruct shallow-to-deep-water hydrography of the Japan Sea during the Mio-Pliocene based on radiolarian assemblages at Sites U1425 and U1430 considering the local tectonism and changes in global/regional climate. Our data suggest that glacioeustatic sea-level changes have probably had an influence on the local paleoceanography between 9.5 and 7.0 Ma. Indeed, warm water probably flowed from the North Pacific into the Japan Sea when sea level was high via shallow central and eastern seaways. In addition, the sill depth of the northern seaway was probably close to 1000 m between 9.5 and 7.8 Ma and had probably allowed inflow of oxygen minimum zone water from the North Pacific to the Japan Sea when sea level was high. In contrast, our data imply that *Cycladophora nakasekoi*, an endemic species to the Japan Sea, dominated between 9.5 and 7.3 Ma when sea level was low. Our data also suggest a progressive shoaling of the sill for the period since 7.8 Ma and that global climatic events such as such the late Miocene cooling (7.5–5.5 Ma) and the early Pliocene warmth have had a sustained influence on the Japan Sea. During the mid-Pliocene, a deep cooling of the subsurface to intermediate water of the Japan Sea likely occurred because species related to subarctic subsurface to intermediate waters were dominant between 5 and 3.8 Ma. The Northern Hemisphere Glaciation (ca. 3.0–2.7 Ma) and Mid-Pleistocene Transition (1.2–0.8 Ma) have both likely intensified the cooling of the Japan Sea.

Keywords: Japan Sea, Integrated Ocean drilling program, Paleoceanography, Sea-level changes, Sill depths, Oxygen minimum zone, Radiolarians

Introduction

The Japan Sea is a back-arc basin opened by continental rifting during the late Oligocene to middle Miocene Epoch, ca. 28–13 million years ago (Ma) (e.g., Tamaki et al. 1992; Jolivet et al. 1994). Currently, the Japan Sea is connected to neighboring seas by four shallow straits: the Tsushima Strait (sill depth of 130 m) connecting to the East China Sea, the Tsugaru Strait (sill depth of 130 m) connecting to the Pacific, and the Soya and Mamiya Straits (sill depths of 55 and 12 m, respectively) connecting to the Sea of Okhotsk (Fig. 1). Therefore, water exchanges are restricted only to the surface waters of the East China Sea, Northwest Pacific, and Sea of Okhotsk, and intrusion of intermediate and deep waters is prevented (e.g., Gamo et al. 2014). This semi-isolation of the Japan Sea leads to formation of its own deep water, the so-called Japan Sea Proper Water, by cooling of the Tsushima Warm Current (TWC) (e.g., Gamo et al. 2014). The Japanese islands and Sakhalin Island, which bound the eastern margin of the Japan Sea, are characterized by

* Correspondence: km.matsuzaki@aori.u-tokyo.ac.jp
[1]Department of Earth and Planetary Science, Graduate School of Science, The University of Tokyo, 7-3-1 Hongo, Bunkyo-ku, Tokyo 113-0033, Japan
[2]Present address: Atmosphere and Ocean Research Institute, The University of Tokyo, 5-1-5 Kashiwanoha, Kashiwa, Chiba 277-8564, Japan
Full list of author information is available at the end of the article

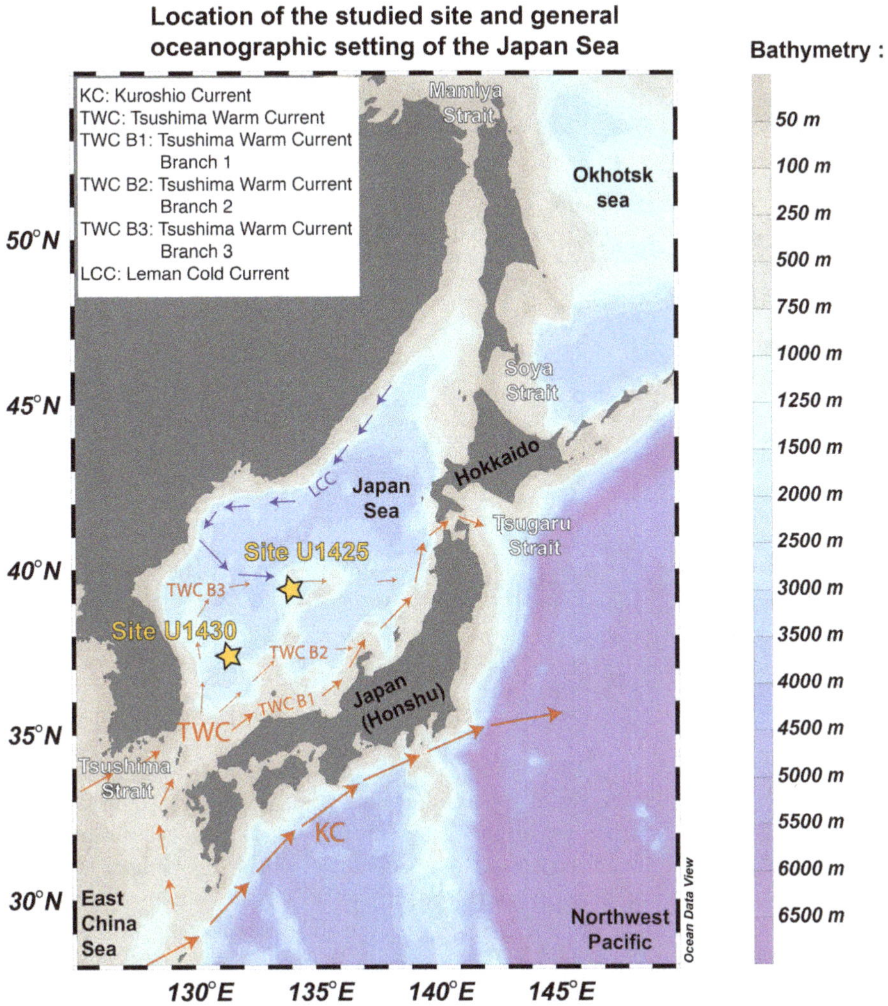

Fig. 1 Location of the study sites and the local oceanographic setting of the Japan Sea. The map was generated by the software Ocean Data View 4 of Schlitzer (2016)

active tectonism, which drastically modified the Japan Sea paleogeography, such as the sill depths of the key straits during the Neogene (e.g., Iijima and Tada 1990; Jolivet et al. 1994; Kosaka et al. 1992; Kano et al. 1991; Suzuki 1989; Taira 2001).

During the middle Miocene, a sedimentary basin called the Fossa Magna was formed in central Japan because of back-arc rifting associated with the opening of the Japan Sea (e.g., Iijima and Tada 1990). The paleobathymetry of this basin evolved from deep water depths in the middle Miocene (middle bathyal sedimentary facies) to shallow water depths in the late Miocene (shallow-water sedimentary facies) (e.g., Fujii et al. 1992; Iijima and Tada 1990; Kobayashi and Tateishi 1992; Kosaka et al. 1992). The beginning of a compressional stress regime in Northeast Japan due to the collision of the Izu-Bonin Arc since the middle Miocene may have caused the shoaling and closure of this seaway ca. 8–7 Ma (e.g., Sato 1994; Sato et al. 2004; Iijima and Tada 1990; Yoshida et al. 2013). In addition, during the late Miocene, a few shallow seaways (about 200 m deep) probably existed in Northeast Japan because of subsidence of the Tohoku-Hokkaido area during the late Miocene (Sato 1994; Yoshida et al. 2013). However, these subsided areas were also likely uplifted because of the compressional stress regime, which started ca. 8–7 Ma (Sato 1994; Yoshida et al. 2013) (Fig. 2). At the same time, the south of the Japan Sea was likely isolated by land from the North Pacific because no marine fossils and microfossils were recorded from the late Miocene (ca. 10 Ma) to the early Pliocene (ca. 4 Ma) along the eastern edges of the Yellow and South China seas (e.g., Iijima and Tada 1990; Kano et al. 1991; Wang et al. 2014) (Fig. 2). In the northern Japan Sea (corresponding to the southern to central part of Hokkaido Island), several geological studies suggested that a bathyal environment prevailed since the beginning of the late Miocene until the late Pliocene (e.g., Fukusawa 1988;

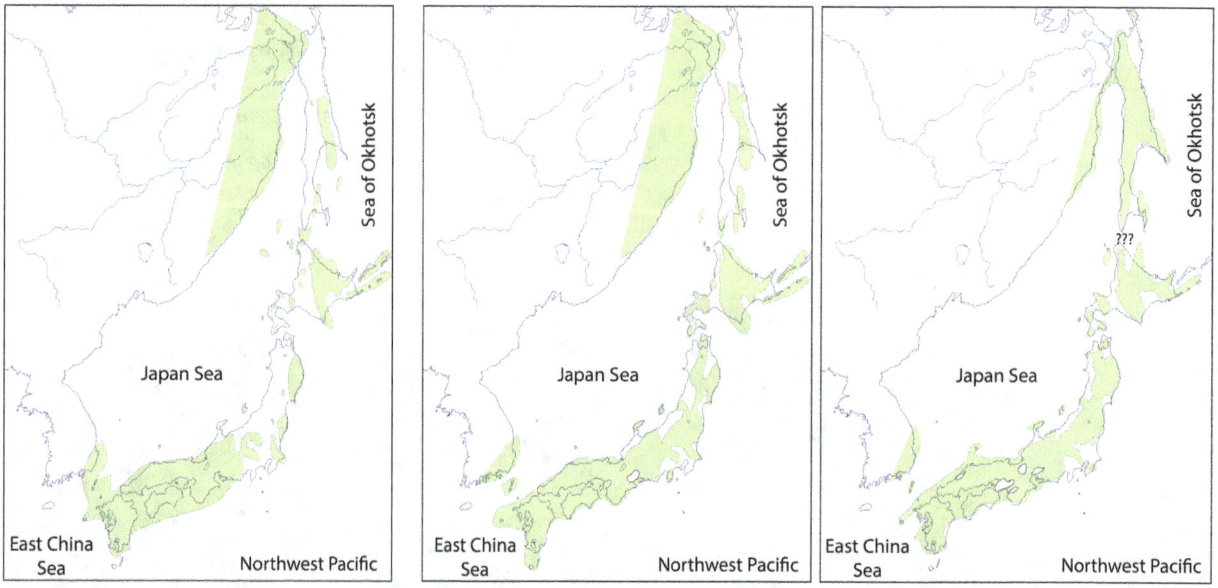

Fig. 2 Evolution of the Japan Sea since 10.5 Ma and its paleomaps (based on Fukusawa 1988; Fujii et al. 1992; Iijima and Tada 1990; Sato 1994; Kano et al. 1991; Kharakhinov 2010; Kobayashi and Tateishi 1992; Kosaka et al. 1992; Sagayama, 2002; Yahata 2002; Yoshida et al. 2013). The base lines of the coasts and hydrography came from http://d-maps.com

Iijima and Tada 1990; Itaki 2016; Kosaka et al. 1992; Kano et al. 1991; Sagayama 2002; Suzuki 1989; Ogasawara 1994; Yahata 2002) (Fig. 2). At the northernmost Japan Sea, in the area presently around the Sakhalin and Tatar straits, tectonic studies suggested that a neritic environment prevailed from the middle Miocene to the end of the late Miocene (Kharakhinov 2010) (Fig. 2). A southward shift of the Okhotsk plate likely caused a southward migration of the Kuril basin and a progressive shoaling and closure of the Sakhalin seaways during the early to late Pliocene (Takeuchi et al. 1999; Kharakhinov 2010).

Paleoceanographic conditions in the Japan Sea are known to have been sensitive to orbital and suborbital forcings and glacioeustatic sea-level changes during the middle to late Pleistocene (e.g., Tada et al. 1999; Itaki et al. 2007; Kido et al. 2007). Currently, the TWC is the unique current flowing into the Japan Sea and controlling its oceanographic condition (Gamo et al. 2014) (Fig. 1). After entering, the TWC bifurcates into three branches at the southwestern part of the Japan Sea near the Tsushima Strait (Hase et al., 1999) (Fig. 1). The first branch of the TWC (TWC B1) flows along the west coast of Honshu Island (Fig. 1). The second branch of the TWC (TWC B2) flows along the Korean peninsula before rejoining TWC B1, and the third branch of the TWC (TWC B3) flows along the Korean Peninsula northward to form the Leman Cold Current (LCC in Fig. 1). During the Pleistocene, the amplitude of glacioeustatic sea-level variations caused by orbital cycles was about 100 m (e.g., Miller et al. 2005; Siddall et al. 2003; Yokoyama et al. 2001; Itaki 2007; Itaki et al. 2004). Because the Tsushima Strait has a shallow sill depth (about 130 m), a sea-level variation of about 100 m restricted flow of the TWC into the Japan Sea, leading to a decrease in surface water salinity and drastically reducing the local deep-water ventilation, which resulted in euxinic bottom-water conditions (e.g., Kido et al. 2007; Oba et al. 1991; Tada et al. 1999).

For the Mio-Pliocene, Tada (1994) compiled the published sedimentological and diatom data of Koizumi (1992a, 1992b) and Tada and Iijima (1992) related to the paleoceanographic history of the Japan Sea and discussed the sill depth of the connecting straits that could have influenced exchanges between the Japan Sea and the North Pacific. Tada (1994) suggested that sediment deposition in the Japan Sea began at ca. 20 Ma, and the sea was open to the south, east, and north until ca. 10.5 Ma. Warm surface water and relatively oxic bottom water conditions prevailed until ca. 15–14 Ma, when a drastic cooling was recorded. Mollusca fossil records also suggest this kind of cooling and indicate that temperate water prevailed since ca. 13 Ma (Chinzei 1978; Ogasawara 1994). According to recent studies, this cooling is closely related to the mid-Miocene global cooling events (15–14 Ma), including the East Antarctic Ice Sheet expansion and stabilization (Holbourn et al. 2013, 2014; Zachos et al. 2001, 2008). The southern strait closed ca. 10.5 Ma, which enhanced the cooling of the Japan Sea (Tada 1994; Ogasawara 1994). In contrast, sedimentary, microfossil facies, and chemical data suggest that the Japan Sea bottom water was oxygen-deficient from

18 to ca. 6.5 Ma (Hanagata 2003, 2006; Tada 1994; Yamamoto et al. 2005). Tada (1994) suggested that the origin of the oxygen-poor water was related to inflow of oxygen minimum zone (OMZ) water of the adjacent North Pacific. The presence of this OMZ in the North Pacific during the Miocene may be controvertible. However, Woodruff and Savin (1989) used measurement of $\delta^{13}C$ in the benthic foraminifer *Cibicidoides* to report a very low $\delta^{13}C$ value for the North Pacific between latitudes 20° and 40° at water depths between ca. 1000 and 3000 m during the middle and late Miocene. This implied the presence of a layer with low dissolved oxygen content in the North Pacific between 1000 and 3000 m at that time (e.g., Woodruff and Savin 1989). Additionally, radiolarian studies were conducted in the context of the Deep-Sea Drilling Program (DSDP) and Ocean Drilling Program (ODP) (namely, legs 31, and 127 and 128). These studies mainly produced biostratigraphic reports (Alexandrovich 1992; White and Alexandrovich 1992; Ling 1992). Kamikuri and Motoyama (2007) showed that, at DSDP Site 302, radiolarian species related to the North Pacific deep water occurred in the Japan Sea between 8 and ca. 3 Ma, which suggests inflow of deep water of the North Pacific into the Japan Sea. These findings showed that the hypothesis proposed by Tada (1994) for that time is plausible, although some concern remains about the ecology of the radiolarian species associated with the North Pacific deep water. For the late Pliocene, few studies have reconstructed paleoceanographic conditions of the Japan Sea. Most of them agree that, during the late Pliocene, cold surface water prevailed in the northern part of the Japan Sea, while warm water from the southern strait (Tsushima Strait) influenced the sea only during interglacial periods (e.g., Kitamura and Kimoto 2006; Kitamura et al. 2001).

Problem addressed here

Our knowledge about global and regional climatic events has improved greatly. Recent studies stressed the importance of the monsoon on the global climate and ocean and also stressed the importance of the uplift of the Tibetan-Himalayan Plateau at the Oligocene/Miocene boundary in establishing the monsoon (e.g., Betzler et al. 2017; Clift 2017; Holbourn et al. 2018; Tada et al. 2016). The monsoon has an influence on the carbon cycle by controlling chemical weathering and thus strongly affecting the global climate (Raymo and Ruddiman 1992; Tada et al. 2016). During the late Miocene between 7 and 5.5 Ma, a period of global cooling was reported for the middle to high latitudes of both hemispheres, the so-called late Miocene cooling event (Herbert et al. 2016). Recent studies suggested that the East Asian winter monsoons (EAWM) were probably stronger at that time, and thus the EAWM probably played an important role in the late Miocene cooling event (Holbourn et al. 2018). Recent studies suggested that the beginning of the shoaling of the Panama Isthmus to a few hundred meters of water depth probably caused changes in the meridional deep ocean circulation (Butzin et al. 2011). Then, during the Pliocene, the final closure of the Panama Isthmus caused drastic changes in ocean salinity/heat distributions that may have triggered the Northern Hemisphere Glaciation (NHG) (e.g., Cortese et al. 2004; Haug et al. 2001; Karas et al. 2017; Motoi et al. 2005; Sandoval et al. 2017; Steph et al. 2006; Raymo 1994; Stroynowski et al. 2015). These changes may have caused an episodic strong Atlantic meridional overturning circulation between 4.5 and 3.8 Ma (Karas et al. 2017). In addition, although it remains under debate, between 16.4 and 8 Ma, the Indonesian seaway, which allowed water mass exchanges between the Northern Pacific and the Indian Ocean, closed progressively and probably caused the formation of a proto-Kuroshio Current (e.g., Kennett et al. 1985; Ogasawara et al. 2008).

Consequently, the reconstruction of changes in the Japan Sea paleoceanography during the Mio-Pliocene is crucial for a better understanding of how the Japan Sea was affected by these global tectonic and climatic events.

Objective and strategy

In this study, we analyzed deep-sea sediments of the Japan Sea collected during Expedition 346 of the Integrated Ocean Drilling Program (IODP) to reconstruct past hydrographic conditions of the Japan Sea during the Mio-Pliocene. Sediment cores were collected from seven sites covering wide latitudinal and depth ranges in the Japan Sea. We selected samples collected at Sites U1425 and U1430 in the central and southcentral parts of the Japan Sea during IODP Expedition 346 because hemipelagic sediments covering the middle Miocene to Pliocene were retrieved (Tada et al. 2015a, 2015b). We analyzed 154 sediment core samples collected at Site U1425 (about 2000 m deep) and 51 samples collected at Site U1430 (about 1000 m deep) to reconstruct past paleoceanography of the Japan Sea since the late Miocene. In the Japan Sea, the present calcite compensation depth (CCD) is about 2000 m, and dissolution of planktic foraminifera begins at a depth of about 1400 m (Ujiie and Ichikura 1973). Lee et al. (2000) reported that the CCD was likely shallower during the Mio-Pliocene. These data show that the use of paleoceanographic proxies based on calcareous microfossils is not necessarily suitable for the Japan Sea, particularly for sites with water depths deeper than 1000 m (Sites U1425 and U1430). Consequently, the use of assemblages of siliceous microfossils appears to be a unique alternative to reconstructing oceanographic conditions of the shallow to deep waters. Because radiolarians are a unique siliceous microfossil group inhabiting shallow to deep waters, we analyzed their assemblages and changes over time to reconstruct hydrographic changes of the shallow to deep waters since the Miocene.

Methods/experimental

Sites studied

In this study, we analyzed sediment core samples collected from Sites U1425 and U1430 drilled during IODP Expedition 346 (2013). However, because a hiatus is present between 7.2 and ca. 4.5 Ma at Site U1430 (Kamikuri et al. 2017; Tada et al. 2015b), most of the analysis was conducted on samples from Site U1425. In this study, we only used data from Site U1430 to provide supporting information.

Site U1425 is located in the central part of the Japan Sea, in the middle of the Yamato Bank (1909 m water depth) at 39° 29.44′ N and 134° 26.55′ E (Tada et al. 2015a). The drilling at Site U1425 recovered a total of 417.5 m of sediment (recovery rate of 98%). The description of the sediment core lithology at Site U1425 given below follows that of Tada et al. (2015a).

The lithology of Mio-Pliocene sediments at this site can be divided into three major lithological units, and each is subdivided into two subunits (e.g., IA and IB) (Figs. 3 and 4). Unit III is the older and deeper lithological unit (> 262 m CCSF-D; > 7.419 Ma) (CCSF-D is an abbreviation for core composite depth below seafloor). This unit consists of clay, diatom-rich clay, and diatom ooze. This unit can be divided into two subunits based on the degree of lithification (IIIA and IIIB). Subunit IIIB corresponds to the depth interval deeper than 356 m CCSF-D (> 9.528 Ma), and it is characterized by gray siliceous claystone with occasional parallel laminations and burrows. Lithological subunit IIIB is not discussed in this study, because it contained no biogenic silica. Subunit IIIA corresponds to the depth interval between ca. 356 and 262 m CCSF-D (9.528–7.419 Ma) (Figs. 3 and 4). This subunit is characterized by decimeter- to meter-scale cyclic alternation of dark gray diatom ooze and moderate to heavily bioturbated clayey diatomaceous ooze. A fine lamination of diatom ooze occurring within a dark organic-rich layer extending to a thickness of about 1.5 m has been recognized at 275 m core depth below seafloor (CSF-A) (Tada et al. 2015a), which corresponds to about 287 m CCSF-D (Irino et al. 2018). Based on the age model of Kamikuri et al. (2017), these laminations occurred about 8 Ma. Lithological unit II corresponds to a depth interval between ca. 262 and

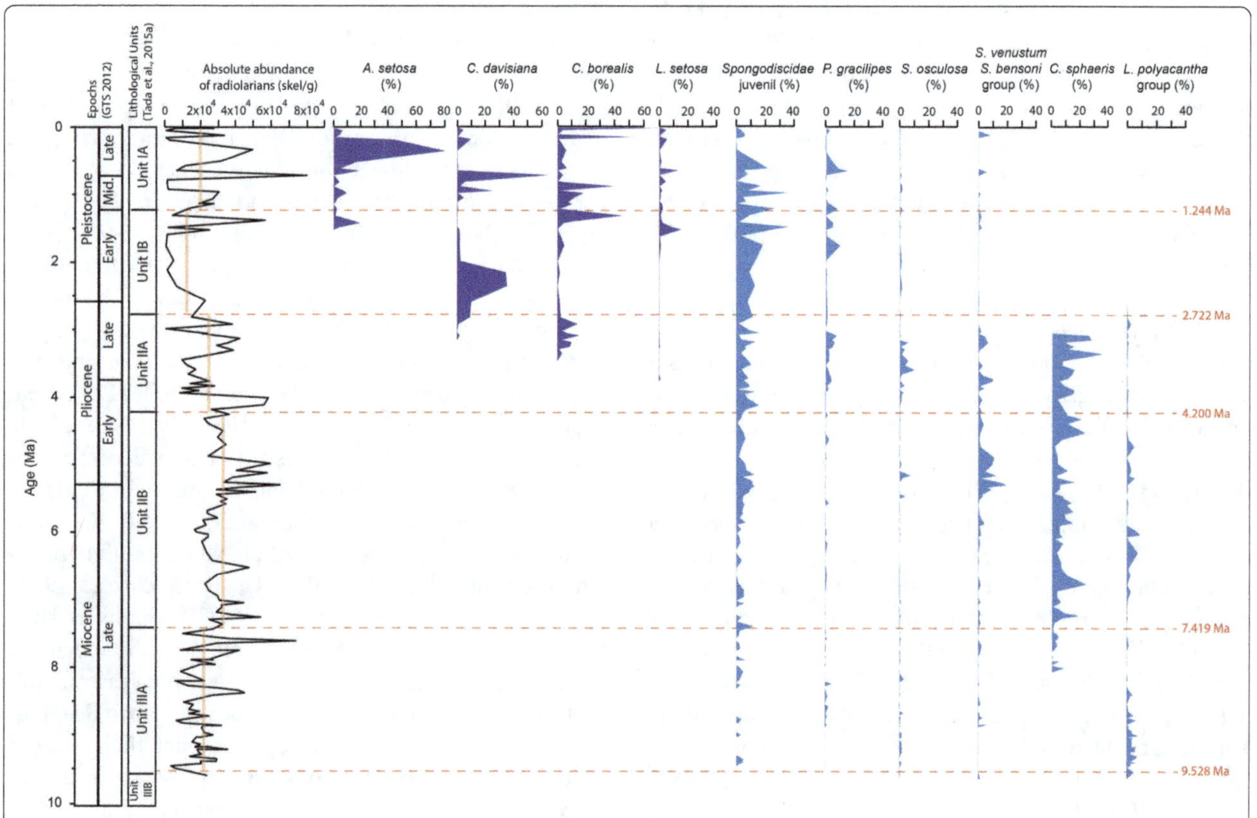

Fig. 3 Radiolarian absolute abundance (AB) (skel/g) and relative abundance (%) of selected radiolarian species (1/2) at Site U1425 plotted by age (Ma) (Kamikuri et al. 2017 depth age model) and compared to the lithological units of Tada et al. (2015a). Narrow red shading on the AB curve corresponds to the mean AB for each lithological subunit

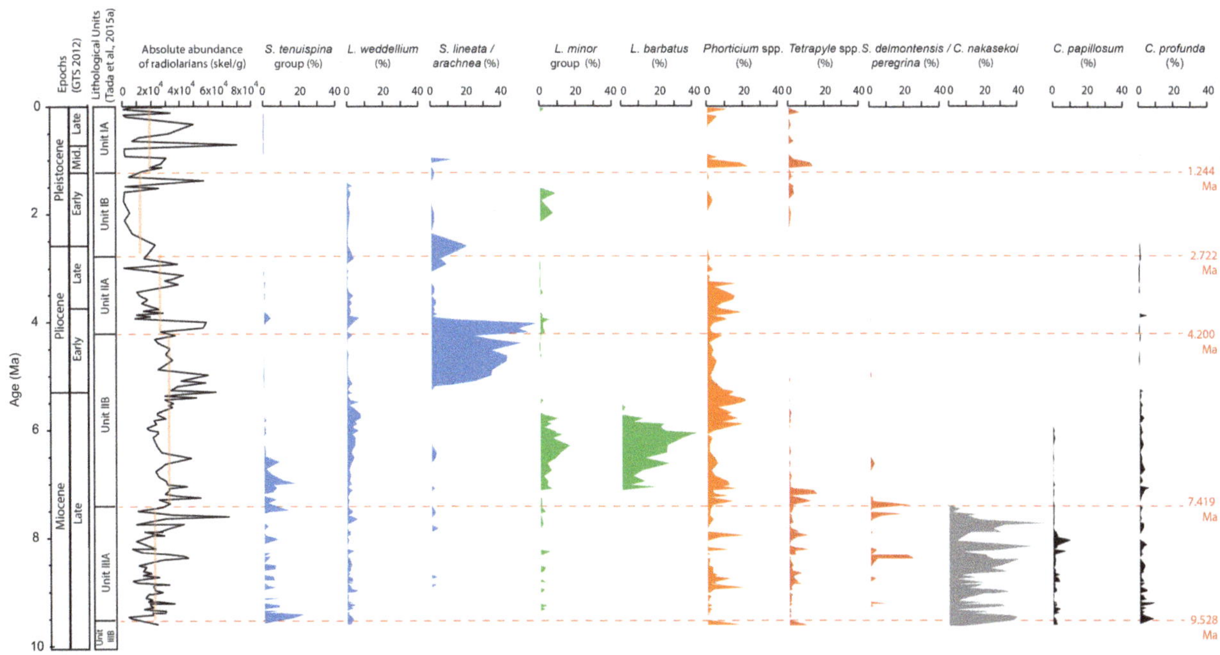

Fig. 4 Radiolarian absolute abundance (AB) (skel/g) and relative abundance (%) of selected radiolarian species (2/2) at Site U1425 plotted by age (Ma) (Kamikuri et al. 2017 depth age model) and compared to the lithological units of Tada et al. (2015a). Narrow red shading on the AB curve corresponds to the mean AB for each lithological subunit

99 m CCSF-D (7.419–2.722 Ma) (Figs. 3 and 4). This unit consists of diatomaceous clay and diatomaceous ooze that is moderately to heavily bioturbated. Unit II can be distinguished from unit III by a decrease in terrigenous matter. This unit is divided into two subunits (IIA and IIB) based on the diatom content and intensity of bioturbation. Subunit IIB (ca. 262–137 m CCSF-D; 7.419–4.200 Ma) is dominated by brownish diatom ooze. Diatom content (70–90% of the smear slide) is a key component for the recognition of this subunit. Subunit IIA (ca. 137–99 m CCSF-D; 4.200–2.722 Ma) mainly consists of brownish-greenish diatom-rich clay. In general, the sediments of subunit IIA are heavily bioturbated. Unit I corresponds to a depth interval between ca. 99 and 0 m CCSF-D (since 2.722 Ma) (Figs. 3 and 4). Unit I consists of clay and silty clay with small amounts of diatom-rich clay, while foraminifer-bearing clay is rare. This unit is divided into two subunits (IA and IB) based on the frequency of the color alternations (dark and light layer) and the intensity of bioturbation. Subunit IB (ca. 99–51 CCSF-D; 2.722–1.244 Ma) consists of light greenish gray clay with a low frequency of dark and light color alternation. Subunit IA (ca. 51–0 m CCSF-D; since 1.244 Ma) is dominated by clay, small amounts of diatomaceous-bearing clay, and tephras. This subunit is also characterized by a pronounced decimeter-scale alternation of light and dark layers.

Site U1430 is located on the southern slope of the eastern South Korean Plateau (1083 m below sea level) at 37° 54.16′ N and 131° 32.25′ E. As above, the lithology of the sediment cores at Site U1430 is described following Tada et al. (2015b) (Fig. 5). The lithology of this site consists of clayey silt, silty clay, nannofossil ooze, diatom ooze, claystone, and sandstone. This site can be divided into four units (I to IV), and each of units I to III is further divided into two subunits based on sediment composition and relative abundances of biosiliceous and siliciclastic fractions. However, subunit IIB and unit IV correspond to glauconite sandstone intervals with no biogenic silica, so these units are not discussed further. Subunit IIIA corresponds to the depth interval between ca. 256 and 103 m CCSF-D (11.714–7.891 Ma) (Fig. 5). This subunit is characterized by decimeter- to meter-scale alternating layers of heavily bioturbated, relatively dark gray diatomaceous ooze and very dark gray diatom-rich clay. Lithological unit II corresponds to a depth interval between ca. 103 and 60 m CCSF-D (7.891–3.003 Ma) (Fig. 5). This unit consists of green silty clay, olive gray diatom-rich silty clay, and dark olive-gray diatom ooze. This unit is divided into two subunits (IIA and IIB) based on the diatom content. Preliminary studies using biostratigraphic data suggest that subunit IIB (103–76 m CCSF-D; 7.891–4.100 Ma) corresponds to a hiatus (Tada et al. 2015b; Kamikuri et al. 2017). Subunit IIA (ca. 76–60 m CCSF-D; 4.100–3.003 Ma) mainly consists of meter-scale alternating heavily bioturbated dark olive-gray diatom-bearing clay and very dark gray diatom ooze. Unit I differs from unit

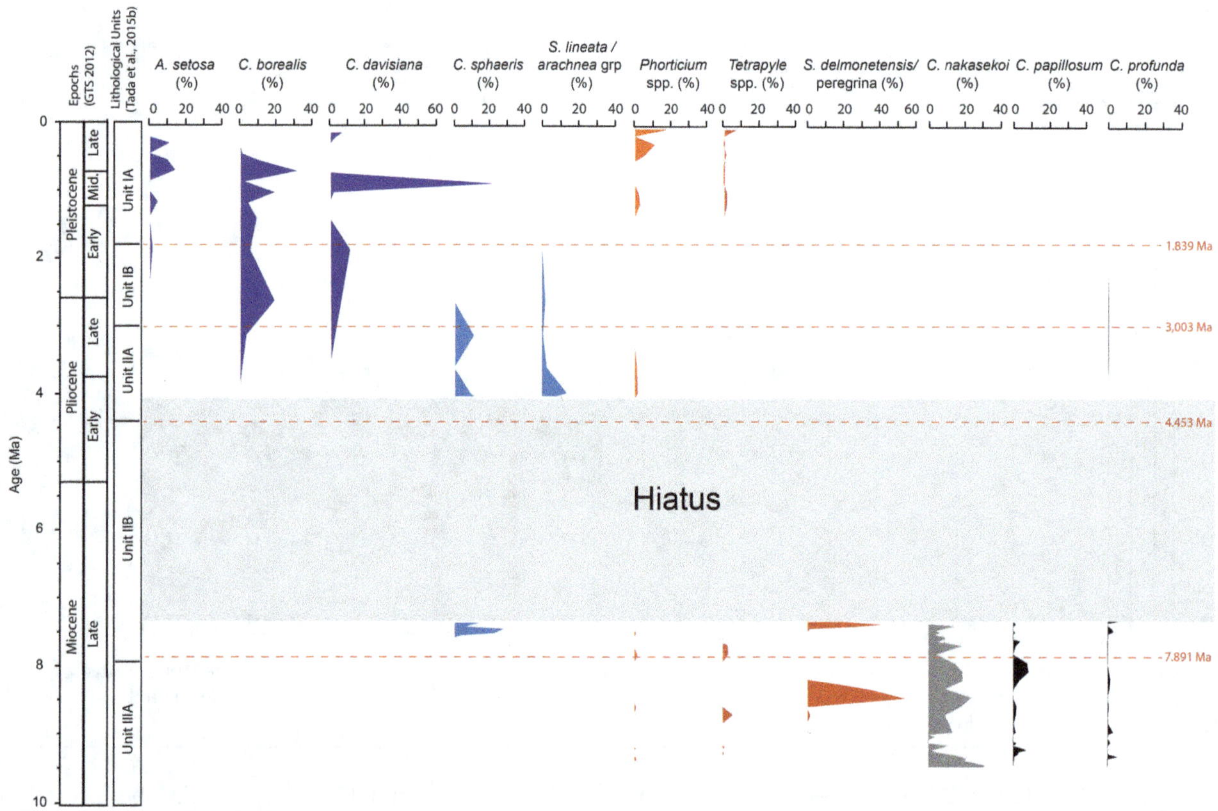

Fig. 5 Relative abundance (%) of selected radiolarian species at Site U1430 plotted by age (Ma) (Kamikuri et al. 2017 depth age model) and compared to the lithological units of Tada et al. (2015b)

II by a decrease in biosiliceous content. This unit consists of alternating interbedded clayey silt, silt clay, foraminifer-rich diatom-bearing clayey silt, nannofossil ooze, and tephras. As for Site U1425, this unit is also divided into two subunits (IA and IB) based on the frequency of the color alternations and the intensity of bioturbation. Subunit IB (60–46 m CCSF-D; 3.003–1.839 Ma) consists of a low-frequency dark and light color alternation and increasing dominance of light gray clayey silt. Subunit IA (< 46 m CCSF-D; since 1.839 Ma) consists of terrigenous grains, biogenic matter, pumice, and volcanic glass.

Assemblage analysis

In this study, we used the depth scale of Irino et al. (2018), who defined the splice and composite depths of all sediment cores retrieved during IODP Expedition 346 (CCSF-D patched-Ver. 2) (see details in Irino et al. 2018). Then, we analyzed changes in the radiolarian assemblage for 154 samples collected at about 3-m intervals from cores drilled at Site U1425 between 0 and 250 m CCSF-D and at about 2-m intervals between 250 and 360 m CCSF-D. To monitor changes in the Japan Sea hydrography, changes in the radiolarian assemblage and radiolarian absolute abundance (AB, in skel/g) are useful tools for reconstructing hydrographic changes throughout the past 10 My. Two types of microscopic slides were prepared: a faunal slide (F-slide) for analyzing changes in the assemblage and a quantitative slide (Q-slide) for accurately estimating the AB of radiolarians in 1 g of dried sediment.

The 154 samples were freeze-dried and treated with diluted hydrogen peroxide (15%) and hydrochloric acid (15%) to remove organic and calcareous matter. The undissolved residue in each sample was washed over a sieve with 45 μm mesh size. For mounting a Q-slide, we followed the procedure proposed by Itaki (2009) and Matsuzaki and Itaki (2016), which starts with transferring the washed undissolved residue to a beaker with 100 mL of water. The obtained solution is homogenized by mixing to settle the radiolarians randomly, after which a 0.5-mL sample is withdrawn (Itaki 2009) and mounted onto an 18 × 24-mm cover slip to make a microscopic slide. All specimens visible on each Q-slide are counted. The AB of radiolarians is calculated as follows (Itaki 2009): $A = 100a/0.5g$, where A is the estimated radiolarian AB (skel/g), a is the number of radiolarian skeletons counted in the Q-slide, and g is the weight of the freeze-dried sample. The remaining undissolved residue in the beaker was also homogenized by mixing to settle the radiolarians randomly and then mounted onto a 22 × 40-mm cover microscopic slide (F-slide) following the procedure of Itaki (2003). Examination of polycystine radiolarians was carried out under an optical microscope at total magnifications of 100–400×. Sample examination continued until at least 400 radiolarians were identified in each sample, or until the sample material was exhausted. The identification followed the nomenclature of Itaki (2009), Matsuzaki et al. (2015a), and Zhang and Suzuki (2017) for extant species. For extinct species, we mainly followed the taxonomic nomenclature of Nigrini and Lombari (1984), Motoyama (1996), and Kamikuri (2010). Microphotographs of selected species are shown in Plate 1. When similar paleoceanographic responses were empirically known for similar species, these radiolarians were grouped for paleoceanographic analysis, as described below. We calculated the relative abundances of the encountered species as a percentage of all radiolarians in the sample. In this study, an average of 412 specimens was counted per sample. We also analyzed the radiolarian assemblage of 51 samples collected at 3.5-m intervals from Site U1430. However, because a hiatus of 2.7 Myr was encountered between 7.2 and ca. 4.5 Ma (Tada et al. 2015b; Kamikuri et al. 2017), only F-slides were mounted, and selected taxa were counted to strengthen the significance of the data obtained from Site U1425.

As mentioned above, a few aspects hamper the paleoceanographic reconstruction when it is based on changes in the microfossil assemblage for a time interval older than the Pleistocene, especially when including extinct species. Generally, knowledge of the vertical and geographical distributions of extinct species is limited. Fortunately, until the late Miocene, there are a few extant radiolarian species whose vertical and geographical distributions are known. Additionally, a few previous studies have clarified geographical distribution of Mio-Pliocene radiolarian species, which helped our reconstruction (Oseki and Suzuki 2009). In this context, we selected 21 extant/extinct species with known geographical distribution from the total assemblage in this study (Figs. 3, 4, and 5).

Age model

We followed the age-depth model of Kamikuri et al. (2017) for Sites U1425 and U1430 for the Mio-Pliocene and the age-depth model of Tada et al. (2018) for the Pleistocene, which relies on correlation of dark and light layers and estimates ages using tephras and magnetostratigraphic events. Details about the tie points and the sedimentation rates used in this study are shown in Tables 1 and 2.

Results

At both sites (U1425 and U1430), the preservation of radiolarian fossils fluctuated between good to moderate for the time interval between the late Miocene and late Pliocene. During the Pleistocene, the preservation of radiolarians fluctuated between poor and good. A similar observation was also made by Kamikuri et al. (2017) and Tada et al. (2015a, 2015b).

Site U1425

Polycystine radiolarian absolute abundance (skel/g)

Because the skeletons of polycystine radiolarians are composed of amorphous opaline silica, their AB can be used to constrain changes in opaline silica (e.g., De Wever et al. 2001) (Figs. 3 and 4).

In lithological subunit IIIA (9.52–7.41 Ma), the mean radiolarian AB is 2.24×10^4 skel/g and the standard deviation (SD) is 1.18×10^4 skel/g ($N = 54$), while the means of radiolarian AB in lithological subunits IIB (7.41–4.20 Ma) and IIA (4.20–2.72 Ma) are 3.26×10^4 skel/g (SD = 1.10×10^4 skel/g; $N = 46$) and 2.50×10^4 skel/g (SD = 1.41×10^4 skel/g; $N = 25$), respectively (Figs. 3 and 4). These data imply that although the mean radiolarian AB is lower in subunit IIIA than in subunits IIB and IIA, the radiolarian AB of subunit IIIA has a higher variability (subunit $\text{IIIA}_{\text{SD}} >$ subunit $\text{IIA}_{\text{SD}} >$ subunit IIB_{SD}). These subunits are all characterized by a diatom ooze or diatom-rich clayey silt, suggesting high biogenic silica content. However, subunit IIIA, which has a lower mean value and higher SD, is characterized by a higher content of terrigenous matter (Tada et al. 2015a). Therefore, it is probable that the higher content of terrigenous matter reduced radiolarian productivity between 9.52 and 7.41 Ma (subunit IIIA). In addition, in subunit IIB (7.41–4.20 Ma), which is characterized by the highest radiolarian AB mean values and lowest SD, the sediments are strongly bioturbated (Tada et al. 2015a). Previous studies demonstrated that highly bioturbated sediments imply oxic bottom water conditions in the Japan Sea during the Pleistocene (e.g., Watanabe et al. 2007). Therefore, it is probable that oxic bottom water conditions prevailed in the Japan Sea between 7.4 and 4.2 Ma, which probably allowed high productivity of radiolarians.

Lithological subunits IA and IB (1.2–0 Ma), which consist of clay and silty clay with small amounts of diatom-rich clay, are characterized by the lowest radiolarian AB. Indeed, the means of radiolarian AB are 1.27×10^4 skel/g (SD = 1.62×10^4 skel/g; $N = 11$) and 2.01×10^4 skel/g (SD = 2.00×10^4 skel/g; $N = 18$) for subunits IB and IA, respectively (Figs. 3 and 4). However, the SD estimated for subunit IA (since 1.2 Ma) is the highest since 9.5 Ma, which suggests high variability in radiolarian productivity since 1.2 Ma, which may be related to the glacioeustatic sea-level variation caused by glacial/interglacial cyclic climate changes (e.g., Elderfield et al. 2012; Lisiecki and Raymo 2005).

Relative abundances of dominant species

We have defined a species or species group as dominant if their relative abundance exceeded 30% of the total

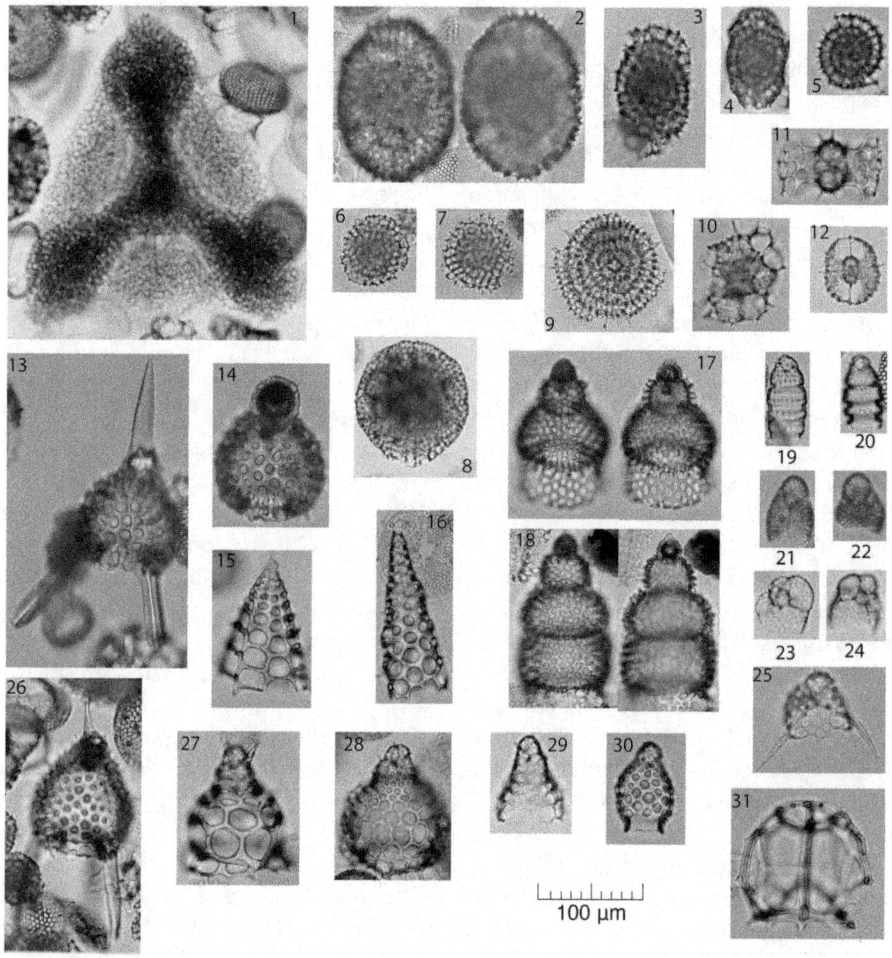

Plate 1 Photomicrographs of select radiolarian species used in this study: 1. *Dictyocoryne profunda* Ehrenberg (sample 346-U1425D-1H1-12–14 cm), 2. *Larcopyle buetschlii* Dreyer (sample 346-U1425D-1H1-12–14 cm), 3. *Lithelius barbatus* Motoyama (sample 346-U1425B-47H1-91-93 cm), 4. *Larcopyle weddellium* Lazarus, Faust and Popova-Goll (sample 346-U1425B-47H1-91-93 cm), 5. *Lithelius minor* Jørgensen group (sample 346-U1425B-29H1-81–83 cm), 6–7. *Spongodiscidae* juvenile (sample 346-U1425B-23H5-129-131 cm), 8. *Spongopyle osculosa* Dreyer (sample 346-U1425B-23H5-129–131 cm), 9. *Stylodictya tenuispina* Jørgensen group, 10. *Phorticium* spp. (sample 346-U1425B-6H5-11–13 cm), 11–12. *Tetrapyle* spp. (Fig. 11: sample 346-U1425B-6H5–11–13 cm, Fig. 12: sample 346-U1425B-34H4–46-48 cm), 13. *Lychnocanoma magnacornuta* Sakai (sample 346-U1425B-51HCC-24-29 cm), 14. *Lipmanella redondoensis* (Campbell and Clark) (sample 346-U1425D-29H1–45-47 cm), 15. *Peripyramis circumtexta* Haeckel group (sample 346-U1425B-47H1–91-93 cm), 16. *Cornutella profunda* Ehrenberg group (sample 346-U1425B-47H1–91-93 cm), 17–18. *Stichocorys peregrina /delmontensis* group (sample 346-U1425B-34H4-46–48 cm), 19–20. *Siphocampe arachnea/lineata* (Ehrenberg) group (sample 346-U1425B-21H6-19–21 cm), 21–22. *Lithomelissa setosa* Jørgensen (sample 346-U1425B-5H6-81–83 cm), 23–24. *Amphimelissa setosa* (Cleve) (sample 346-U1425B-3H4-51–53 cm), 25. *Pseudodictyophimus gracilipes* (Bailey) (sample 346-U1425B-5H6-81–83 cm), 26. *Lychnocanoma parrallelipes* Motoyama (sample 346-U1425B-34H4-46–48 cm), 27. *Cycladophora nakasekoi* Motoyama (sample 346-U1425B-47H1-91–93 cm), 28. *Cycladophora sphaeris* (Popofsky) (sample 346-U1425B-21H6-19–21 cm), 29. *Cycladophora davisiana* (Ehrenberg) (sample 346-U1425B-6H5-11–13 cm), 30. *Carpocanistrum papillosum* (Ehrenberg) group (sample 346-U1425B-34H4-46–48 cm), 31. *Ceratospyris borealis* Bailey (sample 346-U1425B-3H5-12–14 cm)

assemblage at the site. Six species or species groups dominated at Site U1425: *Amphimelissa setosa* (Cleve), *Cycladophora davisiana* (Ehrenberg), *Ceratospyris borealis* Bailey, *Siphocampe lineata/arachnea* sensu Nigrini (1977), *Cycladophora nakasekoi* Motoyama, and the *Lithelius barbatus* Motoyama group (Figs. 3 and 4). Changes in their relative abundances are described below with their inferred paleoceanographic conditions.

The species *A. setosa* is an extant species inhabiting subsurface water depths (50–300 m) in the Arctic Ocean (Bjørklund et al. 2015; Ikenoue et al. 2016; Itaki 2003; Matul and Abelmann 2005). However, this species disappeared from the subarctic North Pacific between marine isotope stages (MISs) 4 and 5 (0.07–0.05 Ma) (e.g., Matul and Abelmann 2005; Ikenoue et al. 2016; Matul et al. 2002; Matsuzaki and Suzuki 2018). In addition, the first occurrence of this species is recorded at 1.48 Ma in the Northeast Pacific (Matsuzaki and Suzuki 2018). In the Japan Sea, this species occurred between 1.47 and 0.05 Ma, although our sampling resolution is relatively

Table 1 Samples used, biostratigraphic tie points, and related sedimentation rates at IODP Site U1425 (cm/ky)

Exp	Site	Hole	Core	Type	Sect	A/W	Offset (cm) Mid	CCSF-A (m) (Tada et al. 2015a)	CCSF-D_Patched (Irino et al. 2018)	Age (Ma)	Sedimentation rates (cm/ky)	Comments
346	U1425	D	1	H	1	W	13	0.130	0.13	0.002		
346	U1425	D	1	H	1	W	103	1.030	1.03	0.017		
346	U1425	D	1	H	3	W	3	3.030	3.03	0.051		
Tie point of Tada et al. (2018)									6.22	0.105	5.924	We used the Tephra & Pmag age of Dark layer No. 1–6 (Additional file 3)
346	U1425	B	1	H	5	W	99	6.992	6.99	0.121		
346	U1425	D	2	H	2	W	14	8.005	8.00	0.143		
346	U1425	D	2	H	3	W	64	10.005	10.00	0.185		
Tie point of Tada et al. (2018)									10.55	0.197	4.707	We used the Tephra & Pmag age of Dark layer No. 2–5 (Addional file 3)
Tie point of Tada et al. (2018)									14.13	0.292	3.776	We used the Tephra & Pmag age of Dark layer No. 3–2 (Additional file 3)
346	U1425	B	2	H	5	W	65	16.006	16.01	0.342		
Tie point of Tada et al. (2018)									17.66	0.386	3.735	We used the Tephra & Pmag age of Dark layer No. 4–2 (Additional file 3)
Tie point of Tada et al. (2018)									21.71	0.483	4.188	We used the Tephra & Pmag age of Dark layer No. 5–3 (Additional file 3)
346	U1425	D	3	H	5	W	42	22.012	22.01	0.493		
346	U1425	B	3	H	4	W	52	24.904	24.90	0.589		
Tie point of Tada et al. (2018)									25.46	0.608	3.002	We used the Tephra & Pmag age of Dark layer No. 6–3 (Additional file 3)
346	U1425	B	3	H	5	W	113	27.014	27.01	0.647		
Tie point of Tada et al. (2018)									29.18	0.700	4.026	We used the Tephra & Pmag age of Dark layer No. 7–2 (Additional file 3)
346	U1425	D	4	H	3	W	28	30.008	30.01	0.720		
346	U1425	D	4	H	5	W	28	33.008	33.01	0.790		
Tie point of Tada et al. (2018)									33.21	0.795	4.251	We used the Tephra & Pmag age of Dark layer No. 9–1 (Additional file 3)
346	U1425	D	4	H	7	W	8	35.808	35.81	0.855		
Tie point of Tada et al. (2018)									38.16	0.909	4.346	We used the Tephra & Pmag age of Dark layer No. 11–1 (Additional file 3)
346	U1425	B	4	H	7	W	32	39.034	39.23	0.937		
346	U1425	D	5	H	3	W	15	40.016	40.21	0.962		
Tie point of Tada et al. (2018)									41.75	1.002	3.869	We used the Tephra & Pmag age of Dark layer No. 13–3 (Additional file 3)
346	U1425	D	5	H	5	W	15	43.016	43.21	1.050		
Tie point of Tada et al. (2018)									44.97	1.109	3.001	We used the Tephra & Pmag age of Dark layer No. 15–2 (Additional file 3)

Table 1 Samples used, biostratigraphic tie points, and related sedimentation rates at IODP Site U1425 (cm/ky) *(Continued)*

Exp	Site	Hole	Core	Type	Sect	A/W	Offset (cm) Mid	CCSF-A (m) (Tada et al. 2015a)	CCSF-D_Patched (Irino et al. 2018)	Age (Ma)	Sedimentation rates (cm/ky)	Comments
346	U1425	D	5	H	6	W	65	45.016	45.21	1.116		
346	U1425	B	5	H	5	W	32	46.026	46.32	1.146		
Tie point of Tada et al. (2018)									47.98	1.192	3.635	We used the Tephra & Pmag age of Dark layer No. 17–1 (Additional file 3)
346	U1425	B	5	H	6	W	82	48.026	48.32	1.204		
346	U1425	D	6	H	3	W	134	50.990	51.28	1.306		
Tie point of Tada et al. (2018)									51.43	1.311	2.892	We used the Tephra & Pmag age of Dark layer No. 19–1 (Additional file 3)
346	U1425	B	6	H	2	W	62	53.017	53.31	1.391		
Tie point of Tada et al. (2018)									54.8	1.455	2.344	We used the Tephra & Pmag age of Dark layer No. 21–2 (Additional file 3)
346	U1425	B	6	H	3	W	112	55.017	55.31	1.468		
346	U1425	B	6	H	5	W	12	57.017	57.31	1.517		
346	U1425	D	7	H	2	W	82	60.013	60.30	1.592		
Tie point of Kamikuri et al. (2017)								64.115	64.63	1.700	4.013	We used the Last Occurrence of A. acquilonium (Radiolaria), which occurred at the mid depth of 64,115 m CCSF-A. This bioevent has no error margin.
346	U1425	D	7	H	5	W	132	65.013	65.30	1.750		
346	U1425	B	7	H	5	W	104	68.016	68.31	1.972		
346	U1425	D	8	H	2	W	99	69.996	70.29	2.119		
346	U1425	D	8	H	4	W	143	73.006	73.30	2.342		
346	U1425	B	8	H	3	W	138	76.007	76.30	2.564		
346	U1425	B	8	H	5	W	138	79.007	79.30	2.787		
Tie point of Kamikuri et al. (2017)								80.005	80.62	2.885	1.350	We used the Last Occurrence of H. parviakitaense (Radiolaria), which occurred at the mid depth of 80,005 m CCSF-A. This bioevent has an error margin of 0,015 Ma.
346	U1425	B	8	H	7	W	39	81.017	81.57	2.905		
346	U1425	B	9	H	3	W	5	85.002	85.55	2.992		
346	U1425	B	9	H	4	W	55	87.002	87.55	3.035		
346	U1425	D	10	H	2	W	65	88.007	88.56	3.057		
346	U1425	D	10	H	4	W	65	91.007	91.56	3.122		
346	U1425	D	10	H	6	W	66	94.017	93.75	3.170		
346	U1425	B	10	H	5	W	94	96.013	95.74	3.213		
346	U1425	D	11	H	3	W	133	99.003	98.73	3.278		
346	U1425	D	11	H	5	W	133	102.003	101.73	3.343		
346	U1425	B	11	H	5	W	62	106.008	105.74	3.430		
346	U1425	D	12	H	3	W	75	109.005	108.73	3.495		
346	U1425	D	13	X	2	W	71	112.011	111.86	3.563		
346	U1425	B	14	H	3	W	59	115.014	114.86	3.628		

Table 1 Samples used, biostratigraphic tie points, and related sedimentation rates at IODP Site U1425 (cm/ky) *(Continued)*

Exp	Site	Hole	Core	Type	Sect	A/W	Offset (cm) Mid	CCSF-A (m) (Tada et al. 2015a)	CCSF-D_Patched (Irino et al. 2018)	Age (Ma)	Sedimentation rates (cm/ky)	Comments
346	U1425	B	14	H	5	W	56	118.014	117.86	3.694		
346	U1425	B	14	H	6	W	106	120.014	119.86	3.737		
346	U1425	D	14	H	2	W	59	122.009	121.86	3.780		
346	U1425	D	14	H	3	W	109	124.009	123.86	3.824		
346	U1425	B	17	H	2	W	124	125.003	124.85	3.845		
346	U1425	D	15	H	3	W	23	127.011	126.86	3.889		
346	U1425	D	15	H	4	W	73	129.011	128.86	3.932		
346	U1425	D	15	H	6	W	73	132.011	131.86	3.997		
Tie point of Kamikuri et al. (2017)								134.43	134.28	4.050	4.606	We used the Last Occurrence of *D. bullatus* (Radiolaria), which occurred at the mid depth of 134.43 m CCSF-A. This bioevent has an error margin of 0.25 Ma.
346	U1425	D	16	H	2	W	4	136.013	135.86	4.089		
346	U1425	D	16	H	4	W	2	139.013	138.86	4.164		
346	U1425	D	16	H	6	W	2	142.013	141.86	4.238		
346	U1425	B	21	H	2	W	21	144.030	143.88	4.289		
346	U1425	B	21	H	4	W	20	147.020	146.87	4.363		
Tie point of Kamikuri et al. (2017)								148.51	148.36	4.400	4.023	We used the First Occurrence of D. bullatus (Radiolaria), which occurred at the mid depth of 148.51 m CCSF-A. This bioevent has an error margin of 0.1 Ma.
346	U1425	B	21	H	6	W	20	150.020	149.87	4.480		
346	U1425	D	18	H	5	W	7	151.996	151.84	4.584		
346	U1425	D	18	H	6	W	60	154.026	153.87	4.691		
346	U1425	B	22	H	3	W	111	156.990	156.84	4.848		
346	U1425	B	22	H	5	W	11	158.990	158.84	4.954		
Tie point of Kamikuri et al. (2017)								160.815	160.66	5.050	1.893	We used the Last Occurrence of L. redondoensis (Radiolaria), which occurred at the mid depth of 160,815 m CCSF-A. This bioevent has an error margin of 0.05 Ma.
346	U1425	D	19	H	3	W	127	161.021	160.87	5.054		
346	U1425	D	19	H	5	W	128	164.031	163.88	5.111		
346	U1425	B	23	H	3	W	130	167.020	166.87	5.167		
346	U1425	B	23	H	5	W	130	170.020	169.87	5.224		
346	U1425	D	20	H	4	W	97	172.014	171.86	5.262		
346	U1425	D	20	H	5	W	146	174.004	173.85	5.299		
346	U1425	B	24	H	2	W	145	176.002	175.85	5.337		
346	U1425	D	21	H	2	W	37	178.013	177.86	5.375		
346	U1425	D	21	H	3	W	87	180.013	179.86	5.413		
346	U1425	D	21	H	5	W	87	183.013	182.86	5.470		
346	U1425	B	25	H	2	W	91	185.001	184.85	5.507		

Table 1 Samples used, biostratigraphic tie points, and related sedimentation rates at IODP Site U1425 (cm/ky) *(Continued)*

Exp	Site	Hole	Core	Type	Sect	A/W	Offset (cm) Mid	CCSF-A (m) (Tada et al. 2015a)	CCSF-D_Patched (Irino et al. 2018)	Age (Ma)	Sedimentation rates (cm/ky)	Comments
346	U1425	B	25	H	4	W	92	188.011	187.86	5.564		
346	U1425	D	22	H	5	W	62	193.011	192.86	5.659		
346	U1425	D	22	H	6	W	112	195.011	194.86	5.696		
346	U1425	D	23	H	2	W	77	198.007	197.86	5.753		
346	U1425	D	23	H	4	W	77	201.007	200.86	5.810		
346	U1425	D	23	H	6	W	77	204.007	203.86	5.867		
346	U1425	B	27	H	4	W	28	207.014	206.86	5.923		
346	U1425	B	27	H	6	W	27	210.004	209.85	5.980		
346	U1425	D	24	H	5	W	6	212.008	211.86	6.018		
346	U1425	D	24	H	6	W	56	214.008	213.86	6.056		
346	U1425	B	28	H	3	W	148	217.010	216.86	6.112		
Tie point of Kamikuri et al. (2017)								219	218.85	6.150	5.290	We used the Last Occurrence of L. parallelipes (Radiolaria), which occurred at the mid depth of 219.00 m CCSF-A. This bioevent has an error margin of 0.25 Ma.
346	U1425	B	29	H	1	W	82	222.021	221.99	6.267		
346	U1425	B	29	H	3	W	82	225.021	224.99	6.378		
346	U1425	B	29	H	5	W	83	228.031	228.00	6.489		
346	U1425	D	27	H	1	W	123	231.023	231.00	6.600		
346	U1425	B	30	H	3	W	32	233.012	234.23	6.720		
346	U1425	B	30	H	5	W	32	236.012	237.23	6.831		
Tie point of Kamikuri et al. (2017)								237.66	239.09	6.900	2.699	We used the Rapid Increase of *L. barbatus* (Radiolaria), which occurred at the mid depth of 237.66 m CCSF-A. This bioevent has an error margin of 0.10 Ma.
346	U1425	D	28	H	4	W	90	238.009	239.23	6.903		
346	U1425	D	28	H	6	W	90	241.009	242.23	6.968		
346	U1425	D	29	H	1	W	46	243.011	244.23	7.012		
346	U1425	D	29	H	2	W	96	245.011	246.23	7.055		
346	U1425	B	33	H	3	W	40	247.002	248.22	7.099		
346	U1425	B	33	H	4	W	90	249.002	250.22	7.142		
346	U1425	B	33	H	5	W	139	250.992	252.21	7.186		
346	U1425	B	33	H	6	W	91	252.012	253.23	7.208		
346	U1425	D	30	H	5	W	76	255.009	256.23	7.273		
Tie point of Kamikuri et al. (2017)								255.58	257.46	7.300	4.593	We used the First Occurrence of L. barbatus (Radiolaria), which occurred at the mid depth of 255.58 m CCSF-A. This bioevent has an error margin of 0.10 Ma.
346	U1425	B	34	H	4	W	47	258.010	259.23	7.342		
346	U1425	B	34	H	5	W	97	260.010	262.23	7.413		
346	U1425	D	31	H	2	W	118	261.012	265.23	7.484		

Table 1 Samples used, biostratigraphic tie points, and related sedimentation rates at IODP Site U1425 (cm/ky) *(Continued)*

Exp	Site	Hole	Core	Type	Sect	A/W	Offset (cm) Mid	CCSF-A (m) (Tada et al. 2015a)	CCSF-D_Patched (Irino et al. 2018)	Age (Ma)	Sedimentation rates (cm/ky)	Comments
346	U1425	D	31	H	4	W	118	264.012	267.23	7.531		
346	U1425	B	35	H	3	W	48	266.008	268.23	7.555		
346	U1425	B	35	H	3	W	148	267.008	270.23	7.602		
346	U1425	B	35	H	5	W	48	269.008	274.23	7.697		
346	U1425	D	32	H	4	W	6	273.007	275.23	7.720		
346	U1425	D	32	H	4	W	106	274.007	277.23	7.768		
346	U1425	D	32	H	6	W	6	276.007	280.23	7.839		
346	U1425	B	37	H	1	W	129	279.011	281.24	7.863		
346	U1425	B	37	H	2	W	80	280.021	281.26	7.863		
346	U1425	D	33	H	4	W	3	282.014	283.23	7.910		
346	U1425	B	38	H	1	W	102	283.006	284.22	7.933		
346	U1425	D	35	H	2	W	48	287.003	288.22	8.028		
346	U1425	B	39	H	2	W	126	290.011	291.23	8.099		
346	U1425	B	39	H	4	W	61	291.991	293.21	8.146		
346	U1425	B	40	X	2	W	59	293.998	294.22	8.170		
346	U1425	B	40	X	3	W	10	295.008	296.22	8.217		
346	U1425	D	37	H	1	W	144	297.010	298.36	8.268		
346	U1425	D	37	H	2	W	94	298.010	299.36	8.292		
346	U1425	D	37	H	4	W	95	300.010	301.36	8.339		
346	U1425	D	37	H	5	W	45	301.010	302.36	8.363		
346	U1425	B	42	H	3	W	67	306.015	307.36	8.481		
346	U1425	D	38	H	2	W	127	307.006	308.35	8.505		
346	U1425	D	39	H	1	W	57	310.003	311.35	8.576		
346	U1425	D	39	H	2	W	107	312.003	313.35	8.623		
346	U1425	D	39	H	3	W	57	313.003	314.35	8.647		
346	U1425	B	44	H	2	W	77	314.012	315.36	8.670		
346	U1425	B	44	H	4	W	5	316.012	317.36	8.718		
346	U1425	D	40	H	2	W	60	318.013	319.36	8.765		
346	U1425	D	40	H	3	W	10	319.013	320.36	8.789		
346	U1425	D	40	H	3	W	110	320.013	321.36	8.812		
346	U1425	B	46	H	1	W	53	321.027	321.85	8.824		
346	U1425	D	41	H	2	W	109	323.018	323.85	8.871		
346	U1425	D	41	H	4	W	37	325.008	325.84	8.918		
346	U1425	B	47	H	1	W	92	326.016	326.84	8.942		
346	U1425	B	47	H	2	W	42	327.016	327.84	8.966		
346	U1425	B	47	H	2	W	143	328.026	328.85	8.990		
346	U1425	B	47	H	3	W	92	329.016	329.84	9.013		
346	U1425	D	43	H	2	W	28	331.016	331.84	9.061		
Tie point of Kamikuri et al. (2017)								332.68	333.51	9.100	4.225	We used the First Occurrence of L. barbatus (Radiolaria), which occurred at the mid depth of 255.58 m CCSF-A. This bioevent has no an error margin.

Table 1 Samples used, biostratigraphic tie points, and related sedimentation rates at IODP Site U1425 (cm/ky) *(Continued)*

Exp	Site	Hole	Core	Type	Sect	A/W	Offset (cm) Mid	CCSF-A (m) (Tada et al. 2015a)	CCSF-D_Patched (Irino et al. 2018)	Age (Ma)	Sedimentation rates (cm/ky)	Comments
346	U1425	D	43	H	3	W	78	333.016	333.84	9.106		
346	U1425	D	43	H	4	W	28	334.016	334.84	9.123		
346	U1425	D	43	H	4	W	128	335.016	335.84	9.140		
346	U1425	D	43	H	6	W	28	337.016	337.84	9.173		
346	U1425	B	49	H	2	W	30	338.016	338.84	9.190		
346	U1425	B	49	H	2	W	130	339.016	339.84	9.207		
346	U1425	D	44	H	2	W	31	341.014	341.84	9.241		
346	U1425	B	50	H	2	W	67	343.994	344.82	9.291		
346	U1425	B	50	H	3	W	118	346.004	346.83	9.325		
346	U1425	D	45	H	2	W	41	347.010	347.84	9.342		
346	U1425	D	45	H	2	W	141	348.010	348.84	9.359		
346	U1425	D	46	H	1	W	76	349.73	350.56	9.393		
346	U1425	D	46	H	2	W	76	351.23	352.06	9.423		
346	U1425	D	46	H	4	W	4	353.23	354.06	9.462		
346	U1425	D	48	H	1	W	26	358.63	359.46	9.568		
Tie point of Kamikuri et al. (2017)								382.925	383.75	9.950	5.911	We used the First Occurrence of C. nakasekoi (Radiolaria), which occurred at the mid depth of 382.925 m CCSF-A. This bioevent has an error margin of 0.25 Ma.

low for the Pleistocene (Fig. 3). During this time interval (1.47–0.05 Ma), the mean of *A. setosa* relative abundance is 10.6% (SD = 18.3; *N* = 22). The highest abundance of this species is 80.55% recorded at 0.34 Ma (Fig. 3 and Additional file 1).

The species *C. davisiana* and *C. borealis* are both subarctic, abundant in areas influenced by the Oyashio Current and in the Sea of Okhotsk (e.g., Matsuzaki and Itaki 2017; Matsuzaki et al. 2014a, 2014b; Okazaki et al. 2004). In the Japan Sea, *C. davisiana* is related to the Japan Sea deep water formed by cooling of the TWC, that is, the Japan Sea Proper Water (Itaki 2003; Itaki et al. 2007), while *C. borealis* prefers cold surface and subsurface water (50–200 m) (Itaki 2003; Itaki et al. 2007; Okazaki et al. 2004). *C. davisiana* is recorded at Site U1425 between 3.21 and 0 Ma, with a mean relative abundance of 6.4% (SD = 13.1; *N* = 37), while *C. borealis* is recorded between 3.56 and 0 Ma with a mean relative abundance of 8.4% (SD = 15.2; *N* = 42) (Fig. 3 and Additional file 1).

The species group *S. lineata/arachnea* is an extant species inhabiting water between 300 and 1000 m deep in the North Pacific (Kling and Boltovskoy 1995). Higher abundances of this species are recorded in the arctic to subarctic areas of the North Pacific (e.g., Boltovskoy and Correa 2016; Kling and Boltovskoy 1995; Takahashi 1991). Additionally, Boltovskoy and Correa (2016) showed that this species is restricted to the Northern Hemisphere high latitudes. Therefore, we assume that this species group is related to the intermediate water of the Northern Hemisphere high latitudes. At Site U1425, this species is encountered intermittently between 9.5 and 0 Ma, with a mean relative abundance of 4.3% (SD = 11.4; *N* = 154) (Additional file 1). This species dominates the radiolarian assemblage (> 30%) between 4.9 and 4.0 Ma (Fig. 4 and Additional file 1).

The *L. barbatus* group is an extinct group (e.g., Motoyama 1996; Kamikuri et al. 2004, 2007) and was described as having relatively wide variation of morphological forms by Motoyama (1996). This species group has been encountered in areas related to the subarctic gyre of the North Pacific (e.g., Motoyama 1996; Kamikuri et al. 2004, 2007), whereas this species group is absent in areas influenced by the subtropical gyre (Kamikuri 2017; Kamikuri et al. 2009). Therefore, we assumed that this species is related to water masses of transitional to subarctic areas. At Site U1425, this species is encountered between 7.55 and 5.51 Ma, with a mean relative abundance of 11.6% (SD = 12.9; *N* = 32) (Fig. 4 and Additional file 1). This species' relative abundance is higher than the mean value between 7.01 and 5.75 Ma, and it dominates the radiolarian assemblage between 6.11 and 6.02 Ma (> 30%) (Fig. 4).

Finally, *C. nakasekoi* is an extinct species endemic to the Japan Sea (Motoyama 1996; Kamikuri et al. 2004, 2007), but its related ecology, such as depth of its habitat and preferred climatic condition, is unknown. This species is recorded between 9.57 and 7.41 Ma at Site U1425, with a mean relative abundance of 20.0% (SD = 12.7; *N* = 55) (Fig. 4 and Additional file 1). The relative abundance of *C. nakasekoi* fluctuates between high (> 30%) and low relative abundances (< 10%) (Fig. 4). Interestingly, the sediments collected at Site U1425 spanning the time interval between 9.57 and 7.41 Ma correspond to lithological subunit IIIA, composed of alternating dark gray diatom ooze and heavily bioturbated clayey diatomaceous ooze (Tada et al. 2015a).

Relative abundance of abundant species

We have defined a species or species group as abundant when its relative abundance is between 10 and 30% of the total assemblage at least once at the site. At Site U1425, abundant species are *Lithomelissa setosa* Jørgensen, *Spongodiscidae* juvenile, *Pseudodictyophimus gracilipes* (Bailey), *Cycladophora sphaeris* (Popofsky), *Phorticium* spp., *Tetrapyle* spp., *Lithelius minor* Jørgensen group, *Stylodictya tenuispina* Jørgensen group, *Stichocorys delmontensis* (Campbell and Clark), *Stylochlanydium venustum/bensoni* group, *Larcopyle polyacantha* group (Campbell and Clark), and *Stichocorys peregrina* (Riedel) (Figs. 3 and 4). The last two species are grouped as *Stichocorys delmontensis/peregrina* in Fig. 4.

Species such as *L. setosa*, *Spongodiscidae* juvenile, and *P. gracilipes* are well-known extant species related to subarctic areas. Based on previous studies, we know that *L. setosa* is related to the shallow water of subarctic to arctic areas of the Northern Hemisphere (Atlantic and Pacific oceans) (Boltovskoy and Correa 2016; Matsuzaki and Itaki 2017; Okazaki et al. 2004). At Site U1425, this species was recorded between 3.69 and 0 Ma, and its mean relative abundance is 1.7% (SD = 3.3; *N* = 44) (Fig. 3 and Additional file 1). *P. gracilipes* is related to the upper intermediate water of areas influenced by the Oyashio Current (300–500 m). Young forms of *Spongodiscidae*, whose taxonomy is still unsettled, are often named as *Spongodiscus resurgens* Ehrenberg group (e.g., Itaki 2009; Matsuzaki et al. 2015a). However, whichever name is used, this form is likely related to the shallow subsurface waters of transitional subarctic areas in the Northwest Pacific (e.g., Kamikuri et al. 2008; Matsuzaki and Itaki 2017; Okazaki et al. 2004). In our study, both the species and species group were found at Site U1425 throughout the period since 9.5 Ma. The mean relative abundance of *P. gracilipes* is 1.2% (SD = 2.2; *N* = 154), while the mean relative abundance of *Spongodiscidae* juvenile is 5.2% (SD = 6.2; *N* = 154) (Fig. 3 and Additional file 1). For *P. gracilipes*, relative abundance higher than the mean value is recorded intermittently between 3.85 and 2.79 Ma, 1.75 and 1.12 Ma, and 0.65 and 0.49 Ma (Fig. 3 and Additional file 1). For juvenile *Spongodiscidae*, progressive increases in its relative abundance are recorded since 7.34 Ma, where its relative abundance exceeds the mean value for the first time (12.47%). From 4.2 and 0.49 Ma, relative abundance higher than 11% is constantly recorded (Fig. 3 and Additional file 1).

C. sphaeris is an extinct species, but it likely inhabited water of transitional-subarctic to subarctic areas of the North Pacific (e.g., Kamikuri et al. 2007; Motoyama 1996; Oseki and Suzuki 2009). In our study, this species occurred between 7.93 and 3.06 Ma at Site U1425 and the mean relative abundance is 9.3% (SD = 6.9; *N* = 79) (Fig. 3 and Additional file 1). Highest abundance of this species was recorded in the late Pliocene at 3.34 Ma (36.62%) (i.e., Fig. 3 and Additional file 1).

Phorticium spp. and *Tetrapyle* spp. are extant species groups, both known as a marker of warm shallow water in the North Pacific (e.g., Lombari and Boden 1985; Itaki et al. 2010; Kamikuri et al. 2008; Matsuzaki et al 2014b; Matsuzaki et al. 2015b; Matsuzaki et al. 2016; Matsuzaki and Itaki 2017). However, the main difference between the two species groups is that while *Tetrapyle* spp. are restricted to the low-middle latitudes, *Phorticium* spp. are also common in temperate areas (Kamikuri et al. 2008; Romine 1985). According to Zhang and Suzuki (2017), our *Phorticium* spp. and *Tetrapyle* spp. might also include morphotypes from different genera because for the Miocene, the taxonomy of both genera is poorly settled and accurate taxonomic study of both genera is required for this period. However, most of the extant taxa, which potentially belong to these two species groups, bear algal symbionts and inhabit warmer shallow waters (Zhang et al. 2018). Therefore, this taxonomic concern does not affect our paleoceanographic reconstruction. At Site U1425, both species groups are recorded continuously since 9.5 Ma, although *Tetrapyle* spp. seem to be absent from 5.05 to 3.06 Ma (Fig. 4 and Additional file 1). *Phorticium* spp. have a mean relative abundance of 5.3% (SD = 6; *N* = 154), while *Tetrapyle* spp. have a mean relative abundance of 1.9% (SD = 3.1; N = 154) (Fig. 4 and Additional file 1). *Tetrapyle* spp. have a relative abundance twice as high as the mean value for the periods between 9.57 and 7.06 Ma and

Table 2 Samples used, biostratigraphic tie points, and related sedimentation rates at IODP Site U1430 (cm/ky)

Exp	Site	Hole	Core	Type	Sect	A/W	Offset (cm) Mid	CCSF-A (m) (Tada et al. 2015b)	CCSF-D_ Patched (Irino et al. 2018)	Age (Ma)	Sedimentation rates (cm/ky)	Comments
346	U1430	A	1	H	CC	W	0	3.68	3.68	0.064		
Tie point of Tada et al. (2018)									6	0.105	5.714	We used the Tephra & Pmag age of Dark layer No. 1–6 (Additional file 3)
346	U1430	A	2	H	4	W	69	9.14	9.14	0.160		
Tie point of Tada et al. (2018)									11.24	0.197	5.677	We used the Tephra & Pmag age of Dark layer No. 1–6 (Additional file 3)
346	U1430	A	2	H	CC	W	0	13.52	13.52	0.264		
Tie point of Tada et al. (2018)									14.46	0.292	3.407	We used the Tephra & Pmag age of Dark layer No. 1–6 (Additional file 3)
Tie point of Tada et al. (2018)									18.78	0.408	3.708	We used the Tephra & Pmag age of Dark layer No. 1–6 (Additional file 3)
346	U1430	A	3	H	4	W	69	19.28	19.28	0.421		
Tie point of Tada et al. (2018)									22.57	0.504	3.981	We used the Tephra & Pmag age of Dark layer No. 1–6 (Additional file 3)
346	U1430	A	3	H	CC	W	0	23.65	23.65	0.530		
Tie point of Tada et al. (2018)									26.88	0.608	4.128	We used the Tephra & Pmag age of Dark layer No. 1–6 (Additional file 3)
346	U1430	A	4	H	4	W	69	29.59	29.59	0.678		
Tie point of Tada et al. (2018)									30.43	0.700	3.855	We used the Tephra & Pmag age of Dark layer No. 1–6 (Additional file 3)
Tie point of Tada et al. (2018)									32.28	0.795	1.945	We used the Tephra & Pmag age of Dark layer No. 1–6 (Additional file 3)
346	U1430	A	4	H	CC	W	0	33.75	33.75	0.842		
Tie point of Tada et al. (2018)									36.11	0.909	3.537	We used the Tephra & Pmag age of Dark layer No. 1–6 (Additional file 3)
346	U1430	A	5	H	4	W	69	39.23	39.23	0.998		
Tie point of Tada et al. (2018)									39.36	1.002	3.502	We used the Tephra & Pmag age of Dark layer No. 1–6 (Additional file 3)
Tie point of Tada et al. (2018)									42.42	1.109	2.852	We used the Tephra & Pmag age of Dark layer No. 1–6 (Additional file 3)
346	U1430	A	5	H	CC	W	0	43.69	43.69	1.145		
Tie point of Tada et al. (2018)									45.39	1.192	3.587	We used the Tephra & Pmag age of Dark

Table 2 Samples used, biostratigraphic tie points, and related sedimentation rates at IODP Site U1430 (cm/ky) *(Continued)*

Exp	Site	Hole	Core	Type	Sect	A/W	Offset (cm) Mid	CCSF-A (m) (Tada et al. 2015b)	CCSF-D_ Patched (Irino et al. 2018)	Age (Ma)	Sedimentation rates (cm/ky)	Comments
												layer No. 1–6 (Additional file 3)
Tie point of Tada et al. (2018)									48.23	1.311	2.381	We used the Tephra & Pmag age of Dark layer No. 1–6 (Additional file 3)
346	U1430	A	6	H	4	W	69	49.15	49.15	1.366		
Tie point of Tada et al. (2018)									50.66	1.455	1.689	We used the Tephra & Pmag age of Dark layer No. 1–6 (Additional file 3)
346	U1430	A	6	H	CC	W	0	53.56	53.56	1.841		
346	U1430	A	7	H	4	W	69	59.2	59.2	2.592		
Tie point of Kamikuri et al. (2017)								61.41	61.41	2.885	0.751	We used the Last Occurrence of H. parviakitaense (Radiolaria), which occurred at the mid depth of 61.41 m CCSF-A. This bioevent has an error margin of 0.015 Ma.
346	U1430	A	7	H	CC	W	0	63.61	63.61	3.073		
346	U1430	A	8	H	4	W	69	69.22	69.22	3.551		
346	U1430	A	8	H	CC	W	0	73.67	73.67	3.931		
346	U1430	A	9	H	1	W	69	74.33	74.33	3.987		
Tie point of Kamikuri et al. (2017)								75.08	75.08	4.050	1.174	We used the First Occurrence of D. bullatus (Radiolaria), which occurred at the mid depth of 75.08 m CCSF-A. This bioevent has an error margin of 0.25 Ma.
346	U1430	A	9	H	2	W	69	75.83	75.83	Hiatus		We defined this depths interval as a hiatus following the findings of Tada et al. (2015b), Kamikuri et al. (2017)
Tie point of Kamikuri et al. (2017)								80.06	80.06	7.300		We used the First Occurrence of L. parallelipes (Radiolaria), which occurred at the mid depth of 80.06 m CCSF-A. This bioevent has an error margin of 0.10 Ma.
346	U1430	A	9	H	5	W	71	80.35	80.35	7.3075186054		
346	U1430	A	9	H	6	W	71	81.85	81.85	7.3457488021		
346	U1430	A	9	H	CC	W	0	83.22	83.22	7.3806657152		
346	U1430	A	10	H	1	W	71	84.75	84.75	7.4196605159		
346	U1430	A	10	H	2	W	71	86.25	86.25	7.4578907126		
346	U1430	A	10	H	3	W	69	87.73	87.73	7.4956111734		
346	U1430	A	10	H	4	W	69	89.23	89.23	7.5338413702		
346	U1430	A	10	H	5	W	71	90.75	90.75	7.5725813029		
346	U1430	A	10	H	6	W	71	92.25	92.25	7.6108114996		
346	U1430	A	10	H	CC	W	0	93.55	93.55	7.6439443368		
346	U1430	A	11	H	4	W	69	99.35	99.35	7.7917677643		

Table 2 Samples used, biostratigraphic tie points, and related sedimentation rates at IODP Site U1430 (cm/ky) *(Continued)*

Exp	Site	Hole	Core	Type	Sect	A/W	Offset (cm) Mid	CCSF-A (m) (Tada et al. 2015b)	CCSF-D_ Patched (Irino et al. 2018)	Age (Ma)	Sedimentation rates (cm/ky)	Comments
346	U1430	A	11	H	CC	W	0	103.71	103.71	7.9028902029		
346	U1430	A	12	H	4	W	69	109.56	109.56	8.0519879702		
346	U1430	A	12	H	CC	W	0	114.02	114.02	8.1656590886		
346	U1430	A	13	H	4	W	69	119.82	119.82	8.3134825161		
346	U1430	A	13	H	CC	W	0	124.17	124.17	8.4243500867		
346	U1430	A	14	H	4	W	69	129.73	129.73	8.5660566826		
346	U1430	A	14	H	CC	W	0	134.19	134.19	8.679727801		
346	U1430	A	15	H	4	W	69	139.65	139.65	8.8188857172		
346	U1430	A	15	H	CC	W	0	144.11	144.11	8.9325568356		
346	U1430	A	16	H	1	W	71	145.44	145.44	8.9664542767		
346	U1430	A	16	H	2	W	71	146.94	146.94	9.0046844734		
346	U1430	A	16	H	3	W	71	148.44	148.44	9.0429146702		
346	U1430	A	16	H	4	W	71	149.94	149.94	9.081144867		
Tie point of Kamikuri et al. (2017)								150.68	150.68	9.100	3.924	We used the Last Occurrence of L. magnacornuta (Radiolaria), which occurred at the mid depth of 150.68 m CCSF-A. This bioevent has no error margin.
346	U1430	A	16	H	5	W	69	151.42	151.42	9.117		
346	U1430	A	16	H	6	W	71	152.94	152.94	9.151		
346	U1430	A	16	H	CC	W	0	154.41	154.41	9.184		
346	U1430	A	17	H	1	W	71	156.81	156.81	9.239		
346	U1430	A	17	H	2	W	71	158.31	158.31	9.273		
346	U1430	A	17	H	3	W	69	159.79	159.79	9.306		
346	U1430	A	17	H	4	W	71	161.33	161.33	9.341		
346	U1430	A	17	H	5	W	69	162.83	162.83	9.375		
346	U1430	A	17	H	6	W	71	164.33	164.33	9.409		
346	U1430	A	17	H	CC	W	0	165.78	165.78	9.442		
Tie point of Kamikuri et al. (2017)								188.24	188.24	9.950	4.418	We used the Last Occurrence of L. magnacornuta (Radiolaria), which occurred at the mid depth of 188.24 m CCSF-A. This bioevent has an error margin of 0.25 Ma

from 1.20 to 0 Ma, while *Phorticium* spp. have relative abundances twice as high as the mean value between 7.27 and 3.28 Ma (Fig. 4 and Additional file 1).

The *L. minor* group and *S. tenuispina* group are extant species, although their ecology is not well understood. The *L. minor* group, which has an unsettled taxonomy and is often named as *Lithelius minor* Jørgensen (e.g., Matsuzaki et al. 2015a), likely inhabits intermediate water in the subtropical North Pacific (e.g., Matsuzaki et al. 2016). This species is also related to subtropical-transitional areas of the North Pacific (Motoyama and Nishimura 2005). The vertical distribution of *S. tenuispina* group is not well understood, but this species is probably related to transitional to subarctic areas in the Northwest Pacific, according to Kamikuri et al. (2008) (named as *Stylodictya stellata* Bailey in Kamikuri et al. 2008). Both species occurred sporadically since 9.5 Ma and their mean relative abundances are 2%

(SD = 3.6; *N* = 154) and 1.8% (SD = 2.8; *N* = 154) for the *S. tenuispina* and *L. minor* groups, respectively (Fig. 4 and Additional file 1).

The *S. venustum/bensoni* groups are extinct, but they probably inhabited the middle to high latitudes of the North Pacific during the Neogene (Kamikuri 2010). The *L. polyacantha* group dominated the radiolarian assemblage between about 15 and 7 Ma at ODP Site 884 (south of the Bering Sea), Site 887 (southern Gulf of Alaska), and Site 1151 (off Northeast Japan) (Kamikuri et al. 2004, 2007), and therefore, this species is also probably related to the middle to high latitudes of the North Pacific during the Neogene. *S. venustum/bensoni* groups occurred throughout the last 9.5 Ma and their mean relative abundance is 1.5% (SD = 3.1; N = 154), with highest relative abundance between 5.3 and 3.06 Ma (Fig. 3 and Additional file 1). The *L. polyacantha* group occurred between 9.57 and 2.3 Ma, and its mean value is 1.6% (SD = 2.1; *N* = 127) (Fig. 3 and Additional file 1).

S. delmontensis/peregrina are extinct species that were widely distributed in the North Pacific during the Mio--Pliocene (Motoyama 1996; Kamikuri 2012; Kamikuri et al. 2007; Oseki and Suzuki 2009; Romine 1985). *S. delmontensis/peregrina* only occur sporadically between 9.57 and 3.89 Ma and their mean relative abundance is 1.5% (SD = 4.4; *N* = 106) (Fig. 4 and Additional file 1). Relative abundance higher than the mean value only occurs between 9.57 and 7.21 Ma (Fig. 4 and Additional file 1). These species are abundant in sediment core samples from the North Pacific (e.g., Romine 1985), while they have only sporadic occurrence in the Japan Sea. The ecology of these species is also under debate. Romine (1985) showed that despite a wide distribution in the North Pacific, these species were probably related to temperate-subarctic areas. Casey (1982) suggested that these species may have inhabited shallow water at high latitudes and intermediate water at low latitudes, while Kamikuri (2012) reported that each species had its own high-latitude and low-latitude forms, which make the diagnostic use of *S. delmontensis/peregrina* problematic. However, Motoyama (1996) suggested that *S. delmontensis/peregrina* are probably a marker of warm water in the Northwest Pacific. In addition, previous studies in the Japan Sea showed that the relative abundances of *S. delmontensis/peregrina* covaried together with those of *Collosphaera* spp. (Kamikuri and Motoyama 2007). *Collosphaera* spp. belong to the order Collodaria, which includes species inhabiting shallow water of subtropical to temperate areas (Suzuki and Not 2015; Biard et al. 2016, 2017). Therefore, in this study, we assumed that the presence of *S. delmontensis/peregrina* suggests subtropical-temperate water of the North Pacific.

Relative abundance of minor species

We have defined species or species groups as minor when their relative abundance is between 0 and 10% of the total assemblage. These minor species are *Spongopyle osculosa* Dreyer; *Larcopyle weddellium* Lazarus, Faust, and Popova-Goll group; and *Cornutella profunda* Ehrenberg and *Carpocanistrum papillosum* (Ehrenberg) group (Figs. 3 and 4).

The species *S. osculosa* is a transitional-subarctic to subarctic species inhabiting intermediate water depths in the North Pacific (Kling and Boltovskoy 1995; Kamikuri et al. 2008; Matsuzaki and Itaki 2017). This species has a sporadic occurrence at Site U1425 and its mean relative abundance is 1% (SD = 1.4; N = 154). This species' relative abundance is higher than the mean value between 4 and 3 Ma, with a maximum of 10.5% at 3.56 Ma (Fig. 3 and Additional file 1).

The species *L. weddellium* is known as a subarctic-arctic species in the North Pacific (Tanaka and Takahashi 2008; Kamikuri et al. 2008; Matsuzaki and Itaki 2017). At the middle to low latitudes of the North Pacific, this species is less abundant and inhabits intermediate water depths only (Matsuzaki et al. 2016). This species has also been abundant at high latitudes of the Southern Hemisphere since 10 Ma (i.e., ODP Sites 689–690 in the Weddell Sea) (Lazarus et al. 2005). Therefore, this species is likely related to subpolar-polar areas of both hemispheres (bi-polar species). This species' mean relative abundance is 1.6% (SD = 1.8; N = 154), although its relative abundance never exceeds 1.0% after 1.47 Ma (Fig. 4 and Additional file 1). This species' relative abundance is generally higher than the mean value between 7.91 and 5.05 Ma (Fig. 4 and Additional file 1).

C. profunda is an extant species that inhabits water deeper than 300 m (Renz 1976; Casey et al. 1979; Takahashi 1991; Takahashi and Honjo 1981), while *C. papillosum*, which is also an extant species, inhabits water deeper than about 1000 m in the North Pacific (Takahashi 1991; Takahashi and Honjo 1981). At Site U1425, *C. profunda* occurs between 9.57 and 2.56 Ma, and its mean relative abundance is 1.8% (SD = 1.7; *N* = 126). The *C. papillosum* group occurs between 9.57 and 3.89 Ma, and its mean relative abundance is 0.9% (SD = 1.5; N = 106) (Fig. 4 and Additional file 1). *C. papillosum* has a higher relative abundance exceeding 2% between 9.57 and 7.91 Ma (Fig. 4 and Additional file 1). Additionally, the species *C. profunda* shows relative abundance of 1–8% from 9.57 to 2.79 Ma (Fig. 4 and Additional file 1).

Site U1430

According to Tada et al. (2015b) and Kamikuri et al. (2017), a hiatus is present at Site U1430 in the upper part of unit IIB (about 83 m CCFS-D according to Tada et al. 2015b) and the stratigraphic interval between the *D. bullatus* (4.3–3.8 Ma) and *L. parallelipes* zones (6.8/7.0 Ma to 7.2/7.4 Ma) is missing (e.g., Kamikuri et al. 2017). Therefore, we only describe the abundance of species that can support the data obtained at Site U1425 for the reconstruction of the paleoceanography of the Pleistocene and late Miocene (Fig. 5). Species related to arctic and subarctic areas, such as *A. setosa*, *C. borealis*, and *C. davisiana*, occur since 3 Ma (Fig. 5). The mean relative abundance of *A. setosa* is 4.04% (SD = 5.03; N = 11), while the means of *C. borealis* and *C. davisiana* are 7.98% (SD = 9.1; N = 14) and 9.27% (SD = 23.41; N = 14), respectively (Fig. 5 and Additional file 2). Relative abundances of *C. sphaeris* and the *S. lineata/arachnea* group could not be estimated between 7.4 and ca. 4 Ma because of the hiatus (Fig. 5). Abundances of *Phorticium* spp. and *Tetrapyle* spp., which are warm-water index species (see above), are also affected by this hiatus. The mean relative abundances of *Phorticium* spp. and *Tetrapyle* spp. are respectively 1.13% (SD = 2.9; N = 51) and 0.68% (SD = 1.48; N = 51) (Fig. 5 and Additional file 2). *Tetrapyle* spp. has a relative abundance higher than the mean value for the periods between 9.24 and 9.15 Ma, 8.93 and 8.68 Ma, 7.64 and 7.54 Ma, and 1.15 and 0 Ma (Fig. 5 and Additional file 2). The *S. delmontensis/peregrina* group occurs only sporadically between 8.41 and 8.31 Ma and 7.38 and 7.30 Ma, with an abundance peak exceeding 30% of the total assemblage (Fig. 5 and Additional file 2). *C. nakasekoi* occurs continuously between 9.44 and 7.38 Ma, and its mean relative abundance is 11.58% (SD = 8.60; N = 32) (Fig. 5 and Additional file 2). *C. papillosum* occurs between 9.44 and 7.38 Ma, and its mean relative abundance is 1.57% (SD = 2.29; N = 33), while *C. profunda* occurs between 9.44 and 2.52 Ma with a mean relative abundance of 0.84% (SD = 1.50; N = 40) (Fig. 5 and Additional file 2).

Discussion

During the Mio-Pliocene, the local paleoceanography of the Japan Sea was closely related to regional tectonic activity and the progressive uplift of the Japanese islands. During the Miocene until 10.5 Ma, the Japan Sea was connected to the North Pacific Ocean via seaways located around the edges of the East China and Yellow seas, Fossa Magna basin, the subsided Tohoku area at mid-latitudes, between northern Honshu and southern Hokkaido islands, and around Sakhalin Island (e.g., Fukusawa 1988; Fujii et al. 1992; Iijima and Tada 1990; Sato 1994; Kano et al. 1991; Kharakhinov 2010; Kobayashi and Tateishi 1992; Kosaka et al. 1992; Sagayama 2002; Yahata 2002; Yoshida et al. 2013). Between ca. 10.5 and 7 Ma, the Japan Sea was connected to the North Pacific via relatively shallow seaways around the Fossa Magna, subsided Tohoku area, and Sakhalin Island, where neritic sedimentary facies (< 200 m) are recognized (e.g., Iijima and Tada 1990; Sato 1994; Yoshida et al. 2013; Kharakhinov 2010; Fukusawa 1988). However, a relatively deep seaway was reported between the northern Honshu and southern Hokkaido islands (e.g., Fukusawa 1988; Iijima and Tada 1990). Tada (1994) suggested that the sill depth of this seaway would have been between 500 and 1000 m based on the sediments and fossil facies reported in Iijima and Tada (1990). Around 8 and 7 Ma, the northeast Japan Islands uplifted because of tectonic activity, which probably closed the Tohoku and Fossa Magna seaways, and thus the Japan Sea was only connected to the North Pacific via seaways located around the southern Hokkaido and Sakhalin islands until 3.1 Ma (e.g., Iijima and Tada 1990; Kano et al. 1991; Sato 1994; Yoshida et al. 2013; Kharakhinov 2010; Fukusawa 1988). At 3.1 Ma, most studies agree that the seaway around the East China and Yellow seas re-opened (Itaki 2016; Kitamura and Kimoto 2006; Kitamura et al. 2001; Tada 1994; Wang et al. 2014). In the discussion below, we propose to reconstruct the paleoceanography of shallow to deep waters with each connecting seaway configuration taken into account.

In the following discussion, we use the following nomenclature: southern seaway is the seaway located around the edge of the East China and Yellow seas, central seaway is the Fossa Magna basin, eastern seaway is the subsided Tohoku area at mid-latitudes, northern seaway is the area between northern Honshu and southern Hokkaido islands, and northernmost seaway is the area around Sakhalin Island.

Evolution of the sill depth of the northern seaway since 9.5 Ma

According to Iijima and Tada (1990), whose paleobathymetrical estimation relied upon sedimentological, micropaleontological (benthic foraminifera), and macrofossil (e.g., bivalves) analyses, the sill depths of the Japan Sea northern seaway are estimated as middle to upper bathyal water depths (500–1000 m) during the late Miocene (ca. 10–6 Ma). Our new dataset provides additional constraints on the estimation of the northern strait sill depths. At Sites U1425 and U1430, the time interval between ca. 9.5 and 7.8 Ma is characterized by occurrences of *C. profunda* inhabiting water below 300 m and *C. papillosum* inhabiting water between 1000 and 3000 m deep in the North Pacific (Figs. 4, 5 and 6) (Takahashi and Honjo 1981; Takahashi 1991). These species are absent in the modern Japan Sea (e.g., Itaki 2003; Itaki 2016; Motoyama et al. 2016a). Itaki (2016) suggested that the absence of these species in the modern Japan Sea is due to the geographical isolation of the intermediate and deep water of the Japan Sea from the North Pacific and thus suggested that these species are related to

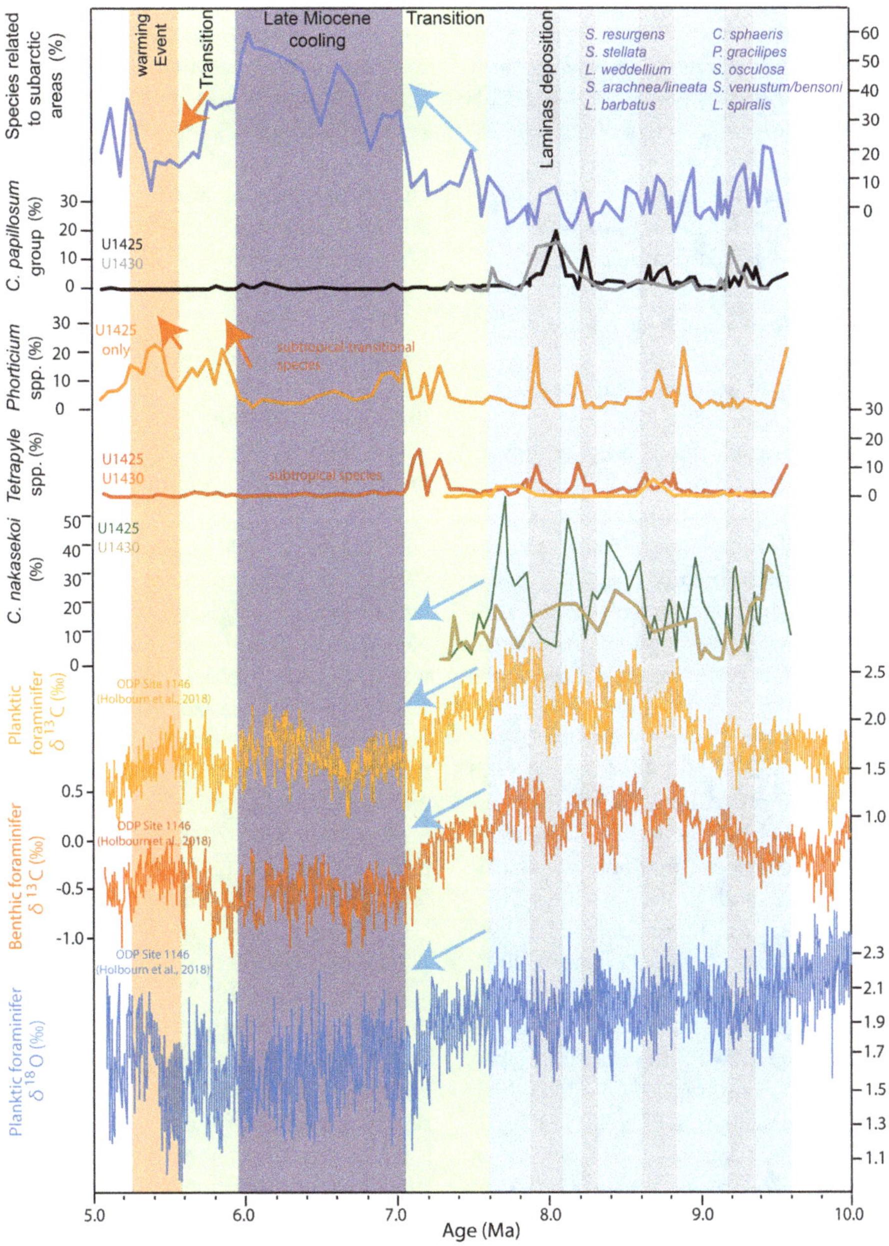

Fig. 6 Details of the late Miocene paleoceanography of the Japan Sea (10–5 Ma). We compare relative abundances of *C. nakasekoi*, *Tetrapyle* spp., and *C. papillosum* from Sites U1425 and U1430 between 9.5 and 7 Ma. The interval between 7.0 and 5.0 Ma shows relative abundance of *Phorticium* spp. and species related to subarctic water (the selected species are listed at the upper right in the figure). These data are compared to oxygen and carbon isotope data for planktic and benthic foraminifer obtained from the South China Sea (ODP 1146, Holbourn et al. 2018)

the deep water of the North Pacific. Therefore, in this study, we assumed that *C. profunda* and *C. papillosum* are indicators of deep water of the North Pacific. Thus, our data imply that between 9.5 and 7.8 Ma intermediate to deep water of the North Pacific probably flowed into the Japan Sea through the northern seaway, which had water depths between 500 and 1000 m, because other seaways open at that time (central, east, and northernmost) were too shallow (< 200 m) to enable deep-water exchanges. Although the passive upwelling of deep-water current may have also had an influence on these inflows of deep water, because our radiolarian data are in agreement with other sedimentological data estimating the sill depth of the northern seaway (e.g. Iijima and Tada 1990), we can consider that the effect of the passive upwelling is minor. Therefore, our new data, combined with the previous data of Iijima and Tada (1990), suggest that the sill depth of the northern seaway was likely close to 1000 m between 9.5 and 7.8 Ma. The very low abundance or absence of *C. papillosum* and the progressive decrease in *C. profunda* relative abundance since 7.8 Ma implies a probable shoaling of the northern seaway between 7.8 and 3 Ma, with the northern seaway probably having a sill depth shallower than 300 m at ca. 3 Ma. This is likely related to regional tectonic activity and progressive uplift of the Japanese islands since 8–7 Ma, when the stress regime in the Japanese islands and eastern Japan Sea changed from extensional to compressional (e.g., Jolivet et al. 1994; Sato 1994; Yoshida et al. 2013).

Paleoceanography of the Japan Sea between 9.5 and 7.0 Ma

During this interval, the Japan Sea was connected to the North Pacific via shallow seaways at the central, east, and northernmost Japan Sea and via a deep seaway in the north. Our data show intermittent high abundances of *Tetrapyle* spp. and *S. delmontensis/peregrina* at Sites U1425 and U1430 and, thus, relatively warm shallow water probably prevailed intermittently (Fig. 6). Because *S. delmontensis/peregrina* had a relatively wide geographical distribution and their ecology remains controvertible, we use mainly *Tetrapyle* spp. to discuss the local paleoceanography.

Tetrapyle spp. occur episodically at ODP Site 302 in the central Japan Sea (Kamikuri and Motoyama 2007) and thus support our findings that the Japan Sea was influenced by warm water of the North Pacific between 9.5 and 7.0 Ma. Taking into account that only the northernmost, northern, eastern, and central seaways were open between 10 and 7 Ma (e.g. Iijima and Tada 1990; Kano et al. 1991; Kharakhinov 2010; Sagayama 2002; Sato 1994), warm current must have flowed into the Japan Sea through one of these seaways. Previous studies did not report *Tetrapyle* spp. flowing the northern Japan outcrops (Motoyama 1992; Motoyama 1993), middle-high latitudes of the Northwest Pacific (ODP Sites 1151/1152) (Kamikuri et al. 2004), the Bering south of Sea/Gulf of Alaska (ODP Sites 884–887) (Kamikuri et al. 2007), or Hokkaido area outcrops (e.g., Shinzawa et al. 2009; Motoyama and Nakamura 2002; Motoyama et al. 2016b). As extant *Tetrapyle* spp. live in warm water and always bear algal symbionts at the subfamily level (e.g., Zhang et al. 2018), shallow-warm waters were not flowing in these regions. Thus, warm water of the North Pacific probably did not flow into the Japan Sea via the northern and northernmost seaways, and the eastern and central Japan Sea seaways are the most plausible pathways for enabling the inflow of warm water into the Japan Sea from the North Pacific. Interestingly, the abrupt decrease in *Tetrapyle* spp. abundances at ca. 7 Ma also corresponds to the timing of emergence of lands in the Tohoku area and the closure of the central seaway (e.g., Fujii et al. 1992; Sato 1994; Yoshida et al. 2013). In consideration of the intermittent occurrence of *Tetrapyle* spp. between 9.5 and 7 Ma, we point out that the northern limit of *Tetrapyle* spp. was probably around the Tohoku area (eastern seaway) during the late Miocene, and thus warm water probably flowed intermittently via the central and eastern seaways between 9.5 and 7 Ma (Fig. 7).

The time interval between 9.5 and 7.0 Ma is also characterized by high abundance of *C. nakasekoi* (about 30–60%) and intermittent abundance of *C. papillosum* fluctuating between 1 and 10% (Fig. 6). The intermittent fluctuation pattern for *C. papillosum* is roughly similar to those of *Tetrapyle* spp. (Fig. 6). We know that *C. papillosum* inhabits water depths between 1000 and 3000 m in the North Pacific and its presence implies that the sill depth of the northern seaway was probably about 1000 m. Previous studies based on measurement of $\delta^{13}C$ in benthic foraminifera imply that a layer with low dissolved oxygen content existed in the North Pacific between 1000 and 3000 m, the so-called OMZ (Woodruff and Savin 1989). If we assume that the habitat zone of *C. papillosum* was similar throughout the Neogene, *C. papillosum* may have inhabited the OMZ water during the late Miocene. This hypothesis is strengthened by the fact that laminations of diatom ooze occurring within a dark organic-rich layer extending to a thickness of about 1.5 m were recognized at about 287 m CCSF-D, which corresponds to an age of ca. 8 Ma (Irino et al. 2018; Kamikuri et al. 2017). Interestingly, the highest relative abundance (about 10%) of *C. papillosum* is recorded at ca. 8 Ma at Sites U1425 and U1430 (Fig. 6), when the thickest lamina were deposited (Tada et al. 2015a). Other laminations are also recorded at depths below 287 m CCSF-D; however, these were not as thick, so their positions were not reported in Tada et al. (2015a). Interestingly, the relative abundance of *C. papillosum* was also intermittent before 8 Ma, but its relative abundance was much lower (Fig. 6). Therefore, it is possible that the occurrence of *C. papillosum* is related to the deposition of laminas. Based on Watanabe et al. (2007), laminas are deposited in the Japan Sea when the bottom water is poor in oxygen. Therefore, it is probable that the *C. papillosum* abundance implies oxygen-poor bottom water in the Japan Sea. Thus, when *C. papillosum* abundance is high, it is probable that OMZ water from the North Pacific was flowing into the Japan Sea, while when the *C. papillosum* relative abundance was lower or close to 0%, OMZ water probably was not flowing. To explain this intermittent fluctuation in *C. papillosum*, one plausible explanation is the glacioeustatic sea-level variation, which was already proposed by Tada (1994), who assumed that orbitally paced glacioeustatic sea-level changes probably caused shoaling and deepening of the northern seaway (ca. 1000 m), although a shoaling/deepening of the OMZ itself cannot be neglected (Fig. 7). However, as said above, the

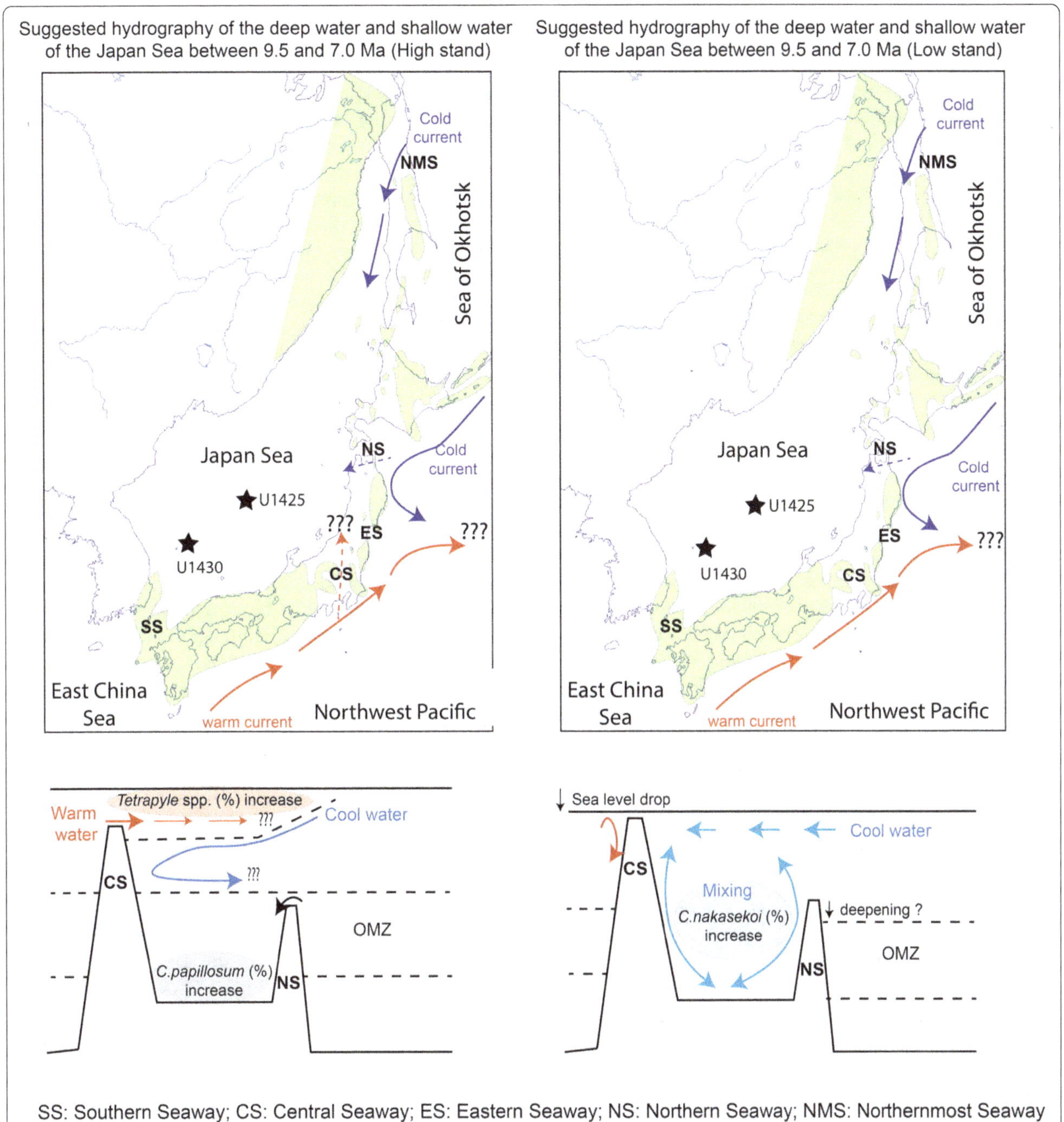

Fig. 7 Suggested hydrography of the deep water and shallow water of the Japan Sea between 9.5 and 7.0 Ma

intermittent fluctuation of *Tetrapyle* spp. is roughly synchronous to that of *C. papillosum*. We have already shown that warm water of the North Pacific probably flowed into the Japan Sea via the eastern and northeastern seaways, which had relatively shallow water depths. Therefore, it is probable that changes in the sill depths of the connecting seaways (northern, eastern, and central) in relation to the glacioeustatic sea-level changes influenced OMZ and warm water inflow from the North Pacific into the Japan Sea at that time.

In contrast, the relative abundance *C. nakasekoi* increased when *C. papillosum* and *Tetrapyle* spp. had a low relative abundance. The concern is that *C. nakasekoi* is a species reported in the Japan Sea only between ca. 10 and 7 Ma (Kamikuri and Motoyama 2007; Kamikuri et al. 2004; Motoyama 1996), and its ecology remains unknown. Previous studies reported that *C. nakasekoi* is absent in the North Pacific Ocean and only constituted about 1% of the total assemblage in samples from outcrops in Hokkaido Island, while this species exceeded 20% of the total assemblage in the Japan Sea (e.g., Kamikuri et al. 2004; Motoyama et al. 2016b; Shinzawa et al. 2009; this study). As we assumed that the sill depth of the northern seaway was ca. 1000 m between 9.5 and 7.8 Ma, one plausible explanation is that *C. nakasekoi* lived in waters deeper than 1000 m, and thus *C. nakasekoi* would have been trapped in the Japan Sea. However, it seems that high abundance of *C. nakasekoi* occurred when the relative abundances of *Tetrapyle* spp. and/or *Phorticium* spp. were low (Fig. 6). This implies that the *C. nakasekoi* relative abundance increased when the relative abundances of species related to warm water decreased. Therefore, although the specific distribution of *C. nakasekoi* suggests that this could be an intermediate-to-deep-water species restricted to the Japan Sea, we cannot assume that this species is not related to a cold marine environment. In fact, the more probable explanation is that *C. nakasekoi* is related to poorly stratified water conditions. Indeed, Tada (1994) suggested that the OMZ was nutrient-rich and probably had a high salinity based on the interpretation of the value of $\delta^{13}C$ reported by Woodruff and Savin (1989). However, warm water is known for having a low density (e.g., Ruddiman 2001). Therefore, when the OMZ and warm water flowed from the North Pacific into the Japan Sea, the water column was likely well stratified (Tada 1994). In such a context, we observed low relative abundance of *C. nakasekoi* (Figs. 6 and 7). In contrast, when there was no inflow of OMZ and warm water from the North Pacific, it is likely that the water masses in the Japan Sea were less stratified, and Tada (1994) suggested that the paleoceanographic condition would have been more or less similar to the present Japan Sea, where deep water is well ventilated because of a vertical mixing of water masses (Fig. 7). Therefore, it is probable that *C. nakasekoi* developed in the Japan Sea when the water column was probably relatively well mixed, and thus it seems that C. *nakasekoi* prefers water masses that are well oxygenated (Figs. 6 and 7).

Paleoceanography of the Japan Sea during the late Miocene and early Pliocene with only the northern strait open (ca. 7–5.0 Ma)

From 7 to 5 Ma, the northern and northernmost seaways were the only gateways enabling water exchanges between the Japan Sea and the North Pacific. As suggested above, the sill depth of the northern strait was likely shallower than 1000 m but deeper than 300 m between ca. 7.8 and 3 Ma because of the presence of *C. profunda* (> 300 m) and low relative abundance/or absence of *C. papillosum* (> 1000 m) (Fig. 6). This suggests that only intermediate to shallow water could have been exchanged between the Japan Sea and North Pacific at that time. Radiolarian data at Site U1425 suggest enhanced cooling conditions in the Japan Sea between 7.0 and 5.8 Ma, which was characterized by an increase in abundances of the subarctic-transitional species, exceeding 30% of the total assemblage during this interval (Fig. 5). Our data therefore suggest that a cold marine environment prevailed in the Japan Sea at that time.

Recently, Herbert et al. (2016) used the alkenone unsaturation index to show that a synchronous cooling occurred in shallow water at high latitudes in the Northern Hemisphere between ca. 7.5 and 5.5 Ma. This cooling was interpreted as a transient glaciation of the Northern Hemisphere and is called the late Miocene cooling (Herbert et al. 2016). Other studies suggested that during this cooling, an increase in C_4 plants was recorded, and thus the climate likely became drier (e.g., Cerling et al., 1997; Holbourn et al. 2018; Huang et al., 2007). Additionally, recent analysis of Asian eolian dust from the Japan Sea also suggested a temporal drying of Central Asia ca. 8–7 Ma (Shen et al. 2017). Holbourn et al. (2018) used $\delta^{18}O$ and $\delta^{13}C$ to measure benthic and planktic foraminifera at ODP Site 1146 (South China Sea) and suggested a cooling at that time because of enhanced EAWM, which probably increased the biological pump and drastically modified the carbon cycle. In Fig. 6, we compare our radiolarian data to the isotope data of Holbourn et al. (2018). The cooling implied by radiolarian assemblages is synchronous with the shift recorded in isotope data from the South China Sea, and thus the cooling implied by radiolarian assemblages is likely related to the late Miocene cooling and the enhanced EAWM. Cooling at high latitudes is generally associated with a southward shift of the subarctic front (e.g., McClymont et al., 2008), while enhanced EAWM generally strengthened winds blowing from north to south because of a Siberian high and Aleutian low (e.g., Tada et al. 2016). Therefore, rather than a local increase in cold radiolarian species productivity because of a cooling climate, it is likely that the influence of cold water currents was stronger during the late Miocene cooling, and thus a larger amount of cold water was transported from high latitudes to the marginal sea. This implies that this cold current influenced the Japan Sea via the northern and northernmost seaways located between northern Honshu and southern Hokkaido islands and around Sakhalin Island because other seaways were closed at that time (Fig. 8).

In this study, at Site U1425, abundances of subarctic--transitional species drastically decreased between 5.8 and 5.2 Ma, while abundances of species related to shallow water of transitional-subtropical (*Phorticium* spp.)

areas increased (by about 20%) (Fig. 6). Radiolarian data imply that cold water continued to flow into the Japan Sea from the northernmost seaway (Sakhalin area) because the relative abundance of transitional-subarctic species was still fluctuating between 10 and 20% (Fig. 6). However, relatively warm water probably flowed at the same time from the northern seaway (northern Honshu and Southern Hokkaido islands) because the relative abundance of subtropical-transitional species was about 20%, while their abundance was less than 5% between 7 and 5.8 Ma (Fig. 8). Previously published data for mollusca fossils confirmed our hypothesis by showing that temperate conditions prevailed in the Japan Sea during that time (Ogasawara 1994). Around 5 Ma (early Pliocene), Earth experienced a warm climate with relatively high atmospheric pCO_2 (e.g., Fedorov et al. 2013). At the same time, the tropical warm pool likely expanded (e.g., Brierley et al. 2009). Therefore, considering that we have used an age model relying on biostratigraphy of radiolarians and an error margin of ca. 0.2 Myr (Kamikuri et al. 2017), the higher abundances of transitional-subtropical species between ca. 5.8 and 5.2 Ma are probably related to the warm conditions prevailing throughout the Earth, which plausibly caused an expansion of the tropical warm pool and a northward shift of the subtropical front, enabling inflow of relatively temperate-warm water into the Japan Sea through the northern strait.

Drastic cooling of the subsurface and intermediate waters during the mid-Pliocene (ca. 5.0–3.8 Ma)

Our radiolarian data from the Japan Sea show that the *S. arachnea/lineata* group, an intermediate-water species restricted to the polar to subpolar areas of the Northern Hemisphere (e.g., Boltovskoy and Correa 2016), dominates our assemblages between ca. 5.0 and 3.8 Ma (Fig. 9). The *S. arachnea/lineata* group reported in previous paleoceanographic and biostratigraphic studies showed a low abundance (< 5%) because this group is relatively small (about 50 μm), and most of the previous studies processed the sediments with 63-μm mesh sieves (e.g., Motoyama 1996; Kamikuri et al. 2004, 2007, 2009). Despite this technical issue, increases and decreases in the abundance of this group have been reported off the northeastern Japanese coast and in the subarctic North Pacific south of the Bering Sea (ODP Sites 1151 and 884) at ca. 4.7 and 4.0 Ma (appendix of Kamikuri et al. 2004, 2007). In the southern Gulf of Alaska (ODP Site 887) and in the Equatorial Pacific (ODP Sites 845 and 1241), this species was absent (Kamikuri et al. 2007; Kamikuri et al. 2009). These findings, combined with our data, suggest that a drastic cooling occurred in the subsurface to intermediate waters of the Japan Sea and the Northwest and Central North Pacific almost simultaneously during the mid-Pliocene. This may have been caused by a southward flow of subarctic subsurface to intermediate water from high latitudes of the Northwest

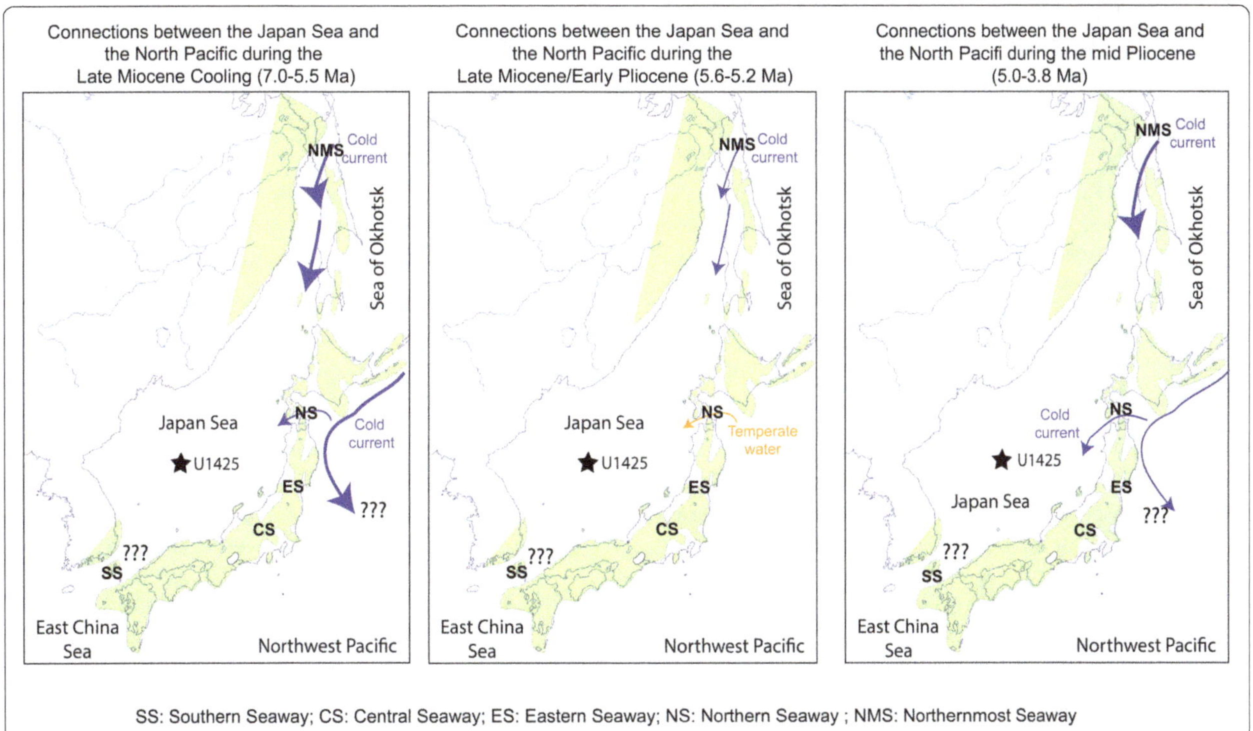

Fig. 8 Connections between the Japan Sea and the North Pacific during the late Miocene to late Pliocene

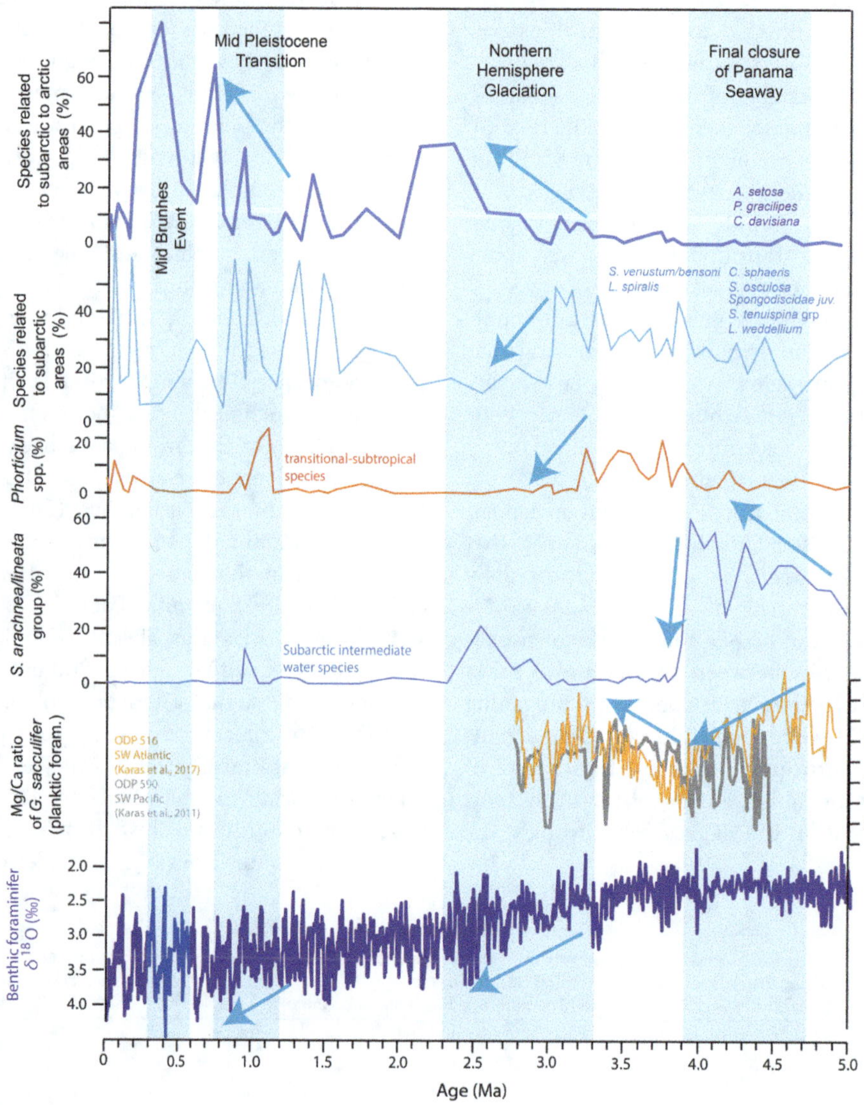

Fig. 9 Details of the paleoceanography of the Japan Sea between 5 and 0 Ma. We compare relative abundances of *S. arachnea/lineata* group, *Phorticium* spp., and species related to subarctic and arctic areas (the selected species are listed at the upper right in the figure) from Site U1425. These data are compared to Mg/Ca ratio data for plankton of the South Pacific and Atlantic (ODP 516/590, Karas et al. 2011, 2017) and benthic oxygen isotopes (Lisiecki and Raymo 2005)

Pacific. At that time, only the northern seaway, which at a water depth deeper than 300 m was deep enough to enable inflow of subsurface to intermediate water of the North Pacific, because *S. lineata/arachnea* inhabits water depths between 300 and 1000 m in the North Pacific (Kling and Boltovskoy 1995). Therefore, we suggest that such intermediate water influenced the Japan Sea via the northern Strait between 5.0 and 3. 8 Ma (Fig. 8). However, causes and mechanisms are difficult to explain. A re-estimation of *S. arachnea/lineata* group presence using 45-µm mesh sieves at other DSDP/ODP sites and comparison with geochemical data would probably provide clarification.

Although it is just speculation, numerous studies have suggested a close link between changes in the North Pacific paleoceanographic conditions and the final shoaling of the Panama Isthmus ca. 5–4 Ma (e.g., Cortese et al. 2004; Haug et al. 2001; Steph et al. 2006; Stroynowski et al. 2015). Steph et al. (2006) showed that the $\delta^{18}O$ measured in *Globigerinoides sacculifer* (Brady) (planktic foraminifera) at sites between the Eastern Equatorial Pacific and the Caribbean Sea began to diverge between ca. 4.7 and 4.2 Ma. This was interpreted as a reduced exchange of water masses through the Panama Isthmus, inhibiting exchanges of shallow water between the North Pacific and North Atlantic (Steph et al. 2006). A coupled atmosphere-ocean climate model experiment by Motoi et al. (2005) suggested that the closure of the Panama Isthmus not only inhibited the export of warm water from the North Atlantic to the North Pacific but also disturbed the water mass density (salinity) balance of both oceans. In addition, Karas et al. (2017) reconstructed temperatures based on the planktic foraminifera Mg/Ca ratio and found a synchronous cooling of the South Atlantic and South Pacific between 5 and 4 Ma. Further, they associated this cooling with an episodic enhanced Atlantic meridional overturning circulation due to water mass density disturbance initiated by the final closure of the Panama Isthmus. In Fig. 9, we plot our radiolarian data against the Mg/Ca ratio obtained by Karas et al. (2017) in the South Atlantic and South Pacific. It seems that the interval where the radiolarian assemblage of the Japan Sea is dominated by the *S. arachnea/lineata* group corresponds to the cooling event reported by Karas et al. (2017). Therefore, it is possible that the cooling recorded in the Japan Sea and Northwest Pacific intermediate water is related to the closing of the Panama Isthmus and its related water mass density perturbation, which may have affected the global thermohaline circulation episodically.

Paleoceanography of the Japan Sea during the late Pliocene (ca. 3.8–2.7 Ma)

The late Pliocene is characterized by a progressive global cooling until the beginning of the NHG (e.g., Haug et al. 1999; Raymo 1994). Abrupt emergence of land is suggested for the Hokkaido and Sakhalin archipelagic water regions based on recent geological studies because of the cooling related to the NHG between 3 and 2 Ma (e.g., Sagayama 2002; Yahata 2002; Kharakhinov 2010). Thus, the exchange of subarctic water via the northernmost seaway would have been affected (Sagayama 2002; Yahata 2002; Kharakhinov 2010). For this interval (3.5–2.7 Ma), our data show that the transitional-subarctic species and transitional-subtropical species had relative abundances varying between 20 and 40% and 10 and 20%, respectively (Fig. 9). Between 3.8 and 2.7 Ma, our data imply that the Japan Sea was probably still influenced by transitional-subarctic and transitional-subtropical water at that time.

Previous studies suggested that the opening of the southern strait at ca. 3 Ma caused only a restricted and localized warming to the southernmost part of the Japan Sea because the channel was probably very narrow and insufficient to supply significant amounts of warm water (e.g., Iijima and Tada 1990; Itaki 2016; Kitamura and Kimoto 2006; Kitamura et al. 2001). Therefore, transitional-subtropical water probably did not flow into the Japan Sea from the southern strait. Koizumi and Yamamoto (2016) showed that at the same latitude, the diatom *Td'*-based sea surface temperature (SST) of the Japan Sea (ODP Site 797) was colder than that recorded in the North Pacific (ODP Site 436) at a similar latitude despite the opening of the seaway into the southern Japan Sea at ca. 3.1 Ma (Iijima and Tada 1990; Tada 1994; Ogasawara 1994). To explain this observation, Itaki (2016) suggested the possibility that subarctic water from the Sea of Okhotsk influenced the Japan Sea via a strait west of Hokkaido Island during the Pliocene. Considering the local geological context, we know that until ca. 3 Ma the area around Sakhalin Island was still characterized by shallow neritic facies, implying that cold water of the high latitudes could still influence the Japan Sea via the northernmost seaway of the Japan Sea (Kharakhinov 2010). Therefore, we can infer that transitional--subarctic species probably continued to enter the Japan Sea via a shallow strait between the Sakhalin Islands and the Japan Sea until its closure between 3 and 2 Ma (Kharakhinov 2010), while transitional-subtropical species probably entered through the northern strait, as during the early Pliocene (Figs. 8 and 9).

Paleoceanography of the Japan Sea during the Pleistocene (ca. 2.5–0 Ma)

At 2.7 Ma, the NHG began, which marked the beginning of the Pleistocene (e.g., Haug et al. 1999; Raymo 1994). At Site U1425, our data show that relative abundances of arctic species increased from 2.7 Ma, while transitional-subarctic species and subtropical-transitional species abundances decreased (Fig. 9). In addition, species related to deep water of the North Pacific disappeared (Figs. 3 and 4). This implies a drastic cooling of the Japan Sea due to the NHG, and the absence of Pacific deep-water species suggests that the deep water of the Japan Sea was isolated from the North Pacific because the sill depth of the northern seaway was too shallow. This shoaling is probably due to the sea-level fall caused by the NHG and tectonic uplift of northeastern Japan (e.g., Kamikuri and Motoyama 2007; Itaki 2016).

The Pleistocene is characterized by cyclical glaciation (e.g., Lisiecki and Raymo 2005), and the relationship between the amplitude of glaciations caused by orbital forcing was not linear (e.g., Lisiecki and Raymo 2005; Elderfield et al. 2012). During the early Pleistocene (2.58–1.2 Ma), weak glaciations with a periodicity of 41 ky characterized the glacial-interglacial cycles, while strong glaciations were repeated with a periodicity of nearly 100 ky since the middle Pleistocene (0.8 Ma) (Lisiecki and Raymo 2005; Elderfield et al. 2012). This periodicity shift, called as the mid-Pleistocene transition (MPT) (Berger and Jansen 1994), occurred from 1.2 to 0.8 Ma. During this event, our data show high abundance peaks of species related to arctic shallow water between 1.3 and ca. 1.0 and Ma and thus imply that the MPT and its related strong glaciation probably intensified the cooling of the Japan Sea. Additionally, the Mid--Brunhes Event (MBE) occurring between 0.45 and 0.3 Ma (Jansen et al. 1986) corresponds to a period when Earth eccentricity was close to 0 (Berger 1978). This likely caused lower pCO_2 and caused a colder spring-summer season, while the autumn/winter season was warmer (Lüthi et al. 2008; Yin and Berger 2010). Our data show high abundance peaks of species related to arctic shallow water between 0.5 and ca. 0.3 Ma (Fig. 9). This suggests intensification of the cooling of the Japan Sea shallow water during these times. Interestingly, this orbital event fit well with our second cooling of the shallow water and thus suggests that the MBE probably intensified the cooling of the Japan Sea.

Conclusions

Radiolarian assemblages analyzed at Sites U1425 and U1430 enabled us to reconstruct the evolution of the Japan Sea hydrographic system since 9.5 Ma. We can conclude that radiolarians in the Japan Sea were sensitive to environmental changes caused by regional tectonism, as well as global climatic events:

1. During the late Miocene, the sill depth of the northern strait was likely close to 1000 m until ca. 7.8 Ma and probably enabled inflow of OMZ water from the North Pacific when the sea level was high. Since 7.8 Ma, a progressive shoaling of the northern strait is implied by the radiolarian data, and at about 3 Ma, the sill depth of the northern strait was probably 300 m. This shoaling is likely due to enhanced tectonic activity on the Japan islands.
2. The central and eastern seaways were likely the key straits enabling inflow of warm-shallow water of the North Pacific into the Japan Sea between 9.5 and 7 Ma, when sea level was high.
3. Our data suggest that *C. papillosum* is probably related to OMZ water of the North Pacific.
4. Although the habitat zone of *C. nakasekoi* remains difficult to assess, it is possible that *C. nakasekoi* was related to poorly stratified water masses formed when the sea level was low during the late Miocene because the inflow of OMZ and warm water were prevented.
5. The Japan Sea recorded a deep cooling between 7 and 5.5 Ma. This cooling is probably related to the late Miocene cooling of Herbert et al. (2016) and enhanced EAWM.
6. A moderate warming occurred between 5.6 and 5.2 Ma in the Japan Sea, and it was probably related to the early Pliocene warming, when an expansion of the tropical warm pool likely occurred (e.g., Fedorov et al. 2013; Brierley et al. 2009).
7. The Japan Sea subsurface to intermediate water drastically cooled between 5.0 and 4.0 Ma and similar cooling likely occurred in the subarctic area of the Northwest Pacific almost synchronously. The reasons and mechanisms of this cooling remain unknown, but because the timing of this cooling was almost simultaneous with the formation of the Panama Isthmus, they may be related.
8. The NHG and Pleistocene major climatic events, such as the MPT and MBE, intensified the cooling of shallow water in the Japan Sea.

Abbreviations

AB: Absolute abundance; CCD: Calcite compensation depth; CCSF: Core composite depth below sea floor; DSDP: Deep-Sea Drilling Program; EAWM: East Asian winter monsoons; F-slide: Faunal slide; IODP: Integrated Ocean Drilling Program; MBE: Mid-Brunhes Event; MIS: Marine isotope stage; MPT: Mid-Pleistocene Transition; NHG: Northern Hemisphere Glaciation; ODP: Ocean Drilling Program; OMZ: Oxygen minimum zone; Q-slide: Quantitative slide; SD: Standard deviation; SST: Sea surface temperature; TWC: Tsushima Warm Current

Acknowledgements

We would like to thank the Integrated Ocean Drilling Program (IODP) Expedition 346 for providing us the samples. We also wish to thank the Kochi Core Center (KCC) and its curators for their sampling assistance. We also wish to thank Mr. Shunsuke Kurokawa for discussions about the manuscript and Mr. Yasuaki Chiba for his help in analyzing samples from Site U1430.

Funding

This work was supported by the Japan Society for the Promotion of Science (JSPS) Research Fellowship for Young Scientists, grant 16J02142 (MKM) and partly founded by the JSPS grant 18H01279 attributed to Takuya Sagawa (Kanazawa University). This work was also partially financed by the IODP Expedition 346 After Cruise Research Program, JAMSTEC.

Authors' contributions

MKM conducted the treatment and analysis on 154 samples from Site U1425, drew figures and tables, and wrote the manuscript. IT and TR proposed the topic and co-designed the study. KSI conducted analysis on samples from Site U1430. All authors read and approved the final manuscript.

Authors' information

MKM is a JSPS Research Fellow at the University of Tokyo and now assistant professor at Atmosphere and Ocean Research Institute of the university of Tokyo; IT is a senior researcher at the Geological Survey of Japan, AIST; TR is a professor of the University of Tokyo, the host researcher of MKM, and the co-chief of IODP Expedition 346; KSI is an associate professor at Ibaraki University.

Competing interests

The authors declare that they have no competing interests.

Author details

[1]Department of Earth and Planetary Science, Graduate School of Science, The University of Tokyo, 7-3-1 Hongo, Bunkyo-ku, Tokyo 113-0033, Japan. [2]Present address: Atmosphere and Ocean Research Institute, The University of Tokyo, 5-1-5 Kashiwanoha, Kashiwa, Chiba 277-8564, Japan. [3]Geological Survey of Japan, AIST, Marine Geology Research Group/Institute of Geology and Geoinformation, Tsukuba Central 7, 1-1-1 Higashi, Tsukuba, Ibaraki 305-8567, Japan. [4]Faculty of Education, Ibaraki University, 2-1-1 Bunkyo, Ibaraki 310-8512, Japan.

References

Alexandrovich JM (1992) Radiolarians from Sites 794, 795, and 797 (Japan Sea). In: Tamaki K, Ingle JCJ et al (eds) Proc ODP Sci Res 127/128(1):291–307. Ocean Drilling Program, College Station

Berger A (1978) Long-term variations of daily insolation and quaternary climatic changes. J Atmos Sci 35(12):2362–2367 https://doi.org/10.1175/1520-0469(1978)035%3C2362:LTVODI%3E2.0.CO;2

Berger WH, Jansen E (1994) Mid-Pleistocene climate shift: the Nansen connection. In: Johannessen OM, Muench RD, Overland JE (eds) The polar oceans and their role in shaping the global environment, 295–311. American Geophysical Union, Washington. https://doi.org/10.1029/ GM085p0295

Betzler C, Eberli GP, Lüdmann T, Reolid J, Kroon D, Reijmer JJG, Swart PK, Wright J, Young JR, Alvarez-Zarikian C, Alonso-García M, Bialik OM, Blättler CL, Guo JA, Haffen S, Horozal S, Inoue M, Jovane L, Lanci L, Laya JC, Hui Mee AL, Nakakuni M, Nath BN, Niino K, Petruny LM, Pratiwi SD, Slagle AL, Sloss CR, Su X, Yao Z (2017) Refinement of Miocene Sea level and monsoon events from the sedimentary archive of the Maldives (Indian Ocean). Prog Earth Planet Sci 5:5 https://doi.org/10.1186/s40645-018-0165-x

Biard T, Bigeard E, Audic S, Poulain J, Gutierrez-Rodriguez A, Pesant S, Stemmann L, Not F (2017) Biogeography and diversity of Collodaria (Radiolaria) in the global ocean. ISME J 11(6):1331–1344 https://dx.doi.org/10.1038%2Fismej.2017.12

Biard T, Stemmann L, Picheral M, Mayot N, Vandromme P, Hauss H, Gorsky G, Guidi L, Kiko R, Not F (2016) In situ imaging reveals the biomass of giant protists in the global ocean. Nat 532(7600):504–507. https://doi.org/10.1038/nature17652

Bjørklund KR, Hatakeda K, Kruglikova SB, Matul AG (2015) *Amphimelissa setosa* (Cleve) (Polycystina, Nassellaria)–a stratigraphic and paleoecological marker of migrating polar environments in the northern hemisphere during the quaternary. Stratigr 12(1):23–37

Boltovskoy D, Correa N (2016) Biogeography of Radiolaria Polycystina (Protista) in the World Ocean. Prog Oceanogr 149:82–105. https://doi.org/10.1016/j.pocean.2016.09.006

Brierley CM, Fedorov AV, Liu Z, Herbert TD, Lawrence KT, LaRiviere JP (2009) Greatly expanded tropical warm pool and weakened Hadley circulation in the early Pliocene. Sci 323:1714–1718 https://doi.org/10.1126/science.1167625

Butzin M, Lohmann G, Bickert T (2011) Miocene Ocean circulation inferred from marine carbon cycle modeling combined with benthic isotope records. Paleoceanogr Paleoclimatol 26(1) https://doi.org/10.1029/2009PA001901

Casey R (1982) *Lamprocyrtis* and *Stichocorys* lineages: biogeographical and ecological perspectives relating to the tempo and mode of polycystine radiolarian evolution. In: Mamet B, Copeland MJ (eds) third American paleontological convention, proceedings vol I. 77:82

Casey R, McMillen K, Reynolds R, Spaw J, Schwarzer R, Gevirtz H, Bauer M (1979) Relict and expatriated radiolarian fauna in the Gulf of Mexico and its implications. Trans Gulf Coast Assoc Geol Soc 29:224–227

Cerling TE, Harris JM, MacFadden BJ, Leakey MG, Quade J, Eisenmann V, Ehleringer JR (1997) Global vegetation change through the Miocene/Pliocene boundary. Nature 389(6647):153

Chinzei K (1978) Neogene molluscan faunas in the Japanese Islands: an ecologic and zoogeographic synthesis. Veliger 21(2):155–170

Clift PD (2017) Cenozoic sedimentary records of climate-tectonic coupling in the western Himalaya. Prog Earth Planet Sci 4:39 https://doi.org/10.1186/s40645-017-0151-8

Cortese G, Gersonde R, Hillenbrand CD, Kuhn G (2004) Opal sedimentation shifts in the World Ocean over the last 15 Myr. Earth Planet Sci Lett 224(3–4):509–527. https://doi.org/10.1016/j.epsl.2004.05.035

De Wever P, Dumitrica P, Caulet JP, Nigrini C, Caridroit M (2001) Radiolarians in the sedimentary record. Gordon and Breach Science Publishers, Amsterdam

Elderfield H, Ferretti P, Greaves M, Crowhurst S, McCave IN, Hodell DA, Piotrowski AM (2012) Evolution of ocean temperature and ice volume through the mid-Pleistocene climate transition. Sci 337(6095):704–709 https://doi.org/10.1126/science.1221294

Fedorov AV, Brierley CM, Lawrence KT, Liu Z, Dekens PS, Ravelo AC (2013) Patterns and mechanisms of early Pliocene warmth. Nat 496:43–49 https://doi.org/10.1038/nature12003

Fujii S, Kaseno Y, Nakagawa T (1992) Neogene paleogeography in the Hokuriku region, Central Japan, based on revised stratigraphic correlation. Mem Geol Soc Jpn 37:85–95

Fukusawa H (1988) Sedimentary mechanism of Neogene bedded siliceous rocks on Late Miocene Wakkanai formation of northern Hokkaido. J Geol Soc Jpn 94:669–688

Gamo T, Nakayama N, Takahata N, Sano Y, Zhang J, Yamasaki E, Taniyasu S, Yamashita N (2014) The sea of Japan and its unique chemistry revealed by time-series observations over the last 30 years. Monogr Environ Earth Planets 2(1):1–22 https://doi.org/10.5047/meep.2014.00201.0001

Hanagata S (2003) Miocene-Pliocene foraminifera from the Niigata oil-fields region, northeastern Japan. Micropaleontol 49(4):293–340 https://doi.org/10.1661/0026-2803(2003)049[0293:MFFTNO]2.0.CO;2

Hanagata S (2006) Foraminiferal proxies of dissolved oxygen and their changes across the Miocene/Pliocene boundary in the Japan Sea. Stratigr 3:285–306

Hase H, Yoon JH, Koterayama W (1999) The current structure of the Tsushima Warm Current along the Japanese coast. J Oceanogr 55(2):217–235

Haug GH, Sigman DM, Tiedemann R, Pedersen TF, Sarnthein M (1999) Onset of the permanent stratification in the subarctic Pacific Ocean. Nat 401(6755): 779–782. https://doi.org/10.1038/44550

Haug GH, Tiedemann R, Zahn R, Ravelo AC (2001) Role of the Panama uplift on oceanic freshwater balance. Geol 29(3):207–210 https://doi.org/10.1130/0091-7613(2001)029%3C0207:ropuoo%3E2.0.co;2

Herbert TD, Lawrence KT, Tzanova A, Peterson LC, Caballero-Gill R, Kelly CS (2016) Late Miocene global cooling and the rise of modern ecosystems. Nat Geosci 9(11):843–847 https://doi.org/10.1038/ngeo2813

Holbourn A, Kuhnt W, Clemens S, Prell W, Andersen N (2013) Middle to late Miocene stepwise climate cooling: evidence from a high-resolution deep water isotope curve spanning 8 million years. Paleoceanogr Paleoclimatol 28(4):688–699. https://doi.org/10.1002/2013PA002538

Holbourn A, Kuhnt W, Clemens SC, Kochhann KG, Jöhnck J, Lübbers J, Andersen N (2018) Late Miocene climate cooling and intensification of southeast Asian winter monsoon. Nat Commun 9(1):1584. https://doi.org/10.1038/s41467-018-03950-1

Holbourn A, Kuhnt W, Lyle M, Schneider L, Romero O, Andersen N (2014) Middle Miocene climate cooling linked to intensification of eastern equatorial Pacific upwelling. Geol 42(1):19–22. https://doi.org/10.1130/G34890.1

Huang Y, Clemens SC, Liu W, Wang Y, Prell WL (2007) Large-scale hydrological change drove the late Miocene C4 plant expansion in the Himalayan foreland and Arabian Peninsula. Geology 35:531–534

Iijima A, Tada R (1990) Evolution of tertiary sedimentary basins of Japan in reference to opening of the Japan Sea. J Fac Sci, U Tokyo Section II 22(2): 121–171

Ikenoue T, Okazaki Y, Takahashi K, Sakamoto T (2016) Bering Sea radiolarian biostratigraphy and paleoceanography at IODP Site U1341 during the last four million years. Deep-Sea Res II Top Stud Oceanogr 125:38–55. https://doi.org/10.1016/j.dsr2.2015.03.004

Irino T, Tada R, Ikehara K, Sagawa T, Karasuda A, Kurokawa S, Seki A, Lu S (2018) Construction of perfectly continuous records of physical properties for dark-light sediment sequences collected during IODP Expedition 346. Prog Earth Planet Sci 5:23 https://doi.org/10.1186/s40645-018-0176-7

Itaki T (2003) Elutriation technique for the extraction of radiolarian skeletons from sandy sediments. Fossils 73:38–41 (in Japanese)

Itaki T (2007) Historical changes of deep-sea radiolarians in the Japan Sea during the last 640 kyrs. Fossils 82:43–51 (in Japanese)

Itaki T (2009) Last glacial to Holocene polycystine radiolarians from the Japan Sea. News Osaka Micropaleontol (NOM) 14:43–89

Itaki T (2016) Transitional changes in microfossil assemblages in the Japan Sea from the late Pliocene to Early Pleistocene related to global climatic and local tectonic events. Prog Earth Planet Sci 3(1):11 https://doi.org/10.1186/s40645-016-0087-4

Itaki T, Ikehara K, Motoyama I, Hasegawa S (2004) Abrupt ventilation changes in the Japan Sea over the last 30 kyr: evidence from deep-dwelling radiolarians. Palaeogeogr Palaeoclimatol Palaeoecol 208(3–4):263–278 https://doi.org/10.1016/j.palaeo.2004.03.010

Itaki T, Kimoto K, Hasegawa S (2010) Polycystine radiolarians in the Tsushima Strait in autumn of 2006. Paleontol Res 14(1):19–32 https://doi.org/10.2517/1342-8144-14.1.019

Itaki T, Komatsu N, Motoyama I (2007) Orbital- and millennial-scale changes of radiolarian assemblages during the last 220 kyrs in the Japan Sea. Palaeogeogr Palaeoclimatol Palaeoecol 247(1–2):115–130 https://doi.org/10.1016/j.palaeo.2006.11.025

Jansen JHF, Kuijpers A, Troelstra SR (1986) A mid-Brunhes climatic event: long-term changes in global atmosphere and ocean circulation. Sci 232:619–622 https://doi.org/10.1126/science.232.4750.619

Jolivet L, Tamaki K, Fournier M (1994) Japan Sea, opening history and mechanism: a synthesis. J Geophys Res 99(B11):22237–22259 https://doi.org/10.1029/93JB03463

Kamikuri S (2010) New late Neogene radiolarian species from the middle to high latitudes of the North Pacific. Rev Micropaleontol 53(2):85–106 https://doi.org/10.1016/j.revmic.2008.06.005

Kamikuri S (2012) Evolutionary changes in the biometry of the fossil radiolarian Stichocorys peregrina lineage in the eastern equatorial and eastern North Pacific. Mar Micropaleontol 90–91:13–28 https://doi.org/10.1016/j.marmicro.2012.04.003

Kamikuri S (2017) Late Neogene radiolarian biostratigraphy of the eastern North Pacific ODP Sites 1020/1021. Paleontol Res 21(3):230–254 https://doi.org/10.2517/2016PR027

Kamikuri S, Itaki T, Motoyama I, Matsuzaki KM (2017) Radiolarian biostratigraphy during the middle Miocene to quaternary, IODP Sites U1422-U1427 and U1430 in the Japan Sea. Paleontol Res 21(4):397–421 https://doi.org/10.2517/2017PR001

Kamikuri S, Motoyama I (2007) Radiolarian assemblage and environmental changes in the Japan Sea since the Late Miocene. Fossils 82:35–42 (in Japanese)

Kamikuri S, Motoyama I, Nishi H, Iwai M (2009) Neogene radiolarian biostratigraphy and faunal evolution rates in the eastern equatorial Pacific ODP Sites 845 and 1241. Acta Paleontol Pol 54(4):713–742. https://doi.org/10.4202/ app.2008.0076

Kamikuri S, Motoyama I, Nishimura A (2008) Radiolarian assemblages in surface sediments along longitude 175°E in the Pacific Ocean. Mar Micropaleontol 69(2):151–172. https://doi.org/10.1016/j.marmicro.2008.07.005

Kamikuri S, Nishi H, Motoyama I (2007) Effects of late Neogene climatic cooling on North Pacific radiolarian assemblages and oceanographic conditions. Palaeogeogr Palaeoclimatol Palaeoecol 249(3–4):370–392 https://doi.org/10.1016/j.palaeo.2007.02.008

Kamikuri S, Nishi H, Motoyama I, Saito S (2004) Middle Miocene to Pleistocene radiolarian biostratigraphy in the Northwest Pacific, ocean drilling program leg 186. Island Arc 13(1):191–226. https://doi.org/10.1111/j.1440-1738.2003. 00421.x

Kano K, Kato H, Yanagisawa Y, Yoshida F (1991) Stratigraphy and geologic history of the Cenozoic of Japan. Report no. 274. Geological survey of Japan, Tokyo

Karas C, Nürnberg D, Bahr A, Groeneveld J, Herrle JO, Tiedemann R (2017) Pliocene oceanic seaways and global climate. Sci Rep 7:39842 https://doi.org/10.1038/srep39842

Karas C, Nürnberg D, Tiedemann R, Garbe-Schönberg D (2011) Pliocene Indonesian Throughflow and Leeuwin current dynamics: implications for Indian Ocean polar heat flux. Paleoceanogr Paleoclimatol 26(2) https://doi.org/10.1029/2010PA001949

Kennett JP, Keller G, Srinivasan MS (1985) Miocene planktonic foraminiferal biogeography and paleoceanographic development of the indo-Pacific region. Geol Soc Am Mem 163:197–236

Kharakhinov VV (2010) Oil-and-gas geology of the Sakhalin region. Scientific World, Moscow Available at http://www.geokniga.org/bookfiles/geokniga-neftegazovaya-geologiya-sahalinskogo-regiona.pdf (in Russian)

Kido Y, Minami I, Tada R, Fujine K, Irino T, Ikehara K, Chun H (2007) Orbital-scale stratigraphy and high-resolution analysis of biogenic components and deep-water oxygenation conditions in the Japan Sea during the last 640 kyr. Palaeogeogr Palaeoclimatol Palaeoecol 247(1):32–49 https://doi.org/10.1016/j.palaeo.2006.11.020

Kitamura A, Kimoto K (2006) History of the inflow of the warm Tsushima current into the sea of Japan between 3.5 and 0.8 Ma. Palaeogeogr Palaeoclimatol Palaeoecol 236(3):355–366 https://doi.org/10.1016/j.palaeo.2005.11.015

Kitamura A, Takano O, Takata H, Omote H (2001) Late Pliocene–early Pleistocene paleoceanographic evolution of the sea of Japan. Palaeogeogr Palaeoclimatol Palaeoecol 172(1):81–98 https://doi.org/10.1016/S0031-0182(01)00272-3

Kling SA, Boltovskoy D (1995) Radiolarian vertical distribution patterns across the southern California current. Deep-Sea Res I Oceanogr Res Pap 42(2):191–231 https://doi.org/10.1016/0967-0637(94)00038-TI

Kobayashi I, Tateishi M (1992) Neogene stratigraphy and paleogeography in the Niigata region, Central Japan. Mem Geol Soc Jpn 37:53–70

Koizumi I (1992a) Biostratigraphy and paleoceanography of the Japan Sea based on diatoms: ODP leg 127. In: Tsuchi R, Ingle JC Jr (eds) Pacific Neogene: environment, evolution and events. Univ. Tokyo Press, pp 15–24

Koizumi I (1992b) Diatom biostratigraphy of the Japan Sea: leg 127. In: Pisciotto KA, Ingle JC Jr, von Breymann MT, Barron J et al (eds) Proc ODP Sci Res 127–128(1): 249–289. https://doi.org/10.2973/odp.proc.sr.127128-1.127.1992

Koizumi I, Yamamoto H (2016) Diatom records in the quaternary marine sequences around the Japanese Islands. Quat Int 397:436–447 https://doi.org/10.1016/j.quaint.2015.03.043

Kosaka T, Midori T, Hoyanagi K, Kubota M, Miyahigashi Y (1992) Late Cenozoic stratigraphy and paleoceanographic changes in the northern Fossa magna. Mem Geol Soc Jpn 37:71–83

Lazarus D, Faust K, Popova-Goll I (2005) New species of prunoid radiolarians from the Antarctic Neogene. J Micropalaeontol 24(2):97–121 https://doi.org/10.1144/jm.24.2.97

Lee GH, Park SC, Kim DC (2000) Fluctuations of the calcite compensation depth (CCD) in the East Sea (Japan Sea). Geo-Mar Lett 20(1):20–26. https://doi.org/10.1007/s003670000029

Ling HY (1992) Radiolarians from the sea of Japan: leg 128. In: Piscotto KA, Ingle JC, von Breymann MT et al (eds) Proc. ODP Sci res 127/128(1):225–236. Ocean drilling program, College Station. https://doi.org/10.2973/odp.proc.sr.127128-1.125.1992

Lisiecki LE, Raymo ME (2005) A Pliocene-Pleistocene stack of 57 globally distributed benthic $\delta^{18}O$ records. Paleoceanogr Paleoclimatol 20(1):PA1003 https://doi.org/10.1029/2004PA001071

Lombari G, Boden G (1985) Modern radiolarian global distributions. Special publication 16A. Cushman Foundation for Foraminiferal Research, Lawrence, pp 68–69

Lüthi D, Le Floch M, Bereiter B, Blunier T, Barnola J, Siegenthaler U, Raynaud D, Jouzel J, Fischer H, Kawamura K, Stocker T (2008) High-resolution carbon dioxide concentration record 650,000–800,000 years before present. Nat 453(7193):379–382. https://doi.org/10.1038/nature06949

Matsuzaki KM, Itaki T (2016) The effect of boron (B) on radiolarian skeletons during extraction from mudstone and soft sediment. News Osaka Micropaleontol (NOM) 16:75–78

Matsuzaki KM, Itaki T (2017) New Northwest Pacific radiolarian data as a tool to estimate past sea surface and intermediate water temperatures. Paleoceanogr Paleoclimatol 32(3):218–245 https://doi.org/10.1002/2017PA003087

Matsuzaki KM, Itaki T, Kimoto K (2016) Vertical distribution of polycystine radiolarians in the northern East China Sea. Mar Micropaleontol 125:66–84 https://doi.org/10.1016/j.marmicro.2016.03.004

Matsuzaki KM, Nishi H, Suzuki N, Cortese G, Eynaud F, Takashima R, Kawate Y, Sakai T (2014b) Paleoceanographic history of the Northwest Pacific Ocean over the past 740 kyr, discerned from radiolarian fauna. Palaeogeogr Palaeoclimatol Palaeoecol 396:26–40 https://doi.org/10.1016/j.palaeo.2013.12.036

Matsuzaki KM, Nishi H, Suzuki N, Kawate Y, Takashima R, Sakai T (2014a) *Cycladophora davisiana* abundances as a paleoceanographic and stratigraphic tool in high latitude siliciceous sediments. Mar Micropaleontol 106:1–9 https://doi.org/10.1016/j.marmicro.2013.11.002

Matsuzaki KM, Suzuki N (2018) Quaternary radiolarian biostratigraphy in the subarctic northeastern Pacific (IODP Expedition 341 Site U1417) and synchroneity of bioevents across the North Pacific. J Micropalaeontol 37:1–10 https://doi.org/10.5194/jm-37-1-2018

Matsuzaki KM, Suzuki N, Nishi H (2015a) Middle to Upper Pleistocene polycystine radiolarians from hole 902-C9001C, Northwestern Pacific. Paleontol Res 19(s1):1–77 https://doi.org/10.2517/2015PR003

Matsuzaki KM, Suzuki N, Nishi H, Hayashi H, Gyawali BR, Takashima R (2015b) Early to middle Pleistocene paleoceanographic history of southern Japan based on radiolarian data from IODP Exp. 314/315 Sites C0001 and C0002. Mar Micropaleontol 118:17–33 https://doi.org/10.1016/j.marmicro.2015.05.001

Matul A, Abelmann A (2005) Pleistocene and Holocene distribution of the radiolarian *Amphimelissa setosa* Cleve in the North Pacific and North Atlantic: evidence for water mass movement. Deep-Sea Res II Top Stud Oceanogr 52(16):2351–2364 https://doi.org/10.1016/j.dsr2.2005.07.008

Matul A, Abelmann A, Tiedemann R, Kaiser A, Nürnberg D (2002) Late Quaternary polycystine radiolarian datum events in the Sea of Okhotsk. Geo-Mar Lett 22(1):25–32 https://doi.org/10.1007/s00367-002-0093-y

McClymont EL, Rosell-Melé A, Haug GH, Lloyd JM (2008) Expansion of subarctic water masses in the North Atlantic and Pacific oceans and implications for mid-Pleistocene ice sheet growth. Paleoceanography and Paleoclimatology 23(4)

Miller KG, Kominz MA, Browning JV, Wright JD, Mountain GS, Katz ME, Sugarman PJ, Cramer BS, Christie-Blick N, Pekar SF (2005) The Phanerozoic record of global sea-level change. Sci 310(5752):1293–1298. https://doi.org/10.1126/science.1116412

Motoi T, Chan WL, Minobe S, Sumata H (2005) North Pacific halocline and cold climate induced by Panamanian gateway closure in a coupled ocean-atmosphere GCM. Geophys Res Lett 32(10):L10618 https://doi.org/10.1029/2005GL022844

Motoyama I (1992) Neogene radiolarian stratigraphy in the Tsugaru peninsula Aomori prefecture, northern Japan. News Osaka Micropaleontol 8:89–100

Motoyama I (1993) Late Miocene and Pliocene radiolarian datum levels from DSDP Sites 192 and 302, and hole 438A of mid-to high-latitude NW Pacific. News Osaka Micropaleontol 9:337–347

Motoyama I (1996) Late Neogene radiolarian biostratigraphy in the subarctic Northwest Pacific. Micropaleontol 42(3):221–262. https://doi.org/10.2307/1485874

Motoyama I, Maruyama T, Nishimura T, Sakurai K (2016b) Geological age of the whale fossil-bearing calcareous float concretion from the Haw River, Hokkaido, Japan, based on radiolarian and diatom analyses. Bull Hobetsu Mus 31:1–6

Motoyama I, Nakamura S (2002) Radiolarian biostratigraphy of the Miocene Masuporo and Wakkanai formations of the Uruyagawa section, Wakkanai, Hokkaido, Japan, with reference to unconformity. J Geol Soc Jpn 108(4):219–231

Motoyama I, Nishimura A (2005) Distribution of radiolarians in the North Pacific surface sediments along the 175°E meridian. Paleontol Res 9(2):95–117. https://doi.org/10.2517/prpsj.9.95

Motoyama I, Yamada Y, Hoshiba M, Itaki T (2016a) Radiolarian assemblages in surface sediments of the Japan Sea. Paleontol Res 20(3):176–206 https://doi.org/10.2517/2015PR030

Nigrini C, Lombari G (1984) A guide to Miocene radiolaria. Special publication 22. Cushman Foundation for Foraminiferal Research, Lawrence

Oba T, Kato M, Kitazato H, Koizumi I, Omura A, Sakai T, Takayama T (1991) Paleoenvironmental changes in the Japan Sea during the last 85,000 years. Paleoceanogr 6(4):499–518 https://doi.org/10.1029/91PA00560

Ogasawara K (1994) Neogene paleogeography and marine climate of the Japanese Islands based on shallow-marine molluscs. Palaeogeogr Palaeoclimatol Palaeoecol 108:335–351 https://doi.org/10.1016/0031-0182(94)90241-0

Ogasawara K, Takano M, Nagato H, Nakano T (2008) Cenozoic molluscan faunas and climatic changes in the Northern Pacific related to Pacific gateways: review and perspective. Bull Geol Surv Jpn 59(7/8):355–364

Okazaki Y, Takahashi K, Itaki T, Kawasaki Y (2004) Comparison of radiolarian vertical distributions in the Okhotsk Sea near the Kuril Island and in the northwestern North Pacific off Hokkaido Island. Mar Micropaleontol 51:57–284. https://doi.org/10.1016/j.marmicro.2004.01.003

Oseki S, Suzuki N (2009) Miocene to Pleistocene paleographic distribution of polycystine radiolarians in the North Pacific. News Osaka Micropaleontol (NOM) 14:183–238

Raymo ME (1994) The initiation of the northern hemisphere glaciation. Annu Rev Earth Planet Sci 22:353–383 https://doi.org/10.1146/annurev.ea.22.050194.002033

Raymo ME, Ruddiman WF (1992) Tectonic forcing of late Cenozoic climate. Nat 359(6391):117–122 https://doi.org/10.1038/359117a0

Renz G (1976) The distribution and ecology of Radiolaria in the Central Pacific: plankton and surface sediments. Bull Scripps Inst Oceanogr U Calif 22:1–267

Romine K (1985) Radiolarian biogeography and paleoceanography of the North Pacific at 8 ma. In: Kennett JP (ed) the Miocene Ocean: paleoceanography and biogeography. Geol Soc Am Mem 163:237–273

Ruddiman W (2001) Earth's climate: past and future, 1st edn. Macmillan, New York

Sagayama T (2002) Middle Miocene to Pliocene sedimentary basin analysis of vertical movements in Hokkaido, Japan. Revista Mexicana de Ciencias Geologicas 19(3):215–225

Sandoval MI, Boltovskoy D, Baxter AT, Baumgartner PO (2017) Neogene paleoceanography of the eastern equatorial Pacific based on the radiolarian record of IODP drill sites off Costa Rica. Geochem Geophys Geosyst 18(3):889–906 https://doi.org/10.1002/2016GC006623

Sato H (1994) The relationship between late Cenozoic tectonic events and stress field and basin development in Northeast Japan. J Geophys Res Solid Earth 99(B11):22261–22274 https://doi.org/10.1029/94JB00854

Sato H, Iwasaki T, Kawasaki S, Ikeda Y, Matsuta N, Takeda T, Hirata N, Kawanaka T (2004) Formation and shortening deformation of a back-arc rift basin revealed by deep seismic profiling, Central Japan. Tectonophysics 388(1–4):47–58 https://doi.org/10.1016/j.tecto.2004.07.004

Schlitzer R (2016) Ocean Data View. http://odv.awi.de

Shen X, Wan S, France-Lanord C, Clift PD, Tada R, Révillon S, Shi X, Zhao D, Liu Y, Yin X, Song Z, Li A (2017) History of Asian eolian input to the sea of Japan since 15 ma: links to Tibetan uplift or global cooling? Earth Planet Sci Lett 474:296–308 https://doi.org/10.1016/j.epsl.2017.06.053

Shinzawa M, Kamikuri S, Motoyama I (2009) Radiolarian fossil from the Miocene series in the Hobetsu area, southern Central Hokkaido, Japan. News Osaka Micropaleontol 14:117–137

Siddall M, Rohling EJ, Almogi-Labin A, Hemleben C, Meischner D, Schmelzer I, Smeed DA (2003) Sea-level fluctuations during the last glacial cycle. Nature 423(6942):853

Steph S, Tiedemann R, Prange M, Groeneveld J, Nurnberg D, Reuning L, Schulz M, Haug GH (2006) Changes in Caribbean surface hydrography during the Pliocene shoaling of the central American seaway. Paleoceanography Paleoclimatol 21:PA4221. https://doi.org/10.1029/2004PA001092

Stroynowski Z, Ravelo AC, Andreasen D (2015) A Pliocene to recent history of the Bering Sea at Site U1430A, IODP Expedition 323. Paleoceanogr 30:1641–1656 https://doi.org/10.1002/2015PA002866

Suzuki N, Not F (2015) Biology and ecology of Radiolaria. In: Ohtsuka S, Suzaki T, Horiguchi T, Suzuki N, Not F (eds) Marine protists. Springer, Tokyo, pp 179–222 https://doi.org/10.1007/978-4-431-55130-0_8

Suzuki U (1989) Geology of Neogene basins in the eastern part of the sea of Japan. Mem Geol Soc Jpn 32:143–183 (in Japanese)

Tada R (1994) Paleoceanographic evolution of the Japan Sea. Palaeogeogr Palaeoclimatol Palaeoecol 108(3–4):487–508 https://doi.org/10.1016/0031-0182(94)90248-8

Tada R, Iijima A (1992) Lithostratigraphy and compositional variation of Neogene hemipelagic sediments in the Japan Sea. In: Tamaki K, Suyehiro K, Allan J, McWilliams M et al (eds) Proc ODP Sci Res 127/128(2):1229–1260. https://doi.org/10.2973/odp.proc.sr.127128-2.188.1992

Tada R, Irino T, Ikehara K, Karasuda A, Sugisaki S, Xuan C, Sagawa T, Itaki T, Kubota Y, Lu S, Seki A, Murray RW, Alvarez-Zarikian C et al (2018) High-resolution and high-precision correlation of dark and light layers in the quaternary hemipelagic sediments of the Japan Sea recovered during IODP Expedition 346. Prog Earth Planet Sci 5:19 https://doi.org/10.1186/s40645-018-0167-8

Tada R, Irino T, Koizumi I (1999) Land-ocean linkages over orbital and millennial timescales recorded in Late Quaternary sediments of the Japan Sea. Paleoceanogr 14(2):236–247 https://doi.org/10.1029/1998PA900016

Tada R, Murray RW, Alvarez Zarikian CA, Anderson WT Jr, Bassetti MA, Brace BJ, Clemens SC, da Costa Gurgel MH, Dickens GR, Dunlea AG, Gallagher SJ, Giosan L, Henderson ACG, Holbourn AE, Ikehara K, Irino T, Itaki T, Karasuda A, Kinsley CW, Kubota Y, Lee GS, Lee KE, Lofi J, Lopes C, Peterson LC, Saavedra-Pellitero M, Sagawa T, Singh RK, Sugisaki S, Toucanne S, Wan S, Xuan C, Zheng H, Ziegler M (2015a) Site U1425. In: Tada R, Murray RW, Alvarez Zarikian CA et al (eds) Proc. IODP vol 346. Ocean drilling program, College Station. https://doi.org/10.2204/iodp.proc.346.106.2015

Tada R, Murray RW, Alvarez Zarikian CA, Anderson WT Jr, Bassetti MA, Brace BJ, Clemens SC, da Costa Gurgel MH, Dickens GR, Dunlea AG, Gallagher SJ, Giosan L, Henderson ACG, Holbourn AE, Ikehara K, Irino T, Itaki T, Karasuda A, Kinsley CW, Kubota Y, Lee GS, Lee KE, Lofi J, Lopes C, Peterson LC, Saavedra-Pellitero M, Sagawa T, Singh RK, Sugisaki S, Toucanne S, Wan S, Xuan C, Zheng H, Ziegler M (2015b) Site U1430. In: Tada R, Murray RW, Alvarez Zarikian CA et al (eds) Proc. IODP vol 346. Integrated Ocean Drilling Program, College Station. https://doi.org/10.2204/iodp.proc.346.110.2015

Tada R, Zheng H, Clift PD (2016) Evolution and variability of the Asian monsoon and its potential linkage with uplift of the Himalaya and Tibetan plateau. Prog Earth Planet Sci 3:4. https://doi.org/10.1186/s40645-016-0080-y

Taira A (2001) Tectonic evolution of the Japanese island arc system. Annu Rev Earth Planet Sci 29:109–134. https://doi.org/10.1146/annurev.earth.29.1.109

Takahashi K (1991) Radiolaria: flux, ecology, and taxonomy in the Pacific and Atlantic. Ocean Biocoenosis Ser 3:1–301

Takahashi K, Honjo S (1981) Vertical flux of Radiolaria: a taxon-quantitative sediment trap study from the western tropical Atlantic. Micropaleontol 7:140–190. https://doi.org/10.2307/1485284

Takeuchi T, Kodama K, Ozawa T (1999) Paleomagnetic evidence for block rotations in Central Hokkaido–South Sakhalin, Northeast Asia. Earth Planet Sci Lett 169(1):7–21 https://doi.org/10.1016/S0012-821X(99)00081-3

Tamaki T, Suyehiro K, Allan J, Ingle JC, Pisciotto KA (1992) Tectonic synthesis and implication of Japan Sea ODP drilling. In: Tamaki K, Suyehiro K, Allan J, McWilliams M et al (eds) Proc ODP Sci res 127/128(2):1333–1348. Ocean Drilling Program, College Station. https://doi.org/10.2973/odp.proc.sr.127128-2.240.1992

Tanaka S, Takahashi K (2008) Detailed vertical distribution of radiolarian assemblage (0-3000 m, fifteen layers) in the central subarctic Pacific, June 2006. Mem Fac Kyushu Univ Ser D, Earth & Planet Sci XXXII(1):49–72

Ujiie H, Ichikura M (1973) Holocene to uppermost Pleistocene in a piston core from off San'in district, sea of Japan. Trans Proc Soc Jpn 91:137–150 https://doi.org/10.14825/prpsj1951.1973.91_137

Wang P, Li Q, Li CF (2014) Geology of the China seas (Vol. 6), 1st edn. Elsevier, Amsterdam

Watanabe S, Tada R, Ikehara K, Fujine K, Kido Y (2007) Sediment fabrics, oxygenation history, and circulation modes of Japan Sea during the Late Quaternary. Palaeogeogr Palaeoclimatol Palaeoecol 247:50–67. https://doi.org/10.1016/j.palaeo.2006.11.021

White LD, Alexandrovich JM (1992) Pliocene and Pleistocene abundance and preservation of siliceous microfossil assemblages from Sites 794, 795, and 797: implications for circulation and productivity in the Japan Sea. In: Pisciotto KA, Ingle JC Jr, von Breymann MT, Barron J et al (eds) Proc ODP Sci res 127/128(1):341–357. Ocean drilling program, College Station. https://doi.org/10.2973/odp.proc.sr.127128-1.140.1992

Woodruff F, Savin SM (1989) Miocene Deepwater oceanography. Paleoceanogr 4(1):87–140 https://doi.org/10.1029/PA004i001p00087

Yahata M (2002) Evolution in space and time of Late Cenozoic hydrothermal activity and ore mineralization in Hokkaido, Japan. Report No. 73, Geological Survey of Hokkaido, pp 151–194

Yamamoto M, Naraoka H, Ishiwatari R, Ogihara S (2005) Carbon isotope signatures of bacterial 28-norhopanoic acids in Miocene-Pliocene diatomaceous and phosphatic sediments. Chem Geol 218(1–2):117–133 https://doi.org/10.1016/j.chemgeo.2005.01.027

Yin Q, Berger A (2010) Insolation and CO_2 contribution to the interglacial climate before and after the mid-Brunhes event. Nat Geosci 3(4):243–246 https://doi.org/10.1038/ngeo771

Yokoyama Y, De Deckker P, Lambeck K, Johnston P, Fifield LK (2001) Sea-level at the Last Glacial Maximum: evidence from northwestern Australia to constrain ice volumes for oxygen isotope stage 2. Palaeogeography, Palaeoclimatology. Palaeoecology 165(3-4):281–297

Yoshida T, Kimura JI, Yamada R, Acocella V, Sato H, Zhao D, Nakajima J, Hasegawa A, Okada T, Honda S, Ishikawa M, Dicky O, Prima A, Kudo T, Shibazaki B, Tanaka A, Ishikawa M (2013) Evolution of late Cenozoic magmatism and the crust–mantle structure in the NE Japan Arc. Geol Soc Lond, Spec Publ 385:SP385–SP315 https://doi.org/10.1144/SP385.15

Zachos J, Dickens GR, Zeebe RE (2008) An early Cenozoic perspective on greenhouse warming and carbon-cycle dynamics. Nat 451(17):279–283 https://doi.org/10.1038/nature06588

Zachos J, Pagani M, Sloan L, Thomas E, Billups K (2001) Trends, rhythms, and aberrations in global climate 65 Ma to present. Sci 292:686–693. https://doi.org/10.1126/science.1059412

Zhang L, Suzuki N (2017) Taxonomy and species diversity of Holocene pylonioid radiolarians from surface sediments of the northeastern Indian Ocean. Palaeontol Electronica 20.3(2017):1–68 https://doi.org/10.26879/718

Zhang L, Suzuki N, Nakamura Y, Tuji A (2018) Modern shallow water radiolarians with photosynthetic microbiota in the western North Pacific. Mar Micropaleontol 139:1–27 https://doi.org/10.1016/j.marmicro.2017.10.007

8

High-resolution and high-precision correlation of dark and light layers in the Quaternary hemipelagic sediments of the Japan Sea recovered during IODP Expedition 346

Ryuji Tada[1*], Tomohisa Irino[2], Ken Ikehara[3], Akinori Karasuda[1], Saiko Sugisaki[3], Chuang Xuan[4], Takuya Sagawa[5], Takuya Itaki[3], Yoshimi Kubota[6], Song Lu[2], Arisa Seki[1], Richard W. Murray[7], Carlos Alvarez-Zarikian[8], William T. Anderson Jr[9], Maria-Angela Bassetti[10], Bobbi J. Brace[11], Steven C. Clemens[12], Marcio H. da Costa Gurgel[13], Gerald R. Dickens[14], Ann G. Dunlea[16], Stephen J. Gallagher[15], Liviu Giosan[16], Andrew C. G. Henderson[17], Ann E. Holbourn[18], Christopher W. Kinsley[19], Gwang Soo Lee[20], Kyung Eun Lee[21], Johanna Lofi[22], Christina I. C. D. Lopes[23], Mariem Saavedra-Pellitero[24], Larry C. Peterson[25], Raj K. Singh[26], Samuel Toucanne[27], Shiming Wan[28], Hongbo Zheng[29] and Martin Ziegler[30]

Abstract

The Quaternary hemipelagic sediments of the Japan Sea are characterized by centimeter- to decimeter-scale alternation of dark and light clay to silty clay, which are bio-siliceous and/or bio-calcareous to a various degree. Each of the dark and light layers are considered as deposited synchronously throughout the deeper (> 500 m) part of the sea. However, attempts for correlation and age estimation of individual layers are limited to the upper few tens of meters. In addition, the exact timing of the depositional onset of these dark and light layers and its synchronicity throughout the deeper part of the sea have not been explored previously, although the onset timing was roughly estimated as ~ 1.5 Ma based on the result of Ocean Drilling Program legs 127/128. Consequently, it is not certain exactly when their deposition started, whether deposition of dark and light layers was synchronous and whether they are correlatable also in the earlier part of their depositional history.

The Quaternary hemipelagic sediments of the Japan Sea were drilled at seven sites during Integrated Ocean Drilling Program Expedition 346 in 2013. Alternation of dark and light layers was recovered at six sites whose water depths are > ~ 900 m, and continuous composite columns were constructed at each site. Here, we report our effort to correlate individual dark layers and estimate their ages based on a newly constructed age model at Site U1424 using the best available paleomagnetic datum and marker tephras. The age model is further tuned to LR04 $\delta^{18}O$ curve using gamma ray attenuation density (GRA) since it reflects diatom contents that are higher during interglacial high-stands. The constructed age model for Site U1424 is projected to other sites using correlation of dark layers to form a high-resolution and high-precision paleo-observatory network that allows to reconstruct changes in material fluxes with high spatio-temporal resolutions.

Keywords: Quaternary sediments, Japan Sea, Inter-site correlation, High-resolution age model, IODP, Expedition 346, U1424, U1425, U1426, U1430

* Correspondence: ryuji@eps.s.u-tokyo.ac.jp
[1]Graduate School of Science, The University of Tokyo, Tokyo, Japan
Full list of author information is available at the end of the article

Introduction

Background

The Japan Sea is one of the marginal seas on the western rim of northern Pacific that were formed more or less simultaneously during the Cenozoic. It is the most isolated marginal sea that is connected to neighboring oceans only with shallow (< 130 m in sill depth) and narrow (< 90 km in width) straits (Fig. 1). Partly because of its semi-isolation, the Japan Sea has its own deep water (called Japan Sea Proper Water), which ventilates the deeper part of the sea with the residence time of ~ 100 years (Gamo et al. 2014), and thus, the sea is occasionally called a miniature ocean (Gamo et al. 2014). The deep water below the water depth of ~ 2000 m is highly homogenous with low and nearly constant temperature of 0.04~0.07 °C, slightly low salinity of 34.07‰, low phosphorous of 1.9~2.2 μmol/kg, and high dissolved oxygen of 200~ 230 μmol/kg (Talley et al. 2006; Gamo et al. 2014). Consequently, the modern Japan Sea is one of the most oxic ocean basins in the world and relatively poor in nutrients. However, the situation was not necessarily the same in the geological past when sea level and the properties of the inflowing waters were different (e.g., Gallagher et al. 2015). We understand that the marginal sea discussed here is known by different names, including the "Japan Sea" and "East Sea". In this paper, we use "Japan Sea" given the order of authorship.

Previous studies

In the summer of 1989, JOIDES Resolution entered the Japan Sea and drilled at six sites during Ocean Drilling Program (ODP) legs 127/128 (Tamaki et al. 1990) to explore its tectonic and paleoceanographic histories. Based on the results of these legs, Tada et al. (1992) reported that Quaternary sediments in the deeper part of the sea are characterized by centimeter- to meter-scale alternations of dark, organic-rich layers and light, organic-poor layers. They further demonstrated that their deposition was synchronous basin-wide and the dark (and light) layers can be correlated in-between sites. Tada et al. (1999) demonstrated that alternation of dark and light layers reflects millennial-scale changes in the contribution of nutrient-rich East China Sea coastal water to the influx through the Tsushima Strait, which probably reflects changes in Yangtze River discharge caused by the changes in East Asian Summer Monsoon intensity in association with the Dansgaard-Oeschger Cycle (DOC) with dark layers deposited during interstadials (see also Tada 2005, 2012). Consequently, the onset timing of the dark and light layer deposition and temporal changes in the amplitude and frequency of their alternation

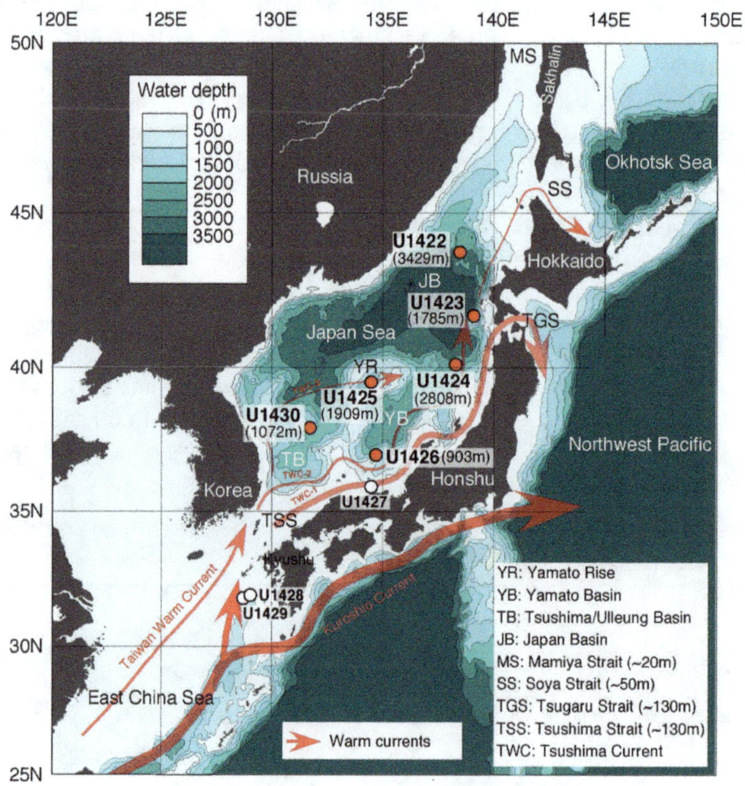

Fig. 1 A map showing locations of studied sites (red circles) together with unstudied sites (white circles) drilled during IODP Expedition 346 in the Japan Sea and northern East China Sea. Surface currents within the sea, water depths of each sites, and sill depths of each straits are also indicated

collected wide attention. Tada (2005) estimated the onset timing of the deposition of dark and light layers at approximately 1.5 Ma. However, since drilling was conducted in a single hole, the correlation gaps existed between cores and correlation among the sites was not perfect. Consequently, correlation went back only to 0.7 Ma (Tada et al. 1992). In addition, the age model was of lower resolution and precision since drilling was in a single hole and only limited paleomagnetic and biostratigraphic events were available at that time (e.g., Tada 1994).

Objective of this study

In order to specify the onset timing of deposition of dark and light layers, to examine the origin and the nature of their alternations, and to explore their temporal changes in frequency and amplitude, Integrated Ocean Drilling Program (IODP) Expedition 346 was conducted in the summer of 2013 and drilled at seven sites in the Japan Sea and two sites in the northern tip of the East China Sea (Fig. 1). To accomplish the objectives described above, high-resolution and high-precision correlation of sedimentary sequences among the sites within the sea is critical, especially for the Quaternary interval. Here, we present a result of group effort to correlate the dark and light layers in the Quaternary interval among six sites located in the deeper (> 900 m) parts of the sea where dark and light color contrast is distinct.

Construction and content

Studied sites

Seven sites were drilled during the IODP Expedition 346. Sites were selected so that they cover wide ranges of latitude, longitude, and water depths to form a "paleo-observatory network". With respect to the present water depth, Site U1427 is the shallowest site at ~ 320 m water depth, Sites U1426 and U1430 are shallower depth sites of around 1000 m, Sites U1425 and U1423 are intermediate depth sites of slightly less than 2000 m, and Sites U1422 and U1424 are the deepest sites in excess of 2800 m water depth (Fig. 1). Sites U1426 and U1430 are on the bank, Site U1425 is on a terrace within the failed rift in the middle of the Yamato Rise, Sites U1423 and U1427 are on gentle slopes, and Sites U1422 and U1424 are on the basin floor. Sites U1427, U1424, and U1423 are under the influence of the first branch of the Tsushima Warm Current (TWC), Site U1426 is under the influence of the second branch, and Sites U1430 and U1425 are under the influence of the third branch of TWC, respectively. Site U1422 is under the influence of the Liman Cold Current. Sites U1422, U1423, and U1424 were influenced by sea ice at least during the last glacial period (Ikehara 2003). Sites U1427, U1426, and U1425, as well as Sites U1424, U1423, and U1422 can form latitudinal transects; Sites U1427, U1426, U1425, and U1424, as well as Sites U1423 and U1422 can form depth transects; and U1430, U1425, and U1424 can form a longitudinal transect, respectively.

Three to five holes are drilled at each site at least to the depth of ~ 100 m so that we can construct a continuous stratigraphic column at each site by core splicing, at least for the Quaternary interval (Tada et al. 2015). Since Sites U1424 and U1426 are drilled approximately at the same locations as ODP Sites 794 and 798, respectively, where previous drilling (legs 127/128) drilled deeper into Miocene, it is possible to use the cores from these sites to complement stratigraphic information on the older part.

Studied methods

Inter-hole correlations and construction of composite columns

It is important to construct perfectly continuous composite columns at all the sites to conduct high-resolution and high-precision correlations of dark and light layers in-between the sites. We conducted inter-hole correlation of all the recovered cores covering the lithological Unit I (Tada et al. 2015) at each site for this purpose. In addition, inter-hole correlation of cores enables us to precisely correlate off-splice parts of cores to the composite columns (spliced sections) so that we can use samples taken from off-splice parts of cores in the same way as those taken from the spliced parts and thus can drastically increase the amount of samples utilized for high-resolution studies.

Preliminary inter-hole correlation of cores was conducted onboard by stratigraphic correlators based on RGB (red-green-blue), GRA (gamma ray attenuation), and NGR (natural gamma ray) profiles taken immediately after core retrieval (Tada et al. 2015). However, there was no time onboard to examine core photographs in detail to cross-check the validity of inter-hole correlations. For this reason, there were possibilities that stratigraphic gaps and/or repetitions of the sequence caused by small faults mostly formed during drilling were overlooked. Also, drilling disturbance may affect physical properties such as GRA and NGR and even color in some cases, but they are easily overlooked just by examining profiles of physical properties and/or color especially if disturbances are minor. If the core sediments are homogeneous or heavily bioturbated, such small faults or disturbances are difficult to identify. However, in the case of the Quaternary part of the Japan Sea sediments, centimeter- to meter-scale dark and light layers serve as good markers to detect such minor faults and disturbances. We went back to high-resolution core photos of Unit I to identify minor faults, core disturbances, and artificial modifications such as plugging sponges in interstitial water (IW) sampling intervals and modified tie points to avoid such intervals. Details of procedures of inter-hole correlation, revision of tie points, and construction of composite sections are presented in Irino et al. (in press), and we adopt their results (revised composite sections) in this study.

Inter-site correlations

Inter-site correlations of dark and light layers are conducted based on visual inspection of composite columns of the six sites. First, we make a continuous composite column at each site using high-resolution core photographs (Irino et al., in press) and pick up relatively thick and distinct dark layers as marker layers. We set the bottom of marker dark layers as correlation tie points since many of the dark layers are laminated especially at their bottoms; thus, the effect of bioturbation is minimal (Watanabe et al. 2007). We use a composite column of Site U1424 as a reference column since its preliminary age model (Tada et al. 2015) suggests that the linear sedimentation rate (LSR) is more or less constant throughout the Quaternary and no significant sedimentation break exists. In addition, magnetostratigraphic events are best identified (Xuan et al., 2018, personal communication) and the largest number of tephra layers is found at this site. Next, we visually correlate marker dark layers at Site U1424 with those at other sites from the top of the column downward. We use marker tephra layers (Sagawa et al., in press) and magnetostratigraphic events (Xuan et al., 2018, personal communication) as guides to correlate marker dark layers among the sites. After inter-site correlation is completed, photographs of composite columnar sections of each sites are adjusted in their lengths at the core top and the Brunhes/Matsuyama boundary so that we can visually check the appropriateness of inter-site correlation of marker dark layers (Additional file 1: Figure S1).

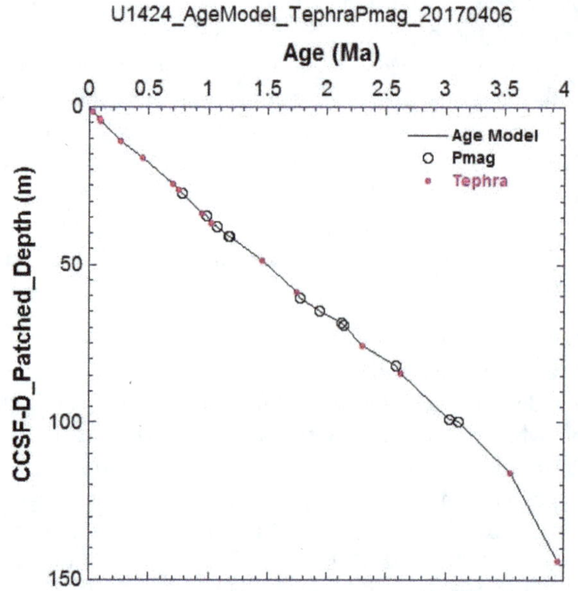

Fig. 2 A diagram showing age versus depth relationship at Site U1424 based on marker tephras and magnetostratigraphic events listed in Table 1. Depth is in CCSFD_patched meter scale

Construction of the age model

The master age model of Unit I is constructed for Site U1424 where sedimentation seems continuous with relatively constant LSR, magnetostratigraphic events are best defined, and the largest number of tephra are identified among the six deeper sites drilled in the Japan Sea during Exp. 346. We use the revised magnetostratigraphy of Xuan et al. 2018 (personal communication) and tephrochronology of Sagawa et al. (in press) and Satoguchi and Nagahashi (2012) to construct an age model, which we call a tephra-p-mag-based age model (Fig. 2). The datum used to construct the age model is listed in Table 1.

A tephra-p-mag-tuned age model is further tuned to LR04 age model (Lisiecki and Raymo 2005) by visually correlating GRA and NGR composite profiles of U1424 to LR04 $\delta^{18}O$ stack of benthic foraminifers. It is well known that GRA decreases with the increase in diatom contents because diatom frustules have high internal porosity (Tada and Iijima 1983). NGR also decreases with the increase in diatom contents because diatoms dilute detrital material. In case of the Quaternary sediments of the Japan Sea, diatom abundance strongly links to glacio-eustatic sea level because surface productivity of the sea is controlled by nutrient influx through the Tsushima Strait, which in turn has been modulated by glacio-eustatic sea level changes (Tada et al. 1992, 1999; Tada 2012). Because the Tsushima Strait is only 130 m deep and its width and depth increase with the rise in sea level, the influx of the TWC is expected to increase drastically with the rise in sea level. Thus, diatom abundance can be used as a sea level proxy.

Tuning of U1424 age to LR04 age is conducted for the last ~ 3 m.y. based on visual correlation of the GRA and NGR composite profiles with LR04 $\delta^{18}O$ stack profile. We put priority on GRA over NGR for correlation because GRA data has higher resolution than NGR. We then cross-check the correlation by comparing LR04 stack profile with NGR profile. For most of glacial-interglacial cycles, GRA and NGR profiles show patterns similar to each other without phase shift and thus confirm the correlation based on GRA. For several intervals where the GRA profile does not show a clear pattern similar to LR04 $\delta^{18}O$ stack profile, NGR profile is used for correlation. The correlation of GRA and NGR with LR04 $\delta^{18}O$ stack is shown in Additional file 2: Figure S2 and CCSF-D_patched depth and age for each correlation point are listed in Table 2 where CCSF-D_patched depth is the depth scale for the revised composite section. A resulted LR04-tuned age model is shown in Fig. 3.

Results and discussion

Age estimation of marker dark layers

In this study, we first constructed perfectly continuous composite columns for the entire Quaternary sequences at the six deeper sites studied, and using these composite columns, we correlated distinct dark layers among the sites

Table 1 A list of tephras and magnetostratigraphic events, their composite depths (CCSFD_patched), ages, and references used to construct a tephra and p-mag based age model shown in Fig. 2

Control No.	Event name	Other name	Age (Ma)	CCSF-D (m)	CCSF-D _Patched (m)
0	Core top		0	0.000	0.000
T01	AT		0.03	1.659	1.350
T02	BT22	Aso-4	0.09	3.960	3.860
T03	JS72	Toya	0.10	4.750	4.650
TK01	Aso-1		0.26	10.852	11.002
TK02	B-Og		0.45	16.675	16.215
T04	Ka1		0.7	24.458	24.358
T05	Imakuma1	Hkd-Ku	0.75	26.527	26.427
PM01	Bot C1n	B/M	0.78	27.542	27.442
T06	Yamada1		0.95	33.871	33.751
PM02	Top C1r.1n	Top Jarallillo	0.99	34.941	34.821
T07	Pink	SS-Pnk	1.02	36.795	37.121
PM03	Bot C1r.1n	Bot Jarallillo	1.07	37.741	38.101
PM04	Top C1r.2n	Top Cobb Mountain	1.17	40.635	40.995
PM05	Bot C1r.2n	Bot Cobb Mountain	1.19	40.845	41.205
T08	Kd18?	Ak-Kd18	1.45	47.586	48.436
T09	Fukuda	Eb-Fkd	1.75	57.812	58.662
PM06	Top C2n	Top Olduvai	1.78	59.692	60.542
PM07	Bot C2n	Bot Olduvai	1.95	63.902	64.752
PM08	Top C2r.1n	Top Reunion	2.13	67.913	68.763
PM09	Bot C2r.1n	Bot Reunion	2.15	68.406	69.256
T10	Mushono	Msn-Jwg4	2.3	74.727	75.577
PM10	Top C2An.1n	M/G	2.58	81.396	82.246
T11	Asashiro	Ass-Tmd2	2.62	83.231	84.451
PM11	Top C2An.1r	Top Kaena	3.03	97.682	98.992
PM12	Bot C2An.1r	Bot Kaena	3.12	98.537	99.847
T12	Souri	Sr-Ity	3.55	114.662	115.972
T16	Ohta	Znp-Ohta	3.95	142.667	143.977

for the interval covering the last 1.5 My corresponding to the lithological Subunit IA (Additional file 1: Figure S1). Then, we constructed a high-resolution and high-precision age model at Site U1424 based on correlation of GRA and NGR with LR04 $\delta^{18}O$ stack (Additional file 2: Figure S2). By using this age model, it is possible to estimate the ages of individual marker dark layers and thus project their ages to other sites.

Coding of dark layers

As is described above, dark layers are useful markers to make a high-resolution correlation between sites and assign precise ages to the sedimentary sequences in the deeper part of the Japan Sea. Consequently, it is useful to give them code numbers so that we can easily identify dark layers of our interest and estimate their ages. As to the last glacial period, thinly laminated (TL) layer numbering was conducted in the way so that TL numbers agree with interstadial numbers of DOC (Tada et al. 1999; Dansgaard et al. 1993). No numbering has been made on dark layers deposited before Marine isotope stage (MIS) 6 except by Tada et al. (1992) who made tentative naming of distinct dark layers at ODP Sites 794, 795, and 797. However, their naming was not systematic and went back to only ca. 0.7 Ma.

In this study, we conduct numbering of the marker dark layers in Subunit IA covering the last 1.45 Ma. We first identify relatively thick and distinct dark layers as marker layers for the purpose of inter-site correlation as is described above. Then, using these marker dark layers and tephra layers as guides, we further identify thinner but distinct dark layers that can be correlated among the deeper six sites and can be used as second order marker layers. The results are listed in Additional file 3: Table S1.

Table 2 A list of correlation point numbers, their depths (CCSFD_patched) on U1424 composite column, and their corresponding ages on LR04 $\delta^{18}O$ stack profile. Correlation points are also shown in Additional file 2: Figure S2

Exp	Site	Hole	Core	Type	Sect	A/W	Offset (cm)	CSF-A (m)	CCSF-A (m)	CCSF-D_Patched (m)	Age Control No.	LR04-tuned Age (Ma)
346	U1424	A	1	H	1	A	97.0	0.97	0.97	0.97	1	0.018
346	U1424	A	1	H	3	A	20.0	3.20	3.20	3.10	2	0.069
346	U1424	B	2	H	1	A	120.0	3.90	6.05	5.95	3	0.135
346	U1424	B	2	H	2	A	6.0	4.26	6.41	6.31	4	0.156
346	U1424	B	2	H	3	A	144.0	7.14	9.29	9.19	5	0.223
346	U1424	B	2	H	4	A	125.0	8.45	10.60	10.50	6	0.246
346	U1424	B	2	H	5	A	78.0	9.48	11.63	11.53	7	0.278
346	U1424	B	2	H	5	A	135.5	10.06	12.21	12.11	8	0.294
346	U1424	A	2	H	5	A	29.0	13.09	13.86	13.76	9	0.341
346	U1424	B	3	H	2	A	28.0	13.98	16.44	16.34	10	0.431
346	U1424	B	3	H	3	A	130.5	16.51	18.97	18.87	11	0.513
346	U1424	B	3	H	4	A	52.5	17.23	19.69	19.59	12	0.536
346	U1424	A	3	H	2	A	127.5	19.08	20.97	20.87	13	0.583
346	U1424	A	3	H	4	A	15.0	20.95	22.85	22.75	14	0.630
346	U1424	B	4	H	1	A	87.5	22.58	25.79	25.69	15	0.718
346	U1424	A	3	H	6	A	74.0	24.54	26.44	26.34	16	0.746
346	U1424	B	4	H	2	A	147.0	24.67	27.89	27.79	17	0.794
346	U1424	B	4	H	4	A	102.5	27.23	30.44	30.34	18	0.872
346	U1424	A	4	H	2	A	104.0	28.34	31.64	31.54	19	0.922
346	U1424	A	4	H	4	A	30.0	30.60	33.90	33.78	20	0.964
346	U1424	A	4	H	4	A	122.5	31.53	34.83	34.71	21	0.986
346	U1424	C	4	H	5	A	119.5	33.10	36.42	36.33	22	1.038
346	U1424	B	5	H	3	A	107.0	34.73	38.13	38.49	23	1.098
346	U1424	B	5	H	4	A	72.5	35.89	39.29	39.65	24	1.122
346	U1424	B	5	H	5	A	79.0	37.45	40.85	41.21	25	1.198
346	U1424	A	5	H	2	A	134.5	38.15	42.14	42.50	26	1.248
346	U1424	C	5	H	3	A	85.0	39.25	43.30	44.06	27	1.290
346	U1424	A	5	H	4	A	102.5	40.83	44.82	45.59	28	1.336
346	U1424	A	5	H	5	A	114.0	42.44	46.44	47.25	29	1.412
346	U1424	B	6	H	2	A	135.5	43.43	47.79	48.64	30	1.456
346	U1424	B	6	H	4	A	17.5	45.25	49.61	50.46	31	1.496
346	U1424	B	6	H	5	A	112.0	47.69	52.06	52.91	32	1.535
346	U1424	A	6	H	3	A	58.0	48.38	52.95	53.80	33	1.575
346	U1424	A	6	H	4	A	62.5	49.93	54.49	55.34	34	1.613
346	U1424	A	6	H	5	A	37.0	51.17	55.74	56.59	35	1.655
346	U1424	A	6	H	5	A	129.5	52.10	56.66	57.51	36	1.700
346	U1424	B	7	H	2	A	137.0	53.07	58.30	59.15	37	1.748
346	U1424	B	7	H	4	A	37.5	55.08	60.31	61.16	38	1.795
346	U1424	B	7	H	5	A	21.5	56.42	61.65	62.50	39	1.863
346	U1424	B	7	H	6	A	36.5	58.07	63.30	64.15	40	1.900
346	U1424	A	7	H	4	A	55.0	59.35	64.48	65.33	41	1.948
346	U1424	A	7	H	5	A	7.0	60.37	65.50	66.35	42	2.000
346	U1424	A	7	H	5	A	142.0	61.72	66.85	67.70	43	2.065

Table 2 A list of correlation point numbers, their depths (CCSFD_patched) on U1424 composite column, and their corresponding ages on LR04 $\delta^{18}O$ stack profile. Correlation points are also shown in Additional file 2: Figure S2 *(Continued)*

Exp	Site	Hole	Core	Type	Sect	A/W	Offset (cm)	CSF-A (m)	CCSF-A (m)	CCSF-D_Patched (m)	Age Control No.	LR04-tuned Age (Ma)
346	U1424	B	8	H	3	A	11.5	62.82	68.88	69.73	44	2.155
346	U1424	B	8	H	3	A	149.0	64.19	70.25	71.10	45	2.200
346	U1424	B	8	H	4	A	117.5	65.38	71.44	72.29	46	2.240
346	U1424	B	8	H	5	A	49.0	66.19	72.25	73.10	47	2.275
346	U1424	A	8	H	4	A	52.5	68.83	73.44	74.29	48	2.313
346	U1424	A	8	H	5	A	52.0	70.32	74.94	75.79	49	2.358
346	U1424	B	9	H	1	A	130.0	70.50	77.14	77.99	50	2.433
346	U1424	B	9	H	2	A	110.5	71.77	78.41	79.26	51	2.488
346	U1424	B	9	H	3	A	74.5	72.91	79.55	80.40	52	2.518
346	U1424	B	9	H	4	A	60.0	74.26	80.90	81.75	53	2.565
346	U1424	A	9	H	3	A	50.5	76.81	82.15	83.09	54	2.600
346	U1424	B	9	H	5	A	139.5	76.56	83.20	84.42	55	2.645
346	U1424	B	9	H	6	A	124.5	77.91	84.55	85.86	56	2.688
346	U1424	A	9	H	5	A	66.5	79.97	85.31	86.62	57	2.715
346	U1424	B	10	H	3	A	122.5	82.93	89.75	91.06	58	2.805
346	U1424	B	10	H	5	A	89.0	85.59	92.41	93.72	59	2.878
346	U1424	A	10	H	4	A	102.5	88.33	94.18	95.49	60	2.920
346	U1424	A	10	H	5	A	117.0	89.97	95.83	97.14	61	2.975

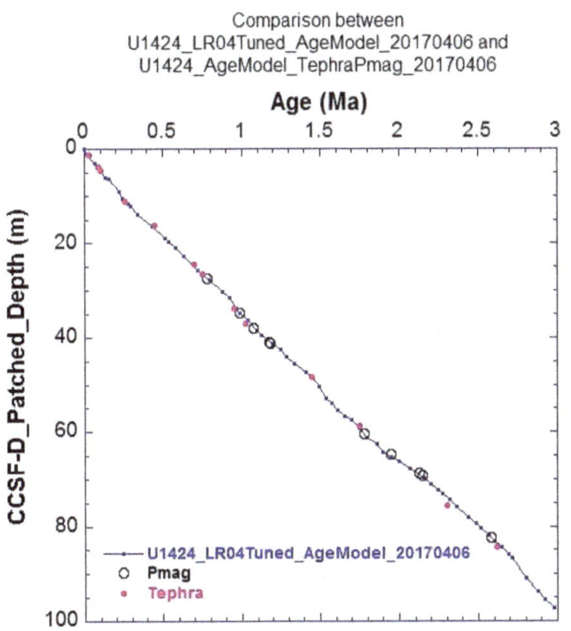

Fig. 3 A diagram showing age versus depth relationship at Site 1424 based on tuning of GRA and NGR profile at U1424 and LR04 benthic $\delta^{18}O$ record of Lisiecki and Raymo (2005). All tuned points fall in an age versus depth relationship based on marker tephras and magnetostratigraphic events within uncertainties

The relatively thick and distinct dark layers occur every several meters and they tend to be associated with quasi-cyclic changes in amplitude and frequency of sediment lightness (L^*) in Subunit IA, which seems to be associated with glacial-interglacial sea level changes as will be discussed later. In brief, thicker dark layers tend to occur during glacial maxima especially when sea level was lowered by more than 100 m and euxinic deep water emerged due to development of low-salinity surface water (Tada et al. 1999; Kido et al. 2007; also Seki 2017). Thick dark layers accumulated under euxinic deep water were found during MIS 2, 6, 10, 12, 16, 20, and 22 (Seki 2017). Such dark layers are overlain by thick light layers which were deposited during interglacial maxima. These characteristics are less clear before MIS 22 (~ 0.9 Ma) when thick diatomaceous dark layers frequently deposited during interglacial maxima.

We define a cycle of such several meter-scale change in amplitude and frequency of lightness as starting from the top of a relatively thick and distinct dark layer downward to the base of the relatively thick light layer so that one cycle basically corresponds to one glacial-interglacial cycle. From the core top downward, cycle number starts from 0 and increases downward. Because we tune changes in sea level proxies such as GRA and NGR to LR04 $\delta^{18}O$ stack so as to construct high-resolution age model for Unit I as described above, and because one cycle basically represents one glacial-interglacial cycle, there is a relation between a cycle number and MIS number. Namely,

$$X = Y/2 - 1$$

where X is a cycle number and Y is an MIS number when the thick dark layer at the top of cycle X deposited. Exceptions are cycle 0 that corresponds to the Holocene when no thick dark layer exists at the top, cycle 1 whose top thick dark layer corresponds to MIS 2 and thus $X = Y/2$, cycle 20 whose thick and dark layer deposited during MIS 45 (interglacial period) and thus $X = Y/2 - 2.5$, and cycle 21 whose thick dark layer deposited at MIS 46 and thus $X = Y/2 - 2$. A relatively thick and distinct dark layer at a top of cycle X is named Dark Layer (DL)-X-1 (Additional file 1: Figure S1; Additional file 3: Table S1). Every cycle contains several thinner but still distinct dark layers that are possible to correlate in-between sites. We put numbers 2, 3, ... on them in descending order and described as suffixes to the cycle number. Namely, Zth dark layer in cycle X is DL-X-Z. In Additional file 3: Table S1, we list ages at the base of each dark layer based on tephra-p-mag-tuned and LR04-tuned age models for the last ~ 3 My. We also list depths and ages at the top of marker dark layer that define the top of each cycle, although it should be noted that their depths and ages may be less well defined and could be slightly diachronous due to the result of bioturbation.

Temporal and spatial changes in linear sedimentation rates (LSRs) in the Japan Sea during the last 1.45 My

As is described in the previous section, we estimated the age at the base of each marker dark layer at Site 1424 and projected its age to other sites based on inter-site correlation described above (Table 2). By using these data, it is possible to calculate linear sedimentation rates (LSRs) at all the six sites with high time resolution (Additional file 4: Table S2). Additional file 5: Figure S3 shows temporal changes in LSRs at all sites. LSRs are more or less constant at ~ 4 and ~ 3 cm/ky at Sites U1422 and U1424 (deeper sites); gradually increasing from ~ 3 to ~ 10 cm/ky, from ~ 2 to ~ 7 cm/ky, and from ~ 2 to ~ 8 cm/ky at Sites U1423, U1425, and U1430 (intermediate depths sites); and relatively high and more or less constant at ~ 10 cm/ky at Site U1426 (shallow site), respectively. Exceptions are between ~ 850 and ~ 1200 ka when LSRs are slightly higher and variable and after ~ 250 ka when LSRs increase upward. At Site U1430, extremely low LSR is observed at ~ 800 ka.

In general, LSR tends to be lower at deeper sites, which probably reflects dissolution of carbonate (and biogenic silica to some extent). The upward increase in LSRs after ~ 250 ka is probably because near-surface sediments are less compacted. The gradually increasing trend of LSR at intermediate depth sites can be partly due to lesser degree of compaction, but the increasing trends started from deeper parts. The upward increasing trend of LSRs could be explained by the upward increase in biogenic carbonate content due to deepening of calcium carbonate compensation depth (CCD) that crossed the intermediate depths of 1000 to 2000 m at ~ 1.0 Ma. Variable LSRs between ~ 850 and ~ 1200 ka could be related to large amplitude changes of CCD during transition from shallower CCD before 1200 ka to deeper CCD after 850 ka. Mass accumulation rate (MAR) calculation of carbonate at these sites based on high-resolution XRF core scanner and gamma ray attenuation (GRA) porosity data combined with LSR data will be necessary to test this possibility.

Establishment of paleo-observatory network

Reconstruction of material cycling such as carbon, sulfur, and phosphorous within the earth surface is critical to understand dynamics and controlling mechanisms of climatic changes in the past and preparing for the future. Especially, changes in centennial to millennial time scales are of prime importance since they are beyond the coverage of observational records, but still, climatic changes of such time scales are relevant to the society. Spatial coverage is also important since spatial changes in mass flux often give clues to understand the processes within the system of interest and their controlling factors. However, it is generally difficult to correlate geological records from different localities with high time precision and resolution. In this respect, Quaternary hemipelagic sediments of the Japan Sea provide a rare opportunity to study spatiotemporal changes of mass flux with high time precision and resolution. Namely, marker dark layers and tephra layers can be correlated with the time precision of less than 100 years, and approximately 250 marker layers are defined during the last 1.45 Ma, providing ~ 6 ky resolution in average. These time precision and resolution are an order of magnitude higher than those for other correlation methods such as biostratigraphy, paleomagnetism, and radiometric dating. In addition, dry bulk density of the sediments can be estimated from GRA data measured onboard. Thus, it is possible to reconstruct MARs with an average time resolution of ~ 6 ky at the six deeper IODP sites that cover the area more than a half of the Japan Sea. Moreover, we can calculate MAR of individual chemical element and/or mineral by combining high-resolution MAR data with high-resolution chemical and/or mineralogical analyses data.

In this study, we propose to utilize the six IODP sites in the Japan Sea as paleo-observatory network since high-precision age data and high-resolution LSRs and GRA data are available; the sites cover wide latitude, longitude, and water depth ranges; proxy data at each site can be correlated with precision of less than 100 years; and good amount of samples are still available from IODP.

Conclusions

The Quaternary hemipelagic sediments in the Japan Sea (Subunit IA) are characterized by centimeter- to decimeter-scale alternation of dark and light layers. Over 250 dark layers are present and 64 distinct dark layers are chosen that are traceable throughout the deeper (> 500 m water depth) part of the basin and correlated in-between the six sites (U1422, 1423, 1424, 1425, 1426, and 1430) drilled by IODP Expedition 346. An age model covering the last 3 My is constructed at Site U1424 based on tephra-chronology and magneto-stratigraphy, which is further tuned to LR04 age using GRA profile, a proxy for sea level change in the Quaternary Japan Sea, correlated to LR0 4 $\delta^{18}O$ stack profile. We estimate ages of marker dark layers at Site U1424, which are projected to other sites based on inter-site correlation of marker dark layers. In this way, we establish paleo-observatory network covering the last 1.45 My in the Southeastern half of the Japan Sea. This network will greatly enhance our understanding on material cycling in the Japan Sea.

Additional files

Additional file 1: Figure S1. Figures showing inter-site correlation of marker dark layers in Subunit IA among six deeper sites in the Japan Sea drilled during Expedition 346. **a** From 0 to 0.8 Ma and **b** from 0.8 to 1.5 Ma. Code numbers of marker dark layers (in Additional file 3: Table S1), marker tephras, and paleomagnetic event boundaries (in Table 1) are also shown.

Additional file 2: Figure S2. a-d Correlation of GRA and NGR profiles at Site U1424 with LR04 benthic δ18O record of Lisiecki and Raymo (2005) during the last 3 My to construct LR04-tuned age model. Correlation points are numbered and listed in Table 2. Also shown at the bottom is L* profile with marker dark layer numbers during the last 1.5 Ma, whose ages and depths are listed in Additional file 3: Table S1.

Additional file 3: Table S1. Code numbers and ages of marker dark layers and their composite depths (CCSFD_patched) at Sites U1422, U1423, U1424, U1425, U1426, and U1430.

Additional file 4: Table S2. A list of linear sedimentation rates (LSRs) between two adjacent marker dark layer horizons at each site during the last 1.4 Ma (corresponding to Subunit IA) based on the two age models.

Additional file 5: Figure S3. Temporal changes in linear sedimentation rates (LSRs) at all the studied sites during the last 1.45 Ma calculated based on ages of marker dark layers whose ages are estimated based on LR04-tuned age model of U1424 (blue) and tephras and paleomagnetic events (orange).

Abbreviations

CCD: Calcium carbonate compensation depth; CCSF: Core composite depth below sea floor; DL: Dark Layer; DOC: Dansgaard-Oeschger Cycle; GRA: Gamma ray attenuation; IODP: Integrated Ocean Drilling Program; IW: Interstitial water; LSR: Linear sedimentation rate; MAR: Mass accumulation rate; MIS: Marine isotope stage; NGR: Natural gamma ray; ODP: Ocean Drilling Program; RGB: Red-green-blue; TL: Thinly laminated; TWC: Tsushima Warm Current; XRF: X-ray fluorescence

Acknowledgements

We thank the technicians and crews of IODP Expedition 346 for their dedicated efforts to help us in retrieving high-quality cores and shipboard data that were used in this study. This research used samples and/or data provided by the Integrated Ocean Drilling Program (IODP).

Funding

This work was supported by a grant from IODP Exp. 346 After Cruise Research Program, JAMSTEC, awarded to TR, IK, Irino T, Itaki T, ST, KY, SS, and KA and from JSPS KAKENHI grant number 16H01765 awarded to TR.

Authors' contributions

Exp. 346 scientists collected basic data for this study during IODP Expedition to the Japan Sea. TR, MRW, and ZCA directed AWT Jr., BMA, BBJ, CSC, GMHC, DGR, GSJ, GL, HACG, HAE, IK, Irino T, Itaki T, KA, KCW, KY, LGS, LKE, LJ, LCICD, PMS, PLC, SRK, SS, ST, TS, WS, XC, ZH, and ZM to collect data. TR, Irino T, and IK proposed the topic and conceived and designed the study. TR, Irino T, IK, TS, KA, LS, and SA conducted detailed inter-site correlation of dark layers, SS and XC conducted revision of magneto-stratigraphy, and Itaki T and KY conducted revision of biostratigraphy. TR, Irino T, Itaki T, KA, LS, and SA drew figures, and TR wrote a draft of the manuscript. All authors read and approved the final manuscript.

Authors' information

TR and MRW were co-chief scientists, ZCA is a staff scientist, and IK, Irino T, Itaki T, ST, KY, SS, KA, XC, AWT Jr., BMA, BBJ, CSC, GMHC, DGR, GSJ, GL, HACG, HAE, KCW, LGS, LKE, LJ, LCICD, PMS, PLC, SRK, TS, WS, ZH, and ZM were shipboard scientists of IODP Expedition 346. KA and SA are students of TR, and LS is a student of Irino T.

Competing interests

The authors declare that they have no competing interests.

Author details

[1]Graduate School of Science, The University of Tokyo, Tokyo, Japan. [2]Faculty of Environmental Earth Science, Hokkaido University, Sapporo, Japan. [3]Research Institute of Geology and Geoinformation, Geological Survey of Japan, AIST, Tsukuba, Japan. [4]National Oceanography Centre Southampton, Southampton, UK. [5]Institute of Science and Engineering, Kanazawa University, Kanazawa, Japan. [6]Department of Geology and Paleontology, National Museum of Nature and Science, Tsukuba, Japan. [7]Department of Earth and Environment, Boston University, Boston, USA. [8]International Ocean Descovery Program, Texas A&M University, College Station, USA. [9]Marine Sciences, Florida International University, Miami, USA. [10]Laboratoire CEFREM, Université de Perpignan, Perpignan, France. [11]Department of Earth and Atmospheric Sciences, University of Nebraska, Lincoln, USA. [12]Earth, Environmental, and Planetary Sciences, Brown University, Providence, USA. [13]Escola de Artes, Ciencias e Humanidades (EACH), Universidade de São Paulo, São Paulo, Brazil. [14]Department of Earth, Environmental, and Planetary Science, Rice University, Houston, USA. [15]School of Earth Sciences, University of Melbourne, Melbourne, Australia. [16]Present Address: Department of Geology and Geophysics, Woods Hole Oceanographic Institution, Woods Hole, USA. [17]School of Geography, Politics and Sociology, University of Newcastle Upon Tyne, Newcastle Upon Tyne, UK. [18]Institute for Geosciences, Christian-Albrechts-Universitat zu Kiel (IFM), Kiel, Germany. [19]Department of Earth, Atmospheric and Planetary Sciences, Massachusetts Institute of Technology, Cambridge, USA. [20]Petroleum and Marine Research Division, Korea Institute of Geoscience and Mineral Resources, Daejeon, South Korea. [21]Department of Ocean Science, Korea Maritime and Ocean University, Busan, South Korea. [22]Géosciences Montpellier, Université Montpellier II, Montpellier, France. [23]IPMA, Lisbon, Portugal. [24]Department of Geosciences, FB5, University of Bremen, Bremen, Germany. [25]Rosenstiel School of Marine and Atmospheric Science, University of Miami, Miami, USA. [26]School of Earth, Ocean and Climate Sciences, Indian Institute of Technology Bhubaneshwar, Bhubaneshwar, India. [27]Centre Bretagne, IFREMER, Plouzané cedex, France. [28]Institute of Oceanology, Chinese Academy of Sciences, Beijing, China. [29]Present Address: Research Center for Earth System Science, Yunnan University, Kunming, China. [30]Present Address: Department of Earth Sciences, Utrecht University, Utrecht, Netherlands.

References

Dansgaard W, Johnsen SJ, Clausen HB, Dahljensen D, Gundestrup NS, Hammer CU, Hvidberg CS, Steffensen JP, Sveinbjornsdottir AE, Jouzel J, Bond G (1993) Evidence for general instability of past climate from a 250-kyr ice-core record. Nature 364(6434):218–220. https://doi.org/10.1038/364218a0

Gallagher SJ, Kitamura A, Iryu Y, Itaki T, Koizumi I, Hoiles PW (2015) The Pliocene to recent history of the Kuroshio and Tsushima Currents: a multi-proxy approach. Prog Earth Planet Sci 2:17. https://doi.org/10.1186/s40645-015-0045-6

Gamo T, Nakayama N, Takahata N, Sano Y, Zhang J, Yamazaki E, Taniyasu S, Yamashita N (2014) The Sea of Japan and its unique chemistry revealed by time-series observations over the last 30 years. Monogr Environ Earth Planets 2(1):1–22. https://doi.org/10.5047/meep.2014.00201.0001

Ikehara K (2003) Late Quaternary seasonal sea-ice history of the northeastern Japan Sea. J Oceanogr 59(5):585–593. https://doi.org/10.1023/B:JOCE.0000009588.49944.3d

Irino T, Tada R, Ikehara K, Sagawa T, Karasuda A, Kurokawa S, Seki A, Lu S (in press) Construction of perfectly continuous records of physical properties for dark-light sediment sequences collected from the Japan Sea during Integrated Ocean Drilling Program Expedition 346 and their potential utilities as paleoceanographic studies. Prog Earth Planet Sci. https://doi.org/10.1186/s40645-018-0176-7.

Kido Y, Minami I, Tada R, Fujine K, Irino T, Ikehara K, Chun J-H (2007) Orbital-scale stratigraphy and high-resolution analysis of biogenic components and deep-water oxygenation conditions in the Japan Sea during the last 640 kyr. Palaeogeogr Palaeoclimatol Palaeoecol 247(1–2):32–49. https://doi.org/10.1016/j.palaeo.2006.11.020

Lisiecki LE, Raymo ME (2005) A Pliocene-Pleistocene stack of 57 globally distributed benthic $\delta^{18}O$ records. Paleoceanography 20(1):PA1003. https://doi.org/10.1029/2004PA001071

Sagawa T, Nagahashi Y, Satoguchi Y, Holborn A, Itaki T, Gallagher SJ, Saavedra-Pellitero M, Ikehara K, Irino T, Tada R (in press) Integrated tephrostratigraphy and stable isotope stratigraphy in the Japan Sea and East China Sea using IODP Sites U1426, U1427 and U1429, Expedition 346 Asian Monsoon. Prog Earth Planet Sci. https://doi.org/10.1186/s40645-018-0168-7.

Satoguchi Y, Nagahashi Y (2012) Tephrostratigraphy of the Pliocene to Middle Pleistocene Series in Honshu and Kyushu Islands, Japan. Island Arc 21 (3):149-169

Seki A (2017) Reconstruction of marine organic matter content in the sediments and deep-water redox condition variability of the Japan Sea during the Quaternary using high-resolution XRF core scanner. Doctoral Thesis, Graduate School of Science, the University of Tokyo, p 198

Tada R (1994) Paleoceanographic evolution of the Japan Sea. Palaeogeogr Palaeoclimatol Palaeoecol 108(3–4):487–508. https://doi.org/10.1016/0031-0182(94)90248-8

Tada R (2005) Onset and evolution of millennial-scale variability in the Asian Monsoon and its impact on paleoceanography of the Japan Sea. Millennial-Scale Variability in the Asian Monsoon, American Geological Union, U.S.A, pp 1–16

Tada R (2012) The Japan Sea sediments and variability of East Asian Monsoon: toward the IODP drilling of the Japan Sea and East China Sea. Quat Res (Tokyo) 51(3):151–164

Tada R, Iijima A (1983) Petrology and diagenetic changes of Neogene siliceous rocks in northern Japan. J Sediment Petrol 53(3):911–930

Tada R, Irino T, Koizumi I (1999) Land-ocean linkages over orbital and millennial timescales recorded in late Quaternary sediments of the Japan Sea. Paleoceanography 14(2):236–247

Tada R, Koizumi I, Cramp A, Rahman A (1992) Correlation of dark and light layers, and the origin of their cyclicity in the Quaternary sediments from the Japan Sea. In Pisciotto, K. A., Ingle, J. C, Jr., von Breymann, M. T., Barron, J., et al. Proc ODP, Sci results, 127/128, Pt. 1: 577–601 College Station(Ocean Drilling Program)

Tada R, Murray RW, Alvarez Zarikian CA, the Expedition 346 Scientists (2015) Proceedings of Integrated Ocean Drilling Program. IODP, vol 346. Integrated Ocean Drilling Program, College Station. https://doi.org/10.2204/iodp.proc.346.101

Talley LD, Min DH, Lobanov VB, Luchin VA, Ponomarev VI, Salyuk AN, Shcherbina AY, Tishchenko PY, Zhabin I (2006) Japan/East Sea water masses and their relation to the sea's circulation. Oceanography 19(3):32–49

Tamaki K, Pisciotto KA, Allan J, et al. (1990) Proceedings of Ocean Drilling Project. Initial Reports, 127, College Station, TX (Ocean Drilling Program). https://doi.org/10.2973/odp.proc.ir.127.1990

Watanabe S, Tada R, Ikehara K, Fujine K, Kido Y (2007) Sediment fabrics, oxygenation history, and circulation modes of Japan Sea during the Late Quaternary. Palaeogeogr Palaeoclimatol Palaeoecol 247(1–2):50–64. https://doi.org/10.1016/j.palaeo.2006.11.021

On the evolution of the water ocean in the plate-mantle system

Takashi Nakagawa[1,2*], Hikaru Iwamori[3,5], Ryunosuke Yanagi[4] and Atsushi Nakao[5]

Abstract

Here, we investigate a possible scenario of surface seawater evolution in the numerical simulations of surface plate motion driven by mantle dynamics, including thermo-chemical convection and water migration, from the early to present-day Earth to constrain the total amount of water in the planetary system. To assess the validity of two hypotheses of the total amount of water inferred from early planetary formation processes and mineral physics, we examine the model sensitivity to the total water in the planetary system (both surface and deep interior) up to 15 ocean masses. To explain the current size of the reservoir of surface seawater, the predictions based on the numerical simulations of hydrous mantle convection suggest that the total amount of water should range from 9 to 12 ocean masses. Incorporating the dense hydrous magnesium silicate (DHMS) with a recently discovered hydrous mineral at lower mantle pressures (phase H) indicates that the physical mechanism of the mantle water cycle would not be significantly influenced, but the water storage region would be expanded in addition to the mantle transition zone. The DHMS solubility field may have a limited impact on the partitioning of water in the Earth's deep mantle.

Keywords: Ocean mass, Water solubility, Dense hydrous magnesium sillicate (DHMS), Plate motion, Mantle dynamics

Introduction

Understanding the water ocean on the Earth's surface is essential for understanding the habitability of Earth-like planets (Maruyama et al. 2013). Geological records suggest that the Earth's surface seawater likely formed between 3.8 Ga and 4.5 Ga based on the appearance of metamorphic rocks derived from sedimentary rocks (one line of evidence that water was present on the Earth's surface) and the initiation of plate tectonics, as suggested by zircon dating (Appel et al. 1998; Maruyama and Komiya 2011; Mojzsis et al. 2001; Valley et al. 2014). However, although constraining the initiation of plate tectonics is very important for determining when water started to be transported into the deep planetary interior, this timing remains debated, with estimates ranging from 2.5 to 4.3 Ga (Condie 2016; Hopkins et al. 2008). To reconcile such geological evidence related to the survival of the water ocean over a period of ~4 billion years, it is crucial to understand the size of the surface seawater reservoir. There are two hypotheses regarding the size of the reservoir that influences the evolution of the water ocean from the early to present-day Earth: (1) estimates based on cosmochemical and geochemical analyses of chondritic material and the modeling of the formation of a water ocean as a result of the solidification of a surface magma ocean (Marty 2012; Hamano et al. 2013) and (2) petrological estimates of the actual water content of the mantle (e.g., Hirschmann 2006). In the early planetary formation hypothesis, the total amount of water in the planetary system after the completion of early planetary formation or the solidification of the surface magma ocean ranges from 5 to 15 ocean masses, which means that a very large amount of water is expected to reside in the mantle (Marty et al. 2012; Hamano et al. 2013). Petrological estimates suggest that the mantle water content may range from 0.2 to 2.3 ocean masses. By adding seawater on the surface, eventually, it turns out that the total present amount of water in the planetary system should be approximately 1.2 to 3.3 ocean masses (Hirschmann 2006).

During the evolution of surface seawater over time, surface plate motions may play an important role, as has

* Correspondence: ntakashi@jamstec.go.jp
[1]Department of Mathematical Science and Advanced Technology, Japan Agency for Marine-Earth Science and Technology, 3173-25, Showa-machi, Yokohama 236-0001, Japan
[2]Department of Earth Sciences, University of Hong Kong, Pokfulam Road, Hong Kong, Hong Kong
Full list of author information is available at the end of the article

been implied by geological records (e.g., Maruyama and Okamoto 2007) and numerical modeling (e.g., Rüpke et al. 2004; Iwamori 2007). The evolution of surface seawater in the plate-mantle system has also been computed in semi-analytical models (Crowley et al. 2011; Sandu et al. 2011; Korenaga 2011). In such models, the partitioning of water between the surface and the deep mantle is determined by the flux balance of water uptake (regassing) by plate subduction and degassing beneath mid-ocean ridges. This modeling results in an estimate of the total water mass in the planetary system of ~2 to 3 ocean masses and consistent with petrological measurements (Hirschmann 2006).

However, our previous studies have indicated that the evolution of the mantle water mass cannot be determined with a simple regassing-degassing balance, which essentially incorporates the excess water transport associated with the dehydration reaction (Nakagawa et al. 2015; Nakagawa and Spiegelman 2017; Nakagawa and Iwamori 2017). Hence, the water solubility limit of each mantle mineral should be included to compute the excess water in the dehydration reaction. By considering this effect, our previous studies have indicated that the mantle water evolution may be strongly regulated by the water solubility limits of upper mantle minerals. In addition, the rheological properties of hydrous mantle rocks may affect the behavior of surface plate motion, with hydrous mantle conditions allowing for more vigorous surface plate motion and a much larger friction coefficient than dry mantle conditions (Nakagawa and Iwamori 2017). Such vigorous surface plate motion induced by a pseudo-plastic rheology may transport large amounts of water (i.e., several ocean masses) into the deep mantle over geologic timescales (approximately 2 billion years), and the mantle transition zone can absorb this amount of water over geologic timescales (Nakagawa and Iwamori 2017). Therefore, the total reservoir size of water at the surface and in the deep interior should be resolved in a plate-mantle system using numerical mantle convection simulations that include the water solubility limits of mantle minerals.

The issues identified in our previous studies include three model assumptions: (1) The boundary condition of mantle water transport is assumed to be an infinite reservoir of water, which is not a realistic situation. To improve the boundary condition of the mantle water transport, the lifetime and amount of water ocean in the planetary system associated with surface plate motions should be evaluated. (2) Defining the water solubility limits of mantle minerals, we only assume those of the upper mantle minerals and assign a constant value to those of the lower mantle minerals. In the lower mantle, the existence of dense hydrous magnesium silicate (DHMS) has previously been established (e.g., Ohtani et al. 2001), and according to recent mineral physics measurements, a new hydrous phase that is stable under lower mantle conditions has been found, which is a DHMS phase called "phase H." This DHMS may affect the evolution of the mantle water mass because, for instance, the solubility limit of phase H is very high (~ 12 wt.%; Nishi et al. 2014; Ohtani et al. 2014; Ohira et al. 2014). (3) Moreover, the pseudo-plastic yielding associated with hydrous mantle rocks is addressed as a reduction in the bulk value of the yield strength of oceanic lithosphere. More realistically, the water-weakening effect on hydrous mantle rocks likely only influences the friction coefficient, not the bulk yield strength of the oceanic lithosphere (Gerya et al. 2008).

The aim of this study is to resolve the issues that arose from our previous study and thereby reveal a possible scenario of the evolution of surface seawater in a plate-mantle system via numerical mantle convection simulations with water migration processes, including the computation of a finite amount of water at the surface, the water solubility limit of DHMS (including phase H), and the improved pseudo-plastic yielding of the oceanic lithosphere associated with hydrous mantle rocks. Using this modeling approach, we can resolve which hypothesis is preferable for understanding the survival time of surface seawater.

Methods/Experimental

Mantle convection simulations

The numerical modeling of global-scale mantle dynamics with water migration has been described by Nakagawa et al. (2015) and Nakagawa and Spiegelman (2017). This process is briefly described here. We assume the thermo-chemical multiphase mantle convection of a compressible and truncated anelastic approximation in a 2D spherical annulus geometry (Hernlund and Tackley 2008). The modeled mantle can be decomposed into depleted harzburgite and enriched basaltic material composed of two-phase transition systems, i.e., olivine-spinel-bridgmanite-post-perovskite and pyroxene-garnet-bridgmanite-post-perovskite, which are associated with changes in the basaltic material. A reference state for each phase transition system is computed as in Tackley (1996). All phase transition parameters can be found in Nakagawa and Tackley (2011). A partial melting effect is included to create an oceanic crust and to allow its segregation. The viscosity of the modeled mantle is dependent on temperature, pressure, water content, and yield strength and is determined via the following equations:

$$\eta_d = A_d \sum_{i,j=1}^{\text{nphase}=3,4} \Delta\eta_{ij}^{\Gamma_{ij} f} \exp\left[\frac{E_d + pV_d}{RT}\right] \quad (1)$$

$$\eta_w = A_w \left(\frac{C_w}{C_{w,\text{ref}}}\right)^{-1} \sum_{i,j=1}^{\text{nphase}=3,4} \Delta\eta_{ij}^{\Gamma_{ij} f_j} \exp\left[\frac{E_w + pV_w}{RT}\right] \quad (2)$$

$$\eta_Y = \frac{\sigma_Y(p, C_w)}{2\dot{e}} \quad (3)$$

$$\eta = \left(\frac{1}{\eta_d} + \frac{1}{\eta_w} + \frac{1}{\eta_Y}\right)^{-1} \quad (4)$$

where $A_{d,w}$ is the prefactor determined by $T = 1600$ K and the ambient pressure at the surface (the subscripts d and w indicate dry and hydrous mantle, respectively), $E_{d,w}$ is the activation energy; $V_{d,w}$ is the activation volume; C_w is the water content in the mantle; $C_{w,\text{ref}}$ is the reference water content and is assumed to be 620 ppm (Arcey et al. 2005); the exponent of the prefactor, which is dependent on the water content, is based on the results of deformation experiments (Mei and Kohlstedt 2000) and is assumed to be 1; Γ_{ij} is the phase function; f is the basaltic composition (varying from 0 to 1); R is the gas constant (8.314 J K mol^{-1}); T is the temperature; p is the pressure; $\sigma_Y(p, C_w)$ is the yield strength of the oceanic lithosphere, which is a function of pressure and water content; $\dot{e}$ is the second invariant of the strain rate tensor; and $\Delta\eta_{ij}$ is the viscosity jump associated with the phase transition, which is assumed to increase by 30 times during the phase transition from ringwoodite or garnet to bridgmanite.

The yield strength of oceanic lithosphere is dependent on pressure and the mantle water content and is determined as follows:

$$\sigma_Y(p, C_w) = C_Y + \mu(C_w)p; \mu(C_w) = \min\left[1, \left(\frac{C_w}{C_{w,\text{ref}}}\right)^{-1}\right]\mu_0 \quad (5)$$

where μ_0 is the Byerlee-type friction coefficient. We include only the water-weakening effect caused by hydrated rocks (Gerya et al. 2008).

Another important influence of hydrated mantle rocks is the variation in density caused by hydrated mantle minerals, as noted by Nakagawa et al. (2015):

$$\rho(T_{ad}, p, C, C_w) = \rho(T_{ad}, C, p)(1 - \alpha(T_{ad}, C, p)(T - T_{ad})) - \Delta\rho_w C_w \quad (6)$$

where $\rho(T_{ad}, C, z)$ is the combined reference density between harzburgite and mid-ocean ridge basalt (MORB) compositions, with a 2.7% density difference, as shown in Fig. 1 (and a 3.6% density difference between olivine- and pyroxene-related phases); T_{ad} is the adiabatic temperature in the mantle; $\Delta\rho_w$ is the density variation due to the water content; and C_w is the water content. Generally, the densities of hydrous minerals are less than those of dry minerals, but the value of $\Delta\rho_w$ is less constrained by high-pressure/high-temperature (high P-T) experiments; the densities of hydrous minerals are generally 0.1 to 1.0% less than those of dry mantle minerals (Richard and Iwamori 2010).

To solve the equations of thermo-chemical mantle convection and to model the chemical composition, we use the numerical code of a finite-volume multigrid flow solver with tracer particles (StagYY; Tackley 2008). To determine the water migration processes, the tracer particle approach is used for the water advection process and degassing process via volcanic eruptions, and a numerical scheme involving a discrete migration velocity approximation is used for the dehydration process (Iwamori and Nakakuki 2013; Nakagawa et al. 2015; Nakao et al. 2016; Nakagawa and Spiegelman 2017). The dehydration process is modeled as the upward migration of excess water, which is defined as the difference between the actual water content and the water solubility at a certain temperature and pressure in a grid (see Fig. 2 in Nakagawa et al. 2015); this upward migration may have a velocity comparable to that of fluid movement in fully two-phase flow modeling (Wilson et al. 2014), which ranges from 0.5 to 50 m/year. Although this may affect the numerical resolution of the model, it should not have a significant influence on the results based on an assessment of the Appendix of Nakagawa and Iwamori (2017). Note that our model of water migration allows for migration only in the vertical direction, whereas in Wilson et al. (2014), the fluid component may migrate appreciably in the horizontal direction over several tens of kilometers near the corner of a mantle wedge. This conventional scheme seems to be valid for global-scale water circulation in a convecting mantle. In addition, we also assume the water partitioning of partially molten material, as in Nakagawa et al. (2015) and Nakagawa and Spiegelman (2017). The partition coefficient of water between solid mantle material and melt is set to 0.01 (Aubaud et al. 2008). In the numerical scheme of material transportation in Nakagawa and Iwamori (2017), two distinct compositional types of tracers are assumed, which can track the water migration for each composition separately; however, here, the chemical composition assigned to each tracer continuously varies with arbitrary melting (see Rozel et al. 2017; Lourenço et al. 2018). Hence, the water migration should be tracked using a bulk composition, which is given as:

$$S(T, P, C) = S_{\text{Peridotite}}(T, P)(1 - C) + S_{\text{Basalt}}(T, P)C \quad (7)$$

where $S_{\text{peridotite}}$ and S_{basalt} are the water solubilities of

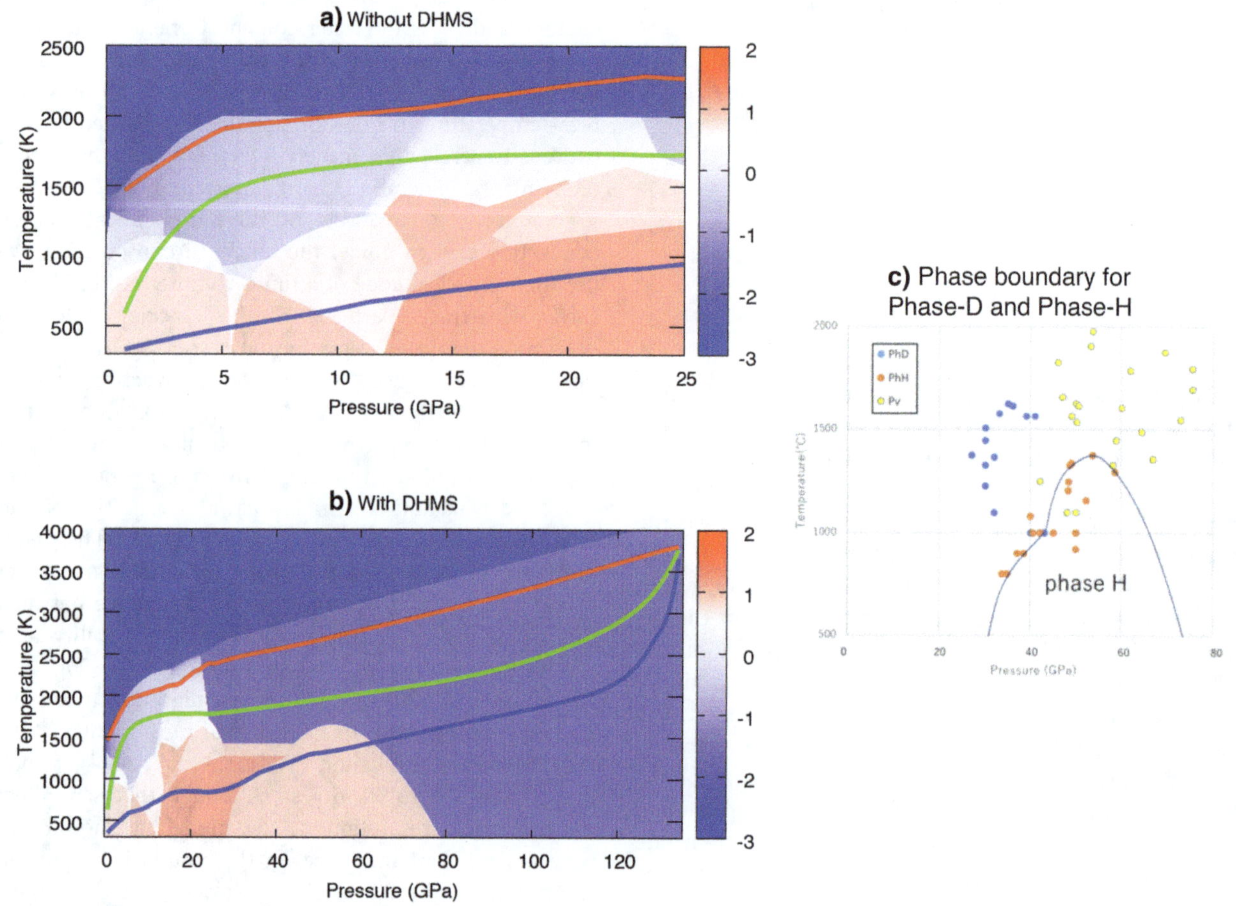

Fig. 1 **a** Water solubility maps of the mantle. Upper mantle only (Iwamori 2007). **b** The full range of mantle temperatures and pressures. Lines Indicate cold (blue), average (green), and hot (red) mantle geotherms for cases of 12 ocean masses of total water without (**a**) or with (**b**) DHMS effects. **c** Fitting for determining the water solubility of phase H included in **b** (Nishi et al. 2014; Ohtani et al. 2014; Ohira et al. 2014; Ohira et al. 2016)

mantle peridotite and oceanic crust as functions of temperature and pressure, respectively, and *C* is the basaltic fraction. This represents a major difference between this study and Nakagawa and Iwamori (2017). The difference between the two different melting approaches is discussed in Appendix.

The numerical setup used in this study is described as follows: 1024 (azimuthal) × 128 (radial) grid points with four million tracers are used to track the chemical composition, melt fraction, and mantle water content. The boundary conditions for temperature are fixed temperatures at the surface (300 K) and the core-mantle boundary (CMB; 4000 K). The initial conditions include an adiabatic temperature of 2000 K at the surface plus a thin thermal boundary layer (to explain the thermal boundary conditions at the surface and CMB), a basaltic composition of 20%, and a dry mantle (zero water content). The composition of the mantle is assumed to be uniform so that partial melting can create heterogeneous features in the mantle. The mantle can become hydrated up to the boundary conditions of the mantle water content (described in the "Computing the water ocean mass") via surface plate motion.

Water solubility maps

Figure 1 shows the maximum H_2O content of the mantle peridotite system based on the work of Iwamori (2004, 2007) and Nakagawa et al. (2015) (Fig. 1a) and includes the stability fields of hydrous phases that are stable at lower mantle conditions (Fig. 1b). These data are used to compute the excess water migration in the convecting mantle. At the lower mantle condition, without DHMS solubility, the water solubility of the lower mantle minerals is set as 100 wt. ppm. With DHMS solubility, in addition to the existing DHMS (phases A to D), because a new hydrous mineral phase has recently been discovered to exist at lower mantle pressures, i.e., "phase H" (Komabayashi and Omori 2006; Nishi et al. 2014; Ohira et al. 2014; Walter et al. 2015; Ohtani 2015), we have added the stability field of DHMS including phase H to the water solubility map of the mantle peridotite system (Fig. 1c). Compared to the water solubility map for pressures of less than 28 GPa (Iwamori 2004, 2007), few experimental results are available for pressures of greater than 28 GPa to constrain the exact phase boundaries and the maximum amount of H_2O in the peridotite system. The effect of aluminum on the stability of DHMS including phase H depends on the partitioning of Al among mantle minerals, including DHMS, phase H, and δ-AlOOH, which is currently poorly constrained. For instance, we assume that the stability of phase H in the natural peridotite system is similar to that of the pure $MgSiO_4H_2$ phase H (Ohira et al. 2016), which can be used to establish a minimum P-T stability range. Considering the bulk peridotite composition and the maximum modal amount of phase H in the peridotite system, we estimate that the maximum H_2O content in the phase H-bearing P-T range is 8 wt.%. At pressures and temperatures higher than the stability field of DHMS with phase H, we set a maximum H_2O content of 100 ppm below the solidus (e.g., Panero et al. 2015) and 0 ppm above the solidus, as in Nakagawa et al. (2015).

Computing the water ocean mass

In a previous study (Nakagawa and Iwamori 2017), we assumed a fixed boundary condition for mantle water migration in terms of the water ocean mass, such as a fixed value of 1.4 × 10^{21} kg (one ocean mass). This is not very realistic for understanding geologic records of the evolution of the water ocean associated with surface plate motion (e.g., Maruyama et al. 2013). To formulate a finite reservoir of surface seawater with a box model assumption, the mass of the water reservoir can be computed as:

$$X_{w,s} = X_{w,\text{total}} - X_{w,m}(F_R, F_H, F_G) \tag{8}$$

where $X_{w,\ s}$ is the mass of the water ocean, $X_{w,\ \text{total}}$ is the mass of the total amount of water in the system, and $X_{w,\ m}$ is the mass of mantle water as a function of regassing (F_R), dehydration (F_H), and degassing (F_G). The boundary condition of mantle water migration at the surface is described as follows:

$$C_w(\text{surface}) = \begin{cases} C_{w,\text{sol}}(300\ K, \text{surface})\ if\ X_{w,s} > 0 \\ 0\ if\ X_{w,s} = 0 \end{cases} \tag{9}$$

where $C_{w,\ \text{sol}}(300K, \text{surface})$ is the water solubility of mantle rocks at the surface.

Results

All physical parameters used in this study are listed in Table 1. This study examines a total of 20 cases that vary with the total amount of water in the planetary system (3 to 15 ocean masses) and the strength of the oceanic lithosphere. The parameters used in each case are listed in Table 2. First, we examine the effect of DHMS in conjunction with a large reservoir of surface seawater. Second, we examine the two hypotheses of the total amount of water in the planetary system. Third, the strength of the oceanic lithosphere is investigated by varying the friction coefficient (μ_0) from 0.1 to 0.6. This range represents the suggested range of lithospheric strength inferred from island loading (0.2–0.75; Zhong and Watts 2013). The minimum friction coefficient is based on the value associated with stable plate-like behavior in a dry mantle convection model (~ 0.13; Moresi and Solomatov 1998). All numerical simulations are performed for approximately 4.6 billion years.

Table 1 Physical parameters

Symbol	Meaning	Value
η_0	Reference viscosity	10^{21} (Pa s)
$\Delta\eta$	Viscosity jump	30
ρ_0	Reference density	3300 (kg m^{-3})
$\Delta\rho_w$	Density anomalies due to mantle water content	4.125 (kg m^{-3})
g_0	Gravity	9.8 (m s^{-2})
α_0	Reference thermal expansivity	5×10^{-5} (K^{-1})
κ_0	Reference thermal diffusivity	7×10^{-7} (m^2 s^{-1})
ΔT_{sa}	Temperature scale	2500 (K)
T_s	Surface temperature	300 (K)
T_{CMB}	Temperature at the CMB	4000 (K)
C_p	Heat capacity	1250 (J kg^{-1} K^{-1})
L_m	Latent heat	6.25×10^5 (J kg^{-1})
H	Internal heating rate	5.2×10^{-12} (W kg^{-1})
C_Y	Cohesion stress	10 (MPa)
E_d	Activation energy of dry mantle	290 (kJ mol^{-1})
V_d	Activation volume of dry mantle	2.4×10^{-6} (m^3 mol^{-1})
E_w	Activation energy of wet mantle	380 (kJ mol^{-1})
V_w	Activation volume of wet mantle	4×10^{-6} (m^3 mol^{-1})
R	Gas constant	8.134 (J K^{-1} mol^{-1})
D	Thickness of mantle	2890 (km)

The Rayleigh number is computed as $\rho_0 g_0 \alpha_0 \Delta T_{sa} D^3/\kappa_0 \eta_0 \sim 1.4 \times 10^8$ with values listed in Table 1. Activation energy and volume for dry mantle are taken from Yamazaki and Karato (2001) and for wet mantle are taken from Korenaga and Karato (2008). Results are not changed with rheological parameters taken from Karato and Wu (1993) for wet mantle. The latent heat shown here is used for temperature feedback caused by partial melting (Xie and Tackley 2004)

Chemical-hydrous structure with effects of DHMS

Figure 2 shows the chemical-hydrous structure of the mantle with and without the water solubility effects of DHMS at $t = 4.6$ billion years and for $\mu_0 = 0.2$, with an assumed total water volume of 12 ocean masses in the planetary system (exosphere plus interior). In both cases, extremely high-viscosity structures are found in the

Table 2 Run summary

Run #	Total water mass	μ_0	Phase H solubility
1	12	0.2	No
2	12	0.2	Yes
3	3	0.2	No
4	6	0.2	Yes
5	9	0.2	Yes
6	15	0.2	Yes
7	12	0.1	Yes
8	12	0.3	Yes
9	12	0.6	Yes

The total water mass is normalized by mass of surface sea water at the present-day Earth (1.4×10^{21}(kg); Genda (2016))

lower mantle due to the high efficiency of heat transfer of hydrous mantle convection (Nakagawa et al. 2015). Without DHMS, we assume that the water solubility of the lower mantle minerals is set as 100 wt. ppm. Without DHMS solubility, the dry mantle transition zone corresponds to upwelling plumes, which can have a temperature that is hotter than 2000 K. As shown in Fig. 6c of Nakagawa and Iwamori (2017), the high temperatures associated with mantle plumes correspond to water contents of zero. Incorporating the DHMS stability field into the water solubility maps (see Fig. 1a, b) reveals that the high-temperature mantle plumes may have a certain water content, unlike in the case without DHMS solubility limits. Another major difference between the cases with and without DHMS solubility is that the hydrated region at a depth of 660 km expands to the uppermost lower mantle in cold downwelling regions because of the high water content of DHMS (see arrows at the bottom of Fig. 2). Since we also assume that a density reduction occurs due to the hydration of the mantle minerals (Nakagawa et al. 2015), smaller-scale basaltic piles are found in the deep mantle compared to those generated by dry mantle convection (e.g., Nakagawa and Tackley 2011).

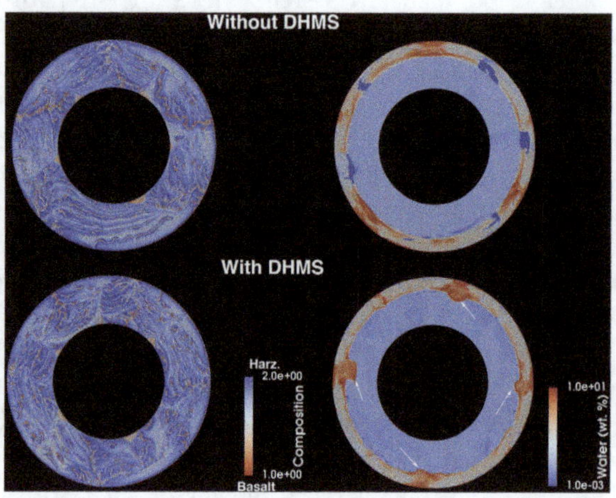

Fig. 2 Compositional (left) and hydrous (right) structures at $t = 4.6$ billion years. Top: without phase H (old look-up database); bottom: with DHMS solubility effects (new look-up database)

To confirm how DHMS affects the mantle water content, Fig. 3 shows the 1D horizontally averaged mantle water content corresponding to the hydrous structure in Fig. 2. Without the effects of DHMS, the mantle transition zone is the main water absorber; however, with the effects of DHMS, an additional region of high water content can be found at depths of up to ~ 1500 km. The water content in the mantle transition zone in these cases ranges from 0.1 to 1.0 wt.%. This range is similar to those inferred from the electrical conductivity structure of the mantle transition zone (e.g., Kelbert et al. 2009) and from seismic imaging (Houser 2016; Schmandt et al. 2014). Moreover, these results also seem to be consistent with the measurements of the water content of diamond inclusions in wadsleyite (Pearson et al. 2014). To confirm the cause of the water enhancement in DHMS found in Fig. 3, Fig. 1a, b shows the water solubility maps on which the 1D vertical structures of temperature have been plotted. Comparing a cold temperature profile with the water solubility map in Fig. 1b appears to indicate that phase H appears in the temperature profile at a depth of ~ 1500 km (corresponding to a pressure of approximately 60 GPa).

Water mass evolution and plate-like behavior: with and without DHMS

Figure 4 shows the temporal variations in the surface water mass and surface mobility that can be computed using the ratio of the surface velocity to the root mean square velocity in the entire mantle with and without DHMS, as shown in Fig. 2. Over 4.6 billion years, the surface seawater comprises 2.2 ocean masses without DHMS and 1.2 ocean masses with DHMS. Both cases are started with 12 ocean masses in total amount of

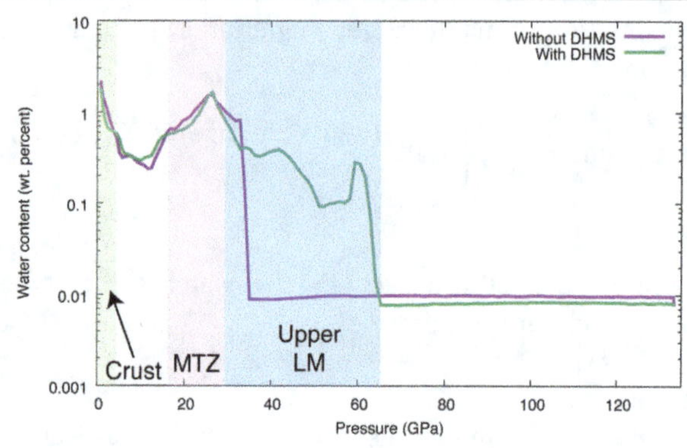

Fig. 3 1D horizontally averaged mantle water content with and without DHMS solubility as a function of pressure scaled as GPa units

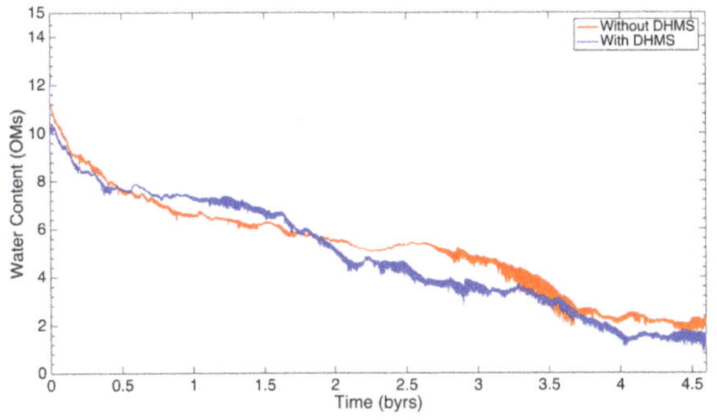

Fig. 4 Temporal variations in the mass of surface seawater with and without DHMS effects

water. This means that, for the mantle water mass, 9.8 ocean masses without DHMS and 10.8 ocean masses with DHMS are absorbed in the mantle. Therefore, when the DHMS effect is included, this water is more likely to be partitioned into the deep mantle to match the constraint on the present-day Earth's ocean mass. In Figs. 1b and 3, with DHMS, more water can be absorbed in the upper lower mantle compared to a case without DHMS, which is caused by phase D and/or phase H. However, as shown in Table 3 of Iwamori (2007), the maximum amount of water that can be stored is 9 to 11 ocean masses, assuming that the maximum H_2O content in the lower mantle is 100 ppm and assuming an average geotherm corresponding to that beneath 60 Ma oceanic lithosphere with a mantle potential temperature of 1300 °C (Iwamori 2007). This may suggest that the DHMS plays a minor role in the mantle water mass, which contributes at most to ~ 1 ocean mass of water mass in the whole mantle because the effects of DHMS only appeared in the cold subducting regions.

Figure 5 shows the mantle water fluxes (ingassing, dehydration, and degassing) as a function of time with and without DHMS. The procedures for computing these water fluxes can be found in Nakagawa and Iwamori (2017) and Nakagawa and Spiegelman (2017). They are given as follows:

$$F_R = \begin{cases} \int_S \rho_m u_z C_w dS \ (\text{if } u_z < 0) \\ 0 \ (\text{if } u_z \geq 0) \end{cases} \quad (10)$$

$$F_G = \dot{M}_{\text{erupt}} C_{w,\text{erupt}} \quad (11)$$

$$F_E = \int_S \rho_m C_{w,\text{ex}} u_f dS \quad (12)$$

where ρ_m is the mantle density; $\dot{M}_{\text{erupt}}$ is the total mass of erupted material; C_w is the water content; u_z is the radial velocity; $C_{w,\text{erupt}}$ is the water content of the erupted material, which is equivalent to the water content of the molten material; $C_{w,\text{ ex}}$ is the excess water content relative to the water solubility; and u_f is the conventional form of fluid migration velocity expressed numerically,

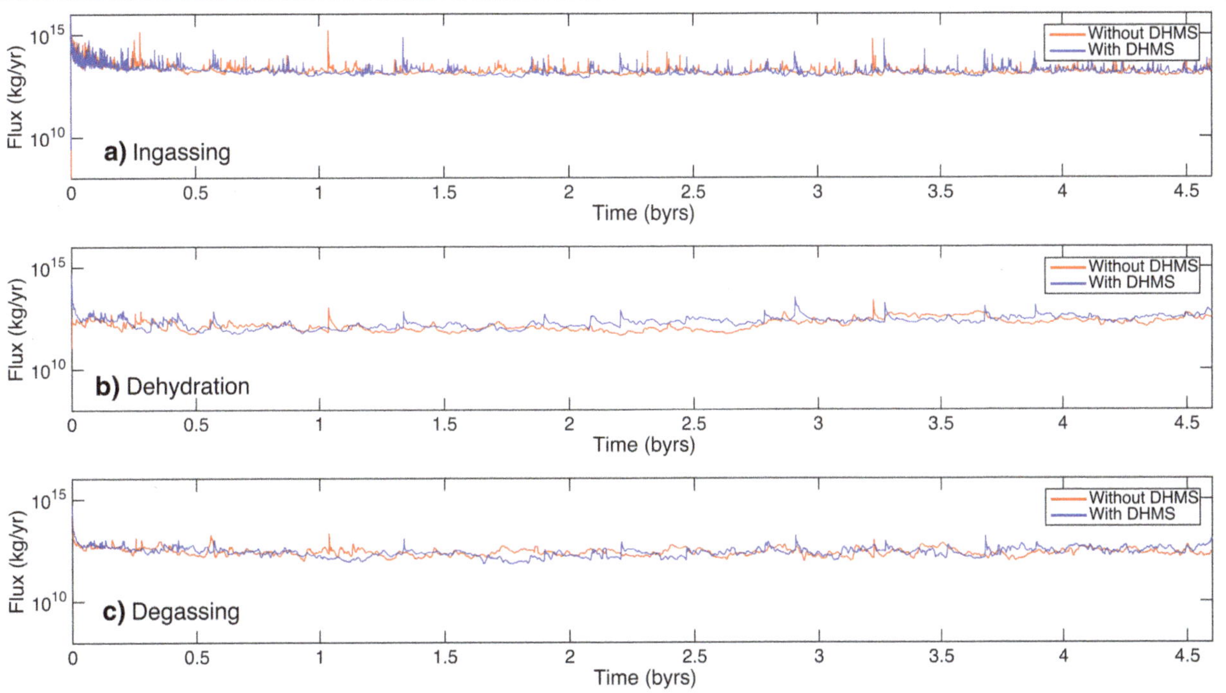

Fig. 5 Water flux diagnostics with and without DHMS solubility as function of time. **a** Ingassing. **b** Dehydration. **c** Degassing

which is computed as the ratio of the radial grid spacing to the time step size but is not very sensitive to the grid spacing shown in the Appendix of Nakagawa and Iwamori (2017), and the subscripts of those fluxes are R = ingassing, G = degassing, and E = dehydration. Basically, the uptake of water from the surface reservoir by plate subduction is the dominant flux process, with a magnitude on the order of 10^{13} kg/year, with the two other fluxes involving the release of water at the surface of approximately an order of magnitude lower than that involved in the uptake process. Thus, the surface seawater is gradually reduced by the ingassing process with surface plate motion. These profiles of water fluxes without DHMS are not very different from those in a case with the DHMS solubility field because the effective depth of the mantle water cycle associated with those fluxes is approximately 150 to 200 km (Nakagawa and Spiegelman 2017). This means that the DHMS solubility field is less sensitive to the physical processes of the mantle water cycle.

To assess the model sensitivities to the total amount of water in the planetary system, Fig. 6 shows the temporal variations in the surface seawater, mantle mass, and surface mobility with DHMS, which indicate that the surface seawater completely dries up before reaching the age of the Earth for a total amount of water of less than nine ocean masses. This implies that the total amount of water in the planetary system should be preferably in about 12 ocean masses, for the surface seawater to be consistent with the surface seawater on the present-day Earth. This preferred amount of water coincides with the storage capacity of the silicate mantle, as mentioned above (Iwamori 2007), although the mantle temperature reproduced in this study (e.g., Fig. 1) is lower than the 60 Ma geotherm assumed by Iwamori (2007). This indicates that (1) DHMS in the lower mantle, which was ignored in Iwamori (2007), plays a minor role, and (2) the present-day mantle could contain a maximum amount of water due to continuous hydration (Fig. 6). As discussed in the the "Introduction," the total amount of water in the Earth system has been estimated to be 1.2 to 3.3 ocean masses, which is based on the water contents of oceanic basalts (mid-ocean ridge basalts (MORB) and ocean island basalts (OIB)) and assumes that the entire mantle is sampled by MORB and OIB (Hirschmann 2006). While the upper mantle comprises the MORB source, OIB likely represents only a part of the crust-mantle cycling system (e.g., White and Hofmann 1982; Christensen and Hofmann 1994). The apparent difference in the estimated amount of water (i.e., 10–12 vs. 1.2–3.3 ocean masses) could be due to the regions that are not sampled by either MORB or OIB.

Regarding the surface mobility, which is one diagnostic for assessing the occurrence of plate-like behavior and is computed as the ratio of surface velocity to the root

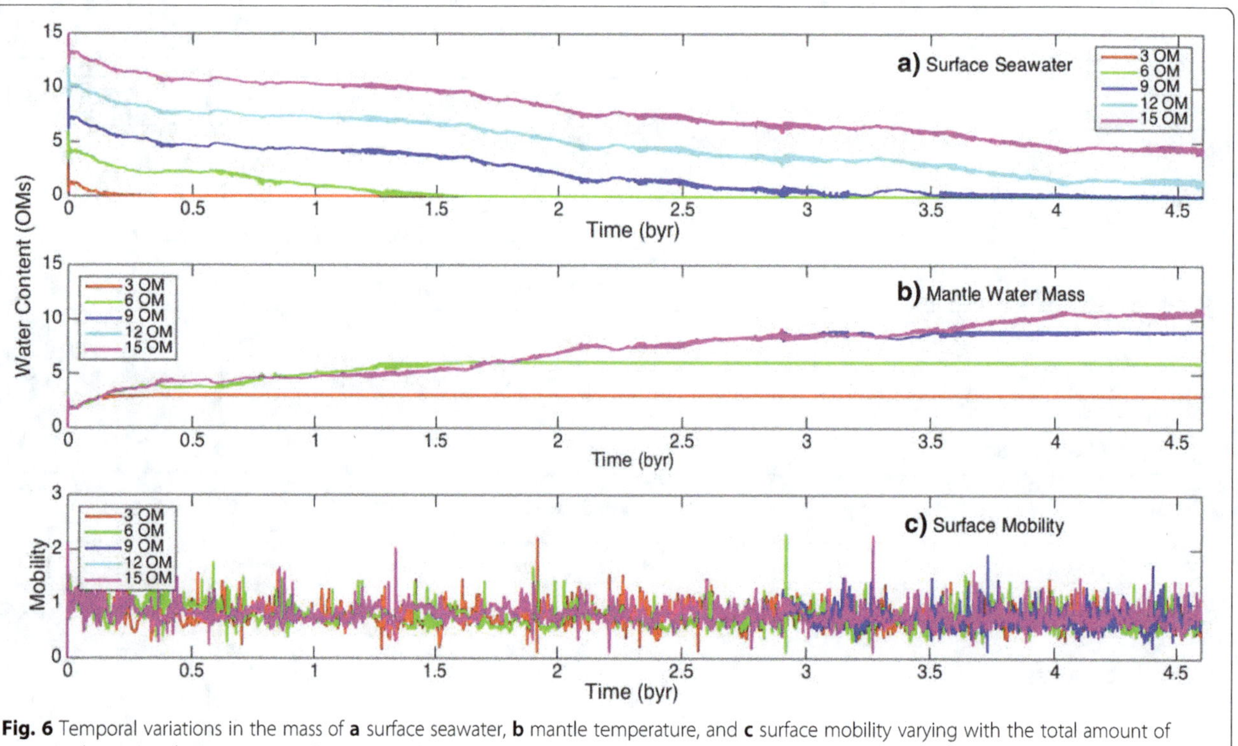

Fig. 6 Temporal variations in the mass of **a** surface seawater, **b** mantle temperature, and **c** surface mobility varying with the total amount of water in the entire planetary system. For cases with 12 and 15 ocean masses, the mantle behaviors (water content and surface mobility) are the same due to the remaining surface seawater

mean square of the convective velocity of the entire mantle (bottom of Fig. 6), it appears to not change very much after all the surface seawater is absorbed into the deep mantle. This suggests that the surface seawater found on the Earth's surface is not strongly correlated with the occurrence of surface plate motion. To confirm this hypothesis, Fig. 7 shows the sensitivity of the total amount of water to water fluxes, which indicates that the degassing flux is still active when the surface seawater is exhausted. This suggests that the surface seawater could be regenerated with mantle degassing, and the water-weakening effect is still valid but that it is immediately returned to the deep mantle via plate subduction. Therefore, surface plate motion will still be very active even if all surface seawater is exhausted.

Water mass evolution: model sensitivity to the strength of the oceanic lithosphere

Figure 8 shows the temporal variations in both the surface and mantle water masses over 4.6 billion years for different friction coefficients (0.1, 0.2, 0.3, and 0.6) corresponding to different strengths of oceanic lithosphere. All of the cases shown here include the solubility effects of DHMS. For weaker oceanic lithosphere ($\mu_0 = 0.1$ and 0.2), the surface seawater can be partitioned with a similar value as the present-day Earth's ocean mass over 4 billion years. On the other hand, for stronger oceanic lithosphere ($\mu_0 = 0.3$ and 0.6), the amount of surface seawater still remains at approximately five ocean masses. This mechanism can be explained by the mass-averaged temperature profile as a function of time as shown in Fig. 7b, c. In weaker oceanic lithosphere, heat transport is more efficient due to the vigorous surface plate motion and large number of plate boundaries (Nakagawa and Iwamori 2017). The mantle temperature is cold enough to pass into the water solubility field of DHMS with phase H. However, stronger oceanic lithosphere cases indicate higher mantle temperatures, so that the mantle temperature does not pass into the water solubility field of DHMS. To confirm this implication, Fig. 9 shows the 1D horizontally averaged mantle water content as a function of pressure. For weaker oceanic lithosphere, a high water content region can be found at both the mantle transition zone and upper lower mantle corresponding to DHMS solubility. In contrast, for stronger oceanic lithosphere, water enhancement is only found in the mantle transition zone. This is caused by the difference in mantle temperatures shown in Fig. 7b and the heat transfer efficiency of plate-mantle dynamics.

Discussion

Findings

The findings from this study are described as follows:

1. The presence of DHMS may have a limited impact

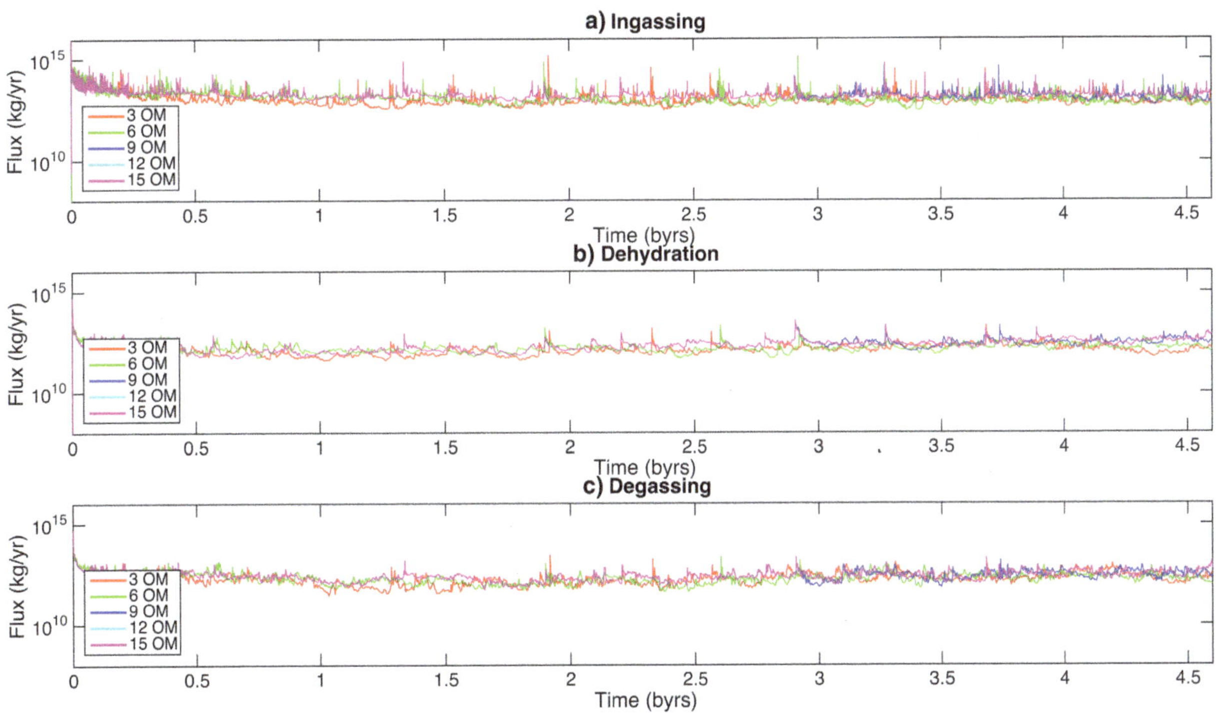

Fig. 7 Water flux diagnostics as a function of time varying with the total amount of water in the entire system as a function of time. **a** Ingassing. **b** Dehydration. **c** Degassing

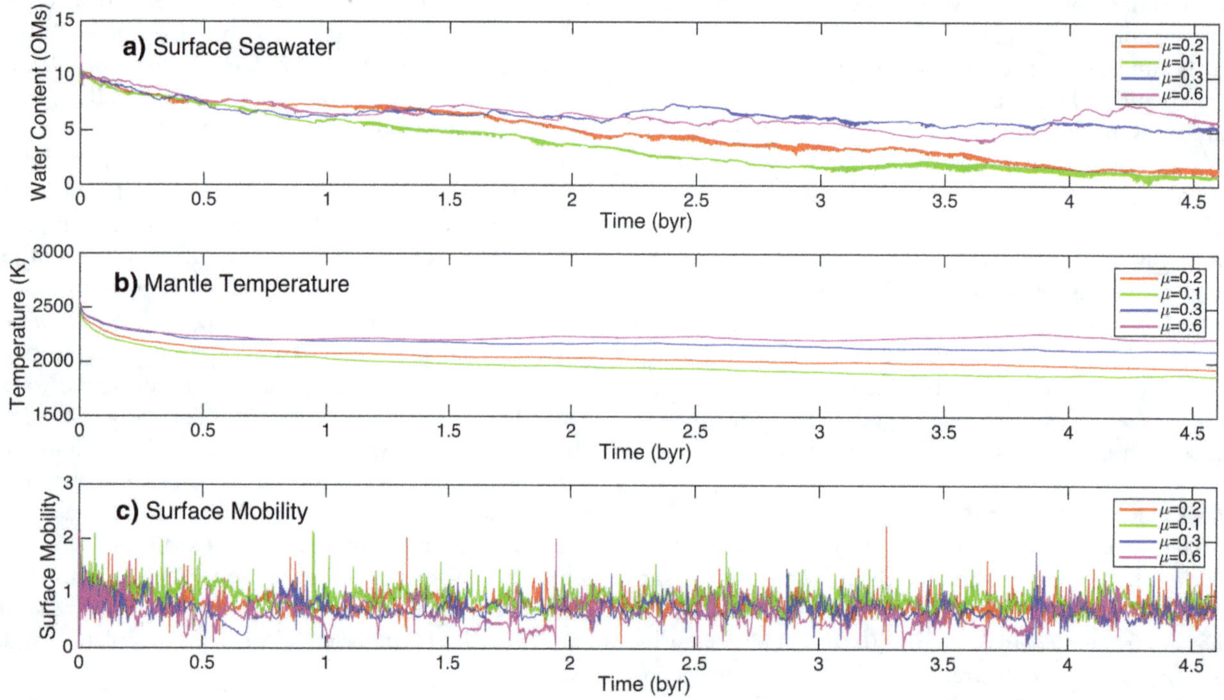

Fig. 8 Temporal variations in the mass of **a** surface seawater, **b** mantle temperature, and **c** surface mobility varying with the friction coefficient of the yield strength of the oceanic lithosphere. The total amount of water is fixed as 12 ocean masses

on the evolution of surface seawater but is not very different from that of the physical processes of the mantle water cycle. The physical mechanism of the mantle water cycle is strongly regulated by the choke point of the water solubility of the mantle minerals (Nakagawa and Spiegelman 2017). However, the DHMS may play some role in expanding the storage region of water transported by plate subduction. Quantitatively, a water reservoir with a water content of 0.1 wt.% is located in the upper lower mantle due to DHMS solubility. To discuss the consistency between this amount and the realistic hydrous

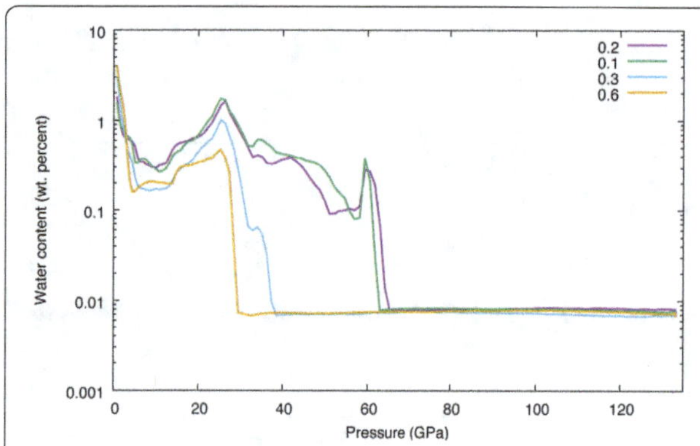

Fig. 9 1D horizontally averaged mantle water content varying with the friction coefficient of the yield strength of the oceanic lithosphere as a function of pressure scaled as GPa units

conditions of the Earth, more observational and experimental measurements are required.

2. The survival time of surface seawater is dependent on the total amount of water in the entire planetary system and the strength of the oceanic lithosphere. To partition water into the surface reservoir with the present-day mass of the surface seawater on the Earth's surface, the entire planetary system should contain ten ocean masses or more, although more accurate knowledge about mantle hydrous phases is required. However, the total amount of water in the planetary system would be reduced if the strength of the oceanic lithosphere was slightly stronger but a large amount of water (~7 ocean masses) was absorbed in the deep mantle. This finding, i.e., that a large amount of water is required for the early Earth, is consistent with an estimate derived from the experimental measurements of chondritic material and inferences based on the solidification of the surface magma ocean during early planetary formation (Marty 2012; Hamano et al. 2013). The total absorbed amount of water in the mantle is also consistent with the estimate obtained based on the material properties of mantle rocks along the realistic geotherm (Iwamori 2007). These results are consistent because the water budget in the mantle is primarily controlled by the water released by dehydration reactions rather than that released by

degassing effects, which has been the main focus of simplified models of hydrous mantle evolution (Franck and Bounama 2001; Rüpke et al. 2004; Crowley et al. 2011; Sandu et al. 2011; Korenaga 2011)

DHMS and other potential mechanisms of water transport into the deep mantle

First, the stability fields of DHMS including phase H that are stable under lower mantle pressure conditions are poorly constrained, as was described earlier in the "Water solubility map" section. In this study, our stability field of phase H corresponds to that of Ohira et al. (2016) and features a minimum stability field of 70 to 80 GPa. If the stability field is raised to more than 100 GPa and higher temperatures, more water can be transported via plate subduction, which may shorten the lifetime of surface seawater; to achieve an Earth-like planet, more water is required in the system.

Second, other potential host minerals of water exist in the deep mantle, such as δ-AlOOH (Ohira et al. 2014) or pyrite-type FeOOH (Nishi et al. 2017). These minerals appear to be stable at typical lower mantle temperatures and pressures. However, the amount of water that these minerals can retain in the deep mantle remains unclear. The possible water solubilities of the lower mantle minerals, including phase H, δ-AlOOH, and pyrite-type FeOOH, which could potentially be accurately determined via experiments, could greatly influence the survival time of surface seawater. The inclusion of additional lower mantle minerals could result in a much shorter seawater survival time than that observed in the cases examined in this study. Note that the volume-averaged water content of the lower mantle is be expected to be ~ 100 ppm or less (Karato 2011; Panero et al. 2015), but these estimates remain highly controversial and are dependent on identifying the possible host minerals of water in the deep lower mantle. Therefore, these estimates should be determined more quantitatively to resolve this issue.

Third, a hydrogen diffusion mechanism is also a significant process in the mantle water cycle (Richard et al. 2002, 2006) and is still unresolved in this study. Numerical modeling results suggest that hydrogen diffusion seems to saturate the mantle with water on a much shorter timescale than that without incorporating hydrogen diffusion (Nakagawa 2017). The incorporation of a water-saturated mantle leads to a steady state in the evolution of surface seawater induced by the mantle water cycle. As a result, the lifetime of surface seawater can be expected to be longer in this study.

Moreover, the rheological properties of the hydrous lower mantle are highly uncertain because of the difficulty of experimental determination, even for dry mantle conditions (e.g., Girard et al. 2016). Under dry mantle conditions, the lower mantle minerals are expected to exhibit the shear localization mechanism with diffusion creep deformation (Girard et al. 2016). However, it is unclear whether this type of deformation mechanism occurs under hydrous lower mantle conditions. If the rheological properties of the lower mantle are similar to those of the upper mantle, more water could be transported into the deep mantle, and the lifetime of surface seawater may be shorter due to the more vigorous convective dynamics in the lower mantle.

Strength of the oceanic lithosphere

As indicated in Fig. 7, the friction coefficient has a great influence on the partitioning of water between the surface and deep mantle reservoir; more surface seawater is absorbed in conjunction with smaller friction coefficients (weaker oceanic lithosphere) because the mantle temperature is sufficiently cold so that both the mantle transition zone and upper lower mantle may be worked as large water reservoirs (see Fig. 8). Hence, a large amount of water should be required when the strength of oceanic lithosphere is very weak; however, using the range of friction coefficients suggested by observational data analysis (up to 0.7; Zhong and Watts 2013), the total amount of water in the entire system could be reduced but still be larger than those of the suggested by simple parameterized convection models (e.g., Sandu et al. 2011). However, it should be noted that a weaker oceanic lithosphere is preferable for understanding the plate-like behavior in a dry mantle convection system ($\mu_0 < 0.1$; Moresi and Solomatov 1998; Crameri and Tackley 2015). Moreover, in this study, we incorporate the effect of "water weakening" to reduce the yield strength of the oceanic lithosphere under a water-saturated situation.

Total amount of water in the planetary system

In most geodynamic models with water circulation, the partitioning of water between surface seawater and the deep mantle at the present is assumed to range from 1:1 to 1:2 (Franck and Bounama 2001; Rüpke et al. 2004; Sandu et al. 2011; Korenaga 2011), which is consistent with the results of the mineral physics experiments (Hirschmann 2006). In particular, Franck and Bounama (2001) also suggested that the lifetime of surface seawater may be dependent on the efficiency of regassing caused by plate subduction. However, in this study, the regassing flux is automatically regulated by the water solubility map, including the "choke point" observed at a depth of 150 to 200 km, which is reduced by up to a few orders of magnitude for the regassing flux at this depth. The most important issue in those geodynamic models is that it is difficult to account for the water solubility of mantle minerals, and the water cycle is assumed to involve only degassing for the water release from the deep interior to the exosphere and water uptake by the regassing process from the surface to the deep mantle. In addition, degassing is assumed to occur only along mid-ocean ridges but should also occur along island arcs. Furthermore, the scaling relationship is based only on heat transfer in steady-state mantle convection, which is only applicable for cases with less vigorous mantle convection. Therefore, simplified mantle dynamics models with water circulation underestimate both the regassing flux and the degassing flux and therefore require a smaller amount of total water in the entire planetary system (~ 3 ocean masses) than the amount inferred from early planetary formation estimates (5 to 15 ocean masses). To avoid these underestimates caused by the assumption of scaling relationships in plate-mantle dynamics, we conducted a series of full mantle convection simulations with water migration, including actual water solubility maps for deep mantle minerals. The results indicate that a large amount of water is needed in the planetary system to achieve a consistent lifetime of surface seawater in the plate-mantle system. This finding is consistent with the estimates of the total amount of water in the entire planetary system inferred from early planetary formation processes.

However, this argument leads to issues related to the sources of the volatile components in the early Earth and their initial amounts (e.g., Albarede 2009). In Nakagawa and Spiegelman (2017), these issues did not affect the initial amount of water in the deep mantle, but the deep mantle should contain a certain amount of water, as illustrated in the water solubility maps of the mantle minerals. These issues remain in models of the early thermal and chemical state of the planetary mantle that do not start from a full magma ocean condition (a fully molten mantle is expected to have been present in the early Earth). To further resolve these issues, the initial state of mantle convection should be seriously examined in a future study that assumes initially fully molten mantle conditions (Lourenço et al. 2016) and then checks the consistency of these conditions with the theoretical estimates of the size of the water reservoir at the surface (Hamano et al. 2013).

Finally, the following evolution of surface seawater associated with a plate-mantle system is proposed; a certain amount of water (volatiles) is delivered to the planet before or after the magma ocean forms (potentially by a giant impact or late veneer accretion, see review by Genda (2016)). After the magma ocean solidifies, surface seawater forms. The partition ratio of water between the surface and the deep mantle is dependent on the water solubility of the deep mantle, which is in turn dependent on the temperature and composition of the early Earth's mantle. Vigorous surface plate motion can transport surface water into the deep mantle; however, because of the high temperatures in the deep mantle, relatively little water stays in the deep mantle. When the mantle is sufficiently cooled by mantle convection, the water transported via plate subduction can be stored in the mantle transition zone and the uppermost lower mantle, thereby gradually reducing the volume of surface water to the present-day amount. Although this conceptual model would be slightly modified by a better understanding of the global-scale mantle water circulation and more accurate water solubility limits of lower mantle mineral assemblage, it is overall a robust model.

Conclusions

In this study, the evolution of surface seawater in the plate-mantle system with the effects of deep mantle water solubility (e.g., DHMS) is investigated to resolve the controversial issue of the total amount of water in the entire planetary system. The conclusions are as follows:

1. The DHMS solubility field may have a small impact on the evolution of surface seawater but is not very sensitive to the physical mechanism of the mantle water cycle because the mantle water cycle is effective up to a depth of 150~200 km, which is much shallower than the DHMS solubility field.
2. In numerical simulations, the total amount of water in the entire planetary system should be greater than at least 7 to 12 ocean masses, which is consistent with the water mass estimate based on early planetary formation (Marty 2012; Hamano et al. 2013) and a petrological estimate that includes a realistic maximum H_2O solubility of mantle material along the slab geotherm (after Iwamori 2007). The main mechanism of the large amount of water required in the entire system is the incorporation of water solubility maps so that the dehydration process can be addressed in numerical mantle convection simulations.
3. DHMS including phase H may represent an additional water reservoir in the deep mantle and may affect the evolution of surface seawater on the present-day Earth. To better estimate the seawater evolution, more accurate constraints on the stability and amount of water stored in DHMS, phase H, and other hydrous minerals (e.g., δ-AlOOH and pyrite-type FeOOH) in the actual mantle are required.

In this study, we successfully computed the evolution of surface seawater in the plate-mantle system caused by hydrous mantle convection and incorporate the solubility fields of hydrous mineral phases in the deep mantle, such as the DHMS. However, it must be noted that due to the global scale of the model, it is very difficult to

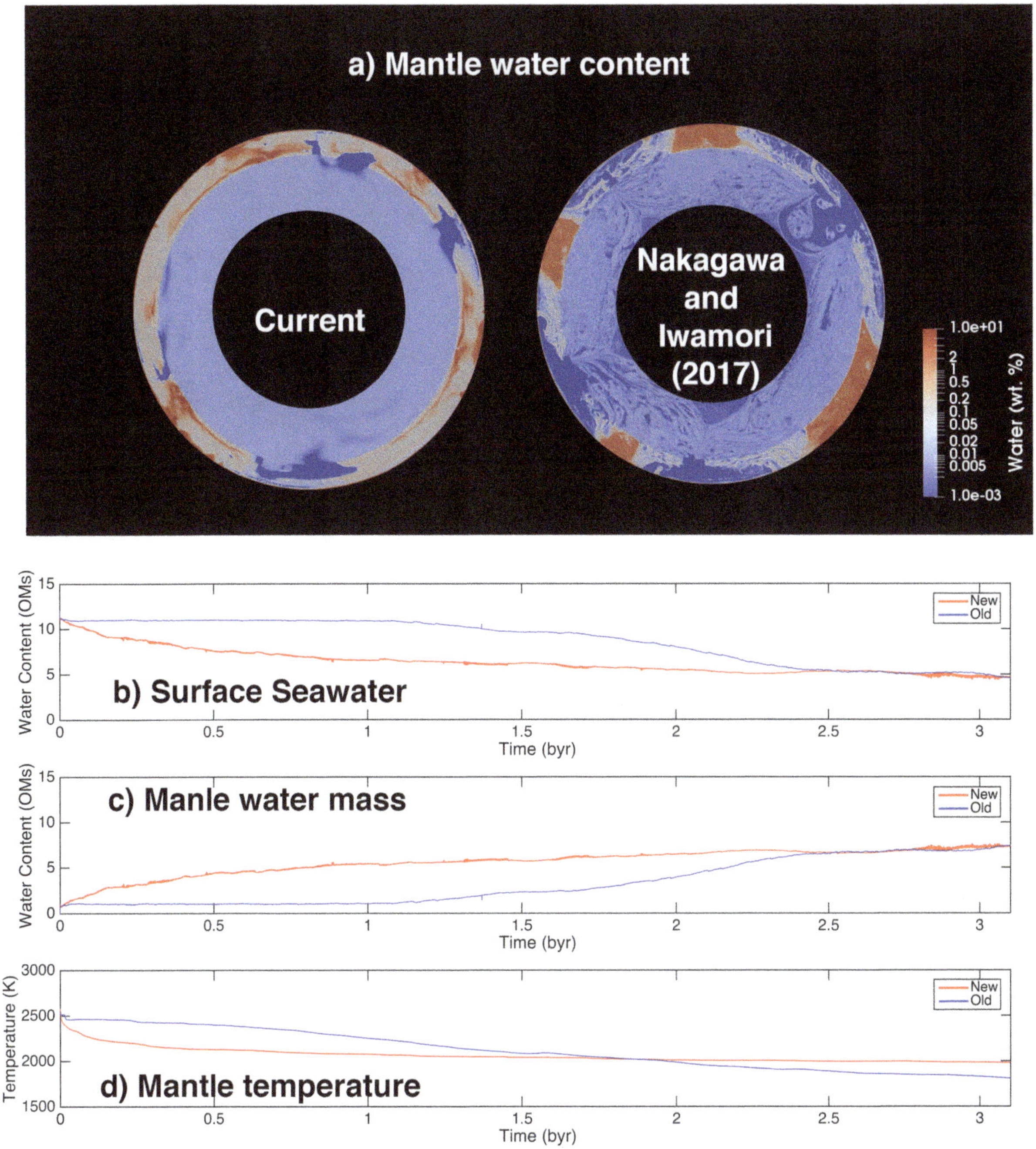

Fig. 10 a Mantle water content taken at 3.1 billion years computed from two different numerical codes without assuming DHMS solubility (left: current version of numerical code; right: the version used in Nakagawa and Iwamori (2017) but corrected for significant issues with the dehydration reaction) and their diagnostics (**b** evolution of surface seawater, **c** mantle water mass, and **d** mantle temperature) as a function of time. On the legend of diagnostics plots, new represents a result from the numerical code used in this study and old represents a result from the numerical code used in Nakagawa and Iwamori (2017)

resolve the detailed physical and chemical processes occurring in the mantle wedge as in Nakao et al. (2016, 2018) and van Keken et al. (2011); therefore, some significant improvements are required to reveal the geological and petrological constraints on mantle water evolution, which requires future investigation.

Appendix A

Comparison between Nakagawa and Iwamori (2017) and the current study

All of the results in this study are very different from those presented in Nakagawa and Iwamori (2017). In particular, as shown in Fig. 6, the sudden change in the mantle water content appears to occur on a shorter time scale than that in Nakagawa and Iwamori (2017). As mentioned in the "Methods/Experimental" section, the major difference in the numerical code between these two studies is the use of arbitrary continuous melting (this study) or discretized melting (previous study). Figure 10 shows a comparison of the mantle water content field and diagnostics (surface seawater, mantle water mass, and mantle temperature) between the numerical code used in this study and that used in Nakagawa and Iwamori (2017), which does not assume DHMS solubility. Unfortunately, the original version of the numerical code in Nakagawa and Iwamori (2017) also includes some incorrect treatments of the numerical procedure of the dehydration reaction, which is related to the data transfer in each numerical domain associated with MPI implementations and resetting an array of excess water due to dehydration reaction. These incorrect stuffs are corrected in this study. Following these corrections, both cases yield similar water evolution results. Using the discretized melting approach, the mantle temperature is expected to be much hotter than that with the arbitrary melting approach, which can allow for continuous compositional variations in each tracer because the discretized melting approach is strongly dependent on the heterogeneous distributions of tracer particles as a result of the migrations caused by numerical simulations, which is determined with a probabilistic approach for the degree of melting and is more likely to indicate a smaller degree of melting than the arbitrary melting approach. Due to this difference, the heat transport caused by melt migration is more efficient in the arbitrary melting approach; therefore, different mantle temperature profiles can be found for each case. The mantle water content profiles that correspond to such a difference between these cases are somewhat different for up to a few billion years, but the final results obtained at 4.6 billion years are not very different.

Abbreviations
CMB: Core-mantle boundary; DHMS: Dense hydrous magnesium silicate

Acknowledgements
The authors thank Guillaume Richard and an anonymous reviewer for significantly improving the original manuscript. We also thank Tomoeki Nakakuki for providing a numerical module of water migration for mantle convection simulations; Masayuki Nishi for providing information about the water contents of deep mantle hydrous phases; Marc Spiegelman for the fruitful discussion of the evolution of finite volumes of water ocean during TN's sabbatical leave to Lamont-Doherty Earth Observatory, Columbia University, New York City; Paul Tackley for providing his numerical code for mantle convection; and Bjorn Mysen for an invitation to write this manuscript based on our presentation at the 2017 JpGU-AGU joint meeting. All numerical computations were performed with the Data Analyzer (DA) system in the JAMSTEC.

Funding
This work was supported by the JSPS KAKENHI Grant Numbers 16K05547 and 18H04467 and by the FLAGSHIP2020 MEXT within the CBSM2 Project "Structure and Properties of Materials in Deep Earth and Planets".

Authors' contributions
TN designed the entire study. HI, RY, and AN provided the parameterized water solubility of hydrated mantle rocks. TN developed a numerical model and carried out all the numerical simulations. TN and HI wrote the manuscript. All authors interpreted the results of the numerical simulations and read and approved the final manuscript.

Competing interests
The authors declare that they have no competing interests.

Author details
[1]Department of Mathematical Science and Advanced Technology, Japan Agency for Marine-Earth Science and Technology, 3173-25, Showa-machi, Yokohama 236-0001, Japan. [2]Department of Earth Sciences, University of Hong Kong, Pokfulam Road, Hong Kong, Hong Kong. [3]Department of Solid Earth Geochemistry, Japan Agency for Marine-Earth Science and Technology, 2-15, Natsushima-cho, Yokosuka 237-0061, Japan. [4]Department of Earth and Planetary Sciences, Tokyo Institute of Technology, 2-12-1, Ookayama, Meguro, Tokyo 152-8551, Japan. [5]Earthquake Research Institute, The University of Tokyo, 1-1-1, Yayoi, Bunkyo, Tokyo 113-0032, Japan.

References
Albarede F (2009) Volatile accretion history of the terrestrial planets and dynamic implications. Nature 461:1227–1233

Appel PWU, Fedo CM, Moorbath S, Myers JS (1998) Recognizable primary volcanic and sedimentary features in a low-strain domain of the highly deformed, oldest known (~3.7–3.8 Gyr) Greenstone Belt, Isua, West Greenland. Terra Nova 10:57–62

Arcey D, Tric E, Doin MP (2005) Numerical simulations of subduction zones: effect of slab dehydration on the mantle wedge dynamics. Phys Earth Planet Int 149:133–153

Aubaud C, Hirschmann MH, Withers AC, Hervig RL (2008) Hydrogen partitioning between melt, clinopyroxene, and garnet at 3 GPa in a hydrous MORB with 6 wt. % H_2O. Contrib Mineral Petrol 156:607–625

Christensen UR, Hofmann AW (1994) Segregation of subducted oceanic crust in the convecting mantle. J Geophys Res 99:19867–19884

Condie KC (2016) A planet in transition: the onset of plate tectonics on Earth between 3 and 2 Ga? Geosci Front. https://doi.org/10.1016/j.gsf.2016.09.001

Crameri F, Tackley PJ (2015) Parameters controlling dynamically self-consistent plate tectonics and single-sided subduction in global models of mantle convection. J Geophys Res Solid Earth 120:3680–3706. https://doi.org/10.1002/2014JB011664

Crowley J, Gérault M, O'Connell RJ (2011) On the relative influence of heat and water transport on planetary dynamics. Earth Planet Sci Lett 310:380–388. https://doi.org/10.1016/j.epsl.2011.08.035

Franck S, Bounama C (2001) Global water cycle and Earth's thermal evolution. J Geodyn 32:231–246

Genda H (2016) Origin of Earth's oceans: an assessment of the total amount history and supply of water. Geochem J 50:27–42

Gerya T, Connolly JAD, Yuen DA (2008) Why terrestrial subduction one-sided? Geology 36:43–46. https://doi.org/10.1130/G24060A.1

Girard J, Amulele G, Farta R, Mohiuddin A, Karato S-I (2016) Shear deformation of bridgmanite and magnesiowüstite aggregates at lower mantle conditions. Science 351:144–147. https://doi.org/10.1126/science.aad3113

Hamano K, Abe Y, Genda H (2013) Emergence of two types of terrestrial planet on solidification of magma ocean. Nature 497:607–610. https://doi.org/10.1038/nature12163

Hernlund JW, Tackley PJ (2008) Modeling mantle convection in the spherical annulus. Phys Earth Planet Int 171:48–54

Hirschmann MH (2006) Water, melting, and the deep Earth H_2O cycle. Annu Rev Earth Planet Sci 34:629–653. https://doi.org/10.1146/annurev.earth.34.031405.125211

Hopkins M, Harrison TM, Manning CE (2008) Low heat flow inferred from >4Gyr zircon suggests Hadean plate boundary interaction. Nature 456:493–496. https://doi.org/10.1038/nature07465

Houser C (2016) Global seismic data reveal little water in the mantle transition zone. Earth Planet Sci Lett 448:94–101. https://doi.org/10.1016/j.epsl.2016.04.018

Iwamori H (2004) Phase relations of peridotites under H_2O saturated conditions and ability of subducting plates for transportation of H_2O. Earth Planet Sci Lett 227:57–71. https://doi.org/10.1016/j.epsl.08.013

Iwamori H (2007) Transportation of H_2O beneath the Japan arcs and its implications for global water circulation. Chem Geol 239:182–198. https://doi.org/10.1016/j.chemgeo.2006.08.011

Iwamori H, Nakakuki T (2013) Fluid processes in subduction zones and water transport to the deep mantle. In: Karato S-i (ed) Physics and chemistry of the deep mantle. John Wiley & Sons, Ltd., Oxford, pp 372–391

Karato S, Wu P (1993) Rheology of the upper mantle: a synthesis. Science 260: 771–778

Karato S-i (2011) Water distribution across the mantle transition zone and its implications for global material circulation. Earth Planet Sci Lett 301:413–423. https://doi.org/10.1016/j.epsl.2010.11.038

Kelbert A, Schultz A, Egbert G (2009) Global electromagnetic induction constraints on transition-zone water content variations. Nature 460:1003–1006. https://doi.org/10.1038/nature08257

Komabayashi T, Omori S (2006) Internally consistent thermodynamics data set for dense hydrous magnesium silicates up to 35 GPa, 1600C: implications for water circulation in the Earth's deep mantle. Phys Earth Planet Int 156:89–107. https://doi.org/10.1016/j.pepi.2006.02.002

Korenaga J (2011) Thermal evolution with a hydrating mantle and the initiation of plate tectonics in the early Earth. J Geophys Res 116:B12403. https://doi.org/10.1029/2011JB008410

Korenaga J, Karato S-I (2008) A new analysis of experimental data on olivine rheology. J Geophys Res 113:B02403. https://doi.org/10.1029/2007JB005100

Lourenço D, Rozel A, Tackley PJ (2016) Melting and crustal production helps plate tectonics on Earth-like planets. Earth Planet Sci Lett 439:18–28. https://doi.org/10.1016/j.epsl.2016.01.024

Lourenço DL, Rozel A, Gerya TV, Tackley PJ (2018) Efficient cooling of rocky planets by intrusive magmatism. Nat Geosci 11:322–327. https://doi.org/10.1038/s41561-018-0094-8

Marty B (2012) The origins and concentrations of water, carbon, nitrogen and noble gases on Earth. Earth Planet Sci Lett 313-314:56–66. https://doi.org/10.1016/j.epsl.2011.10.040

Maruyama S, Ikoma M, Genda H, Hirose K, Yokoyama T, Santosh M (2013) The naked planet Earth: most essential pre-requisite for the origin and evolution of life. Geosci Front 4:141–165

Maruyama S, Komiya T (2011) The oldest pillow lavas, 3.8–3.7 Ga from the Isua supracrustal belt, SW Greenland: plate tectonics already begun by 3.8 Ga. J Geogr 120:869–876

Maruyama S, Okamoto K (2007) Water transportation from the subducting slab into the mantle transition zone. Gonwana Res 11:148–165. https://doi.org/10.1016/j.gr.2006.06.001

Mei S, Kohlstedt DL (2000) Influence of water on plastic deformation of olivine aggregates 1. Diffusion creep regime. J Geophys Res 105:21457–31469

Mojzsis SJ, Harrison TM, Pidgeon RT (2001) Oxygen-isotope evidence from ancient zircons from liquid qater at the Earth's surface 4,300 Myr ago. Nature 409:178–81

Moresi L, Solomatov V (1998) Mantle convection with a brittle lithosphere: thoughts on the global tectonic styles of the Earth and Venus. Geophys J Int 133:669–682

Nakagawa T (2017) On the numerical modeling of the deep mantle water cycle in global-scale mantle dynamics: the effects of the water solubility limit of lower mantle minerals. J Earth Sci 28:563–577. https://doi.org/10.1007/s12583-017-0755-3

Nakagawa T, Iwamori H (2017) Long-term stability of plate-like behavior caused by hydrous mantle convection and water absorption in the deep mantle. J Geophys Res Solid Earth 122. https://doi.org/10.1002/2017JB014052

Nakagawa T, Nakakuki T, Iwamori H (2015) Water circulation and global mantle dynamics: insight from numerical modeling. Geochem Geophys Geosyst 16: 1449–1464. https://doi.org/10.1002/GC005071

Nakagawa T, Spiegelman MW (2017) Global-scale water circulation in the Earth's mantle: implications for the mantle water budget in the early earth. Earth Planet Sci Lett 464:189–199. https://doi.org/10.1016/j.epsl.2017.02.010

Nakagawa T, Tackley PJ (2011) Effects of low-viscosity post-perovskite on thermos-chemical mantle convection in a 3-D spherical shell. Geophys Res Lett 38:L04309. https://doi.org/10.1029/2010GL046494

Nakao A, Iwamori H, Nakakuki T (2016) Effects of water transportation on subduction dynamics: roles of viscosity and density reduction. Earth Planet Sci Lett 454:178–191. https://doi.org/10.1016/j.epsl.2016.08.016

Nakao A, Iwamori H, Nakakuki T, Suzuki YJ, Nakamura H (2018) Role of hydrous lithospheric mantle in deep water transportation and subduction dynamics. Geophys Res Lett. https://doi.org/10.1029/2017GL076953

Nishi M, Irifune T, Tsuchiya J, Tange Y, Nishihara Y, Fujino K, Higo Y (2014) Stability of hydrous silicate at high pressures and water transport to the deep lower mantle. Nat Geosci:224–227. https://doi.org/10.1038/NGEO2074

Nishi M, Kuwayama Y, Tsuchiya J, Tsuchiya T (2017) The pyrite-type high-pressure form of FeOOH. Nature 547:205–208. https://doi.org/10.1038/nature22823

Ohira I, Ohtani E., Kamada S., Hirao N (2016) Formation of phase H-δ-AlOOH solid solution in the lower mantle, Goldschmidt Conference Abstract, 2346

Ohira I, Ohtani E, Sakai T, Miyahara M, Hirao N, Ohishi Y, Nishijima M (2014) Stability of a hydrous δ-phase, AlOOH-$MgSiO_2(OH)_2$, and a mechanism for water transport into the base of lower mantle. Earth Planet Sci Lett 401:12–17. https://doi.org/10.1016/j.epsl.05.059

Ohtani E (2015) Hydrous minerals and the storage of water in the deep mantle. Chem Geol 418:6–15. https://doi.org/10.1016/j.chemgeo.2015.05.005

Ohtani E, Amaike Y, Kamada S, Sakamaki T, Hirao N (2014) Stability of hydrous phase H $MgSiO_4$ under lower mantle conditions. Geophys Res Lett. https://doi.org/10.1002/2014GL061690

Ohtani E, Toma M, Litasov K, Kubo T, Suzuki A (2001) Stability of dense hydrous magnesium silicate phases and water storage capacity in the transition zone and lower mantle. Phys Earth Planet Int 124:105–117. https://doi.org/10.1016/S0031-9201(01)00192-03

Panero WR, Pigott JS, Reaman DM, Kabbes JE, Liu Z (2015) Dry $(Mg,Fe)SiO_3$ perovskite in the Earth's lower mantle. J Geophys Res Solid Earth 120. https://doi.org/10.1002/2014JB011397

Pearson DG, Brenker FE, Nestola F, McNeill J, Nasdala L, Hutchison MT, Mateev S, Mather K, Silversmit G, Schmitz S, Vekemans B, Vincze L (2014) Hydrous mantle transition zone indicated by ringwoodite included within diamond. Nature 507:221–224. https://doi.org/10.1038/nature13080

Richard G, Bercovici D, Karato S-I (2006) Slab dehydration in the Earth's mantle transition zone. Earth Planet Sci Lett 251:156–167. https://doi.org/10.1016/j.epsl.2006.09.006

Richard G, Monneraeu M, Ingrin J (2002) Is the transition zone an empty water reservoir? Influence from numerical model of mantle dynamics. Earth Planet Sci Lett 205:37–51

Richard GC, Iwamori H (2010) Stagnant slab, wet plumes and Cenozoic volcanism in East Asia. Phys Earth Planet Int 183:280–287. https://doi.org/10.1016/j.pepi.2010.02.009

Rozel A B, Golabek G J, Jain C, Tackley P J, Gerya T V (2017) Continental crust formation on early Earth controlled by intrusive magmatism. Nature 545: 332–335. doi: https://doi.org/10.1038/nature22042

Rüpke LH, Morgan JP, Hort M, Connolly JAD (2004) Serpentine and the subduction zone water cycle. Earth Planet Sci Lett 223:17–34

Sandu C, Lenardic A, McGovern P (2011) The effects of deep water cycling on planetary thermal evolution. J Geophys Res 116:B12404. https://doi.org/10.1029/2011JB008405

Schmandt B, Jacobsen SD, Becker TW, Liu Z, Dueker KG (2014) Dehydration melting at the top of the lower mantle. Science 344:1265–1268. https://doi.org/10.1126/sciecnce.1253358

Tackley PJ (1996) Effects of strongly variable viscosity on three-dimensional compressible convection in planetary mantles. J Geophys Res Solid Earth 101:3311–3322

Tackley PJ (2008) Modelling compressible mantle convection with large viscosity contrast in a three-dimensional spherical shell using the yin-yang grid. Phys Earth Planet Int 171:7–18

Valley JW, Cavosie AJ, Ushikubo T, Reinhard DA, Lawrence DF, Larson DJ, Clifton PH, Kelly TF, Wilde SA, Moser DE, Spicuzza MJ (2014) Hedean age for a post-magma-ocean zircon confirmed by atom-probe tomography. Nat Geosci 7: 219–223

van Keken PE, Hacker BR, Syracuse EM, Abers GA (2011) Subduction factory: 4. Depth-dependent flux of H2O from subducting slabs worldwide. J Geophys Res 116:B01401. https://doi.org/10.1029/2010jb007922

Walter MJ, Thomson AR, Wang W, Lord OT, Ross J, McMahon SC, Baron MA, Melekhova E, Kleppe AK, Kohn SC (2015) The stability of hydrous silicates in Earth's lower mantle: Experimental constraints from the systems MgO–SiO2–H2O and MgO–Al2O3–SiO2–H2O. Chem Geol 418:16–29. https://doi.org/10.1016/j.chemgeo.2015.05.001

White WM, Hofmann AW (1982) Sr and Nd isotope geochemistry of oceanic basalt and mantle evolution. NAature 296:821–825

Wilson CR, Spiegelman M, van Keken PE, Hacker BR (2014) Fluid flow in subduction zones: the role of solid rheology and compaction pressure. Earth Planet Sci Lett 401:261–274. https://doi.org/10.1016/j.epsl.2014.05.052

Xie S, Tackley PJ (2004) Evolution of U–Pb and Sm–Nd systems in numerical models of mantle convection. J Geophys Res 109:B11204. https://doi.org/10.1029/2004JB003176

Yamazaki D, Karato S (2001) Some mineral physics constraints on the rheology and geothermal structure of Earth's lower mantle. Am Mineral 86:385–301

Zhong S, Watts AB (2013) Lithospheric deformation induced by loading of the Hawaiian Islands and its implications for mantle rheology. J Geophys Res Solid Earth 118:6025–6048. https://doi.org/10.1002/2013JB010408

10

Hydrogen-rich hydrothermal environments in the Hadean ocean inferred from serpentinization of komatiites at 300 °C and 500 bar

Takazo Shibuya[1,2,3*], Motoko Yoshizaki[4], Masahiko Sato[5], Kenji Shimizu[2,6], Kentaro Nakamura[7], Soichi Omori[8], Katsuhiko Suzuki[2,3], Ken Takai[1,2,3,9], Hideo Tsunakawa[4] and Shigenori Maruyama[4,10]

Abstract

Serpentinization of Al-depleted and Al-undepleted komatiites (and olivine for comparison) was experimentally characterized under high-temperature and high-pressure conditions of 300 °C and 500 bar to evaluate the H_2 generation potential in komatiite-hosted hydrothermal systems in the early Earth. From the results, the steady-state H_2 concentrations of fluids were estimated to be approximately 20 and 0.05 mmol/kg during the serpentinization reactions for Al-depleted and Al-undepleted komatiites, respectively (60 mmol/kg in the case of olivine). The H_2 concentration of hydrothermal fluid generated from the serpentinization of Al-depleted komatiite is lower than that from olivine but is comparable to that of typical modern peridotite-hosted hydrothermal systems (~16 mmol/kg). The relatively low H_2 concentration from Al-undepleted komatiite is similar to the levels in modern basalt-hosted hydrothermal fluids. Considering that the generation of Al-depleted komatiite melt requires a hotter mantle upwelling (plume) than the generation of Al-undepleted komatiite melt and that the temperature of the mantle has gradually decreased throughout Earth's history, Al-depleted komatiite may have constituted ultramafic volcanism in Hadean oceanic islands/plateaus. Furthermore, it seems unlikely that seafloor exposure of mantle peridotites occurred frequently in the Hadean because the oceanic crust of that time was presumably much thicker than the modern equivalent. Therefore, the serpentinization of Al-depleted komatiites may have been the main process that provided abundant H_2-rich seafloor hydrothermal environments in the Hadean ocean, which potentially acted as a nursery for the prebiotic chemical evolution and the emergence and early evolution of life on Earth.

Keywords: Serpentinization, Komatiite, Hydrogen, Hydrothermal fluid, Hydrothermal experiment, Hadean, Origin of life, Early evolution of life

Background

Serpentinization of seafloor ultramafic rocks is one of the most important geological processes that could have been involved in the emergence and early evolution of life on Earth. During serpentinization, molecular hydrogen (H_2) is characteristically generated by the reaction of water with ferrous oxide in rocks. This process would have sustained H_2-rich seafloor hydrothermal environments where organic materials necessary for the emergence of life could have been synthesized and preserved (Amend and McCollom 2009), and energetically preferred catabolic and anabolic metabolisms of early life could have been established (Takai et al. 2006; Martin et al. 2008; Shibuya et al. 2010; Russell et al. 2014; Sleep et al. 2011).

In modern oceans, serpentinization of ultramafic rocks and the resulting H_2-rich hydrothermal activity often occur in slow-spreading ridges without sufficient magmatic supply, such as the Mid-Atlantic Ridge (MAR)

* Correspondence: takazos@jamstec.go.jp
[1]Laboratory of Ocean-Earth Life Evolution Research (OELE), Japan Agency for Marine-Earth Science and Technology (JAMSTEC), 2-15 Natsushima-cho, Yokosuka 237-0061, Japan
[2]Research and Development (R&D) Center for Submarine Resources, Japan Agency for Marine-Earth Science and Technology (JAMSTEC), 2-15 Natsushima-cho, Yokosuka 237-0061, Japan
Full list of author information is available at the end of the article

(Charlou et al. 2002) and the Central Indian Ridge (Kumagai et al. 2008; Nakamura et al. 2009; Morishita et al. 2015) and exceptionally in off-axis mantle-exposed regions such as the Lost City hydrothermal field away from the MAR (Kelley et al. 2001, 2005; Lang et al. 2010). These peridotite-associated hydrothermal systems have been a subject of great research interest because of their geological, geochemical, and biological uniqueness in comparison with conventional basalt-hosted mid-ocean ridge hydrothermal systems (Takai et al. 2006; Nakamura and Takai 2014). Furthermore, an experiment simulating hydrothermal reactions between peridotite and seawater also revealed the great potential of serpentinization to generate H_2-rich hydrothermal fluid (Seyfried et al. 2007). Therefore, such peridotite-associated hydrothermal systems have been perceived as modern analogues of the Hadean ultramafic-hosted systems that could have been possible nurseries for the emergence and early evolution of life.

In the early Earth, the oceanic crust was likely much thicker (approximately three times thicker) than the modern equivalent (6–7 km thickness) owing to the hotter mantle at that time and has become thinner as the temperature of the mantle has decreased with time, as suggested by the geological occurrence of ophiolites (Moores 2002) and the compositional evolution of greenstones derived from mid-ocean ridges (Komiya 2004). Therefore, it has been theoretically hypothesized that the thick lid of oceanic crust probably limited exposure of mantle peridotite on the seafloor; thus, komatiite rather than peridotite would have been the predominant ultramafic rock in the early Archean ocean floor (Takai et al. 2006). Regarding the genesis of komatiite, there was a controversy as to whether komatiites were derived from wet or dry mantle (Arndt et al. 1998; Grove and Parman 2004) since komatiites were first recognized by Viljoen and Viljoen (1969). It is now known that komatiites can be generated not only by high-temperature and high-pressure partial melting in dry mantle (Takahashi and Scarfe 1985) but also under much lower temperature and pressure conditions in hydrous mantle plumes (Inoue and Sawamoto 1992; Inoue et al. 2000). Grove and Parman (2004) even interpreted komatiites as being formed by shallow-level melting of hydrous mantle wedge above subduction zone. However, the low water contents in komatiitic melt inclusions (Shimizu et al. 2001; Berry et al. 2008) and the similar oxygen fugacity of komatiite magmas to the modern normal upper mantle (Canil 1997, 1999; Puchtel et al. 2013) are not consistent with the hydrous melting model. Therefore, most researchers have agreed that komatiites were produced by extraordinarily hot melting in mantle upwellings (plumes) under nearly dry conditions (e.g., Takahashi and Scarfe 1985; Campbell et al. 1989; Herzberg et al. 2010).

Furthermore, Archean komatiites have generally been classified into two types by their Al_2O_3 content, such as Al-depleted (Barberton-type) and Al-undepleted (Munro-type) komatiites, which mainly occur in the early and late Archean greenstone belts, respectively (Nesbitt et al. 1979). Melting experiments under dry conditions revealed that Al-depleted komatiite (ADK) was formed by high-temperature, ultra-deep, high-degree mantle partial melting, whereas Al-undepleted komatiite (AUK) was generated under relatively lower pressure and temperature conditions than ADK (e.g., Wei et al. 1990; Herzberg 1992). Therefore, this compositional change of komatiite likely reflects the decrease in mantle temperature through geologic time (Herzberg et al. 2010).

Of course, komatiite-associated hydrothermal activity does not occur in the modern ocean; thus, the H_2 generation potential of komatiites during serpentinization can only be experimentally estimated through simulated hydrothermal fluid–rock reactions under high-temperature and high-pressure conditions, as previously conducted to reconstruct modern (Seyfried 1987; Seyfried et al. 2007; McCollom et al. 2010; Kato et al. 2013; Suzuki et al. 2015a, b), ancient (Yoshizaki et al. 2009; Lazar et al. 2012; Shibuya et al. 2013), and even extraterrestrial subseafloor hydrothermal systems (Hsu et al. 2015; Sekine et al. 2015). Yoshizaki et al. (2009) confirmed hydrogen generation by a preliminary, ongoing experiment using komatiite and pure water (Yoshizaki et al. 2009). However, the H_2 generation potential of not only ADK but also AUK, during serpentinization, has not yet been evaluated by completed experiments.

In this study, we conducted experiments to simulate the reactions between komatiite and seawater (NaCl solution) at 300 °C and 500 bar, using a batch-type (closed system) hydrothermal reactor (Yoshizaki et al. 2009). For these experiments, both ADK and AUK were prepared to ascertain the H_2 generation potential in komatiite-hosted hydrothermal system through geologic time. The results allowed us to estimate the H_2 concentration of hydrothermal fluids in Hadean komatiite-hosted hydrothermal systems, which can be used to obtain further insights into the possible energetics and kinetics of prebiotic chemical evolution and the emergence and early evolution of life on the early Earth.

Methods

Preparation of the starting solid materials

Olivine (San Carlos) and two types of synthetic komatiite were prepared for use as the starting solid materials in the experiments. Two types of komatiites were synthesized from reagents. The composition of the reagent powders was adjusted to produce synthetic Al-depleted (Al_2O_3 = c. 5 wt%) and Al-undepleted (Al_2O_3 = c. 10 %) komatiites based on the compositions of natural ADK and AUK (Arth et al. 1977; Wei et al. 1990) (Table 1).

Table 1 Composition of the starting materials used in the experiments and the natural Al-depleted and Al-undepleted komatiites (wt%)

	Starting material			Natural komatiite			
	Olivine[a]	Al_2O_3-5 % komatiite[b]	Al_2O_3-10 % komatiite[b]	ADK[c]	ADK[d]	AUK[e]	AUK[f]
SiO_2	41.25	46.92	44.30	46.50	47.16	45.90	48.96
TiO_2	0.01	0.30	0.29	0.30	0.38	0.41	0.47
Al_2O_3	0.02	4.78	10.11	4.74	4.09	7.97	9.87
FeO_{total}	9.20	11.16	10.53	11.90	12.01	11.07	10.96
MnO	0.10	0.21	0.20	0.21	0.20	0.21	0.19
MgO	49.92	29.61	27.95	29.30	27.97	26.38	19.50
CaO	0.09	6.55	6.19	6.32	6.61	7.74	8.91
Na^2O	0.01	0.35	0.33	0.35	0.37	0.43	1.06
K_2O	0.00	0.11	0.11	0.11	0.04	0.09	0.08

[a]Determined by EPMA analysis
[b]Calculated from the amounts of reagents
[c]An Al-depleted komatiite collected from Barbertone Greenstone Belt (South Africa) (Yoshizaki et al. 2009)
[d]An average composition of Al-depleted komatiites from Barbertone greenstone belt (South Africa) (Wei et al. 1990)
[e]An average composition of Al-undepleted komatiites from Munro/Newton Township (Canada) and Belingwe greenston belt (Zimbabwe) (Wei et al. 1990)
[f]An average composition of Al-undepleted komatiites from Munro Township (Canada) (Arth et al., 1977)

Approximately 30 g of sample powder was placed in a Pt-Rh crucible and heated at 1000 °C for 1 h in an electric furnace to decarbonate the initial reagents. Subsequently, the sample was fused at 1600 °C for 1 h while regulating oxygen fugacity at a QFM (quartz–fayalite–magnetite) buffer under an H_2–CO_2 mixed gas atmosphere. To create the spinifex texture of olivine, the temperature was lowered to 1450 °C over 1.5 h and then to 1350 °C over 18 h under QFM conditions, after which the sample was immediately quenched, yielding a fresh komatiite.

The San Carlos olivine crystals and synthesized olivine-spinifex komatiite (with minor spinel phase) was crushed in a tungsten mortar and sieved to obtain the <90 μm fraction. To remove possible contamination by organic matter during sample preparation, the powdered komatiite was ultrasonically washed with acetone and pure water several times and then dried at 50 °C for 12 h prior to the experiment. The composition of olivine was determined using an electron probe microanalyzer, and the composition of synthetic komatiites was also confirmed using X-ray fluorescence.

Experimental system

An autoclave based on Seyfried et al. (1979) was used for the hydrothermal reaction experiment in this study. The autoclave is made of Inconel-alloy, which is corrosion-resistant and possesses adequate strength at elevated temperatures and pressures of up to 600 °C and 600 bar, respectively (Fig. 1; modified after Yoshizaki et al. 2009). The reaction cell is made of a gold bag with a titanium head because these materials are inert with respect to high-temperature water. In addition, as gold is flexible, the water inside the reaction cell can be pressurized by the surrounding water. Although the Ti head is corrosion-resistant, it is known that metallic Ti can react with water to produce hydrogen; thus, the surface of the Ti head was oxidized prior to use. The flexible gold reaction cell allows on-line sampling of the aqueous fluid at almost constant temperature and pressure simply by adding a small amount of pressurized water to the space surrounding the reaction cell in the autoclave. In this way, fluid samples can be obtained from the reaction cell through a gold-lined sampling tube at any time during an ongoing experiment. All materials that would come into contact with the reaction fluid in the experiment were

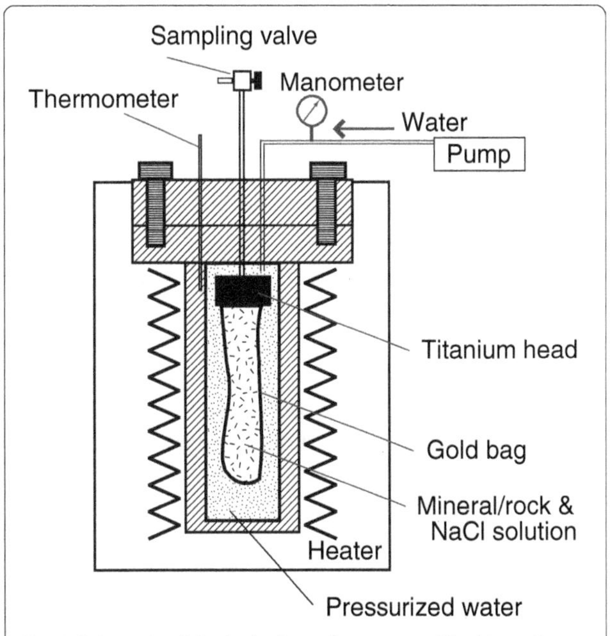

Fig. 1 Schematic of the hydrothermal reactor used in this study, modified after Yoshizaki et al. (2009)

baked in a muffle furnace at 500 °C for 3 h prior to use to eliminate organic matter.

The NaCl solution (approximately 6–7 mol/kg) was prepared from NaCl and pure water to simulate the Hadean seawater, which was potentially saltier than that of the present day owing to the absence of continental crust and associated salt deposits in the Hadean (Knauth 2005). To avoid contamination with organic matter, the NaCl was baked prior to use. The water/rock mass ratio was adjusted to four at the beginning of the experiment (approximately 60 g of solution and 15 g of rock powder) based on the results of Wetzel and Shock (2000) that the water/rock mass ratio in the high-temperature region in the subseafloor hydrothermal system is limited to values less than five. The overall background level of hydrogen concentration in hydrothermal fluid is lower than 0.002 mmol/kg, which was estimated from a 3-month experiment using pure silica powder and NaCl solution under the same temperature and pressure conditions.

Sampling and analysis

Fluid samples were obtained from the reaction cell through a gold-lined sampling tube several times during the ongoing experiment. For the analysis of H_2 concentration, approximately 0.5 ml of fluid was directly introduced into an Ar-purged, sealed vial without air contamination at room temperature. After equilibration between gas and liquid phases in the vial, the H_2 concentration in the gas phase was quantitatively analyzed by gas chromatography (GC) at JAMSTEC. The analytical reproducibility was better than 5 % (1σ). The solid materials of alteration products were dried in an oven immediately after the experiments and then preserved in a vacuum desiccator. The alteration products were analyzed by X-ray diffraction (XRD) at JAMSTEC and by a magnetometer (see below) after the experiment.

Magnetic measurements

For the starting solid material and alteration product of the experiments with olivine, Al_2O_3-5 %, and Al_2O_3-10 % komatiites, the magnetic hysteresis loops at room temperature were measured to estimate the amount of magnetite generated through the experiments using a MicroMag 2900 Alternating Gradient Magnetometer (AGM, Princeton Measurements Corporation) at Kyushu University with a maximum field of 500 mT. The strong-field thermomagnetic curve of the alteration product of the olivine experiment was measured between 50 °C and 780 °C using a MicroMag 3900 Vibrating Sample Magnetometer (VSM, Princeton Measurements Corporation) at the Tokyo Institute of Technology. The magnetic hysteresis loops at low temperatures (10, 15, 20, 25, 30, 35, 40, 45, 50, 55, 60, and 65 K) were measured for the starting materials and alteration products of the komatiite experiments using an MPMS-XL5 Magnetic Property Measurement System (MPMS, Quantum Design) at the Center for Advanced Marine Core Research (CMCR), Kochi University. In the low-temperature measurements, the maximum field in the hysteresis loop measurement was set to be 3 T.

Results and discussion

The powdered olivine, Al_2O_3-5 %, and Al_2O_3-10 % komatiites were reacted with NaCl solution at 300 °C and 500 bar for 2112–2688 h to assess their H_2 generation potential in seafloor hydrothermal systems. In this section, we discuss the alteration of solid materials and its relation to the H_2 concentration in hydrothermal fluid, which yields implications for H_2-rich hydrothermal environments in the Hadean ocean.

Alteration products

The XRD analyses of the alteration products revealed that all the experiments generated a certain amount of serpentine during the hydrothermal reactions (Fig. 2), indicating that serpentinization occurred during these experiments. However, peaks of smectite were observed in the alteration products after the experiments with both Al_2O_3-5 % and Al_2O_3-10 % komatiites. The peaks of smectite in the alteration products were larger for Al_2O_3-10 % komatiite than for Al_2O_3-5 % komatiite; this suggests that smectite becomes more stable as the Al_2O_3 content of the starting material is increased. In addition, CaO originally contained in the starting material of both komatiites was converted to xonotlite.

In general, H_2 generation during serpentinization is caused by the reduction of water as a result of oxidation of ferrous iron in fresh ultramafic rocks (e.g., Janecky and Seyfried 1986). This process can be written as

$$2(FeO)_{rock} + H_2O \rightarrow (Fe_2O_3)_{rock} + H_2,$$

where $(FeO)_{rock}$ refers to the ferrous oxide component in silicate minerals (and glass) while $(Fe_2O_3)_{rock}$ indicates the ferric oxide component of minerals in a serpentinized rock (e.g., McCollom and Bach 2009). Because the ferric oxide typically precipitates as magnetite (ferric/ferrous oxide: Fe_3O_4), the amount of magnetite generated through reactions is strongly related to the H_2 concentration in hydrothermal fluid. In the XRD analyses of the alteration products, no clear peaks of magnetite were identified owing to its small quantities (below the detection limit). However, the magnetic measurements provided constraints on the amounts of effective magnetite generated in the experiments. Figure 3 shows the hysteresis loops of the starting material and alteration product of all experiments. From the hysteresis loops, the saturation magnetization (M_s) values were also

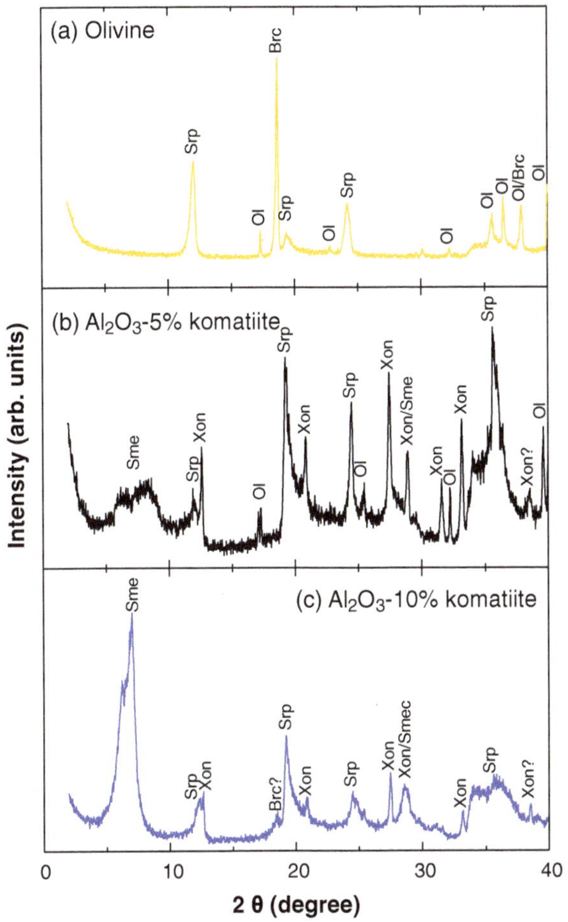

Fig. 2 Results of the XRD analysis (CuKα) of the alteration products. XRD pattern of the alteration products from the experiments with (**a**) olivine, (**b**) Al_2O_3-5 % komatiite, and (**c**) Al_2O_3-10 % komatiite. Mineral abbreviations and ideal chemical formula: Ol = olivine ($(Mg, Fe)_2SiO_4$), Sme = smectite ($(Na, Ca)_{0.3}(Al, Fe, Mg)_{2-3}(Si, Al)_4O_{10}(OH)_2 \cdot nH_2O$), Srp = serpentine ($(Mg, Fe)_3Si_2O_5(OH)_4$), Brc = brucite ($(Mg, Fe)(OH)_2$), and Xon = xonotlite ($Ca_6Si_6O_{17}(OH)_2$)

calculated (Table 2). In the olivine experiment, the M_s values of the alteration product and starting material were estimated from the room-temperature hysteresis loops to be 3.69 Am^2/kg and less than 0.01 Am^2/kg, respectively (Fig. 3a, b, and Table 2). Furthermore, the strong-field thermomagnetic curve of the alteration product of the olivine experiment shows that the Curie temperature (T_C) was 538 °C (Fig. 4a), indicating the formation of nearly pure magnetite during the experiment. Using the well-known magnetite M_s value of 92 Am^2/kg (Hunt et al. 1995), the alteration product of the olivine experiment is estimated to contain 4.0 wt% magnetite.

In both komatiite experiments, the M_s values of the starting material and alteration product were less than 0.05 Am^2/kg at room temperature (Table 2), which indicates that these solid materials contain no or only minute quantities of ferromagnetic minerals with T_C higher than room temperature. In contrast, the M_s values calculated from the hysteresis loops measured at 10 K for these samples were greater than 0.80 Am^2/kg (Fig. 3c–f, Table 2). These results revealed that the solid materials mainly contained a magnetic mineral with T_C lower than room temperature. The temperature dependences of M_s were also calculated from their hysteresis loops (Fig. 4b), which yielded T_C values of these materials of approximately 30 K. Although it is difficult to determine the composition of these magnetic minerals, the most probable candidate with a T_C value of approximately 30 K is a solid solution between magnetite and spinel phases such as $(Fe_3O_4)_x(MgAl_2O_4)_{1-x}$ (Harrison and Putnis 1996). The starting materials showed a certain level of M_s value (Table 2), which is consistent with the starting materials originally containing spinel. In addition, the normalized M_s of the starting material of the Al_2O_3-5 % komatiite experiment is slightly higher than that of the alteration product at temperatures above 30 K (Fig. 4b), indicating that a tiny amount of magnetite is probably contained in the starting material in addition to the magnetite–spinel solid solution. On the other hand, the M_s values of the alteration products are clearly higher than those of the starting materials in both experiments. More importantly, the increment of the M_s value in the Al_2O_3-5 % komatiite experiment is higher than that in the Al_2O_3-10 % komatiite experiment (Table 2). This trend strongly suggests that the total amount of effective magnetite component in the magnetite–spinel solid solution that was newly generated during serpentinization is greater in the Al_2O_3-5 % komatiite experiment than in the Al_2O_3-10 % komatiite experiment.

Even considering the results of all experiments, the increment of the M_s value decreases with increasing Al_2O_3 content in the starting material, which is likely derived from the decrease in the amount of effective magnetite generated with increasing Al_2O_3 content in the starting material. It is therefore suggested that the increase in Al_2O_3 content in ultramafic rock elevates the amount of smectite and reduces the amount of magnetite in the alteration minerals. The formation of smectite presumably inhibited magnetite formation because the FeO originally contained in the starting material was incorporated directly into smectite.

Effect of Al_2O_3 content in rocks on H_2 concentration in hydrothermal fluid

The three experiments revealed that the final, steady-state H_2 concentrations of fluids were approximately 60 mmol/kg (with olivine), 20 mmol/kg (with Al_2O_3-5 % komatiite), and 0.05 mmol/kg (with Al_2O_3-10 % komatiite) (Figs. 5 and 6; Table 3). The H_2 concentration in the olivine

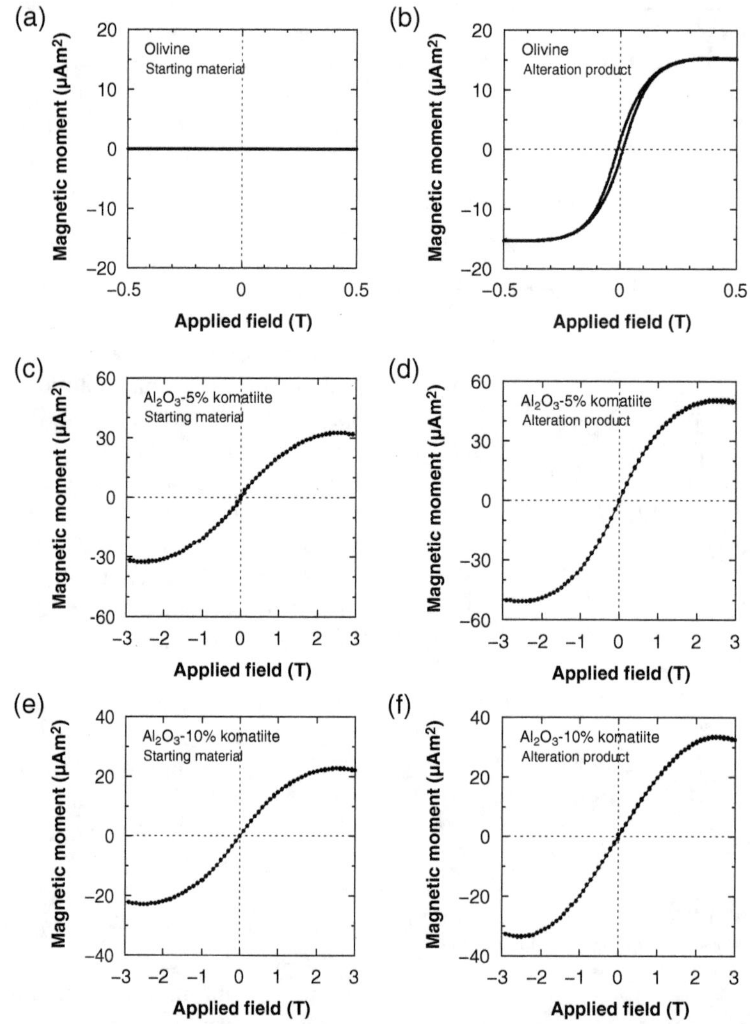

Fig. 3 Magnetic hysteresis loops of (**a**) the starting material and (**b**) the alteration product of the olivine experiment, (**c**) the starting material and (**d**) the alteration product of the Al_2O_3-5 % komatiite experiment, and (**e**) the starting material and (**f**) the alteration product of the Al_2O_3-10 % komatiite experiment. The hysteresis loops were corrected by subtracting diamagnetic/paramagnetic slopes. Note that the loops were measured at 10 K for both komatiite samples and at room temperature for the olivine samples

Table 2 Summary of the magnetic measurements

		Room temperature			10 K			
Experiment	Sample state	m (mg)	M_s (μAm^2)	M_s (Am^2/kg)	m (mg)	M_s (μAm^2)	M_s (Am^2/kg)	T_C (K)
Olivine	Starting material	9.22	0	0	–	–	–	–
	Alteration product	4.14	15.3	3.69	–	–	–	811
Al_2O_3-5 % komatiite	Starting material	11.98	0.62	0.05	33.6	32.1	0.95	~30
	Alteration product	6.58	0.01	0	18.6	50.2	2.70	~30
Al_2O_3-10 % komatiite	Starting material	5.96	0.01	0	41.1	33.0	0.80	~30
	Alteration product	2.70	0.04	0.01	13.1	22.5	1.72	~30

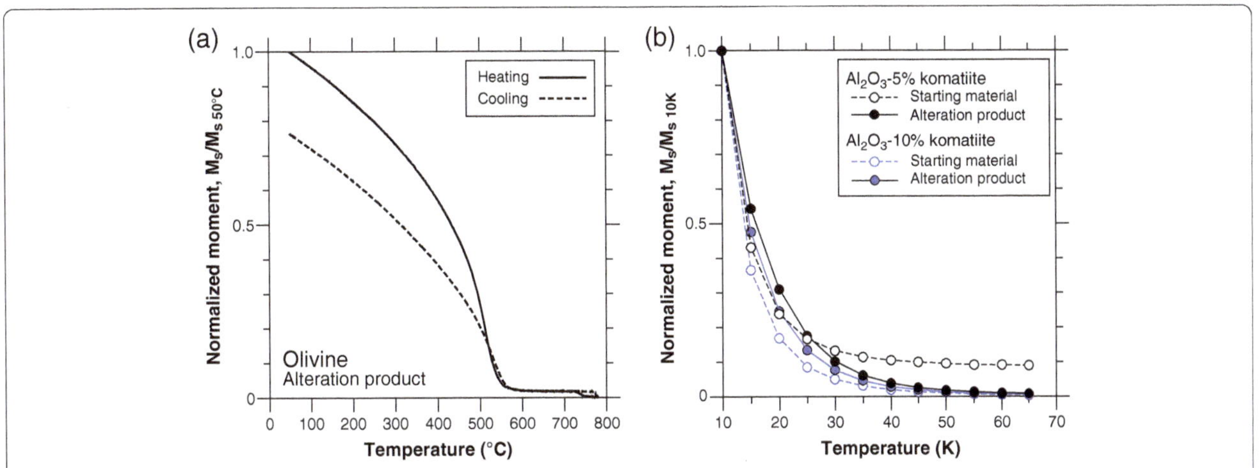

Fig. 4 Results of the Curie temperature measurements. **a** Strong-field thermomagnetic curve for the alteration product of the olivine experiment. **b** Temperature dependency of saturation magnetization of the starting material and alteration product from the Al_2O_3-5 % and Al_2O_3-10 % komatiite experiments, which were calculated from their hysteresis loops measured at various temperatures (e.g., Fig. 3)

experiment is significantly higher than that obtained in a previous experiment with olivine at 400 °C (Allen and Seyfried 2003) and is rather comparable with that generated by an experiment simulating serpentinization of lherzolite at 200 °C (Seyfried et al. 2007). This is consistent with the thermodynamic analysis of serpentinization; olivine becomes stable above approximately 315–390 °C, which limits magnetite formation and H_2 generation in hydrothermal fluid (McCollom and Bach 2009). The steady-state H_2 concentration in the Al_2O_3-5 % komatiite experiment is clearly higher than that obtained from the previous ongoing experiment using Al_2O_3-5 % komatiite and pure water (2.4 mmol/kg) (Yoshizaki et al. 2009). This may be due to the difference of initial solution because NaCl solution is generally much more reactive with rocks than pure water. More importantly, the experiments in this study showed that the steady-state H_2 concentration in fluid increases as the Al_2O_3 content in the starting material is reduced (Fig. 7). This pattern is also consistent with the amount of generated magnetite, which increases with decreasing Al_2O_3 content in the starting material. Therefore, it is suggested that the Al_2O_3 content in the starting material strongly affects the generated H_2 concentration in fluid because the Al_2O_3 level controls the formation of smectite, which potentially inhibits magnetite formation.

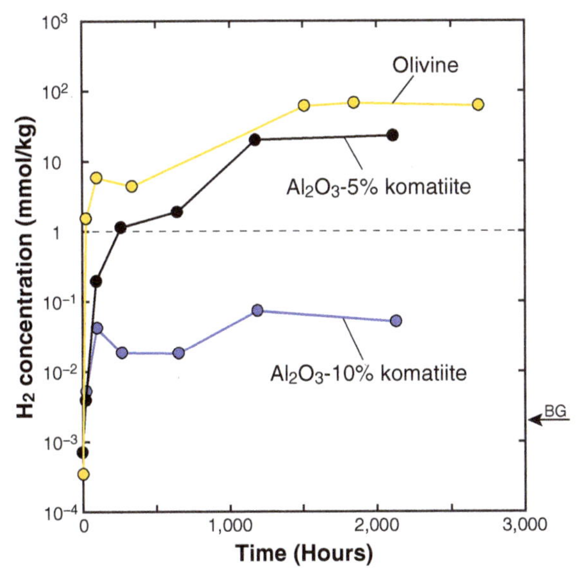

Fig. 5 Changes in H_2 concentration during the experiments with olivine, Al_2O_3-5 % and Al_2O_3-10 % komatiites. The *arrow* indicates the maximum background H_2 concentration after c. 2000 h

Comparison with known H_2 concentrations in modern seafloor hydrothermal vent fluids shows that the concentration in the Al_2O_3-10 % komatiite experiment is comparable to those observed in typical MORB-hosted systems (generally 0.1–1 mmol/kg) (Fig. 6). In contrast, natural peridotite-associated high-temperature hydrothermal fluids have higher H_2 concentrations of up to 16 mmol/kg (Charlou et al. 2002; Gallant and Von Damm 2006; Kumagai et al. 2008), which was successfully reconstructed in experiments using orthopyroxene ± clinopyroxene ± olivine at 400 °C (Allen and Seyfried 2003). Thus, the Al_2O_3-5 % komatiite has the potential to generate an H_2 concentration similar to or higher than those observed in modern peridotite-associated hydrothermal fluids.

Potential role of komatiite serpentinization in the origin of life and early ecosystems in the Hadean ocean

Theoretical predictions for the thermodynamic state of prebiotic chemical evolution have highlighted that the H_2-rich hydrothermal environment is energetically advantageous

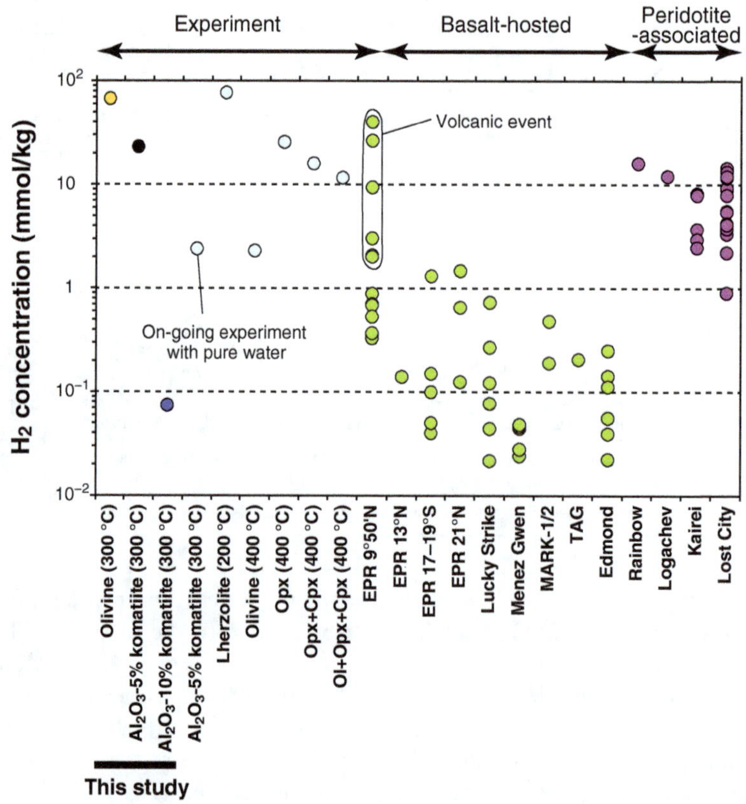

Fig. 6 Comparison of H_2 concentrations in hydrothermal fluid from experiments and modern natural hydrothermal vent systems. Data sources: Al_2O_3-5 % komatiite experiment with pure water at 300 °C and 500 bar (Yoshizaki et al. 2009), lherzolite experiment at 200 °C and 500 bar (Seyfried et al. 2007), four experiments with Ol, Opx, and/or Cpx at 400 °C and 500 bar (Allen and Seyfried 2003), EPR 9°50'N (Lilley et al. 2003; Von Damm and Lilley 2004), EPR 13°N (Von Damm 1995), EPR 17–19°S (Charlou et al. 1996b), EPR 21°S (Lilley et al. 1983), Lucky Strike and Menez Gwen (Charlou et al. 2000), MARK-1/2 (Charlou et al. 2002), TAG (Charlou et al. 1996a), Logatchev and Rainbow (Charlou et al. 2002), Kairei (Gallant and Von Damm 2006; Kumagai et al. 2008), and Lost City (Proskurowski et al. 2006). The H_2 concentrations of seafloor hydrothermal fluids were mainly obtained from the compilation of Nakamura and Takai (2014). Mineral abbreviations: Ol = olivine, Opx = orthopyroxene, Cpx = clinopyroxene

for synthesis and preservation of organic molecules such as amino acids and fatty acids (Amend and McCollom 2009; Amend et al. 2011). Furthermore, it has been hypothesized that the most plausible energy metabolisms to support the emergence and early evolution of life may have been H_2-driven hydrogenotrophic methanogenesis/acetogenesis and/or methanotrophic acetogenesis (Russell et al. 1989; Takai et al. 2006; Martin et al. 2008; Russell et al. 2010, 2014). In this section, we discuss the possible geological settings that may have generated H_2-rich hydrothermal systems in the Hadean ocean.

Considering the relationship between the decreasing mantle temperature during the history of Earth and the

Table 3 Changes in H_2 concentration and pH of fluid samples through time

Olivine			Al_2O_3-5 % komatiite			Al_2O_3-10 % komatiite		
Time (h)	$pH_{25\ °C}$	H_2 (mmol/kg)	Time (h)	$pH_{25\ °C}$	H_2 (mmol/kg)	Time (h)	$pH_{25\ °C}$	H_2 (mmol/kg)
0	6.8	3.5×10^{-4}	0	7.0	7.1×10^{-4}	0	7.0	n.d.
24	10.3	1.5	24	7.5	3.9×10^{-3}	24	7.7	0.005
96	9.6	5.9	96	8.2	0.20	96	7.3	0.042
336	10.9	4.4	264	8.2	1.1	264	7.8	0.019
1512	10.8	61.2	648	7.7	1.9	648	8.5	0.018
1848	10.9	67.4	1176	8.1	20.0	1176	7.5	0.074
2688	10.5	61.8	2112	8.0	23.0	2112	7.7	0.052
						2668	8.0	0.030

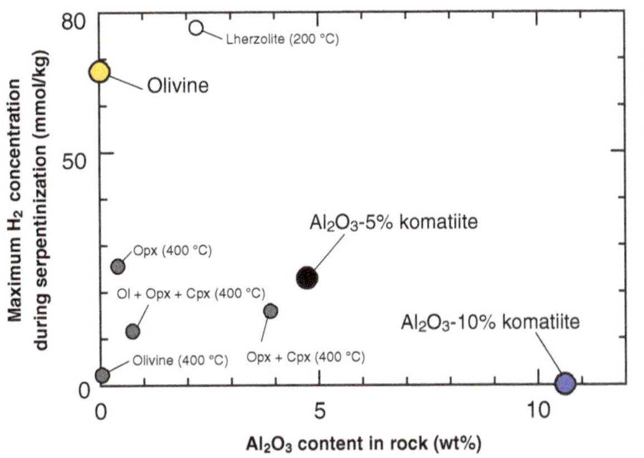

Fig. 7 Relationships between fluid H_2 concentration during serpentinization and Al_2O_3 content in the starting solid material. For comparison, the results of the lherzolite experiment at 200 °C (Seyfried et al. 2007) and the experiments with Ol, Opx, and/or Cpx at 400 °C (Allen and Seyfried 2003) are also shown

chemical composition/production process of komatiites, the different H_2 generation potentials of ADK and AUK imply that komatiite-hosted hydrothermal fluid evolved from H_2-enriched to H_2-depleted through geologic time. Previously, high-pressure experiments on the generation of komatiite melts revealed that ADK melt is generated under high-temperature and high-pressure (>7 GPa) conditions, leaving majoritic garnet as a liquidus phase of mantle, whereas AUK may form owing to high degrees of mantle melting at shallower depths where olivine is a liquidus phase (e.g., Herzberg 1992; Arndt et al. 2008). Therefore, it is believed that the formation of ADK requires a higher mantle plume temperature than for the generation of AUK. This indicates that ADK was produced more abundantly in the oceanic-island/plateau volcanism of the earlier Earth with the hotter mantle than AUK, which is consistent with the geological records of ~3.5 Ga Barberton-type (Al-depleted) and c. 2.7 Ga Munro-type (Al-undepleted) komatiites (Nesbitt et al. 1979; Herzberg 1992). As the Al_2O_3-5 % (Al-depleted) komatiite has a greater potential to enrich H_2 in hydrothermal fluid than Al_2O_3-10 % (Al-undepleted) komatiite (Figs. 6 and 7), the H_2-rich hydrothermal environments hosted by ADK were present abundantly and ubiquitously on the ocean floor prior to the early Archean.

In contrast, the exposure of mantle peridotite at the ocean floor was probably rare in the Archean (Takai et al. 2006). Modern abyssal peridotite is frequently observed near slow-spreading ridges without sufficient magmatic supply because large-scale normal faults (e.g., detachment faults) penetrate deeply, and the mantle peridotite is dragged up to the seafloor (Escartín et al. 2003). However, it has been revealed that the Archean oceanic crust was much thicker than the modern equivalent owing to the higher potential mantle temperature at that time (Moores 2002; Ohta et al. 1996; Komiya 2004; Shibuya et al. 2007, 2012). The thicker ocean crust would have limited the development of deep normal faults reaching to the mantle and exposure of mantle peridotite, which has led to the hypothesis that the early Archean H_2-rich hydrothermal environment was mainly driven by komatiite volcanism instead of much less abundant fault-related peridotite (Takai et al. 2006). This hypothesis is further substantiated by the results obtained in this study and can be applied to the Hadean. The serpentinization of ADK may have generated abundant H_2-rich (ca. 20 mmol/kg) hydrothermal fluid and proximal H_2-rich hydrothermal fluid–seawater mixing zones in the Hadean ocean, which could potentially have served as nurseries for prebiotic chemical evolution and the subsequent emergence and early evolution of life on Earth. Furthermore, considering that the oceanic crust was probably sufficiently thick throughout the Hadean and Archean to limit the exposure of peridotite on the seafloor and that ADK disappeared after the early Archean, the Hadean era was more favorable for the emergence of life than the Archean era in terms of H_2-rich hydrothermal environments.

Conclusions

The H_2 generation potential of ADK and AUK during serpentinization was estimated by the experimental hydrothermal reactions of komatiites at 300 °C and 500 bar. The experiments revealed that the H_2 concentration (ca. 20 mmol/kg) in hydrothermal fluid generated from the serpentinization of ADK is comparable to that of the most H_2-rich hydrothermal systems in modern oceans. Furthermore, considering the limited exposure of mantle peridotite owing to the thick lid of oceanic crusts in the Hadean ocean floor, ADK was likely the most ubiquitous seafloor ultramafic rock that could host H_2-rich hydrothermal system. Such H_2-rich hydrothermal systems would have provided a favorable environment (redox, electrochemical, and thermal gradients in the seawater/hydrothermal fluid mixing zone) for the acquisition of bioavailable free energy and for prebiotic chemical evolution in the Hadean. Further experiments and thermodynamic calculations to simulate the mixing between Hadean seawater and H_2-rich hydrothermal fluid should be carried out to test this hypothesis.

Competing interests
The authors declare that they have no competing interests.

Authors' contributions
TS, KN, KT, and SM proposed and designed the study. TS and MY carried out the experiments. TS, MY, and KS prepared the starting materials. TS, MY, MS, and HT analyzed the fluid samples and solid materials. TS, MY, MS, KN, SO, KS, and KT interpreted the data. All authors read and approved the final manuscript.

Acknowledgements
We thank the three anonymous reviewers for their helpful comments and H Kawahata for careful editorial handling. This study was partially supported by the Trans-crustal Advection and In-situ biogeochemical processes of Global sub-seafloor Aquifer (TAIGA) project and the Grants-in-Aid for Scientific Research (No. 22740333) from the Japanese Ministry of Education, Culture, Sports, Science and Technology. This work was also performed under the cooperative research program of Center for Advanced Marine Core Research (CMCR), Kochi University (Accept No. 14A007 and 14B005).

Author details
[1]Laboratory of Ocean-Earth Life Evolution Research (OELE), Japan Agency for Marine-Earth Science and Technology (JAMSTEC), 2-15 Natsushima-cho, Yokosuka 237-0061, Japan. [2]Research and Development (R&D) Center for Submarine Resources, Japan Agency for Marine-Earth Science and Technology (JAMSTEC), 2-15 Natsushima-cho, Yokosuka 237-0061, Japan. [3]Project Team for Development of New-generation Research Protocol for Submarine Resources, Japan Agency for Marine-Earth Science and Technology (JAMSTEC), 2-15 Natsushima-cho, Yokosuka 237-0061, Japan. [4]Department of Earth and Planetary Sciences, Tokyo Institute of Technology, 2-12-1 Ookayama, Meguro-ku, Tokyo 152-8551, Japan. [5]Geological Survey of Japan, National Institute of Advanced Industrial Science and Technology, 1-1-1 Higashi, Tsukuba 305-8567, Japan. [6]Kochi Institute for Core Sample Research, Japan Agency for Marine-Earth Science and Technology (JAMSTEC), 200 Monobe, Nankoku, Kochi 783-8502, Japan. [7]Department of Systems Innovation, The University of Tokyo, 7-3-1 Hongo, Bunkyo-ku, Tokyo 113-8656, Japan. [8]The Open University of Japan, 2-11 Wakaba, Mihama-ku, Chiba 261-8586, Japan. [9]Department of Subsurface Geobiological Analysis and Research (SUGAR), Japan Agency for Marine-Earth Science and Technology (JAMSTEC), 2-15 Natsushima-cho, Yokosuka 237-0061, Japan. [10]Earth-Life Science Institute, Tokyo Institute of Technology, 2-12-1 Ookayama, Meguro-ku, Tokyo 152-8551, Japan.

References

Allen DE, Seyfried Jr WE. Compositional controls on vent fluids from ultramafic-hosted hydrothermal systems at midocean ridges: an experimental study at 400 °C, 500 bars. Geochim Cosmochim Acta. 2003;67:1531–42.

Amend JP, McCollom TM. Energetics of biomolecule synthesis on early Earth. In: Zaikowski L, Friedrich JM, Seidel SR, editors. Chemical evolution II: from the origins of life to modern society. Washington, D.C.: American Chemical Society; 2009. p. 63–94.

Amend JP, McCollom TM, Hentscher M, Bach W. Catabolic and anabolic energy for chemolithoautotrophs in deep-sea hydrothermal systems hosted in different rock types. Geochim Cosmochim Acta. 2011;75:5736–48.

Arndt N, Ginibre C, Chauvel C, Albarède F, Cheadle M, Herzberg C, et al. Were komatiites wet? Geology. 1998;26:739–42.

Arndt N, Lesher CM, Barnes SJ. Komatiite. Cambridge: Cambridge University Press; 2008.

Arth JG, Arndt NT, Naldrett AJ. Genesis of Archean komatiites from Munro Township, Ontario: trace-element evidence. Geology. 1977;5:590–4.

Berry AJ, Danyushevsky LV, O'Neill HSC, Newville M, Sutton SR. Oxidation state of iron in komatiitic melt inclusions indicates hot Archaean mantle. Nature. 2008;455:960_3.

Campbell IH, Griffiths RW, Hill RI. Melting in an Archean mantle plume—heads its basalts, tails its komatiites. Nature. 1989;339:697–9.

Canil D. Vanadium partitioning and the oxidation state of Archaean komatiite magma. Nature. 1997;389:842–5.

Canil D. Vanadium partitioning between orthopyroxene, spinel and silicate melt and the redox states of mantle source regions for primary magmas. Geochim Cosmochim Acta. 1999;63:557–72.

Charlou JL, Donval JP, Jean-Baptiste P, Dapoigny A, Rona PA. Gases and helium isotopes in high temperature solutions sampled before and after ODP Leg 158 drilling at TAG hydrothermal field (26°N, MAR). Geophys Res Lett. 1996a;23:3491–4.

Charlou JL, Fouquet Y, Donval JP, Auzende JM, Jean-Baptiste P, Stievenard M, et al. Mineral and gas chemistry of hydrothermal fluids on an ultrafast spreading ridge: East Pacific Rise, 17° to 19°S (Naudur cruise, 1993) phase separation processes controlled by volcanic and tectonic activity. J Geophys Res. 1996b; 101:15899–919.

Charlou JL, J. P. Donval a ED, Jean-Baptiste P, Radford-Knoery J, Fouquet Y, Dapoigny A, et al. Compared geochemical signatures and the evolution of Menez Gwen (37°50′N) and Lucky Strike (37°17′N) hydrothermal fluids, south of the Azores Triple Junction on the Mid-Atlantic Ridge. Chem Geol. 2000;171:49–75.

Charlou JL, Donval JP, Fouquet Y, Jean-Baptiste P, Holm N. Geochemistry of high H_2 and CH_4 vent fluids issuing from ultramafic rocks at the Rainbow hydrothermal field (36°14′N, MAR). Chem Geol. 2002;191:345–59.

Escartín J, Mével C, MacLeod CJ, McCaig AM. Constraints on deformation conditions and the origin of oceanic detachments: the Mid-Atlantic Ridge core complex at 15°45′N. Geochem Geophy Geosy. 2003;4:1067. doi:10.1029/2002GC000472.

Gallant RM, Von Damm KL. Geochemical controls on hydrothermal fluids from the Kairei and Edmond Vent Fields, 23°–25°S, Central Indian Ridge. Geochem Geophy Geosy. 2006;7:Q06018. doi:10.01029/02005GC001067.

Grove TL, Parman SW. Thermal evolution of the Earth as recorded by komatiites. Earth Planet Sci Lett. 2004;219:173–87.

Harrison RJ, Putnis A. Magnetic properties of the magnetite-spinel solid solution: curie temperature, magnetic susceptibilities, and cation ordering. Am Mineral. 1996;81:375–84.

Herzberg C. Depth and degree of melting of komatiites. J Geophys Res. 1992;97:4521–40.

Herzberg C, Condie K, Korenaga J. Thermal history of the Earth and its petrological expression. Earth Planet Sci Lett. 2010;292:79–88.

Hsu H-W, Postberg F, Sekine Y, Shibuya T, Kempf S, Horányi M, et al. Ongoing hydrothermal activities within Enceladus. Nature. 2015;519:207–10.

Hunt CP, Moskowitz BM, Banerjee SK. Magnetic properties of rocks and minerals. In: Ahrens TJ, editor. Rock physics & phase relations: a handbook of physical constants. Washington, D. C.: American Geophysical Union; 1995. p. 189–204.

Inoue T, Sawamoto H. High pressure melting of pyrolite under hydrous condition and its geophysical implication. In: Syono Y, Manghnani MH, editors. High-pressure research: application to earth and planetary sciences. Washington D. C.: Terra, Tokyo and AGU; 1992. p. 323–31.

Inoue T, Rapp RP, Zhang J, Gasparik T, Weidner DJ, Irifune T. Garnet fractionation in a hydrous magma ocean and the origin of Al-depleted komatiites: melting experiments of hydrous pyrolite with REEs at high pressure. Earth Planet Sci Lett. 2000;177:81–7.

Janecky DR, Seyfried Jr WE. Hydrothermal serpentinization of peridotite within the oceanic crust: experimental investigations of mineralogy and major element chemistry. Geochim Cosmochim Acta. 1986;50:1357–78.

Kato S, Shibuya T, Nakamura K, Suzuki K, Rejishkumar VJ, Yamagishi A. Elemental dissolution of basalts with ultra-pure water at 340 °C and 40 MPa in a newly developed flow-type hydrothermal apparatus. Geochem J. 2013;47:89–92.

Kelley DS, Karson JA, Blackman DK, Früh-Green GL, Butterfield DA, Lilley MD, et al. An off-axis hydrothermal vent field near the Mid-Atlantic Ridge at 30°N. Nature. 2001;412:145–9.

Kelley DS, Karson JA, Früh-Green GL, Yoerger DR, Shank TM, Butterfield DA, et al. A serpentinite-hosted ecosystem: the Lost City hydrothermal field. Science. 2005;307:1428–34.

Knauth LP. Temperature and salinity history of the Precambrian ocean: implications for the course of microbial evolution. Palaeogeogr Palaeoclimateol Palaeoecol. 2005;219:53–69.

Komiya T. Material circulation model including chemical differentiation within the mantle and secular variation of temperature and composition of the mantle. Phys Earth Planet Inter. 2004;146:333–67.

Kumagai H, Nakamura K, Toki T, Morishita T, Okino K, Ishibashi J-i, et al. Geological background of the Kairei and Edmond hydrothermal fields along the Central Indian Ridge: implications of their vent fluids' distinct chemistry. Geofluids. 2008;8:239–51.

Lang SQ, Butterfield DA, Schulte M, Kelley DS, Lilley MD. Elevated concentrations of formate, acetate and dissolved organic carbon found at the Lost City hydrothermal field. Geochim Cosmochim Acta. 2010;74:941–52.

Lazar C, McCollom TM, Manning CE. Abiogenic methanogenesis during experimental komatiite serpentinization: implications for the evolution of the early Precambrian atmosphere. Chem Geol. 2012;326–327:102–12.

Lilley MD, Baross JA, I GL. Reduced gases and bacteria in hydrothermal fluids: the Galapagos spreading center and 21°N East Pacific Rise. In: Rona PA, Bostrom K, Laubier L, Smith Jr KL, editors. Hydrothermal processes at seafloor spreading centers. Marine Sciences: NATO Conference Series IV; 1983. p. 411–49.

Lilley MD, Butterfield DA, Lupton JE, Olson EJ. Magmatic events can produce rapid changes in hydrothermal vent chemistry. Nature. 2003;422:878–81.

Martin W, Baross J, Kelley D, Russell MJ. Hydrothermal vents and the origin of life. Nat Rev Microbiol. 2008;6:805–14.

McCollom TM, Bach W. Thermodynamic constraints on hydrogen generation during serpentinization of ultramafic rocks. Geochim Cosmochim Acta. 2009;73:856–75.
McCollom TM, Sherwood Lollar B, Lacrampe-Couloume G, Seewald JS. The influence of carbon source on abiotic organic synthesis and carbon isotope fractionation under hydrothermal conditions. Geochim Cosmochim Acta. 2010;74:2717–40.
Moores EM. Pre-1 Ga (pre-Rodinian) ophiolites: their tectonic and environmental implications. Geol Soc Am Bull. 2002;114:80–95.
Morishita T, Nakamura K, Shibuya T, Kumagai H, Sato T, Okino K, et al. Petrology of peridotites and related gabbroic rocks around the Kairei hydrothermal field in the Central Indian Ridge. In: Ishibashi J, Okino K, Sunamura M, editors. Subseafloor biosphere linked to hydrothermal systems. Tokyo: Springer Japan; 2015. p. 177–93.
Nakamura K, Morishita T, Bach W, Klein F, Hara K, Okino K, et al. Serpentinized troctolites exposed near the Kairei Hydrothermal Field, Central Indian Ridge: insights into the origin of the Kairei hydrothermal fluid supporting a unique microbial ecosystem. Earth Planet Sci Lett. 2009;280:128–36.
Nakamura K, Takai K. Theoretical constraints of physical and chemical properties of hydrothermal fluids on variations in chemolithotrophic microbial communities in seafloor hydrothermal systems. Prog Earth Planet Sci. 2014;1:5. doi:10.1186/2197-4284-1-5.
Nesbitt RW, Sun SS, Purvis AC. Komatiites: geochemistry and genesis. Can Mineral. 1979;17:165–86.
Ohta H, Maruyama S, Takahashi E, Watanabe Y, Kato Y. Field occurrence, geochemistry and petrogenesis of the Archean mid-oceanic ridge basalts (AMORBs) of the Cleaverville area, Pilbara Craton, Western Australia. Lithos. 1996;37:199–221.
Proskurowski G, Lilley MD, Kelley DS, Olson EJ. Low temperature volatile production at the Lost City hydrothermal field, evidence from a hydrogen stable isotope geothermometer. Chem Geol. 2006;229:331–43.
Puchtel IS, Blichert-Toft J, Touboul M, Walker RJ, Byerly GR, Nisbet EG, et al. Insights into early Earth from Barberton komatiites: evidence from lithophile isotope and trace element systematics. Geochim Cosmochim Acta. 2013;108:63–90.
Russell MJ, Hall AJ, Turner D. In vitro growth of iron sulphide chimneys: possible culture chambers for origin-of-life experiments. Terra Nova. 1989;1:238–41.
Russell MJ, Hall AJ, Martin W. Serpentinization as a source of energy at the origin of life. Geobiology. 2010;8:355–71. doi:10.1111/j.1472-4669.2010.00249.x.
Russell MJ, Barge L, Bhartia R, Bocanegra D, Bracher P, Branscomb E, et al. The drive to life on wet and icy worlds. Astrobiology. 2014;14:308–43.
Sekine Y, Shibuya T, Postberg F, Hsu H-W, Suzuki K, Masaki Y, et al. High-temperature water–rock interactions and hydrothermal environments in the chondrite-like core of Enceladus. Nat Commun. 2015;6:8604. doi:10.1038/ncomms9604.
Seyfried Jr WE, Gordon PC, Dickson FW. A new reaction cell for hydrothermal solution equipment. Am Mineral. 1979;64:646–9.
Seyfried Jr WE. Experimental and theoretical constraints on hydrothermal alteration processes at mid-ocean ridges. Annu Rev Earth Planet Sci. 1987;15:317–35.
Seyfried Jr WE, Foustoukos DI, Fu Q. Redox evolution and mass transfer during serpentinization: an experimental and theoretical study at 200 °C, 500 bar with implications for ultramafic-hosted hydrothermal systems at mid-ocean ridges. Geochim Cosmochim Acta. 2007;71:3872–86.
Shibuya T, Kitajima K, Komiya T, Terabayashi M, Maruyama S. Middle Archean ocean ridge hydrothermal metamorphism and alteration recorded in the Cleaverville area, Pilbara Craton, Western Australia. J Metamorph Geol. 2007;25:751–67.
Shibuya T, Komiya T, Nakamura K, Takai K, Maruyama S. Highly alkaline, high-temperature hydrothermal fluids in the early Archean ocean. Precambrian Res. 2010;182:230–8.
Shibuya T, Tahata M, Kitajima K, Ueno Y, Komiya T, Yamamoto S, et al. Depth variation of carbon and oxygen isotopes of calcites in Archean altered upper oceanic crust: implications for the CO_2 flux from ocean to oceanic crust in the Archean. Earth Planet Sci Lett. 2012;321–322:64–73.
Shibuya T, Yoshizaki M, Masaki Y, Suzuki K, Takai K. Reactions between basalt and CO_2-rich seawater at 250 and 350 °C, 500 bars: implications for the CO_2 sequestration into the modern oceanic crust and the composition of hydrothermal vent fluid in the CO_2-rich early ocean. Chem Geol. 2013;359:1–9.
Shimizu K, Komiya T, Hirose K, Shimizu N, Maruyama S. Cr-spinel, an excellent micro-container for retaining primitive melts – implications for a hydrous plume origin for komatiites. Earth Planet Sci Lett. 2001;189:177–88.
Sleep NH, Bird DK, Pope EC. Serpentinite and the dawn of life. Philos Trans R Soc Lond B Biol Sci. 2011;366:2857–69.
Suzuki K, Kato S, Shibuya T, Hirose T, Fuchida S, Kumar YR, et al. Development of hydrothermal and frictional experimental systems to simulate sub-seafloor water–rock–microbe interactions. In: Ishibashi J, Okino K, Sunamura M, editors. Subseafloor biosphere linked to hydrothermal systems. Tokyo: Springer Japan; 2015a. p. 71–85.
Suzuki K, Shibuya T, Yoshizaki M, Hirose T. Experimental hydrogen production in hydrothermal and fault systems: significance for habitability of subseafloor H_2 chemoautotroph microbial ecosystems. In: Ishibashi J, Okino K, Sunamura M, editors. Subseafloor Biosphere Linked to Hydrothermal Systems. Tokyo: Springer Japan; 2015b. p. 87–94.
Takahashi E, Scarfe CM. Melting of peridotite to 14 GPa and the genesis of komatiites. Nature. 1985;315:566–8.
Takai K, Nakamura K, Suzuki K, Inagaki F, Nealson KH, Kumagai H. Ultramafics-Hydrothermalism-Hydrogenesis-HyperSLiME (UltraH3) linkage: a key insight into early microbial ecosystem in the Archean deep-sea hydrothermal systems. Paleontol Res. 2006;10:269–82.
Viljoen MJ, Viljoen RP. Evidence for the existence of a mobile extrusive peridotitic magma from the Komati Formation of the Onverwacht Group. Geol Soc S Afr Spec Publ. 1969;2:87–113.
Von Damm KL. Controls on the chemistry and temporal variability of seafloor hydrothermal fluids. In: Humphris SE, Zierenberg RA, Mullineaux LS, Thomson RE, editors. Seafloor hydrothermal systems: physical, chemical, biological, and geological interactions. Washington DC: Geophysical Monograph. American Geophysical Union; 1995. p. 222–47.
Von Damm KL, Lilley MD. Diffuse flow hydrothermal fluids from 9°50'N East Pacific Rise: origin, evolution and biogeochemical controls. In: Wilcock WSD, DeLong EF, Kelley DS, Baross JA, Cary SC, editors. The subseafloor biosphere at mid-ocean ridges, Geophysical Monograph, vol. 144. Washington DC: American Geophysical Union; 2004. p. 245–68.
Wei K, Tronnes RG, Scarfe CM. Phase relations of aluminum-undepleted and aluminum-depleted komatiites at pressures of 4–12 GPa. J Geophys Res. 1990;95:15817–27.
Wetzel LR, Shock EL. Distinguishing ultramafic- from basalt-hosted submarine hydrothermal systems by comparing calculated vent fluid compositions. J Geophys Res. 2000;105:8319–40.
Yoshizaki M, Shibuya T, Suzuki K, Shimizu K, Nakamura K, Takai K, et al. H_2 generation by experimental hydrothermal alteration of komatiitic glass at 300 °C and 500 bars: a preliminary result from on-going experiment. Geochem J. 2009;43:e17–22.

11

Lg wave propagation in the area around Japan: observations and simulations

Takashi Furumura[1*], Tae-Kyung Hong[2] and Brian LN Kennett[3]

Abstract

Regional wavefields are strongly influenced by crustal structure heterogeneity variations along their propagation paths. Observations of such effects between the Asian continent and the Japanese subduction zone across the Sea of Japan (East Sea) have been strongly assisted recently by the development of high-density seismic networks in Japan and Korea, as well as the supercomputer-based three-dimensional finite difference method seismic wave propagation simulations using detailed heterogeneous crustal models. In this study, an *Lg* propagation map derived from 289,000 ray paths connecting sources to observation stations reveals efficient *Lg* wave propagation from continental Asia to Kyushu through the Korean Peninsula, and to Hokkaido, thus indicating a laterally consistent crustal structure extending from continental Asia to Japan. However, the *Lg* wave propagation to the Japanese main island of Honshu is totally blocked as it crosses the continental-oceanic boundary surrounding the Sea of Japan. Three-dimensional (3-D) finite-difference method seismic wave propagation simulations performed using a detailed crustal structural model allow us to clearly visualize the way in which an *Lg* wave develops from a shallow source in the crust and its propagation in the crustal waveguide by means of multiple post-critical *S* wave reflections in the continental structure. The sudden thinning of the continental crust at the edge of the Asian continent adjacent to the oceanic crust in the Sea of Japan, which involves a thickness change from 30 to 10 km within a 100-km distance (with the thinner crust extending over 600 km), decreases the *Lg* wave energy by 10%. It has been confirmed that 50% of this *Lg* wave energy loss occurs during wave passage along the thinner crust and that the other 50% results from conversion into *P* energy in the overlying seawater.

Keywords: Computational seismology; Crustal structure; Finite-difference method simulation; *Lg* wave; Regional wave; *Sn* wave; Wave propagation

Background

An *Lg* wave is a crustally guided *S* wave composed of a superposition of post-critical reflections in the crust such as that between a free surface and the Mohorovičić discontinuity (Moho), which can also be interpreted as a combination of higher-mode surface waves. In the short-period seismograms, an *Lg* wave is usually a prominent feature at regional distances from 150 km to well over 1,000 km from the hypocenter of the event in continental structures and can be identified on three-component seismograms by the amplitude similarities of each component. The phase is dominated by relatively high frequencies between 0.2 and 5 Hz. Therefore, *Lg* phases have been used to magnitude estimates of shallow events ($m_b(Lg)$, see, e.g., Nuttli 1973; Herrmann and Nuttli 1982) and to map crustal heterogeneities by estimating *Lg* attenuation variations (Q_{Lg}; e.g., Phillips et al. 2000; Xie et al. 2006; Chung et al. 2007; Hong et al. 2008). The *Lg* phase is also utilized for discriminating manmade sources from natural earthquakes, such as by using the *Lg/Pg* and *Lg/Pn* amplitude ratio (e.g., Kennett 1993; Walter et al. 1995; Taylor 1996; Fisk 2006; Kim and Richards 2007).

It is well recognized that *Lg* wave propagation is extremely sensitive to lateral crustal structure variations. For example, although *Lg* waves propagate efficiently from most areas of the Asian and Eurasian continents, strong attenuation and/or *Lg* wave blockage has been identified in a number of regions, such as southern Eurasia (Ruzaikan et al. 1977; Ni and Barazangi 1983), northwestern Russia (Baumgardt 1985), the Tibetan Plateau (Xie 2002; Fan and

* Correspondence: furumura@eri.u-tokyo.ac.jp
[1]Earthquake Research Institute, The University of Tokyo, 1-1-1 Yayoi, Bunkyo-ku, Tokyo 113-0032, Japan
Full list of author information is available at the end of the article

Lay 2003), the Himalaya-Tibet collision zone (Singh et al. 2012), and the Alpine Mountains (Campillo et al. 1987). In these cases, attenuation and/or blockage has been traced to the geological and tectonic environments of each region and to features such as abrupt lateral variations in the crustal thickness, thick sediment cover, low-*Q* materials in the crust, or partial crustal melting.

The *Lg* phase can also be dramatically attenuated by the sudden crustal structural change from thick continental crust to thinner oceanic crust, and a distance as short as 100 km across oceanic crust is sufficient to eliminate *Lg* wave propagation (Press and Ewing 1952). *Lg* wave propagation is also inefficient in areas with thinner crust, such as the Mediterranean, Black Sea, southern Caspian Sea (Press and Ewing 1952; Kadinsky-Cade et al. 1981; Maupin 1989; Rodgers et al. 1997), and the North Sea Central Graben (Gregersen 1984; Kennett et al. 1985).

Such variations in the *Lg* wave propagation in different parts of the world have been studied extensively within the contexts of investigating crustal heterogeneity, monitoring nuclear explosion tests, and discriminating between such tests and natural shallow earthquakes, etc. However, considerably less study has been done in the areas surrounding Japan.

The Japanese Islands are unique in that they are situated on a continental crust surrounded by oceanic crust, which means that *Lg* waves develop along limited paths. However, the amplitude behavior of the *Lg* waves in such combined environments has not yet been fully investigated. Utsu (1956) identified the arrival of *Lg* waves in the Galitzin and Benioff long-period seismograms recorded in Japan and later (1958) examined *Lg* wave propagation from the epicenters of Asian earthquakes to Japan. His 1958 study demonstrated that the *Lg* waves propagate to Japan from continental Asia across the Yellow Sea and the northern part of the East China Sea, but he also noted that *Lg* wave propagation is totally blocked across the Sea of Japan (East Sea). Next, Shima (1962) examined the regional variations of *Lg* wave propagation in Japan and noticed that the *Lg* wave propagation speed in northeastern Japan (3.57 km/s) is slightly faster than the propagation speed in southwestern Japan (3.44 km/s).

In the 40 years that followed, little work on *Lg* propagation in Japan was undertaken until the studies of Furumura and Kennett (2001) and Kennett and Furumura (2001), who developed an *Lg* wave propagation map for Japan using short-period seismograms from the *J-Array* short-period seismic network that extends across Japan (Yomogida 1996). Their studies demonstrated regional variations in *Lg* wave propagation in Japan that are characterized by efficient *Lg* wave propagation in western Japan and relatively inefficient *Lg* wave propagation in northern Japan. They also demonstrated that the *Lg* wave propagation along the Japanese Islands is attenuated as the waves traverse the Pacific Ocean and Sea of Japan in areas where water depths exceed approximately 1,000 m. In 2001, Kennett and Furumura conducted a two-dimensional (2-D) finite-difference method (FDM) simulation of regional wavefields along the Japanese subduction zone in order to investigate the nature of *Lg* wave propagation in the Japanese Islands across the continent-ocean transition area.

More than 10 years have passed since our previous studies (Furumura and Kennett 2001; Kennett and Furumura 2001), and during this interval, there has been considerable progress in both seismic monitoring and available computational capabilities. Therefore, in this study, we revisit our former studies into the nature of *Lg* wave propagation in Japan, with special emphasis on *Lg* wave propagation from outside Japan to the Japanese Islands across the Sea of Japan using a large amount of waveform data obtained from recently deployed high-density seismic networks in Japan and Korea, and supercomputer-based three-dimensional (3-D) FDM simulation of regional wave propagation by the use of a high-resolution crustal structure model.

In this paper, we first revise and extend earlier works (Furumura and Kennett 2001; Hong 2010) into *Lg* wave propagation and then present a detailed *Lg* propagation map of the area around Japan, including Korea and parts of continental Asia. We then conducted a 3-D FDM simulation of the regional seismic wavefield in the area around Japan using the Earth Simulator supercomputer (The Earth Simulator Center, Yokoyama, Japan), which is operated by the Japan Marine Science and Technology Center (JAMSTEC) and using the recently released *CRUST 1.0* crust model (Laske et al. 2013), in order to visualize the properties of the *Lg* wave propagation along the continental waveguide and the dissipation of the *Lg* waves caused by crustal barriers.

Tectonic environment in the area around Japan

The Japanese Islands are located in a tectonically complicated area where oceanic plates (the Pacific and Philippine Sea Plates) and continental plates (the North American and the Eurasian Plates) collide. As a result, the nation is subject to large numbers of earthquakes occurring along the plate boundaries and in the various plates. The northern half of the Japanese Islands are located on the North American Plate with the Pacific Plate subducting from the Japan Trench beneath northern and central Japan, which extends from Hokkaido to Honshu (Figure 1). The volcanic front associated with this subducting plate lies across the central Japanese island of Honshu, and a large number of active volcanoes extend from Hokkaido to central Honshu (Figure 1). The western part of the Japanese Islands is located on the Eurasian Plate of the Asian continent, where the Philippine Sea Plate subducts at a

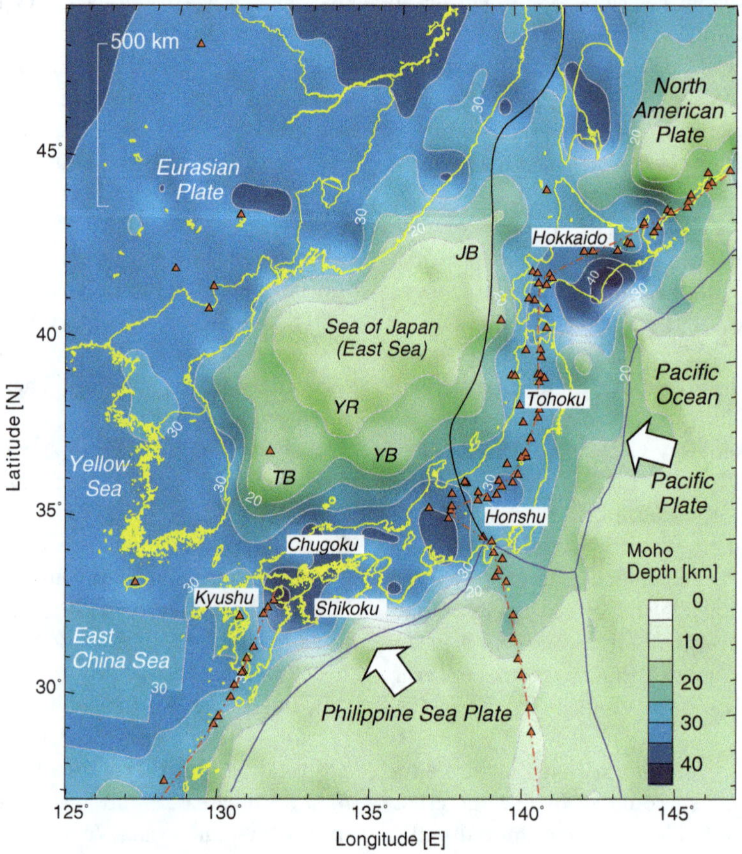

Figure 1 Tectonic environment and crust thickness in the area around Japan. Tectonic environment in the area around Japan. The blue lines denote the plate boundaries, and the gray lines illustrate the isodepth contours of the Moho depth in kilometers derived from the *CRUST 1.0* model (Laske et al. 2013). The red triangles and lines denote active volcanoes and the volcanic front, respectively. The major features in the Sea of Japan are marked as follows: JB, Japan Basin; YB, Yamato Basin; TB, Tsushima Basin; YR, Yamato Rise.

relatively gentle (<10° to 20) angle beneath Shikoku and Chugoku Islands, but at a much steeper (>50°) angle beneath Kyushu. The line of the volcanic front in western Japan lies across Kyushu and the Ryukyu Islands, where many active volcanoes are present; however, no volcanic front exists on the Shikoku or Chugoku Island. Therefore, the crustal structure of western Japan is more like a continental structure where efficient *Lg* wave propagation can be expected, similar to most areas of the Asian and Eurasian continents.

The Sea of Japan is a marginal sea surrounded by continental Asia, the Korean Peninsula, and the Japanese Islands. Here, an oceanic basin developed due to continental rifting during the Oligocene to mid-Miocene periods, at which time the Japanese Islands separated from the Eurasian Plate. However, the exact formation and evolution of the Sea of Japan remains with the subject of various proposed models, including an explanation involving a fan-shaped opening with different arc rotations (e.g., Otofuji et al. 1991), and a hypothesis that the Sea of Japan's formation was due to a pull-apart opening along the share zone in the eastern and western margins of the sea (e.g., Lallemand and Jolivet 1985). Based on an analysis of paleo-magnetic data (Otofuji et al. 1991), the age of the opening of the Sea of Japan has been estimated at 12 to 21 Ma.

The isodepth contour of the Moho depth derived from a recently released crustal model (*CRUST 1.0*) (Laske 2013) shows a transitional structure for the Sea of Japan with a thicker (>10 km) crust than is normal for ocean areas (about 5 to 7 km), but which is much thinner than the crust of the surrounding continental regime (about 30 to 35 km). There is a rapid transition in the Moho depth from continental structure (around 30 km) to approximately 20 km along the Sea of Japan border. The average and maximum water depths in the Sea of Japan are about 2 and 3.6 km, respectively, which are still much shallower than the depth of the Pacific Ocean to the east and south of the Japanese Islands (Figure 2).

In contrast to the thick (>35 km) and laterally homogeneous crustal structure of continental Asia, lateral changes in crustal thickness are dramatic along the Japanese Islands, particularly in the direction perpendicular across the width. The isodepth contours of the Moho depth in Figure 1

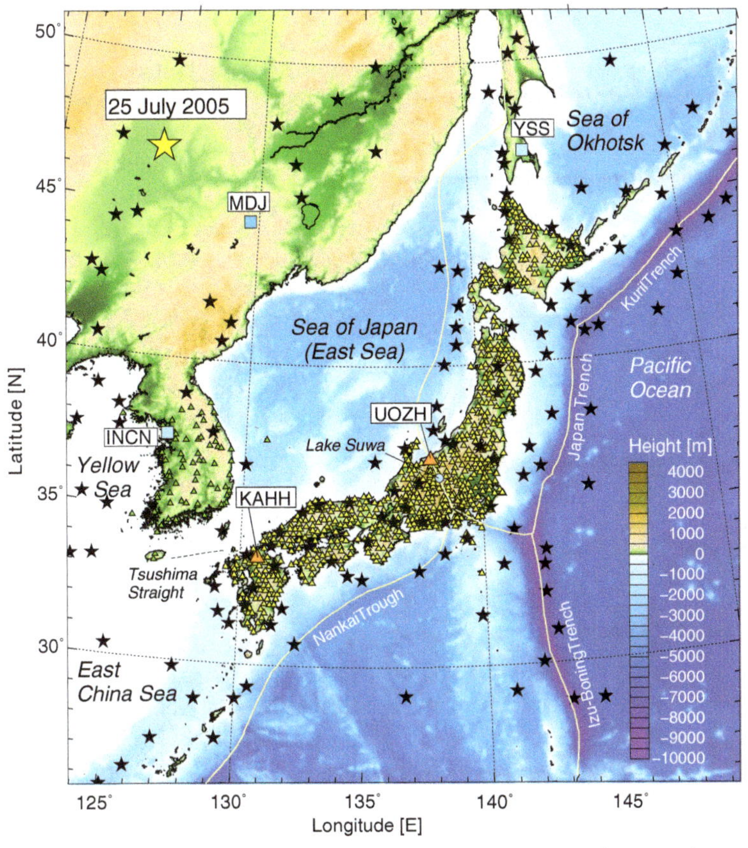

Figure 2 Distribution of seismic stations and earthquake epicenters used in this study. Map illustrating the stations in Hi-net in Japan (yellow triangles), the Korean network (green triangles), and Global Seismic Network (GSN) of the Incorporated Research Institutions for Seismology (IRIS) broadband stations (INCN, MDJ, YSS; blue squares) together with event epicenters (red stars) for the period from April 2001 to September 2013. The sea bottom topography is shown with a color scale.

display some deepened (>35 km) Moho depth zones in southern Hokkaido, Honshu, central Japan, and the area between Shikoku and Kyushu. Here, it can be seen that crustal thickness decreases rapidly with distance from the coast off the east and southwestern shorelines of the Japanese Islands towards the Pacific Ocean.

Methods and Results

Waveform analysis

Figure 2 shows the distribution of the Hi-net high-gain seismic network across Japan, which is operated by the National Research Institute for Earth Science and Disaster Prevention (NIED). This network consists of more than 800 stations with an average separation of approximately 20 km. Each station consists of a three-component short-period (f_0 = 1 Hz) velocimeter, and seismic data are continuously recorded at a sampling rate of 100 Hz (Okada et al. 2004; Obara et al. 2005). We also used the seismic stations the Global Seismic Network (GSN) of the Incorporated Research Institutions for Seismology (IRIS) in the study area, such as YSS in Sakhalin, MDJ in northeastern China, and INCN in Korea, as well as short-period and broadband stations of the Korean seismological network operated by the Korea Institute of Geoscience and Mineral Resources (KIGAM) and the Korean Meteorological Administration (KMA) (Park et al. 2009) to improve the ray path coverage in this study area. The stations employed and the epicenters of the events studied are plotted in Figure 2.

Figure 3 shows examples of three-component seismograms of transverse (T), radial (R), and vertical (Z) component ground motions recorded by the Hi-net stations. The upper panel (Figure 3a) shows the record of a Hi-net station in Honshu (UOZH; see Figure 2), and the lower panel (Figure 3b) shows the records of a shallow (h = 12 km) intermediate scale (Mw 5.0) earthquake that occurred in northeastern China on 25 July 2005 as recorded by a station in Kyushu (KAHH). The seismogram recorded by this station shows a clear *Lg* wave arrival as the largest signal on all three components, with a dominant frequency between 0.5 and 2 Hz. The *Lg* wave also has an extended wave train spanning the group velocity range from 3.5

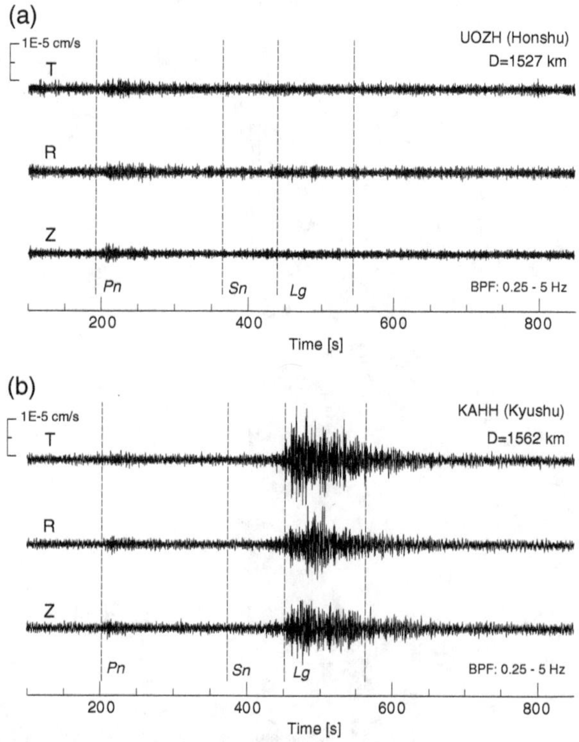

Figure 3 Example of Japan Hi-net record from a shallow crustal earthquake in northeastern China. Three-component seismograms for a 25 July 2005 event in northeastern China recorded at stations **(a)** UOZH in Honshu where the *Lg* wave is missing and **(b)** KAHH in Kyushu where *Lg* is seen clearly (see Figure 2 for locations). A band-pass filter with a passband between 0.25 and 5 Hz is applied to all records.

to 2.8 km/s due to the superposition of multiple crustal-reflected shear waves. Thus, this *Lg* wave can be recognized as having developed from a superposition of multiple post-critical *S* wave reflections propagating in the crust. Such clear *Lg* wave observations from a shallow *Mw* 5.0 event traveling more than 1,500 km might seem rather surprising, but the same *Lg* wave is very difficult to see in the record of a Honshu station (UOZH), which is at the same epicentral distance, even though the *Pn* phase amplitude is almost the same. These observations demonstrate the strong variation in both the crustal structure and nature of *Lg* wave propagation between continental Asia and Japan, through the Korean Peninsula, and across the Sea of Japan.

Usually, the mantle lid *S* wave (*Sn*) phase is detected in the regional seismograms of shallow events, but it is not clearly seen in the seismograms of both the above-mentioned stations. This is probably because of strong *Sn* wave attenuation during the long propagation of the lower *Q*s mantle structure in the back-arc side of tectonically active subduction zones.

Properties of the *Lg* waves around the Japanese Islands

Figure 4 shows a vertical (Z) component of seismograms recorded by 30 Hi-net stations in Japan, which extend along the coast of the Sea of Japan from the north (Hokkaido) to the south (Kyushu) through Honshu. Here, we show the features of *Lg* wave propagation to Japan from three shallow events that occurred in (a) northeastern China, (b) the Yellow Sea, and (c) central Japan. The travel-time curves for the *Pn*, *Sn*, and *Lg* phases are shown on the record sections by assuming group velocities of 7.7 km/s for *Pn*, 4.2 km/s for *Sn*, and 3.5 to 2.8 km/s for the *Lg* phases, along with the later arrival for the *Rg* phase (2.8 km/s). The amplitude of each trace is normalized by using the later *S* wave coda (specifically, the coda normalization technique; Aki and Chouet 1975) in order to compensate for localized site amplification at each station. In this figure, we can see strong variations in the regional wavefield for the different events, especially for the efficiency and inefficiency of *Lg* wave propagation along the paths from the sources to the stations across different tectonic environments. For example, it can be seen that there is efficient *Lg* wave propagation along the path from continental Asia to Kyushu and Hokkaido, but *Lg* wave propagation is suppressed for paths across the Sea of Japan. Additionally, *Lg* wave propagation is very efficient in southwestern Japan, but much less effective in northeastern Japan, even though the mantle *Pn* phases for all three events could be clearly observed at most stations. Although the mantle *Sn* phase was clearly observed with propagation in western Japan (Figure 4c) and from the Korean Peninsula to Kyushu (Figure 4b), *Sn* phase attenuation was very strong in northern Japan and along the path across the Sea of Japan (Figure 4a).

We measured the efficiency of the *Lg* wave propagation along each path, from the sources to the monitoring stations, by taking the ratio of average *Lg* amplitude to average *Sn* amplitude in the group-velocity windows 4.6 to 4.2 km/s for *Sn* and 3.5 to 2.8 km/s for the *Lg* wave, and over the three components of the seismograms. In Figure 4, we show ray paths that have good *Lg* propagation paths ($Lg/Sn > 1.5$) in green and less efficient ($1.0 < Lg/Sn < 1.5$) propagation paths in orange. The propagation paths with very weak or missing *Lg* ($Lg/Sn < 1.0$) are plotted in red. Note that in the area where the *Sn* wave is very small and comparable to the background noise level, the *Lg*/*Sn* ratio measures the strength of the *Lg* wave relative to the noise occurring just prior to the *Lg* wave arrival.

We can immediately see that the *Lg* propagation is efficient along the paths from the Asian continent to Kyushu and Hokkaido that do not pass through the Sea of Japan. However, there is a sharp *Lg* wave signal cutoff in the record section of Figure 3a for stations in the east of Kyushu, even though the corresponding decline in *Lg* efficiency in

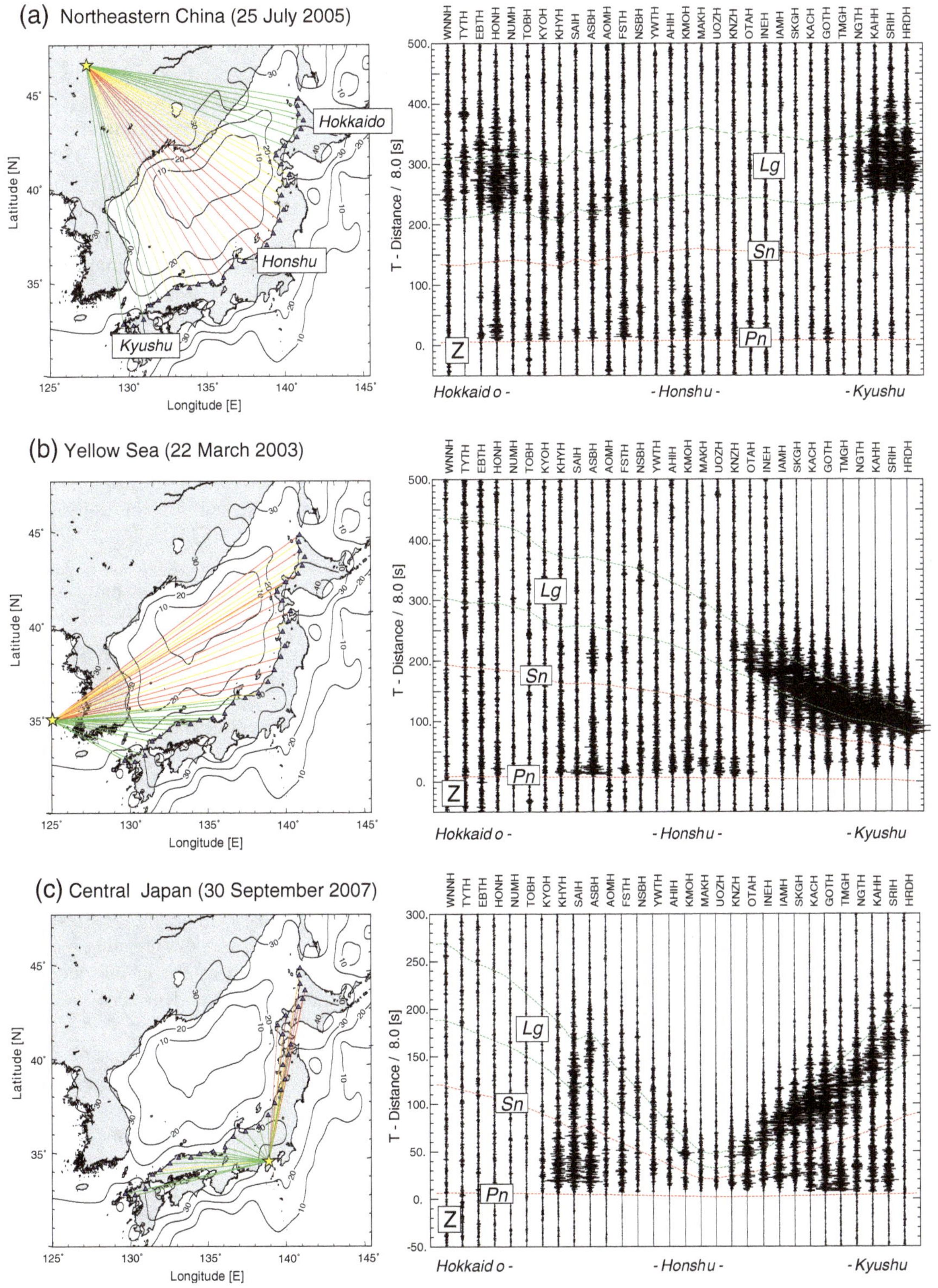

Figure 4 *Lg* wave propagation from the Asian continent to Japan and along the Japanese Islands. Variation in the *Lg/Sn* ratio along the paths from the event to recording stations and the corresponding vertical component seismograms recorded at Hi-net stations from Hokkaido to Kyushu. Isodepth contours of the Moho depth for 10, 20, 30, and 40 km are shown in the *Lg* propagation map. Efficient *Lg* paths (*Lg/Sn* > 1.5) are shown in green, inefficient *Lg* (1.0 < *Lg/Sn* < 1.5) are shown in orange, and no *Lg* (*Lg/Sn* < 1.0) shown in red. The frames display events in **(a)** northeastern China, **(b)** the Yellow Sea, and **(c)** central Japan. The travel times of *Pn*, *Sn*, and *Lg* are marked. The Hi-net station code is shown at the top of the seismogram sections.

southern Hokkaido is not as drastic. This suggests that the transition from the continental to oceanic crustal structure along the northern margin of the Sea of Japan is not as sharp as that along the east close to the Korean Peninsula. Recent marine survey investigations involving refraction and reflection experiments have revealed that the crustal thickness beneath the northeastern side of the Sea of Japan (Japan Basin; see Figure 1) is roughly 10 to 12 km (Hirata et al. 1992), which is thicker than a typical ocean crust (about 5 to 7 km) and that there is a large lateral crustal structure variation in the center of the Sea of Japan where the crustal thickness sudden grows to 22 km beneath the Yamato Bank (Kurashimo et al. 1996). The intermediate variations between the ocean and continent crustal structures beneath the Sea of Japan are associated with the lithospheric deformation that occurred during the opening of the sea, which resulted from continental rifting during the Oligocene and Miocene.

The efficient propagation of *Lg* waves across the Yellow Sea to Japan, which was first noted by Utsu (1958), is seen in Figure 4b. This observation is consistent with the continuity of continental Asia to the Korean Peninsula, the extension of the continental structure to the Yellow Sea, and to the north of the East China Sea.

In Japan, the *Lg* wave propagation is very clear in the western part of the country, but is not clearly seen from central to northern Honshu Japan, Hokkaido, and in southern Kyushu (Figure 4c), as pointed out by Furumura and Kennett (2001) and Kennett and Furumura (2001). These areas of inefficient *Lg* propagation correlate well to the zone of low-Q anomaly in the crust (Nakamura and Uetake 2002; Sekine 2005). The *Lg* waves traveling to Hokkaido across the Pacific Ocean and the Sea of Japan for more than 100 km dissipate completely.

Tomographic inversion of *Lg* propagation efficiency

We chose 149 events in the area spanning 125°E to 145°E and 25°N to 55°N, with source depths less than 33 km and magnitudes greater than Mw 4.5 that occurred during the period between April 2001 and September 2013 (Figure 2). The coverage of 289,000 ray paths in total, from sources to stations, provides dense sampling over the study area. More than 500 of the ray paths sampled cross the Japanese Islands entirely, and more than 100 ray paths (covering most parts of the study area) travel across the Sea of Japan. However, relatively few ray paths (<20) are found in the outer zone of the study area, such as in northeast China, the Yellow Sea, the East China Sea, the Sea of Okhotsk, and the eastern and southern Pacific Ocean. This might cause uncertainties when attempting to map *Lg* wave propagation properties in those outer areas (Figure 5).

We then applied a conventional tomography technique, following the procedure described by Furumura and Kennett (2001), in order to construct a detailed *Lg* propagation map across the study area. This tomographic inversion is based on the simultaneous iterative reconstruction technique (SIRT) such as that described by Kak and Slaney (1987) in their textbook, which was introduced for early medical computer tomography (CT) usage. Here, it is used to map the *Lg*/*Sn* ratio value along a path with inversion for the local properties of the *Lg* wave transmission and blockage using 0.2° × 0.2° cells. The iteration is repeated 10 times in situations where the assignment of *Lg/Sn* ratio value to the cells along the ray path from the hypocenter to the station, and smoothing values in the neighboring eight cells are applied to stabilize the inversion result and create a smooth reconstruction image (see Furumura and Kennett (2001) for details). This SIRT-based inversion procedure has a modest conversion rate but works well with the large sparse system associated with *Lg* paths. Many other least-squares inversion approaches can be used, but those require a suitable damping procedure to stabilize the results.

We first assess the inversion resolution by looking for the reconstruction of a single-cell (0.2° × 0.2°) anomaly distributed across the study area on a 5° × 5° grid. Figure 6 shows the results obtained from this point-spreading test, which indicates that even with the strong path coverage, there are possibilities for preferential anomaly smearing in certain directions. The source and station distribution offers fairly good coverage over the Japanese Islands from northeast to southwest, but there are fewer ray paths in the perpendicular direction. In consequence, there is a tendency for northeast-to-southwest resolution smearing to appear on the inversion map. For the same reason of abovementioned dominant ray path direction, there is a trend towards northwest-to-southeast smearing for *Lg* propagation anomalies on continental Asia. However, resolution in the Sea of Japan and in the Korean Peninsula is well controlled, with anomaly spreading recorded at less than 0.5° × 0.5°. In northeastern China, where ray path coverage is rather sparse (less than 200; see Figure 5), the point anomalies spread out to 1.5° × 1.5°.

The *Lg* propagation efficiency map results are shown in Figure 7 together with the crust thickness (the distance between surface/sea bottom and Moho), which provides a summary of the *Lg* wave propagation from continental Asia to Japan and around the Korean Peninsula. The green color on the map indicates areas where *Lg* wave propagation is efficient, and the red color indicates the least efficient *Lg* wave propagation area. Compared with the earlier study (Furumura and Kennett 2001), which employed data obtained from *J-Array* stations, the resolution of the present map, which uses the Hi-net and Korean stations, represents a considerable improvement, and it can be seen that there is intense and almost uniform ray path coverage from continental Asia to Japan and across the Japanese Islands. Nevertheless, the *Lg* wave

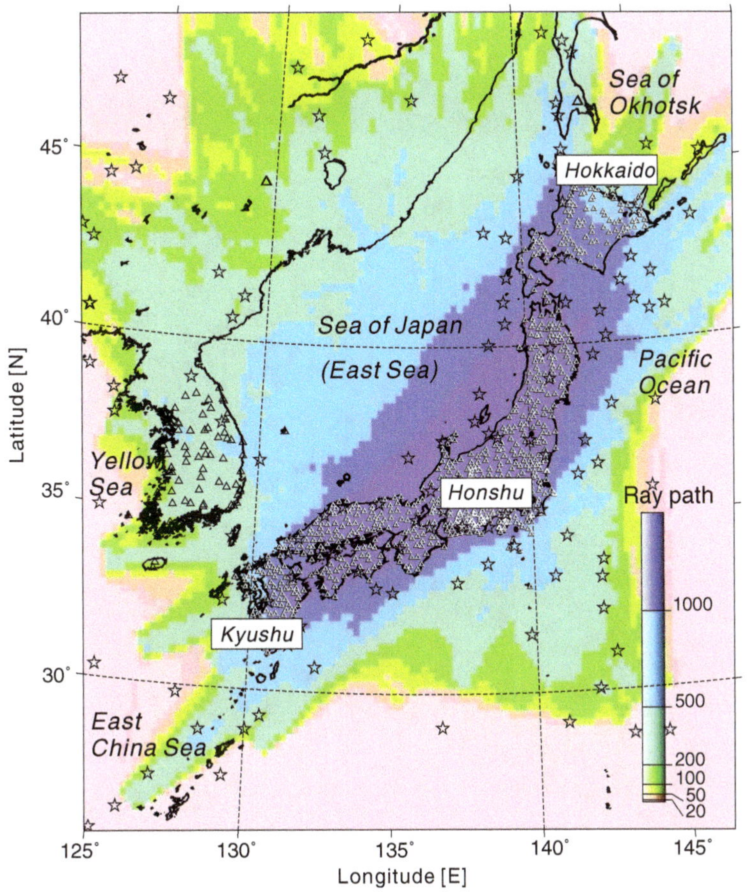

Figure 5 Ray density map of the study area. Map of ray path density counted in each 0.2° × 0.2° cell showing the coverage and density of the ray paths in the study area. Source (squares) and stations (triangles) are shown.

propagation patterns shown are in good agreement with former studies (Furumura and Kennett 2001; Hong 2010) in their areas of coverage.

Also in Figure 7, we can see that the area of efficient *Lg* propagation continues from Asia to Kyushu through the Korean Peninsula across the Tsushima Strait and extends further west to the Yellow Sea and to the north of the East China Sea. In addition, an extension of relatively good *Lg* propagation is seen in the area from continental Asia to Hokkaido.

For the areas surrounding Japan, it can be seen that there is a sharp *Lg* wave propagation cutoff along the edge of the Sea of Japan where the crust thickness suddenly changes from a thick (>30 km) continental crust to a thinner (<10 km) oceanic crust (Kurashimo et al. 1996), and it has been confirmed that the *Lg* wave dissipation in the Sea of Japan starts when the crustal thickness decreases below 30 km. The lighter red color in the northern half of the Sea of Japan indicates relatively weaker attenuation of the *Lg* wave than in the south, which is shown in dark red.

We also find that along the continent to oceanic boundary on the Pacific Ocean side of Japan (which corresponds roughly to the 1,000-m isodepth contour of the water depth), the *Lg* wave propagation is totally blocked. The darker red color in the Pacific Ocean indicates that the *Lg* attenuation across the Pacific Ocean is much stronger than that across the Sea of Japan.

On the west sides of the Japanese Islands, there is a good *Lg* propagation area that runs from central Honshu to northern Kyushu and then connects to continental Asia through the Korean Peninsula. In contrast, the *Lg* wave propagation is poor in northern Japan from central Honshu to southern Hokkaido, as well as in southern Kyushu. These findings are consistent with the earlier results of Furumura and Kennett (2001), and the area corresponds to the anomalously low Q_{Lg} (<200) zone derived in the inversion by Hong (2010).

Such areas of inefficient *Lg* wave propagation in Japan might relate to low-*Q* anomalies in the crust beneath active volcanoes (shown as triangles in Figure 7), irregular topography at the surface and the Moho, and thick sediment cover. The net results of their effects are to disrupt coherent *Lg* reflection patterns in the crustal waveguide.

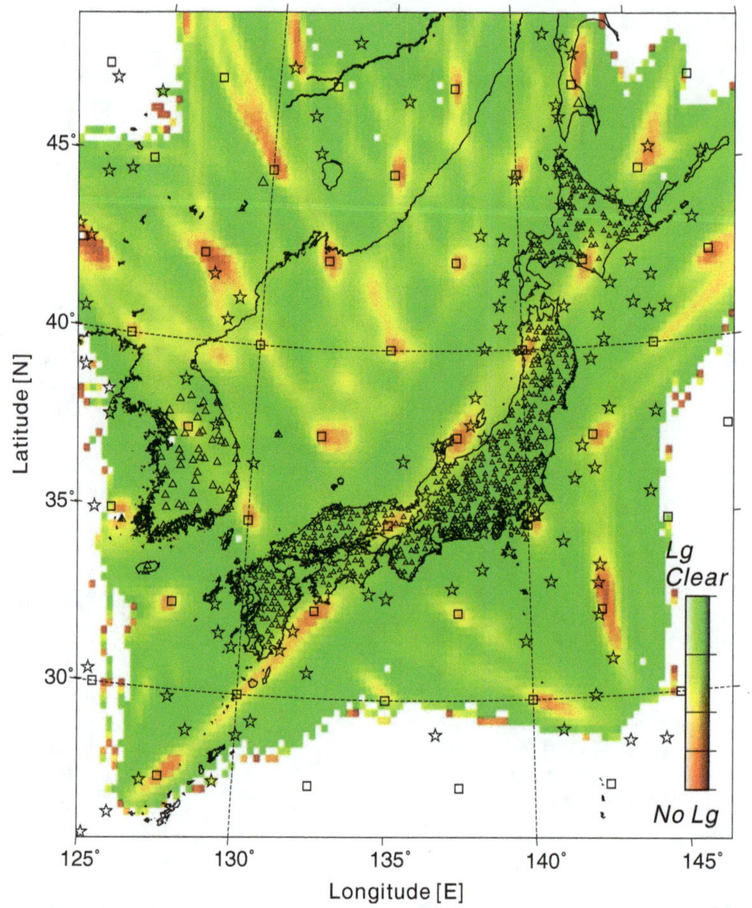

Figure 6 Resolution test for the inversion analysis of the *Lg* wave propagation. Resolution test for *Lg* wave propagation map shown in Figure 5, illustrating the smearing pattern of the point (0.2° × 0.2°) anomalies placed at every 5° × 5° (blue squares).

Numerical investigation

Numerical simulation of Lg wave propagation

In order to understand the processes by which an *Lg* wave is able to travel large distances (more than 1,000 km) in the continental crustal waveguide, yet can be totally blocked by crustal barriers, we have produced a 3-D FDM simulation of seismic wave propagation in the study area using a heterogeneous crust and topography model based on the recently released *CRUST 1.0* structural model (Laske et al. 2013). We then considered a wave propagation from the 25 July 2005 earthquake in northeastern China discussed in Figures 2, 3, 4a. The simulation area covers a horizontal zone of 1,900 × 2,000 km and 96 km in depth, which has been discretized with a uniform grid size of 0.5 km in the horizontal direction and 0.25 km in the vertical direction (Figure 8). The heterogeneous crust and upper-mantle structure of the FDM simulation was constructed by assigning Lame's constants that relate to *P* and *S* wave speed (*Vp* and *Vs*, respectively), density (*r*), and the anelastic attenuation constants for *P* and *S* waves (*Qp* and *Qs*, respectively) at each grid point of the 3-D simulation model. The seismic wave propagation at each grid point is calculated explicitly with increasing time steps and by solving the equation of motions using an eighth-order staggered-grid FDM in the horizontal directions and a fourth-order staggered-grid FDM in the vertical direction, where the smaller (0.25 km) grid is applied. The *CRUST 1.0* structural model (Laske et al. 2013) provides depth distributions for topography, sedimentary layer, upper crust, middle crust, and lower crust, as well as the *P* and *S* wave speed and density at a resolution of 1° × 1°. The *CRUST 1.0* model is expected to provide significant improvements over earlier crustal models in the area around the Sea of Japan, along the Japanese Islands (Ryoki 1999) and around the Korean Peninsula (Chang and Baag 2005) with the results of recent marine and land geophysical experiments. However, we did not adopt the sedimentary layer function of the *CRUST 1.0* model since we had confirmed that the sediments in the study area were not uniformly covered by the present *CRUST 1.0* model, but were instead restricted to those areas where marine experiments have

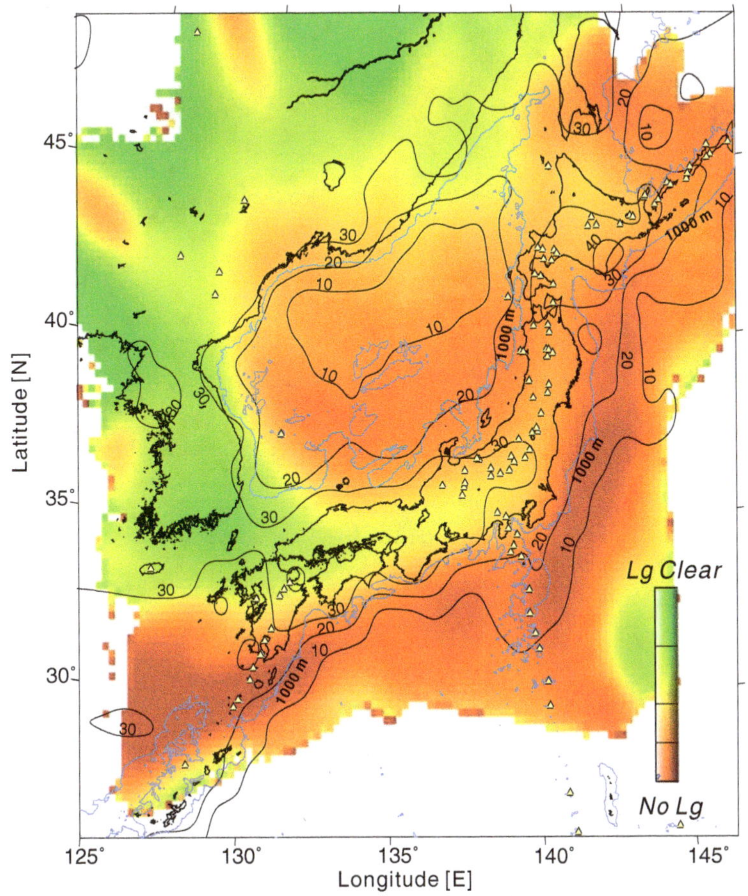

Figure 7 ***Lg*** **wave propagation map in the area around Japan.** Summary map of the variation in the *Lg* wave propagation in the area around Japan derived from the inversion of the *Lg/Sn* ratio. Green marks efficient *Lg* propagation, and red marks indicate inefficient *Lg* propagation. Triangles denote major active volcanoes. Black lines denote isodepth contours of the crustal thickness at 10-km intervals, and blue lines correspond to the 1,000-m contour of sea depth.

been carried out. Although low-wave speed and low-*Q* sediments also have the potential to significantly effect *Lg* wave attenuation, the present simulation will only examine how changes in crustal thickness and sea depth affect *Lg* wave propagation.

The crustal model (Figure 8) shows strong lateral crustal structure changes with the crust thinning from a thick (30 to 35 km) continental structure to a relatively thin (10- to 20-km-thick) oceanic structure along the outer boundary of the Sea of Japan, and a much thinner (5-to 10-km-thick) crust along the coastal edge of the Pacific Ocean. Along the Japanese Islands, a somewhat thicker crust (30 to 35 km) extends from the northern island of Hokkaido to the southern island of Kyushu (with some bumps along the Moho) and then continues to the continental crust beneath the Yellow Sea, the northern part of East China Sea, and continental Asia, through the Korean Peninsula. There is some particularly noticeable Moho topography at the northeastern margin of the Sea of Japan between continental Asia and Hokkaido.

In the FDM simulation, the model has a minimum *P* wave speed ($Vp = 1.5$ km/s) in seawater and a minimum *S* wave speed ($Vs = 3.5$ km/s) in the upper crust, which allows for *S* wave propagation simulations at frequencies less than 1.5 Hz, with a sampling rate of 3.5 grid points per minimum *S* wavelength.

Since the *CRUST 1.0* structural model was believed to be too simple to represent the scattering of high-frequency waves, which play an important role in shaping the character of high-frequency seismic signals above 0.5 to 1 Hz, we superimposed stochastic random heterogeneity to deviate *P* and *S* speeds of the *CRUST 1.0* model of the crust and upper-mantle. We used a von Karman stochastic random heterogeneity distribution function with a longer horizontally correlation distance ($a_x = 10$ km) than vertical ($a_z = 0.5$ km) and a 2% standard deviation of *P* and *S* wave speeds. In the upper mantle, we use a somewhat larger scale of heterogeneity ($a_x = 20$ km, $a_z = 1$ km, and 2% standard deviation) than that in the crust.

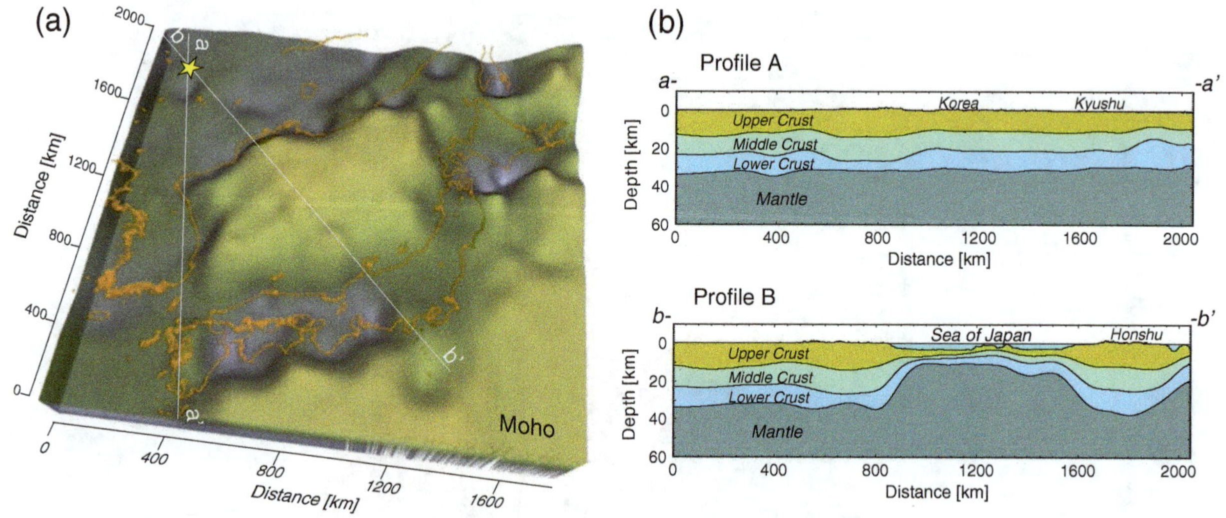

Figure 8 3-D model of crustal structure used in the FDM simulations. (a) 3-D view of Moho depth. **(b)** Vertical profiles of crustal structure along profile A and profile B showing lateral crustal structure variations.

Anelastic attenuation for both *Qp* and *Qs* is employed in the FDM simulation using the frequency-independent Q model implemented for the FDM simulation created by Robertsson et al. (1994). After some preliminary experiments, we finally selected the anelastic parameters of $Qs = 400$ for the crust and $Qs = 350$ for the mantle, with $Qp = 2 \times Qs$ calculated by comparing the relative strength of the *Pn*, *Sn*, and *Lg* phases in the observed seismograms. The relatively low *Qs* (equal to 350 to 400) compared with that of the normal value (e.g., $Qs = 1{,}000$) in the continental structure might correspond to the low-*Q* mantle lid below the Sea of Japan. The lower *Qs* in the crust and mantle are consistent with those of the Japan Seismic Hazard Information Station (J-SHIS) standard velocity model (Azuma et al. 2013) corresponding to the tectonics of the young and tectonically active Japanese Islands. We assigned a large *Qp* (equal to 10,000) for seawater.

Following the Harvard centroid moment tensor (CMT) catalog solution (Harvard Seismology Group, Cambridge, MA, USA), a double-couple reverse-fault point source for the 25 July 2005 northeastern China earthquake (see Figure 4a) is placed in the upper left of the simulation model at a depth of 12 km. This source radiates seismic wave into the crust with a maximum frequency of $f = 1.5$ Hz.

The 3-D FDM simulation was conducted on the Earth Simulator supercomputer (Earth Simulator Center) while employing 1.02 TByte of memory for a single-precision calculation with a wall-clock time of 6.5 h using 32 supercomputer nodes (256 vector processors) and 50,000 time steps in order to carry wave propagation out to 750 s after initiation. Snapshots of seismic wave propagation derived from the FDM simulation for the northeastern China earthquake that display the way in which the regional wavefield spreads from a shallow source in the continent are shown in Figure 9. The simulated wavefield is separated into the *P* wave contribution by taking the divergence of the 3-D wavefield plotted in red, and the *S* wave contribution is plotted by extracting the curl (rot.) of the 3-D wavefield shown in green. The wave amplitudes in the snapshot are multiplied by the hypocentral distance to each grid point in order to compensate for the geometrical attenuation of the body waves and to improve the visibility of the seismic phases in the later timeframes.

In the first frame of the snapshot taken 180 s after the earthquake initiation (Figure 9a), the spreading of the *P* and *S* waves from the source shows an isotropic distribution. The original radiation pattern of the *P* and *S* waves is largely eliminated by multiple reflections and scattering in the heterogeneous subsurface structure. As time passes, the *Lg* wave builds up as a sequence of multiple post-critical *S* wave reflections in the crust, which then appear in the snapshot along with a spread of *S* wave energy over continental Asia for more than 400 km.

For the *Pg* wave, which is a sum of multiple wide-angle *P* wave reflections in the crust (somewhat like the *Lg* wave for *S*), the attenuation is rather strong and difficult to see in the next ($T = 280$ s) time frame of the snapshots. This is because the *P* wave reflection is not totally occurred when the *P* waves encounter the free surface and Moho, so some of the *P* wave energy is successively transmitted to the mantle by *Pg*-to-*S* conversion when bouncing from the Moho.

As the *Lg* wave emerges from continental Asia and propagates into the Sea of Japan, its progression suddenly attenuates along the thinner crust (Figure 9c; $T = 380$ s). Following the continental to oceanic transition, the circular wavefront of the *Lg* wave cuts sharply along the eastern coast of Korean Peninsula and between the Asian

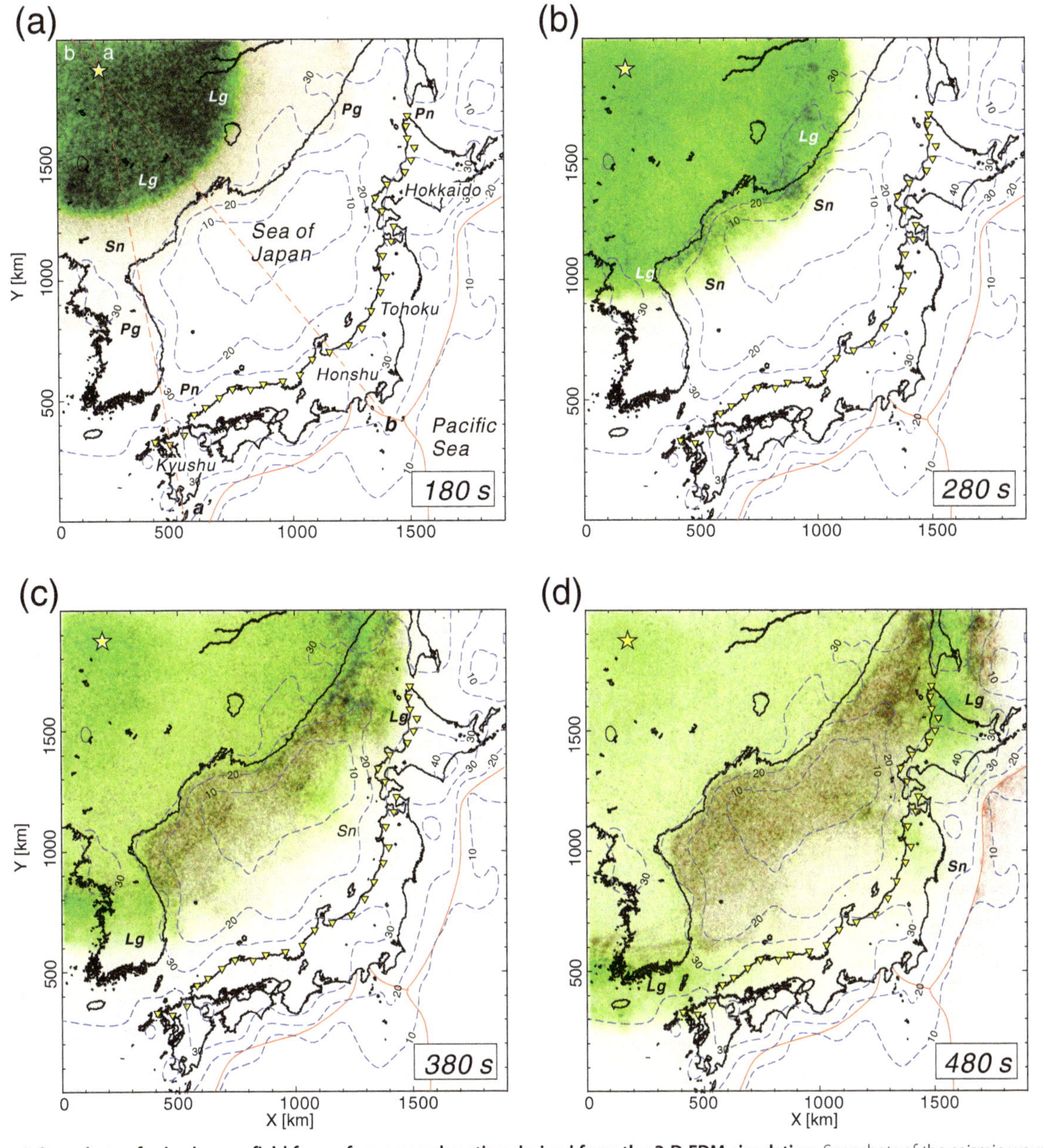

Figure 9 Snapshots of seismic wavefield for surface ground motion derived from the 3-D FDM simulation. Snapshots of the seismic wavefield of the surface ground motion at time $T = 180$ s **(a)**, 280 s **(b)**, 380 s **(c)**, and 480 s **(d)** from the earthquake start. *P* waves are shown in red, and *S* waves are shown in green. Isodepth contours of the crustal thickness are shown by purple lines. Red lines denote plate boundaries. Major phases are marked.

continent and Hokkaido (between the Eurasian and North American plates). We conjecture that the transmission of *Lg* wave energy into the mantle from the slope of the thinning crustal boundary and that multiple reflections of the *Lg* wave within the thinner crust of irregular seafloor and Moho boundaries cause the *Lg* wave to attenuate rapidly. The spread of red color (*P* wave) across the Sea of Japan with a long duration in the sea indicates that *Lg*-to-*P* conversion has occurred at the crust-sea interface, which also removes seismic energy from the *Lg* wave in the thinner crust.

Since Press and Ewing (1952), it has been recognized that anything more than 100 km of propagation along a purely oceanic path is sufficient to totally eliminate an *Lg* wave. Observations of weak *Lg* waves crossing 300 km of the middle of the Sea of Japan occur because the crust is much thicker (10 to 20 km) than a normal ocean crust (7 km).

The last frame of the snapshot ($T = 480$ s) illustrates the remainder of the *Lg* wavefront traveling in northern Hokkaido and arriving in western Kyushu with a long tail. This confirms our observations of efficient *Lg* wave propagation from continental Asia to Japan (with a propagation distance of more than 1,500 km) and inefficient propagation to Honshu across the Sea of Japan. There is also a pronounced area extending through Honshu from central Japan to the northern Tohoku region in which the *Lg* wave propagation is inefficient.

A segment of the *S* wave signal with a long coda can be found in northern Honshu (Tohoku) in the last frame ($T = 480$ s). This was developed by strong *S* wave focusing along the concave crustal structure at the northwestern margin of the Sea of Japan and propagation in the mantle as an *Sn* wave to the Tohoku region.

In Figure 10, synthetic seismograms of the vertical-component ground velocity from the simulation are compared with Hi-net record observations for the northeastern China event, along with the application of a band-pass filter with corners at 0.5 to 1.5 Hz. In the simulation, we can see the arrival of a clear *Lg* wave train in Kyushu and Hokkaido, but no *Lg* wave can be seen in the Honshu stations, just as we have seen in the Hi-net data. However, the amplitude of the simulated *Lg* wave in Kyushu is relatively small when compared with the actual observation. We tried to solve this problem by taking into consideration uncertainties in the estimated focal mechanism and depth of the source which might modifying slightly the *S* wave radiation pattern, but the relative weakness of the *Lg* wave in Kyushu did not show significant improvement. We believe that the problems arise from subtle issues in the structural model, such as the delicate curvature of the Moho along the continental to oceanic boundary.

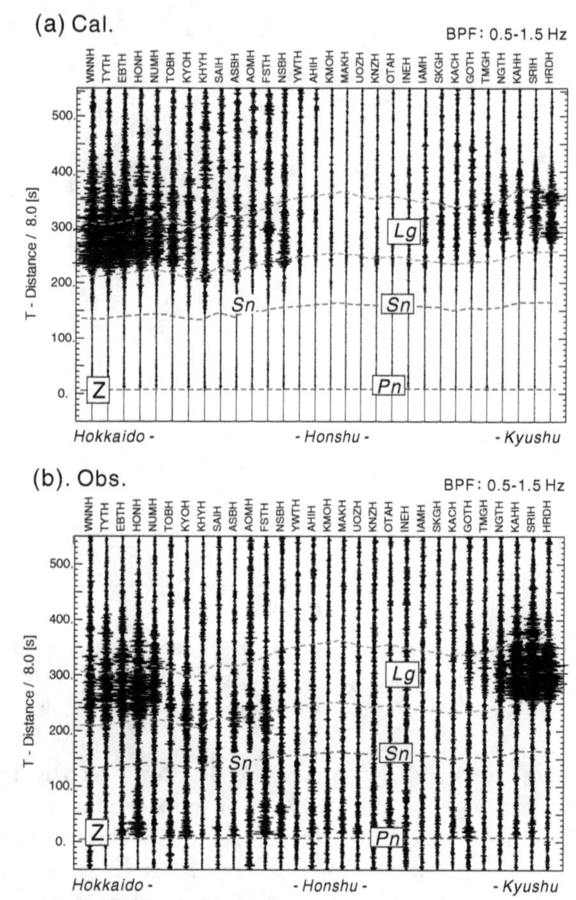

Figure 10 Record sections derived by the 3-D FDM simulation and Hi-net observations. Comparison of record sections of vertical-component ground motion for station from Hokkaido (left) to Kyushu for **(a)** synthetic seismogram derived by the FDM simulation and **(b)** observed seismograms at Hi-net stations. A band-pass filter with a pass band frequency between 0.5 and 1.5 Hz is applied to both records to facilitate comparison.

A larger *Sn* wave with a long coda that was developed by *Lg*-to-*Sn* wave conversion, which results from the strong focusing and defocusing effect imposed by the irregular Moho interface and heterogeneous crustal structures, is seen in the snapshots (Figure 9c,d). This wave appears across numerous stations between Hokkaido and Tohoku within the time window expected for the *Sn* and *Lg* waves. These long-tailed converted *Sn* waves are more clearly seen in the high-frequency seismograms of the Hi-net (see Figure 4a) in Hokkaido (e.g., KHYH) and in the Tohoku region (e.g., ASBH).

The present FDM simulation results obtained using the *CRUST 1.0* model well explain the broad features of the observed regional wavefields that develop in continental Asia and travel to Japan across the Sea of Japan. However, further refinement of the simulation model is likely to be necessary in order to match the observations more consistently. For example, it would appear that there are problems with our understanding of the continent-ocean transition because the continental margin off the east coast of the Korean Peninsula appears to be too sharp and cuts out *Lg* energy that should arrive in Kyushu (snapshot Figure 9d). On the other hand, the transition at the northeast margin of the Sea of Japan may need to be sharper than in the current model in order to match the rapid drop in *Lg* amplitude seen in southern Hokkaido.

Features of regional phases

In order to show how *Lg* wave propagation patterns are drastically modified by interaction with crustal heterogeneities and can be totally blocked by crustal barriers such as the central Sea of Japan, we show a set of snapshots in a vertical section by cutting the FDM model along a profile from the source to a station, together with a record section of synthetic seismograms along the profile.

Figure 11 shows a snapshot of the seismic wavefield along the profile from continental Asia to Kyushu (where the *Lg* propagation is very efficient) together with synthetic seismograms of vertical-component ground motions along the profile (profile A; see Figures 8a,b and 9c). Each seismogram trace is multiplied by the epicentral distance in order

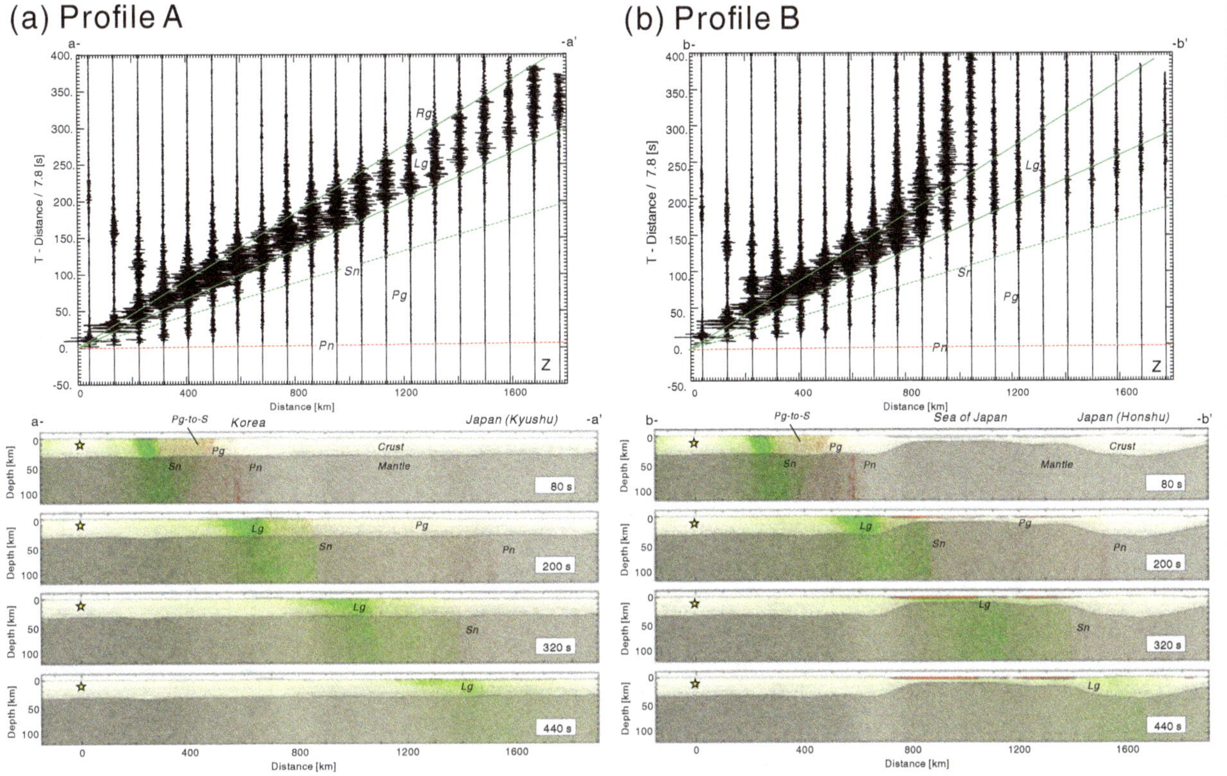

Figure 11 Record section and snapshot derived by FDM simulation along profiles A and B. Record section of vertical-component seismogram along **(a)** profile A (to Kyushu) and **(b)** profile B (to Honshu across the Sea of Japan) and the snapshots of *P* (red) and *S* (green) wavefields at 80, 200, 320, and 440 s from earthquake start. Theoretical travel time curves for the major phases are shown.

to compensate for the geometrical spreading of body waves.

The first frame of the snapshot ($T = 80$ s) displays the *Lg* wave train build up resulting from multiple post-critical *S* wave reflections between the free surface and Moho. As can be seen in the figure, *Lg* wave propagation is very efficient in the continental structure, which is able to carry *S* wave energy over distances in excess of 1,500 km. The multiple forward scattering of the high-frequency waves in the crust from the horizontally elongated configuration of the small-scale heterogeneities also helps trap high-frequency ($f > 1$ Hz) *Lg* wave energy in the crustal waveguide. In the second frame of the snapshot ($T = 200$ s), the *Sn* wave separates from the *Lg* wave due to a faster *S* wave speed in the mantle.

A weak *Pn* wave can be found propagating in the crust and mantle as a head wave while keeping nearly the same amplitude in all frames. The attenuation of the *Pg* phase propagating in the crust is very strong because large *Pg*-to-*S* conversion occurs at the free surface and the Moho (shown in green between *Pg* and *Sn* in the $T = 80$- and 200-s frames), which removes the *Pg* wave energy from the crust down to the mantle. In the $T = 200$-s time frame, it is hard to see the *Pg* wave signal. It is also recognized that the *Pg* wave often develops with a low-wave speed layer cover below the surface (see Olsen et al. 1983), but it might be underestimated in the present simulation where no sedimentary layer is present. The record section of the vertical-component ground motion along the path from the Asian continent to Kyushu along the Korean Peninsula illustrates the band of *Lg* waves developed by a superposition of the *S* wave reflections in the crust, which leads to an extended wave train with a group velocity range 3.5 to 2.8 km/s. Some *Lg* wave attenuation occurs beyond 1,300 km when the waves cross the sea between Korea and Kyushu.

Lg crustal barrier blockage

We will now look at the process by which the crustal barrier beneath the Sea of Japan blocks the *Lg* wave propagation. In Figure 11b, we show snapshots of a vertical section of the simulated seismic wavefield, in addition to synthetic seismograms along the profile from continental Asia to Honshu across the Sea of Japan (profile B; see Figures 8a,b and 9c).

These snapshots demonstrate the process by which the sudden upheaval of the Moho from 35 to 10 km at the continental to oceanic margin (over a distance of 100 km) and the extension of the thinner (10 km thick) crust over 600 km blocks the *Lg* wave propagation. This blockage occurs primarily by transferring the *Lg* wave energy into the mantle as a consequence of the rising Moho slope (see the $T = 200$- and 320-s frames), but the multiple *Lg* wave reflections in the thinner crust between the sea bottom and at the rough topography of the Moho top also leads to a dramatic decline of the *Lg* wave energy in the waveguide. In addition, significant *Lg*-to-*P* conversion occurs at the sea bottom due to the multiple reflections in the thinner crust, which also leads to a larger removal of *Lg* wave energy from the waveguide (see the red color in the sea in $T = 320$-s frame). The transmitted *S* wave propagates in the mantle as an *Sn* wave with a faster speed, and some of this energy is returned to the crust when the *Sn* waves impinge on the thick continental structure beneath Honshu (see the $T = 440$-s frame). This converted *Sn*-to-*Lg* energy arrives earlier than would be expected for the *Lg* wave itself. Some *Lg* wave energy dissipates into the water of the Sea of Japan due to *Lg*-to-*P* conversion at the sea bottom, and some of this energy can also be transferred to the crustal waveguide at a later time and results in the development of a long *Lg* wave coda. The overall attenuation of the *Lg* wave is very significant because of its interaction with the crustal barrier, which results in a sudden thinning of the crustal waveguide from 35 to 10 km over a distance of 100 km, and due to the more than 600 km extent of a thinner (10 km) waveguide, which results in *Lg* wave energy dissipation into the seawater by *Lg*-to-*P* conversion. Consequently, very weak *Lg* waves can be seen in the right corner of the snapshot in the last ($T = 440$ s) time frame.

The synthetic seismograms of the vertical ground motion component along the profile across the crustal barrier demonstrate the dramatic attenuation of the *Lg* wave energy as it crosses the Sea of Japan over a 1,400-km distance from the hypocenter onwards. The loss of the *Lg* wave energy as it propagates across the crustal barrier is significantly enhanced when compared with that of an *Lg* wave propagating in a predominantly continental environment (Figure 11a). The long tail of the *Lg* coda is the result of *P*-to-*Lg* conversions at the sea bottom and is caused by multiple *P* wave reverberations in the seawater, as we observed in the seismic wavefield snapshots. We also note an enhancement of the *Sn* wave and its coda at points above the crustal barrier (1,100 km from the hypocenter) as *Lg* wave energy is transmitted into the mantle.

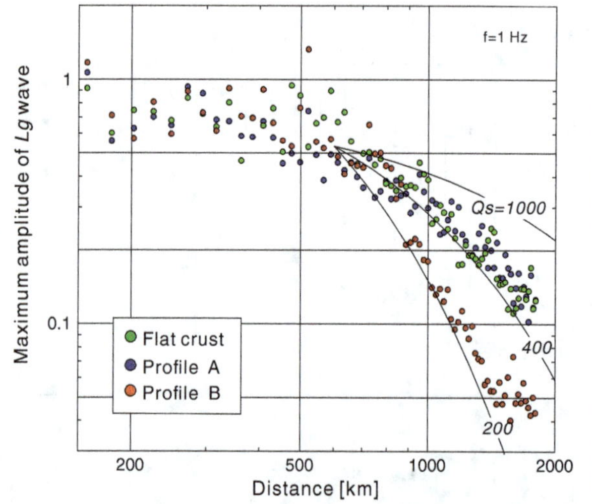

Figure 12 Attenuation functions of *Lg* wave along profiles A and B. Comparison of the attenuation functions of the maximum amplitude of the *Lg* wave of 1 Hz for the synthetic seismogram derived from the FDM simulation as a function of distance from the hypocenter along profile A (blue circles) and along profile B (red circles). The attenuation function for a laterally homogeneous flat-crust model is also shown (green circles). The theoretical attenuation curves for the *Lg* wave in anelastic media ($Qs = 1000$, 400, and 200) is also displayed by solid lines.

Lg attenuation

In Figure 12, we compare the apparent attenuation of *Lg* waves along paths crossing the continental crust (profile A) and across the Sea of Japan (profile B) as a function of epicentral distance. To accomplish this, we measure the averaged root-mean-square (RMS) amplitude of the *Lg* wave on the seismograms for frequencies around 1 Hz by applying a band-pass filter with a pass band frequency between 0.75 and 1.5 Hz. We also examine *Lg* wave attenuation based on an FDM simulation using a uniform 35-km-thick crust (flat-crust model) as a reference. Figure 12 plots the RMS amplitude of the *Lg* wave for the models of profiles A and B with the theoretical attenuation functions of the *Lg* wave, assuming the anelastic coefficients of $Qs = 200$, 400, and 1,000 and the geometrical spreading factor of the *Lg* wave ($r^{-0.83}$, where r is the hypocentral distance) (Nuttli 1973).

The apparent attenuation of the *Lg* waves for the profile A model is comparable to that for the flat-crust model with a laterally homogeneous crustal structure and fits the theoretical attenuation function with $Qs = 400$, which is the value assigned to the crust in the present FDM simulation. For the laterally heterogeneous waveguide of the profile B model, the *Lg* wave attenuation is very strong and tends to follow the apparent attenuation function for $Qs = 200$ and geometrical spreading factor of $r^{-0.83}$.

It is obvious that such drastic *Lg* wave attenuation results from an increase in the effective geometrical spreading factor associated with the laterally heterogeneous and thinner crustal waveguide, rather than a decrease of the *Qs* value in the crust. We note that the seismic waves lose amplitude with propagation due to spreading of seismic wave energy (geometrical spreading effect), which, for body waves, are often represented as r^{-1} (and $r^{-0.83}$ for *Lg* wave), where r is

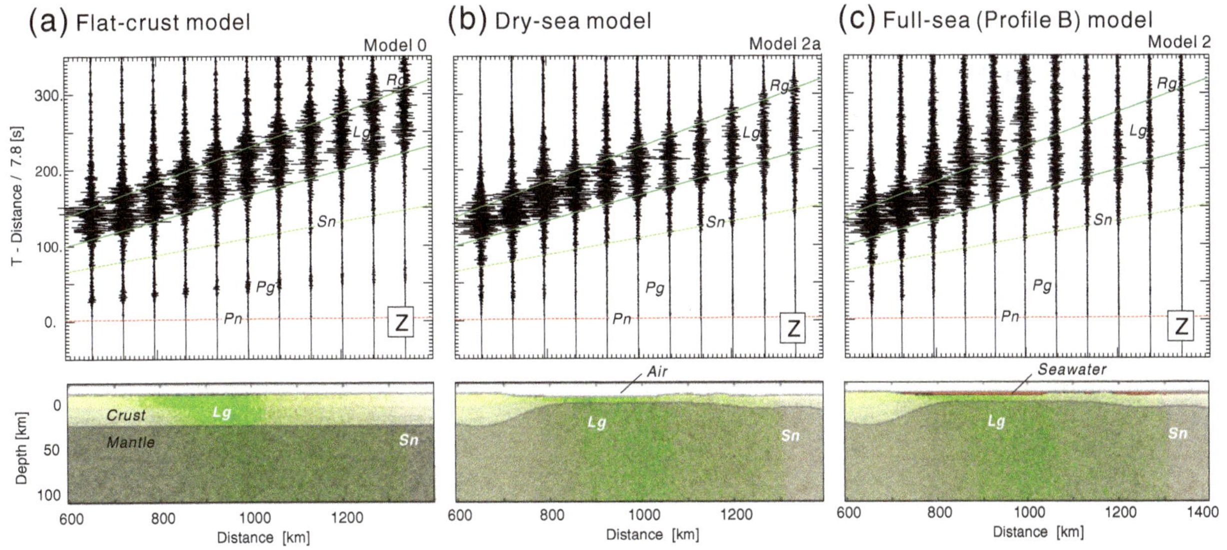

Figure 13 Comparison of *Lg* wave propagation for three models for structure across the Sea of Japan. Record section of vertical-component seismogram and partial snapshots around the crustal pinch derived by the FDM simulation at $T = 320$ s for modified Sea of Japan models with **(b)** drained seawater (dry-sea model) and **(c)** full-sea model (same as the profile B model in Figure 11) and **(a)** a 35-km-thick flat-crust model for reference.

the epicentral distance. This is also due to anelastic attenuation (Qs), which takes the form $\exp(-\pi^*r/(Vs^*Qs))$, where Vs is the shear wave speed. The geometrical spreading factor of r^{-1} corresponds to the isotropic spreading of the seismic wave in a homogeneous structure. Since the seismic wavefront in a heterogeneous structure has directional characteristics, the rate at which seismic wave energy spreads is also heterogeneous. The directional energy spreading variation results in smaller (or larger) apparent Qs along the propagation path when the actual geometrical spreading of the S wave is stronger (or weaker) than the conventional geometrical spreading factor (r^{-1}) in all directions. The drop in the apparent Qs from 400 to 200 corresponds to an increase in the effective geometrical spreading factor from $r^{-0.83}$ to $r^{-1.15}$ in the heterogeneous crustal waveguide at a distance of 1,000 km (assuming $Vs = 3.5$ km/s).

Nuttli's (1973) geometrical spreading factor of $r^{-0.83}$ for *Lg* wave was derived from a theoretical model of an Airy phase trapped in a homogeneous crustal waveguide with a strong contrast at the Moho. In our case, the *Lg* wave attenuation is much stronger in the heterogeneous and thinner crustal waveguide due to the loss of *Lg* wave energy from the crustal waveguide into the mantle.

As we noted in Figure 11b, the partitioning of the *Lg* wave energy into seawater accelerates the attenuation and elongation of the *Lg* wave train. We examined such effects by comparing the seismogram and attenuation function of the *Lg* wave elastic energy using a modified dry-sea model that excluded seawater (Figure 13b; dry-sea model) and compared the result with the simulation mentioned above (Figure 13c; full-sea model), which is equivalent to the profile B model shown in Figure 10b, and the result for the reference flat-crust model (Figure 13a). We observed *Lg* attenuation and *Sn* development during propagation across a thin crust due to the *Lg*-to-*Sn* phase conversion (Figure 13b). Strong *Lg*-to-*P* conversions at the bottom of the sea and the trapping of *P* wave energy into seawater (red color in Figure 13c) drastically elongate the later *Lg* wave coda by converting it back into the *S* wave.

The strength of the *Lg* wave is examined by calculating the strain energy (i.e., $E = 1/2 \times \rho \times v^2$, where ρ is density and v is particle velocity). Figure 14 plots the *Lg* wave train energy measurements derived by the square average sum of *Lg* wave ground velocities for horizontal and vertical motions over the time window for group velocities between 3.5 and 2.8 km/s. The results show that a large amplitude *Lg* wave develops for epicentral distances above 150 km and that the energy can be sustained over large distances in a homogeneous crustal waveguide up to an epicentral distance of 700 km, where crust thinning starts. Once crust thinning starts, the *Lg* wave strain energy increases over the epicentral distance range of 700 to 800 km. This is due to the initial concentration of *Lg* wave energy in the thinning crustal structure.

Next, dramatic *Lg* wave attenuation starts with consistent propagation in the thinner crust for epicentral distances from 800 to 1400 km. Thus, after crossing the crustal barrier, the strain energy of the *Lg* wave drops to 10% of

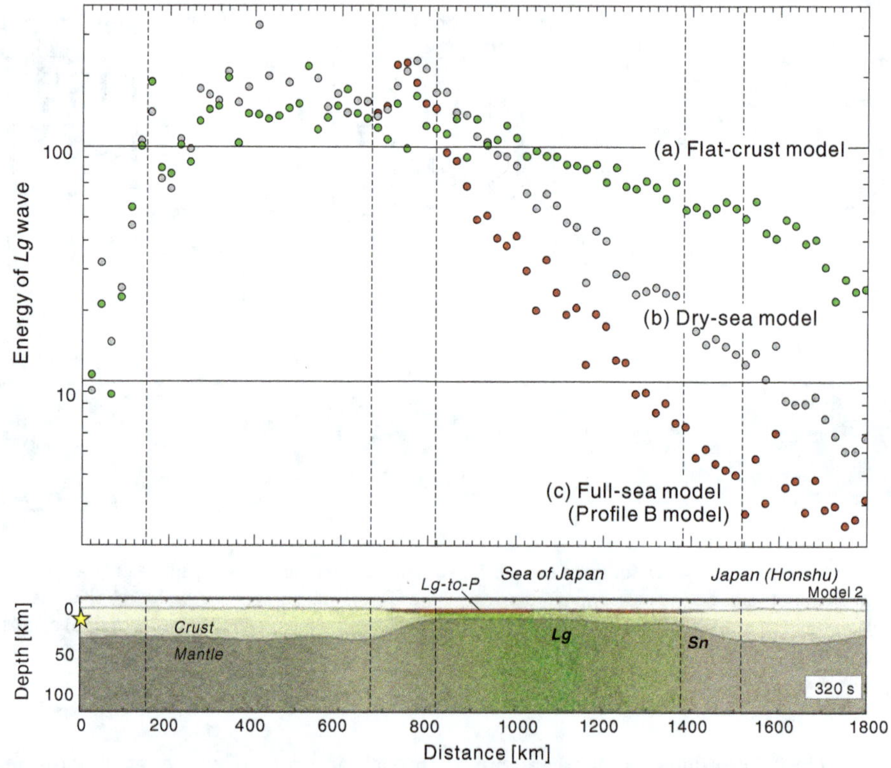

Figure 14 Attenuation function of the *Lg* wave energy for the Sea of Japan structure (Figure 13). Comparison of the attenuation function of the *Lg* wave energy calculated from the synthetic seismograms for the three models in Figure 13a,b,c.

the flat-crust model at the distance of 1,500 km. Half of the *Lg* wave energy loss occurs as a result of escaping energy into the mantle during propagation in the thinner crustal waveguide (see the gray points in Figure 14 for the dry-sea model), and the other half of the energy attenuation is associated with transfer of *Lg* wave energy into *P* waves in seawater (see the red points for the full-sea model). Therefore, we have concluded that both the effects of thinning crust and the presence of seawater are important components of the dramatic loss of *Lg* wave energy that occurs when such waves propagate across the complex crustal barrier.

Discussion

Magnitude estimation based on *Lg* wave

The strong variations seen in the *Lg* wave propagation efficiency and inefficiency in the area around Japan, as shown in Figure 7, may cause significant bias for *mb*(*Lg*) magnitude estimates that are based on *Lg* amplitudes. As we show in Figure 12, the *Lg* amplitude drops across the continental to oceanic transition zone are roughly twice as great as those along the continental path, which would cause underestimation of *mb*(*Lg*) by a 0.3 magnitude unit when using a magnitude scale function based on log(*A*), where *A* is the amplitude of the *Lg* wave, such as in the scale derived by Nuttli (1973). Hong (2012) pointed out that the estimated *mb*(*Lg*) for an underground nuclear explosion test conducted by North Korea on 9 October 2006 recorded in stations across the continental margin using an *Lg* wave set produced a smaller *mb*(*Lg*) = 2.6 to 3.4 than was recorded at stations along continental paths (*mb*(*Lg*) = 3.8 to 4.2). Hong (2012) also proposed a correction function for the *mb*(*Lg*) estimation by employing a ray path-dependent *Qs*, with much lower *Qs* along the path from the Japanese subduction zone to the Korean Peninsula, rather than using a constant *Qs* for all earthquakes.

A similar discussion on local magnitude overestimations was made in Furumura and Kennett (2001) and Kennett and Furumura (2002) for the 2000 western Tottori (western Japan) earthquake. In those studies, they pointed out that the Japan Meteorological Agency (JMA) provided a relatively high local magnitude value (M_{JMA} 7.3) compared to the moment magnitude estimated by various other agencies (*Mw* 6.6 to 6.8) and explained this discrepancy as the result of *Lg* and *Rg* wave development in western Japan resulting from the western Tottori earthquake, which carried a large-amplitude seismic wave over long distances with significantly less attenuation than a body *S* wave. This explanation is consistent with the efficient *Lg* wave propagation noted in the regional distances around Tottori, as shown in Figure 7. They also noted that the magnitude function used by the JMA was derived by Tsuboi (1954) based on the analysis of earthquakes that had occurred mostly in northern Japan, where the *Lg* wave propagation is not as efficient (see Figure 7).

Potential *Lg* wave seismic hazard

In some areas where the *Lg* propagation is efficient, such as in western Japan, *Lg* waves have the potential ability to carry strong ground motions over significant distances from the source region. The dominant frequency of an *Lg* wave, ranging 0.2 to 5 Hz, causes significant resonance with wooden-frame houses and low-rise buildings with fewer than ten floors. The JMA shaking intensity scale is designed to emphasize the strength of ground motions in this frequency range.

Furumura and Kennett (1998) and Furumura and Singh (2002) analyzed seismograms in Mexico City from Mexican subduction zone earthquakes, such as for the 1985 Michoacán earthquake (*Mw* 8.0) and showed that the *Lg* wave carried large-amplitude ground motions very effectively in the 0.2- to 0.3-Hz frequency band from the coast to more than 300 km inland. Therefore, it has been recognized that the dominance in the seismic wave in frequencies around 0.2 to 0.3 Hz radiating from the source of the 1985 Michoacán earthquake (e.g., Campillo et al. 1989; Singh et al. 1988, 1990) and a dramatic amplification of ground motions in these frequencies due to resonance with the soft reclaimed sediments below the lake-bed zone of Mexico City (e.g., Sánchez-Sesma et al. 1993; Kawase and Aki 1989; Chávez-García and Cuenca 1996) was the major cause of the destructive damage suffered by Mexico City during the 1985 Michoacán earthquake. We also note that in this frequency band, the path effect of efficient seismic wave transportation from the coast to inland for significant distance due to the efficient propagation of *Lg* wave is an important contributor to the large and long-time duration strong ground motions developed in Mexico City. Using FDM simulations, Furumura and Kennett (1998) and Furumura and Singh (2002) demonstrated that the shallowly subducting Cocos plate beneath Mexico reinforced the post-critical *S* wave reflections between the free surface and the top of the dipping plate interface, thus directing a large amount of *S* wave energy inland, which then developed a large and elongated *Lg* wave train.

A similar sort of distant and localized damage occurred beside Lake Suwa in central Honshu (see Figure 2) during the 1944 Tonankai earthquake (*M* 7.9) that ruptured along the eastern half side of the Nankai Trough at a hypocentral distance of approximately 300 km (Usami 1996). There is no doubt that strong ground motion amplification in the soft sediments of this lake, like in Mexico City, is the primary cause of the significant damage suffered, but an additional ground motion reinforcement effect due to the shallowly dipping Philippine Sea Plate beneath the Japanese Islands helped direct strong ground motions toward the inland direction in a similar way to that seen in Mexico City (Kennett and Furumura 2002). Furthermore, as we saw in Figure 7, there is a zone of good *Lg* wave propagation from the Nankai Trough to central Japan.

Additional factors affecting *Lg* propagation

Thus far, this study has focused primarily on the influence of the major features of crustal structure on *Lg* propagation and blockage. However, the influence of thick sediment cover over heterogeneous crust could well cause additional *Lg* attenuation effects, and some components of the *Lg* wave propagation variation across Japan might be associated with such thick sediment covers in addition to the crustal thickness changes. In the present large-scale (1,900 km × 2,000 km) 3-D FDM simulation, the implementation of the low-wave speed (e.g., $Vs < 0.5$ km/s) sediments while retaining the same span of frequencies is somewhat difficult, even when the most advanced supercomputers are used. However, the steadily improving state of computer technology should enable such simulations in the near future. In such high-resolution simulations, we will also need a detailed model of sedimentary layers that can be produced by combining different data derived by reflection experiments, tomography, etc. Furthermore, so far, we have assumed a constant *Qs* value in the crust and mantle, but lateral variation of attenuation structure could also cause a strong effect on high-frequency *Lg* wave propagation, such as when the waves cross the volcanic zone in the Tohoku region of Japan.

Conclusions

We have analyzed a large number of regional waveform records obtained from dense seismic stations in Japan (Hi-net) and in Korea and coupled these observations to large-scale 3-D FDM simulations of regional wavefields using the Earth Simulator supercomputer (The Earth Simulator Center). With the aid of the recently developed *CRUST 1.0* model (Laske et al. 2013), we have successfully described the main features of the regional wavefield arising via the crustal structure transitions from continental Asia to the Korean Peninsula, the Japanese subduction zones, and across the Sea of Japan.

Efficient *Lg* wave propagation was found from continental Asia to Kyushu, through the Korean Peninsula, and to Hokkaido. However, the *Lg* propagation to the Japanese main island of Honshu was totally blocked because it crosses the continental to oceanic boundary surrounding the Sea of Japan. The visualized regional wavefield derived from these simulations demonstrates the process of the *Lg* wave suppression by transmission of the *S* waves into the mantle through the Moho interface and the dissipation of the *Lg* energy into seawater by conversion to *P* wave. Crustal thinning and varying seawater cover were found to extract significant effects on eliminating *Lg* by extracting its energy from the crustal waveguide and by elongating the weaker *Lg* wave train. An investigation into the regional *Lg* wavefield can be expected to provide a key for improving the knowledge about detailed crustal structures and local tectonics.

Efficient *Lg* propagation in relatively homogeneous waveguides, such as from the Asian mainland to Kyushu and Hokkaido and along western Japan, has the potential to carry strong ground motions and produce disasters associated with amplification of ground motion in soft sediments beneath populated cities. Thus, detailed evaluations of the regional variations in *Lg* wave propagation are necessary for evaluating for the causes of strong ground motion damage in past earthquakes and for seismic hazard mitigation efforts in anticipation of future earthquakes.

Competing interests

The authors declare that they have no competing interests.

Authors' contributions

TF proposed the topic and carried out the main data analysis and numerical simulations. TKH analyzed data, and BLNK helped with its interpretation. TKH and BLNK collaborated with the corresponding author in the preparation of the manuscript. All authors have read and approved the final manuscript.

Authors' information

TF is a Professor of the Earthquake Research Institute and the Center for Integrated Disaster Information Research, the University of Tokyo. He received his Ph.D. in Geophysics from Hokkaido University in 1992 and has been working in the field of computer simulations of seismic wave and tsunami propagation, data assimilation, and the visualization of ground motion in order to further understanding of seismic wavefields in heterogeneous structures. He has been a board member of the Japan Geoscience Union (JpGU) since 2008. TKH is an Associate Professor of Geophysics in the Department of Earth System Sciences at Yonsei University. He received his Ph.D. in Geophysics from The Australian National University in 2003. His research interests focus primarily on the investigation of crustal heterogeneities by processing seismic waveform data and analyzing source properties, along with the development and application of numerical techniques for the analysis of seismic wave propagation and scattering in heterogeneous structures. BLNK is a Professor of Seismology at the Research School of Earth Sciences, Australian National University and is the director of the ANSIR National Research Facility of Australia. He is the past President of the International Association of Seismology and Physics of the Earth's Interior (IASPEI) and received his Ph.D. in 1973 from the University of Cambridge. His research interests include the development of interpretational techniques for seismic records with the stated objective of extracting detailed information about the nature of the structure within the Earth and the characteristics of seismic sources.

Acknowledgements

We would like to thank the National Research Institute for Earth Science and Disaster Prevention (NIED) of Japan for providing Hi-net data, the Korea Meteorological Administration (KMA), the Korea Institute of Geoscience and Mineral Resources (KIGAM), and the US IRIS Data Management Center in Seattle for providing seismic data. We also extend our thanks to the Earth Simulator Center of the Japan Marine Science and Technology Center (JAMSTEC) for providing central processing unit (CPU) time on the Earth Simulator. This study is supported by the Grant-in-Aid for Scientific Research (c) 23540490 from the Japan Society for the Promotion of Science and the Observation and Research Program for Prediction of Earthquakes and Volcanic Eruptions from the Ministry of Education, Culture, Sports, Science, and Technology. The constructive comments received from the three anonymous reviewers and the Editor were very helpful for revising this manuscript.

Author details

[1]Earthquake Research Institute, The University of Tokyo, 1-1-1 Yayoi, Bunkyo-ku, Tokyo 113-0032, Japan. [2]Department of Earth System Sciences, Yonsei University, 50 Yonsei-ro, Seodaemun-gu, Seoul 120-749, South Korea. [3]Research School of Earth Sciences, The Australian National University, Canberra ACT 0200, Australia.

References

Aki K, Chouet B (1975) Origin of coda waves: source, attenuation, and scattering effects. J Geophys Res 80:3322–3342

Azuma H, Kawai S, Fujiwara H (2013) Development of J-SHIS and applications using API. J Disast Res 8:869–877

Baumgardt DR (1985) Attenuation, blockage, and scattering of teleseismic *Lg* from underground nuclear explosions in Eurasia. AFGL-TR-85-0332, ENSCO, Inc, Virginia

Campillo M (1987) *Lg* wave propagation in a laterally varying crust and the spatial distribution of the quality factor in central France. J Geophys Res 92:12604–12614

Campillo M, Gariel JC, Aki K, Sánchez-Sesma FJ (1989) Destructive strong ground motion in Mexico City: source, path and site effects during great 1985 Michoacán earthquake. Bull Seism Soc Am 79:1718–1735

Chang SJ, Baag CE (2005) Crustal structure in southern Korea from joint analysis of teleseismic receiver functions and surface-wave dispersion. Bull Seism Soc Am 95:1516–1534

Chávez-García FJ, Cuenca J (1996) Site effects in Mexico City urban zone: a complementary study. Soil Dyn Earthq Eng 15:141–146

Chung TW, Noh MH, Kim JK, Park YK, Yoo HJ, Lees JM (2007) A study of the regional variation of low-frequency Q_{Lg} around the Korean peninsula. Bull Seism Soc Am 97:2190–2197

Fan GW, Lay T (2003) Strong *Lg* attenuation in the Tibetan plateau. Bull Seism Soc Am 93:2264–2272

Fisk MD (2006) Source spectral modeling of regional *P/S* discriminants at nuclear test sites in China and the former Soviet Union. Bull Seism Soc Am 96:2348–2367

Furumura T, Kennett BLN (1998) On the nature of regional seismic phases III: the influence of crustal heterogeneity on the wavefield for subduction earthquakes: the 1985 Michoacán and 1995 Copala, Guerrero, Mexico earthquakes. Geophys J Int 135:1060–1084

Furumura T, Kennett BLN (2001) Variations in regional phase propagation in the region around Japan. Bull Seism Soc Am 91:668–682

Furumura T, Singh SK (2002) Regional wave propagation from Mexican subduction zone earthquakes: the attenuation functions for interplate and inslab events. Bull Seism Soc Am 92:2110–2125

Gregersen S (1984) *Lg* wave propagation and crustal structure differences near Denmark and the North Sea. Geophys J R Astron Soc 79:217–234

Herrmann RB, Nuttli OW (1982) Magnitude: the relation of M_L to mb_{Lg}. Bull Seism Soc Am 72:389–397

Hirata N, Karp BY, Yamaguchi T, Kanazawa T, Suyehiro K, Kasahara J, Shiobara H, Shinohara M, Kinoshita H (1992) Oceanic crust in the Japan Basin of the Japan Sea by the 1990 Japan-USSR expedition. Geophys Res Lett 19:2027–2030

Hong TK (2010) *Lg* attenuation in a region with both continental and oceanic environments. Bull Seism Soc Am 100:851–858

Hong TK (2012) *Lg* body-wave magnitude scaling for the continental margin around Korea and Japan. Pure Appl Geophys 169:641–658

Hong TK, Baag CE, Choi H, Sheen DH (2008) Regional seismic observations of the 9 October 2006 underground nuclear explosion in North Korea and the influence of crustal structure on regional phases. J Geophys Res 113:B03305. doi:10.1029/2007JB004950

Kadinsky-Cade K, Barazangi M, Oliver J, Isacks B (1981) Lateral variations in high-frequency seismic wave propagation at regional distances across the Turkish and Iranian plateaus. J Geophys Res 86:9377–9396

Kak AC, Slaney M (1987) Principles of computerized tomographic imaging. IEEE, Piscataway

Kawase H, Aki K (1989) A study on the response of a soft basin for incident *S*, *P*, and Rayleigh waves with spectral reference to the long duration observed in Mexico City. Bull Seism Soc Am 79:1361–1382

Kennett BLN (1993) The distance dependence of regional phase discriminants. Bull Seism Soc Am 83:1155–1166

Kennett BLN, Furumura T (2001) Regional phases in continental and oceanic environments. Geophys J Int 146:562–568

Kennett BLN, Furumura T (2002) The influence of 3-D structure on the propagation of seismic waves away from earthquakes. Pure Appl Geophys 159:2113–2131

Kennett BLN, Gregersen S, Mykkeltveit S, Newmark R (1985) Mapping of crustal heterogeneity in the North Sea basin via the propagation of *Lg* waves. Geophys J R Astron Soc 83:299–306

Kim WY, Richards PG (2007) North Korean nuclear test: seismic discrimination at low yield. Eos 14:157–161

Kurashimo E, Shinohara M, Suyehiro K, Kasahara J, Hirata N (1996) Seismic evidence for stretched continental crust in the Japan Sea. Geophys Res Lett 23:3067–3070

Lallemand S, Jolivet L (1985) Japan Sea: a pull-apart basin? Earth Planet Sci Lett 76:375–389

Laske G, Masters G, Ma Z, Pasyanos M (2013) Update on CRUST 1.0: a 1-degree global model of Earth's crust. Geophys Res Abstr 15:EGU2013–EGU2658

Maupin V (1989) Numerical modelling of *Lg* wave propagation across the North Sea Central Graben. Geophys J Int 99:273–283

Nakamura R, Uetake T (2002) Three-dimensional attenuation structure and site amplification inversion by using a large quantity of seismic strong motion records in Japan. Zisin II 54:475–488 (in Japanese with English abstract)

Ni J, Barazangi M (1983) High frequency seismic wave propagation beneath the Indian shield, Himalayan Arc, Tibetan Plateau, and surrounding regions: high uppermost mantle velocities and efficient *Sn* propagation beneath Tibet. Geophys J R Astron Soc 72:655–689

Nuttli OW (1973) Seismic wave attenuation and magnitude relations for eastern North America. J Geophys Res 78:876–885

Obara K, Kasahara K, Hori S, Okada Y (2005) A densely distributed high-sensitivity seismograph network in Japan: Hi-net by National Research Institute for Earth Science and Disaster Prevention. Rev Sci Instrum 76:021301. doi:10.1063/1.1854197

Okada Y, Kasahara K, Hori S, Obara K, Sekiguchi S, Fujiwara H, Yamamoto A (2004) Recent progress of seismic observation networks in Japan—Hi-net, F-net, K-NET and KiK-net. Earth Planets Space 56:15–28

Olsen KH, Braile LW, Stewart JN (1983) Modeling short-period crustal phases (*P*, *Lg*) for long-range refraction profiles. Phys Earth Planet Int 32:334–347

Otofuji Y, Itaya Y, Matsuda T (1991) Rapid rotation of southwest Japan: paleomagnetism and K-Ar ages of Miocene volcanic rocks of southwest Japan. Geophys J Int 105:397–405

Park JH, Chi HC, Lim IS, Kim GY (2009) Korea seismic networks and Korea Integrated Seismic System (KISS). Geophys Res Abstr 11:EGU2009–EGU9191

Phillips WS, Hartse HE, Taylor SR, Randall GE (2000) 1 Hz *Lg* tomography in Central Asia. Geophys Res Lett 27:3425–3428

Press F, Ewing M (1952) Two slow surface waves across North America. Bull Seism Soc Am 42:219–228

Robertsson JOA, Blanch JO, Symes WW (1994) Viscoelastic finite-difference modeling. Geophysics 59:1444–1456

Rodgers AJ, Ni JF, Hearn TM (1997) Propagation characteristics of short-period *Sn* and *Lg* in the Middle East. Bull Seism Soc Am 87:396–413

Ruzaikan AI, Nersesov IL, Khalturin VI, Molnar P (1977) Propagation of *Lg* and lateral variation in crustal structure in Asia. Geophys J R Astron Soc 82:307–316

Ryoki K (1999) Three-dimensional depth structure of the crust and uppermost mantle beneath southwestern Japan and its regional gravity anomalies. Zisin II 52:51–63 (in Japanese with English abstract)

Sánchez-Sesma FJ, Pérez-Rocha LE, Reinoso E (1993) Ground motion in Mexico City during the April 25, 1989, Guerrero earthquake. Tectonophysics 218:127–140

Sekine S (2005) Tomographic inversion of ground motion amplitudes for the 3-D attenuation structure beneath the Japanese islands. Report Nat Res Institute Earth Sci Disaster Prev 68:137–174

Shima H (1962) On the velocity of *Lg* waves in Japan. Quart J Seismol 27:1–6 (in Japanese with English abstract)

Sing SK, Mena E, Castro R (1988) Some aspects of source characteristics of the 19 September 1985 Michoacán earthquake and ground motion in and near Mexico City from strong motion data. Bull Seism Soc Am 78:451–477

Singh SK, Mori A, Mena E, Krüger F, Kind R (1990) Evidence for anomalous body-wave radiation between 0.3 and 0.7 Hz from the 1985 September 19 Michoacán, Mexico earthquake. Geophys J Int 101:37–48

Singh C, Shekar M, Singh A, Chadha RK (2012) Seismic attenuation characteristics along the Hi-CLIMB profile in Tibet from *Lg Q* inversion. Bull Seism Soc Am 102:783–789

Taylor S (1996) Analysis of high-frequency *Pg*/*Lg* ratios from NTS explosions and western U.S. earthquakes. Bull Seism Soc Am 86:1042–1053

Tsuboi C (1954) Determination of the Gutenberg-Richter's magnitude of earthquakes occurring in and near Japan. Zisin II 7:185–193 (in Japanese with English abstract)

Usami T (1996) Materials for comprehensive list of destructive earthquakes in Japan. Revised and enlarged edition. University of Tokyo Press, Tokyo, pp 294–295 (in Japanese)

Utsu T (1956) On some remarkable phases on seismograms of near earthquakes (part 2). Quart J Seismol 21:107–111 (in Japanese with English abstract)

Utsu T (1958) On the *Lg* phase of seismic waves observed in Japan. Quart J Seismol 23:61–76 (in Japanese with English abstract)

Walter WR, Mayeda K, Patton HJ (1995) Phase and spectral ratio discrimination between NTS earthquakes and explosions: part 1: empirical observations. Bull Seism Soc Am 85:1050–1067

Xie J (2002) *Lg Q* in the eastern Tibetan Plateau. Bull Seism Soc Am 92:871–876

Xie J, Wu Z, Liu R, Schaff D, Liu Y, Liang J (2006) Tomographic regionalization of crustal *Lg Q* in eastern Eurasia. Geophys Res Lett 33:L03315. doi:10.1029/2005GL024410

Yomogida K (1996) J-array: deep structure of the Earth. J Phys Earth 44:655–656

12

Deep-float salinity data synthesis for deep ocean state estimation: method and impact

Shuhei Masuda*, Satoshi Osafune and Tadashi Hemmi

Abstract

The importance of deep ocean observations has been recognized with regard to changes in the deep ocean such as global bottom-water warming. Therefore, sustainable deep ocean monitoring networks that use autonomous profiling floats have been widely proposed, and a number of deep-float deployment initiatives have begun around the world. Deployed floats promise to provide unprecedented deep ocean information. However, present deep-float data are known to have biases. In particular, a depth-dependent bias in salinity data is a major issue that prevents us from constructing global deep ocean monitoring networks. This paper proposes a new approach to utilize ongoing deep-float salinity data to reduce the bias in estimates of the global full-depth ocean state. It reports results from comparative experiments with and without deep-float data by using the proposed approach to examine the impact of data from currently operating deep floats on ocean state estimates. The results demonstrate that available float data possibly contribute local corrections to the modeled climate ocean state. Furthermore, we clarify how interannual basin-scale estimations are controlled by available deep-float salinity data in two specific regions of the Southern and Indian Oceans.

Introduction

The deep ocean has gained attention from climate researchers since Fukasawa et al. (2004) documented bottom-water warming in the abyssal North Pacific Ocean. The warming was detected by repeated high-accuracy ship-based observations by the World Ocean Circulation Experiment (WOCE) Hydrographic Program on its P1 survey line across the subarctic Pacific along 47° N, and Johnson et al. (2008) reported similar bottom-water warming in the Indian basin, also from revisit cruises. Subsequent researchers have detected global bottom-water warming (e.g., Purkey and Johnson 2010, Kouketsu et al. 2011).

Recent observational studies have indicated that Antarctic Bottom Water (AABW) has been freshening in some observational sections (e.g., Aoki et al. 2005, Rintoul 2007, Jacobs and Giulivi 2010). Purkey and Johnson (2013) quantified the freshening of AABW around Antarctica in a straightforward way. Nevertheless, deep ocean observations remain too sparse to resolve the temporal evolution of the deep ocean state.

* Correspondence: smasuda@jamstec.go.jp
Research and Development Center for Global Change, Japan Agency for Marine-Earth Science and Technology (JAMSTEC), Yokosuka, Japan

At the beginning of the 2000s, dramatic progress was made in ocean observations with the introduction of Argo profiling floats capable of continuously monitoring ocean properties in the upper 2000 m (e.g., Argo Science Team, 2001). The success of the monitoring network for the upper ocean (e.g., Riser et al. 2016) motivates the construction of a similar global monitoring network for the deep ocean.

Recently, several types of deep floats have been developed or released. Among these, Deep NINJA is a reliable deep float that can continuously monitor depths to 4000 m as an extension of the conventional Argo float (Kobayashi et al. 2013). Deep NINJAs have been deployed in the North Pacific, South Indian, and Southern Oceans. Although interesting information about the deep ocean state has been obtained, it is known that the salinity measurements are sometimes biased by O(0.01) mainly because of the conductivity sensor measurements, which makes it difficult to exploit them (Kobayashi 2016). Such bias is also found in measurements of other types of deep floats (e.g., Zilberman and Maze 2015).

In this paper, we propose a practical technique to blend these problematic but scientifically valuable data into full-depth ocean state estimates (e.g., Osafune et al.

2015) by reducing their bias through a dynamical approach. Then, we examine the impact of available Deep NINJA data on deep ocean state estimation by comparing two data sets from a set of comparative experiments with and without the deep-float data.

Methods experimental

Data and assimilated system

The deep-float data that we used were 316 profiles obtained from available Deep NINJA floats collected and released by the Japan Agency of Marine-Earth Science and Technology (JAMSTEC: http://www.jamstec.go.jp/ARGO/deepninja/). Figure 1 shows the 17 Deep NINJA trajectories that were used in this study, covering the Southern (11 floats), North Pacific (4), and South Indian (2) Oceans (Table 1), each profile including temperature, salinity, and pressure. The equipped sensor accuracy was 0.002 °C for temperature and 0.005 for salinity. The data were processed through basic quality control in the same way as for a conventional Argo float (Wong et al., 2018).

Deep NINJA salinity data are known to have a bias that depends on pressure. Kobayashi (2018) thoroughly examined the bias in Appendix B of his paper. The problem is that the bias is different for each float; the values range from – 0.005 to – 0.02 at depths of 3500–4000 m based on analysis of float data for which there are also comparable ship-based Conductivity Temperature Depth profiler (CTD) data obtained at the same time as the float deployments (Fig. 2; Kobayashi 2016). We assume the bias value to be time independent since float lifetime is relatively short with a maximum of approximately 2 years (Table 1). This assumption may be premature since it has not yet been validated, so there could be some kind of time-dependent bias. Nevertheless, here, we assume there is not as a first step toward improved deep ocean state estimation.

The data synthesis system is based on the work of Osafune et al. (2015). It generates a dynamically self-consistent long-term ocean state estimation, "Estimated STate of global Ocean for Climate research" (ESTOC), by applying a four-dimensional variational (4D-VAR) adjoint approach (e.g., Sasaki 1970, Stammer et al., 2002, Wunsch and Heimbach 2007). The ocean general circulation model we used has a horizontal resolution of 1° in both latitude and longitude, and it includes 46 vertical levels for the global ocean basin. The control variables are surface forcing data and initial conditions for three-dimensional temperature and salinity distributions.

The ESTOC system is quite suitable for data synthesis of relatively sparse observations because the 4D-VAR adjoint method can generate a dynamically self-consistent state estimation through backward tracking of adjoint sensitivity (e.g., Köhl and Stammer 2004, Masuda et al. 2010), which takes advantage of both statistical and dynamical connections between the model and observations (e.g., Wunsch and Heimbach 2013). Thus, we used this system to examine the impact of available and locally distributed deep-float salinity data on deep ocean state estimation through a comparative twin experiment.

The other observations that were synthesized are similar to those in Osafune et al. (2015): subsurface temperature and salinity from the ENSEMBLES (EN4) dataset compiled by the Hadley Centre of the UK Meteorological Office (Good et al. 2013), sea surface temperature (SST) from historical 10-day datasets compiled from the Optimally Interpolated SST (OISST) dataset (Reynolds et al. 2002), and the historical 10-day sea surface dynamic height anomaly (SSDHA) derived

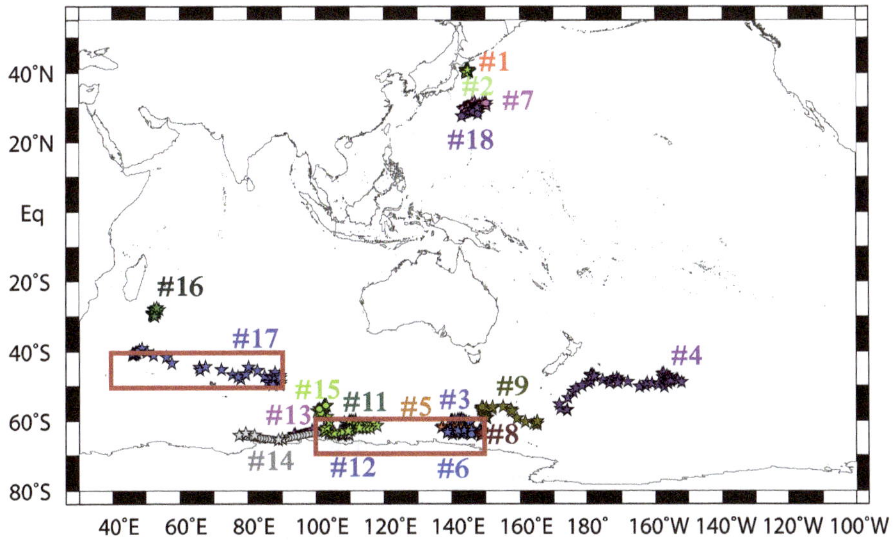

Fig. 1 Trajectories of the Deep NINJA floats listed in Table 1. Red rectangles show the analytical areas for Figs. 4 and 5 (see the text for more details)

Table 1 Inventory of Deep NINJA floats used in this study including information on the deployment and total number of observed profiles

Serial no.	Date of deployment	Date of the last observation	Area of deployment	Position of deployment	No. of profiles
1	2012/8/29	2012/8/30	Subarctic, North Pacific	40° 40.00′ N, 114° 46.18′ E	1
2	2012/8/29	2012/8/29	Subarctic, North Pacific	40° 40.00′ N, 114° 46.18′ E	16
3	2012/12/19	2013/5/31	Off the Adélie Coast, Antarctica	60° 00.62′ S, 139° 56.12′E	9
4	2012/12/5	2012/12/5	South of NZ, Southern Ocean	56° 59.87′ S, 173° 59.42′ E	25
5	2012/12/17	2013/7/14	Off the Adélie Coast, Antarctica	62° 00.38′ S, 137° 00.42′ E	10
6	2012/12/16	2014/8/22	Off the Adélie Coast, Antarctica	62°45.73′ S, 143° 36.05′ E	23
7	2013/7/17	2013/7/17	Subtropical, North Pacific	30° 03.91′ N, 144° 58.09′ E	33
8	2014/3/6	2014/3/27	Off the Adélie Coast, Antarctica	63° 40.55′ S, 148° 35.65′ E	3
9	2014/3/7	2014/3/7	Off the Adélie Coast, Antarctica	59° 14.34′ S, 150° 08.85′ E	23
11	2014/1/19	2014/1/19	Off the Budd Coast, Antarctica	60° 03.50′ S, 109° 53.32′ E	10
12	2014/1/22	2014/1/22	Off the Budd Coast, Antarctica	62° 00.42′ S, 110° 00.19′ E	5
13	2014/1/27	2014/1/27	Off the Budd Coast, Antarctica	64° 15.90′ S, 107° 42.00′ E	24
14	2014/1/24	2014/1/24	Off the Budd Coast, Antarctica	64° 05.36′ S, 109° 51.01′ E	40
15	2014/12/6	2014/12/6	Off the Budd Coast, Antarctica	59° 59.96′ S, 109° 51.92′ E	43
16	2016/1/31	2016/1/31	Western Indian Ocean	29° 59.98′ S, 52° 32.01′ E	18
17	2016/2/2	2016/2/2	Western Indian Ocean	40° 41.49′ S, 45° 54.43′ E	20
18	2016/3/2	2016/3/2	Subtropical, North Pacific	29° 56.31′ N, 147° 27.26′ E	12

from sea surface height data in the Copernicus Marine Environment Monitoring Service (CMEMS) dataset. In addition, we used a global mean sea level reconstructed through the method of Church and White (2011), which is updated and provided by the Commonwealth Scientific and Industrial Research Organisation (CSIRO). The EN4 dataset includes some Deep NINJA data only for water temperature.

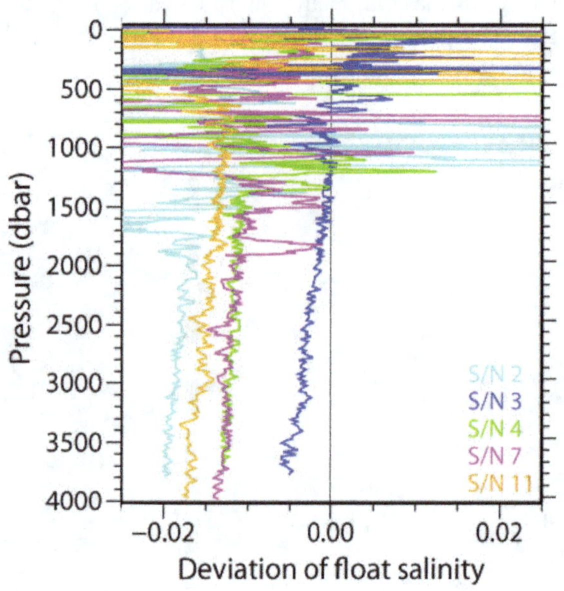

Fig. 2 Depth profiles of the deviation of float salinity from shipboard reference CTD observations cited from Kobayashi (2016). The colors identify the Deep NINJA floats with serial numbers 2, 3, 4, 7, and 11 listed in Table 1

Dynamical approach to bias reduction of salinity data

There are many studies on observation bias correction (e.g., Eyre 1992, Zhu et al. 2014, Eyre 2016). Here, we propose a pre-process for biased Deep NINJA salinity data. This is a kind of "static scheme" in which the statistics of the differences between observations and model background equivalents are evaluated to remove biases (Eyre 2016). The advantage of our method is the use of time-dependent ocean states based on an archived ocean observation dataset to calculate the statistics.

First, the Deep NINJA data were compiled as monthly gridded data averaged over each model grid volume. Then, we compared the salinity profiles from each gridded Deep NINJA with the observation-based background ocean state of a time-varying (monthly) reference field. The state was made by a conventional optimal interpolation of a monthly gridded EN4 dataset based on the World Ocean Atlas 2013 (WOA13) produced by the National Oceanographic Data Center. The EN4 dataset included Argo float data together with quality-controlled available observations. In particular, the number of profiles in the Southern Ocean greatly increased after 2008 (not shown).

Since deep floats drift freely with the ocean currents, the reference ocean state should be determined region-by-region, depth-by-depth, and month-by-month. The depth-dependent salinity bias value b can thus be defined as a function dependent only on depth (z) for each float as

$$b(z) = \overline{y(t, r(t)) - y_{\text{ref}}(\text{mon}(t), r(t))},$$

where y is the salinity profile value of the float observation, t the time, $r(t)$ the three-dimensional float position, y_{ref} the monthly background ocean state, and mon(t) the month relevant to the observation time t. The overbar denotes the time-mean procedure for the period during which each float operates. The procedure means we calculate the average value of the differences between the reference ocean state and float observation along the float trajectory for each depth from the surface to 4000 m under the assumption that both observational mean values match each other.

The depth-dependent salinity bias was determined by the above approach for each float. The values at 4000 m depth range from − 0.044 to 0.003 (Fig. 3 a, b) shows the estimated mean bias profile with standard deviations. The values at depths of 3500 and 4000 m are − 0.015 and − 0.017, respectively. The standard deviation values are 0.008 and 0.009, respectively for each depth. The magnitude of the bias increases as the depth increases beyond 2000 m.

Besides bias, there are other factors that could contribute to the differences between the reference ocean states and float observations, for example, sampling error and representativeness error. Although we cannot evaluate either of them, the gross features of Fig. 3a are largely consistent with those of Fig. 2 (and Kobayashi 2018) where negative biases of order 0.01 with standard deviations of order 0.005 were detected at 3500–4000 m depth when compared with shipboard CTD observations performed when several Deep NINJAs were deployed. This partly validates our methodology. Further examination by using more in situ data will be required to test and improve our approach. In addition, our scheme has some limitations due to the fact that the statistics depend only on a priori available information. Some update of the bias correction scheme, for instance, by using a variational scheme which are adaptive rather than static ones (e.g., Derber and Wu, 1998; Zhu et al. 2014), should lead to further refinement.

We used the salinity values for each float, with the individual depth-dependent bias values removed, as observational values in the ESTOC system. The ESTOC system is based on an anomaly data assimilation method that automatically determines the weight of observational mean values to observational time-varying components by using the statistical values of model variables (Eq. (2) of Osafune et al. 2015). We assume that corrected Deep NINJA data are reliable with regard to variability. Then, the assimilation approach of Osafune et al. (2015) reduces to conventional anomaly assimilation for the compiled Deep NINJA dataset. In addition, we place weight on synthesizing the Deep NINJA data by taking into consideration the observational number included in each profile and the accuracy of the observation. We assume 80 observation levels are included in one deep-ocean grid whose height is approximately 400 m. Deep NINJA sensors have about 2.5 times greater accuracy than gridded EN4 salinity data, which we estimate from averaged EN4 standard deviation values from 2012 to 2016 in the world's oceans at a depth of 4000 m. The total weight may be somewhat overestimated due to differences in the statistics between EN4 and Deep NINJA data. We discuss this point below.

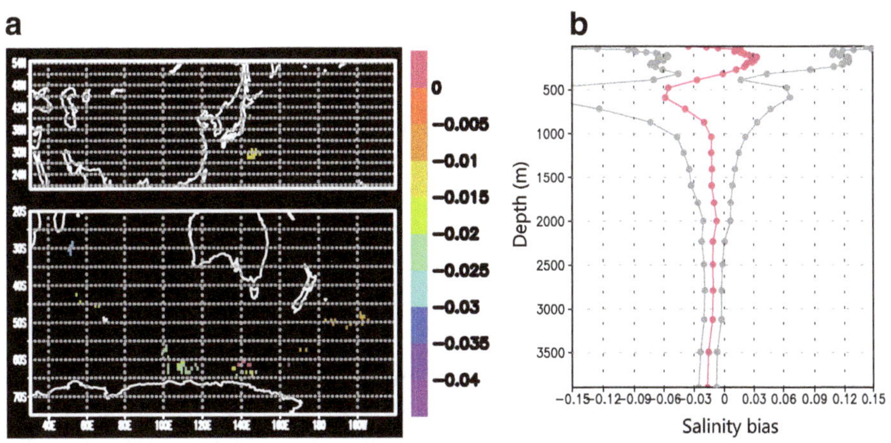

Fig. 3 a Estimated salinity bias at 4000 m depth for each float in Fig. 1. **b** Depth profile of the estimated mean salinity bias for all 17 Deep NINJA floats. The error area is defined by one standard deviation (gray lines) calculated from the 17 profiles

Comparative state estimations to examine deep-float impact

We carried out two data synthesis experiments on the Earth Simulator parallel supercomputer of JAMSTEC to obtain ocean state estimations with and without deep-float data. The case without deep-float data, hereafter called the reference case, is the same as in Osafune et al. (2015), except for the length of the assimilation window (2000–2016) and implementation of global mean sea-level data assimilation. The deep-float case (hereafter, DN case) is the same as the reference case but includes Deep NINJA salinity data for the period 2012–2016 (Table 1), synthesized through the proposed data process and anomaly assimilation described in the "Dynamical approach to bias reduction on salinity data" section. The ocean state estimation was obtained after 50 iterations. The period of analysis in this paper is from 2007 to 2016 to focus on the impact of deep float observations.

Results and discussion

Local impact of deep-float salinity data

To examine the temporal development of the estimated subsurface ocean state, we focus on two areas of 50° longitude by 10° latitude (Fig. 1): the Southern Ocean area off Antarctica's Wilkes Land (100–150° E, 60–70° S) and the South Indian Ocean area (40–90° E, 40–50° S). The former is an area of concentrated monitoring over 5 years (Table 1), while the latter is an area covered by only two floats for a 1-year period.

Figure 4 shows the temporal development of salinity at 3500 m depth for the period 2012–2016. The area

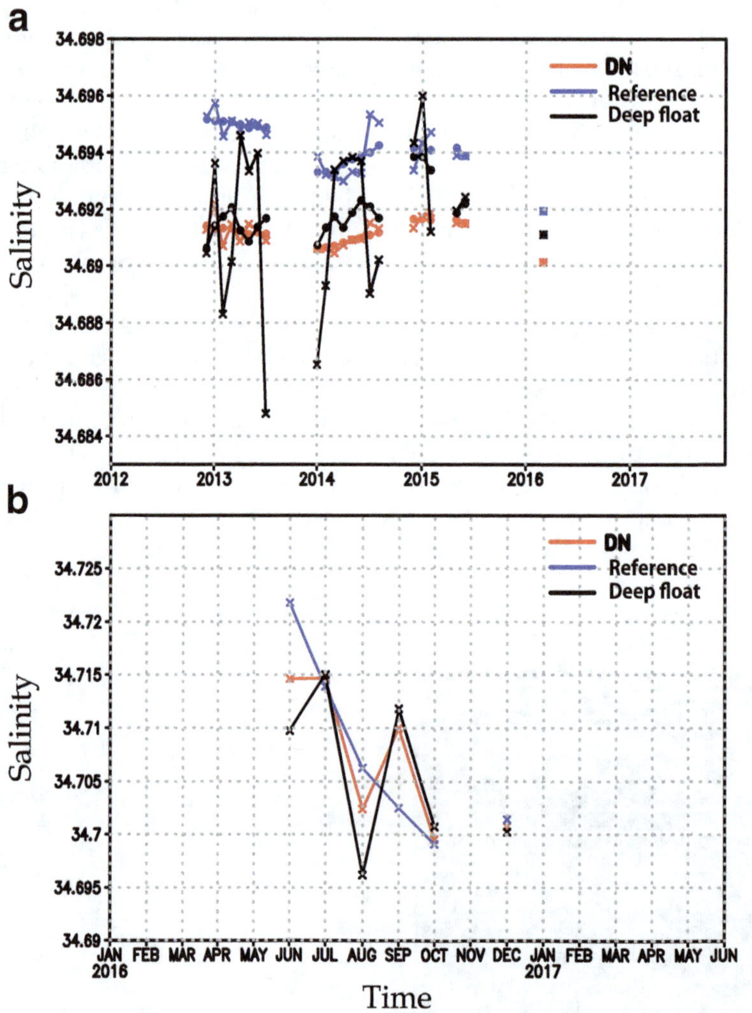

Fig. 4 Temporal development of area-mean salinity at 3500 m depth for **a** the Southern Ocean area (100–150° E, 60–70° S) and **b** the South Indian Ocean area (40–90° E, 40–50° S), outlined in Fig. 1. The black lines indicate deep-float observations, and the blue and red lines denote the reference and DN cases, respectively, averaging only where deep-float observations exist (see the text for more details). Crosses denote monthly values and dots denote 6-month running mean values

average is determined for the grid points and times where deep-float observations exist. For the Southern Ocean, the results for the DN case tend to follow the deep-float data closely. The root mean square difference values are reduced to 0.0013 compared to 0.003 for the reference case values for the 6-month running mean (black dots in Fig. 4a). The raw values of the deep-float observations (crosses in Fig. 4a) generally have larger variances than the state estimations. This is mainly because these include variability at a scale smaller than the model grid size of 1°. For the Indian Ocean, the root mean square difference is reduced to 0.0033 compared to 0.0075 for the reference case values (crosses in Fig. 4b). Although the differences between the results for the DN case and deep-float observations are less than the initial sensor accuracy of 0.005, these corrections can be meaningful since data synthesis reduces errors through integration of other information. These subtle but systematic reductions of the differences demonstrate the potential importance of deep-float observations for estimating the tempo-spatially "local" deep ocean state. That is, available deep-float salinity data can contribute to local ocean state estimation by providing a control variable, although our approach is an empirical one that includes some assumptions.

Representativeness of float observations

We further investigated the tempo-spatial representativeness of float observations for most of the above areas and time-windows by using estimated states. Figure 5 is the same as Fig. 4, but the area average of the reference and DN cases is shown for all grid points included in the area, resulting in temporally continuous curves.

For the Southern Ocean, the estimated deep ocean state shows a salinity trend of 0.002 per decade (red line in Fig. 5a), which appears consistent with bias-corrected

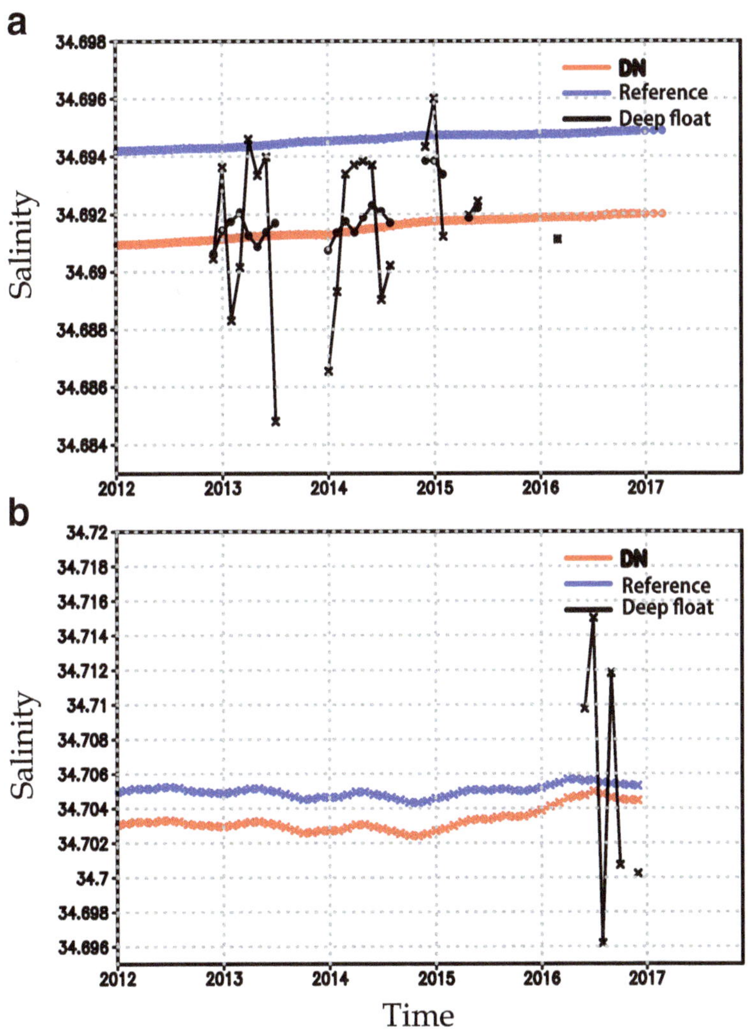

Fig. 5 The same as Fig. 4a and b for **a** and **b**, except that the area averages of the reference and DN cases are determined for all the grid points within each analytical area

deep-float observations (black dots). This result suggests that monthly mean gridded salinity values from sparse deep-float data can represent the salinity in the entire target area to some extent, which demonstrates that it is possible for ongoing deep-float observation to control the modeled deep ocean state even for a relatively large area. Note that the well-known freshening in this region (e.g., Purkey and Johnson 2013; Kobayashi 2018) is visible when the time series are replotted with a neutral density coordinate as in Kobayashi (2018; Fig. 6), although our anomaly data assimilation makes neutral density values of the mean state shift to lighter ones.

For the Indian Ocean, the impact of deep-float data is limited to the past 2 years, before which only the baseline is shifted. This result implies that two deep-float deployments during 1-year periods for this area can contribute only one more year of correction by adjoint retro-chronic temporal development. Longer-term deep ocean climate state estimation requires more data even for short periods of 1-year monitoring since the adjoint signal itself can be traced back for a period of 50 years (e.g., Masuda et al. 2010).

It should be noted that this observational density (number of deep floats and monitoring periods: ten deployments during 5 years for the Southern Ocean and two deployments during 1 year for the Indian Ocean) does not provide precise indices for a global monitoring scheme for some ocean state estimations. This is because the obtained quantitative impacts can depend highly on the local ocean state, period, model representativeness of each region, data assimilation schemes, and other factors. In addition, some issues remain to be solved in the data processing. For example, the monthly background ocean state used as a reference field in the data processing basically works to exclude the detection of long-term trends.

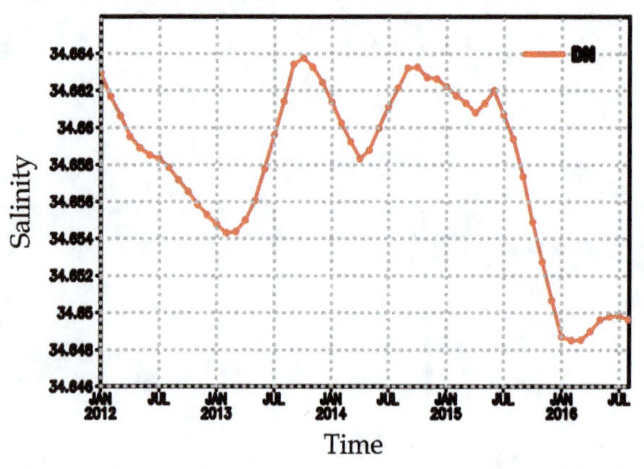

Fig. 6 Temporal development of area-mean salinity at the 28.32 neutral density surface for the Southern Ocean area (100–150° E, 60–70° S) for the DN case. The curve is smoothed out through a 6-month running mean procedure

Another problem is that of weight validation in data synthesis. We made another state estimation in which the weight of the Deep NINJA data synthesis was reduced by 95%. The root mean square difference values for the Southern Ocean area, discussed in the "Local impact of deep-float salinity data" section, change to 0.0018 from 0.0013 for the case of Fig. 4a, consistently reducing compared to 0.003 for the reference case values for the 6-month running mean (not shown). For the Indian Ocean, the root mean square difference changes to 0.0059 from 0.0033 for the case of Fig. 4b, also reducing compared to 0.0075 for the reference ones (not shown). This shows that a relatively small weight can still result in certain corrections and may suggest that further iteration possibly leads to a closer solution. Nevertheless, how we should determine the weight remains to be solved.

Our estimation is a practical first step toward a better deep-float observation scheme by using the present deep floats. Further research along these lines will provide indispensable information for the construction of future deep-ocean monitoring systems and, in addition, the improvement of the deep floats themselves inclusive of equipped sensors as well as data assimilation systems.

Conclusions

We have proposed a practical technique of deep-float salinity data processing for deep ocean state estimation to explore deep ocean climate change. The method consists of preprocessing for reduction of bias with a dynamical approach suitable for anomaly data assimilation under some assumptions.

We examined the impact of deep-float salinity data on ocean state estimation by using a 4D-VAR data synthesis system and proposed data processing in target regions in the Southern Ocean and Indian Ocean. The results show that corrected deep-float data can help constrain the modeled ocean state around the specific localities of float profiles. Regarding basin-scale climate change, in the case of the Southern Ocean, where we made concentrated observations with ten floats over 5 years, a slight salinization trend was estimated for the entire 50° longitude by 10° latitude area, basically in agreement with bias-corrected deep-float data. In the case of the South Indian Ocean where only two floats were available for a 1-year period, no significant change in time development was found until 1 year before data input.

Although these results do not provide concrete indices for either the required float number or the deployment period due to the limited sampling, they can contribute to global deep-float network planning in the near future by demonstrating the controllability of the modeled deep ocean state.

Acknowledgements

We thank Dr. Kanako Satoh, Mr. Fumihiko Akazawa, Ms. Mizue Hirano, and Mr. Tetsuharu Iino for their technical support. We are also thankful for the valuable comments and suggestions of Dr. Taiyo Kobayashi, Dr. Shigeki Hosoda, Dr. Nozomi Sugiura, and Mr. Toshimasa Doi. Figure 2 was kindly provided by Dr. T. Kobayashi. Numerical calculations were performed on the Earth Simulator of JAMSTEC. This work was partly supported by a Grant-in-Aid for Scientific Research on Innovative Areas (MEXT KAKENHI-JP15H05817/JP15H05819) and JSPS KAKENHI Grant 17H04579. WOA13 data were provided by the NOAA National Centers for Environmental Information from its website at https://www.nodc.noaa.gov/OC5/woa13/. The CMEMS altimeter products were obtained from http://marine.copernicus.eu/.

Funding

This work was done by management expenses grants for Japan Agency for Marine-Earth Science and Technology (JAMSTEC), and was partly supported by a Grant-in-Aid for Scientific Research on Innovative Areas (MEXT KAKENHI-JP15H05817/JP15H05819) and JSPS KAKENHI Grant 17H04579.

Authors' contributions

SM analyzed the data and wrote the paper. SO proposed the basic concept of quality control and helped in the coding of FORTRAN. TH compiled the float data and supported numerical experiments. All authors read and approved the final manuscript.

Competing interests

The authors declare that they have no competing interests.

References

Aoki S, Rintoul SR, Ushio S, Watanabe S, Bindoff NL (2005) Freshening of the Adélie Land Bottom Water near 140°E. Geophys Res Lett 32:L23601. https://doi.org/10.1029/2005GL024246

Argo Science Team (2001) In: Koblinsky CJ, Smith NR (eds) Argo: the global array of profiling floats, in observing the oceans in the 21st century. GODAE Project Office, Bureau of Meteorology, Melbourne, pp 248–258

Church JA, White NJ (2011) Sea-level rise from the late 19th to the early 21st century. Surv Geophys 32:585–602

Derber JC, Wu WS (1998) The use of TOVS cloud-cleared radiance in the NCEP SSI analysis system. Mon Weather Rev 126:2287–2299. https://doi.org/10.1175/1520-0493(1998)126 <2287:TUOTCC>2.0.CO;2

Eyre JR (1992) A bias correction scheme for simulated TOVS brightness temperatures, Technical Memorandum. ECMWF, Reading, Uk, p 176

Eyre JR (2016) Observation bias correction schemes in data assimilation systems: a theoretical study of some of their properties. Q J R Meteorol Soc 142(699):2284–2291. https://doi.org/10.1002/qj.2819

Fukasawa M, Freeland H, Perkin R, Watanabe T, Uchida H, Nishina A (2004) Bottom water warming in the North Pacific Ocean. Nature 427:825–827. https://doi.org/10.1038/nature02337

Good SA, Martin MJ, Rayner NA (2013) EN4: quality controlled ocean temperature and salinity profiles and monthly objective analyses with uncertainty estimates. J Geophysical Res Oceans 118:6704–6716. https://doi.org/10.1002/2013JC009067

Jacobs SS, Giulivi CF (2010) Large multidecadal salinity trends near the Pacific-Antarctic continental margin. J Clim 23:4508–4524. https://doi.org/10.1175/2010JCLI3284.1

Johnson GC, Purkey SG, Bullister JL (2008) Warming and freshening in the abyssal southeastern Indian Ocean. J Clim 21:5351–5363

Kobayashi T (2016) Deep NINJA observation. Seventeenth Argo Steering Team Meeting Document 7(2):1–12 http://www.argo.ucsd.edu/DeepNINJA_AST17.pdf

Kobayashi T (2018) Rapid volume reduction in Antarctic Bottom Water off the Adélie/George V Land coast observed by deep floats. Deep Sea Res., in press. https://doi.org/10.1016/j.dsr.2018.07.014

Kobayashi T, Watanabe K, Tachikawa M (2013) Deep NINJA collects profiles down to 4,000 meters. Sea Technol 54(2):41–44

Köhl A, Stammer D (2004) Optimal observations for variational data assimilation. J Phys Oceanogr 34(3):529–542. https://doi.org/10.1175/2513.1

Kouketsu S, Doi T, Kawano T, Masuda S, Sugiura N, Toyoda T, Igarashi H, Kawai Y, Katsumata K, Uchida H, Fukasawa M, Awaji T (2011) Deep ocean heat content changes estimated from observation and reanalysis product and their influence on sea level change. J Geophys Res 116:C03012. https://doi.org/10.1029/2010JC006464

Masuda, S., T. Awaji, N. Sugiura, J. P.Matthews, T. Toyoda, Y.Kawai, T. Doi, S. Kouketsu, H. Igarashi, K. Katsumata, H. Uchida, T. Kawano, M. Fukasawa, 2010: Simulated rapid warming of abyssal North Pacific waters, Science, 329, \–322, doi:https://doi.org/10.1126/science.1188703

Osafune S, Masuda S, Sugiura N, Doi T (2015) Evaluation of the applicability of the Estimated Ocean State for Climate Research (ESTOC) dataset. Geophys Res Lett 42(12):4903–4911

Purkey SG, Johnson GC (2010) Warming of global abyssal and deep Southern Ocean waters between the 1990s and the 2000s: contributions to global heat and sea level rise budgets. J Clim 23:6336–6351

Purkey SG, Johnson GC (2013) Antarctic Bottom Water warming and freshening: contributions to sea level rise, ocean freshwater budgets, and global heat gain. J Clim 26:6105–6122. https://doi.org/10.1175/JCLI-D-12-00834.1

Reynolds RW, Rayner NA, Smith TM, Stokes DC, Wang W (2002) An improved in situ and satellite SST analysis for climate. J Clim 15(13):1609–1625

Rintoul SR (2007) Rapid freshening of Antarctic Bottom Water formed in the Indian and Pacific oceans. Geophys Res Lett 34:L06606. https://doi.org/10.1029/2006GL028550

Riser SC, Freeland HJ, Roemmich D et al (2016) Fifteen years of ocean observations with the global Argo array. Nat Clim Chang 6:145–153

Sasaki Y (1970) Some basic formalisms in numerical variational analysis. Mon Weather Rev 98:875

Stammer D, Wunsch C, Fukumori I, Marshall J (2002) State estimation in modern oceanographic research. EOS 83(27):289 294–295

Wong, A., Keeley, R., Carval, T., and Argo Data Management Team: Argo Quality Control Manual for CTD and Trajectory Data 2018, doi:https://doi.org/10.13155/33951

Wunsch C, Heimbach P (2007) Practical global oceanic state estimation. Physica D-Nonlinear Phenomena 20:197–208. https://doi.org/10.1016/j.physd.2006.09.040

Wunsch C, Heimbach P (2013) Dynamically and kinematically consistent global ocean circulation and ice state estimates. In: Siedler G, Church J, Gould J, Griffies S (eds) Ocean circulation and climate: a 21st century perspective. Chapter, vol 21. Elsevier, pp 553–579. https://doi.org/10.1016/B978-0-12-391851-2.00021-0

Zhu Y, Derber J, Collard A, Dee D, Treadon R, Gayno G, Jung JA (2014) Enhanced radiance bias correction in the National Centers for Environmental Prediction's Gridpoint Statistical Interpolation data assimilation system. Q J R Meteorol Soc 140:1479–1492. https://doi.org/10.1002/qj.2233

Zilberman, N., and G. Maze, 2015: Report on the Deep Argo Implementation Workshop, http://www.argo.ucsd.edu/DAIW1report.pdf

13

High-resolution mapping and time-series measurements of ^{222}Rn concentrations and biogeochemical properties related to submarine groundwater discharge along the coast of Obama Bay, a semi-enclosed sea in Japan

Shiho Kobayashi[1*], Ryo Sugimoto[2], Hisami Honda[3], Yoji Miyata[4], Daisuke Tahara[2], Osamu Tominaga[2], Jun Shoji[5], Makoto Yamada[3], Satoshi Nakada[6] and Makoto Taniguchi[3]

Abstract

High-resolution mapping along the coast and time-series measurements of the radon-222 (^{222}Rn) concentrations in the shallow zone in a semi-enclosed sea, Obama Bay, Japan, were undertaken in 2013. The temporal and spatial variations in the ^{222}Rn concentrations were analyzed in parallel with meteorological conditions, physical–biogeochemical characteristics, and the submarine groundwater discharge (SGD) flux measured with a seepage meter. These data indicate that the groundwater influences the water properties of the bay and that the groundwater supply pathways are not limited to the local SGD. The concentrations of ^{222}Rn flowing into the bay from rivers was known to be relatively high because groundwater seeps from the river bed. High-^{222}Rn water was almost always present around the river mouth, and northward advection of the water affected the distribution of ^{222}Rn concentrations in the bay. The southward wind suppressed the advection of the high-^{222}Rn water and largely controlled the temporal variations in ^{222}Rn concentrations at a station located on the north side of the river mouth, whereas the local SGD affected the short-term changes in the ^{222}Rn concentrations. The concentrations of ^{222}Rn and chlorophyll-a, an indicator of phytoplankton biomass, show a significant positive correlation in the surface layer along the coastline in seasons when the nutrient supply was the main factor limiting primary productivity.

Keywords: Submarine groundwater discharge (SGD), ^{222}Rn monitoring, Chlorophyll-a, Nutrients, Wind-driven advection, Coastal seas

Introduction

It is challenging to obtain evidence of the effects of submarine groundwater discharges (SGDs) on marine ecosystems, partly because groundwater is an invisible source of the freshwater supplied to the sea. Fresh groundwater discharged from the sea bed often reduces the salinity of the bottom water and increases the concentrations of dissolved nutrients that are essential for primary producers. The correspondence between the low-salinity area associated with groundwater discharge and the high concentration of primary producers, such as benthic algae (Bruce 1925; Sanders 1979), seagrass (Kohout and Kolipinski 1967), and salt marsh plants (Nestler 1977), has been recognized since the 1920s. Based on those results, the ecological significance of groundwater discharge into coastal seas was discussed by Johannes (1980), but the distribution of salinity alone was an insufficient index with which to describe the

* Correspondence: shihok@kais.kyoto-u.ac.jp
[1]Field Science Education and Research Center, Kyoto University, Kitashirakawaoiwake, Sakyo-ku, Kyoto 606-8502, Japan
Full list of author information is available at the end of the article

influence of groundwater (Johannes 1980) because it does not distinguish SGD from other freshwater.

Since the 1980s, various methods have been developed to estimate the influence of SGD on marine primary production. The influence of SGD on primary producers and coastal ecosystems was quantified with the nitrogen budget method (Valiela and Costa 1988; Valiela et al. 1992). The effects of SGD on benthic microalgae and red tides were also quantified using the groundwater flow rate, which was calculated from the hydraulic gradient of groundwater on land (Campbell and Bate 1996; Laroche et al. 1997; Gobler and Sanudo-Wilhelmy 2001). A correspondence between the distributions of SGD and benthic microalgae was suggested with a visualization method using thermal imaging, based on the difference between the temperature of the seawater and that of SGD (Miller et al. 2004).

In recent years, the geochemical tracers radium and radon-222 (^{222}Rn) have been used to trace groundwater (Moore 1996; Burnett et al. 2006). ^{222}Rn is a short-lived radioisotope (half-life, 3.83 days) that is present in much higher concentrations in groundwater than in surface flow (Ellins et al. 1990). Using this powerful tracer, the influence of SGD on benthic microalgae was evaluated in the intertidal zone of the Yellow Sea (Waska 2011; Waska and Kim 2010). The nutrient fluxes caused by SGD and their potential effects on primary production have also been demonstrated in several coastal seas around Japan using ^{222}Rn (Shiokawa et al. 2013; Sugimoto et al. 2016).

A method of continuously monitoring the active concentrations of ^{222}Rn in water has been developed and used to visualize the distribution of groundwater in coastal seas (Burnett et al. 2001; Burnett and Dulaiova 2003). This method was first applied in tracking lines of over 100 km and successfully revealed the spatial distribution of SGD (Santos et al. 2008; Stieglitz et al. 2010). The applicability of this approach within a site of only a few kilometers in size was confirmed in a recent study using multiple ^{222}Rn detectors (Hosono et al. 2012).

SGD typically displays significant spatial and temporal variability (Burnett et al. 2006). Therefore, high-resolution mapping and time-series measurements of the ^{222}Rn concentrations and biogeochemical properties in the area of interest are required to quantify the influence of SGD on primary production along the coastline. In this study, we monitored the active concentrations of ^{222}Rn with multiple detectors from a boat (Stieglitz et al. 2010) to determine the distribution of ^{222}Rn concentrations in Obama Bay, Japan, within a study site of only a few kilometers in size (Fig. 1). We also monitored ^{222}Rn and the physical–biogeochemical properties of a site for a month and measured SGD directly with seepage meters (Lee 1977; Rutkowski et al. 1999; Taniguchi and Iwakawa 2004) to investigate the source of the ^{222}Rn. We then compared the distributions of SGD and chlorophyll-a (Chl-a), an indicator of the phytoplankton biomass. In Obama Bay, a correspondence between the high ^{222}Rn concentrations in the bottom layer and the elevated Chl-a concentrations were observed, suggesting that SGD influences phytoplankton growth (Honda et al. 2016). In this study, we investigated this relationship in more detail, focusing on the phytoplankton biomass in the shallow zone along the coastline.

Methods/Experimental

Study site

Obama Bay is a semi-enclosed embayment in Fukui Prefecture, Japan (Fig. 1a). The bay is one of the tributaries of Wakasa Bay, which is connected to the Japan Sea, and the tidal range is less than 20 cm. The surface area, volume, and mean depth of the bay are 58.7 km^2, 0.74 km^3, and 13 m, respectively. The annual precipitation in the watershed of Obama Bay is over 2000 mm/year. The discharge of two major rivers, the Kita and Minami rivers, is usually less than 10 m^3/s for each river between May and August (data on river discharge were provided by the Ministry of Land, Infrastructure, Transport and Tourism, Japan). There are significant groundwater resources in the basin, particularly on the Obama Plain (Sasajima and Sakamoto 1962), and there are more than 100 flowing artesian wells on this alluvial plain (Sugimoto et al. 2016).

The SGD rates (m^3/d) flowing into Obama Bay were estimated based on the monthly ^{222}Rn data and a steady-state mass balance model and were relatively high in spring (March–April), when snowmelt water was predominant, and in the rainy summer season (June–September) (Sugimoto et al. 2016). The annual mean concentration of ^{222}Rn in the river water flowing into Obama Bay was about 59.6 dpm/L, which was lower than that in the groundwater (mean 660 dpm/L) but was not negligible (Sugimoto et al. 2016).

The rainfall, snowfall, and wind velocity data, measured at the weather station designated "Obama" from 20 April to 15 May 2013, were obtained from the Automated Meteorological Data Acquisition System (AMeDAS) of the Japan Meteorological Agency. The mean monthly river discharge data from the largest river (Kita River) for 2013 were obtained from the Ministry of Land, Infrastructure, Transport and Tourism of Japan. The monthly changes in the meteorological conditions in 2013 are shown in Fig. 2. The total monthly rainfall was relatively high in January but gradually decreased and was lowest in May. It increased again in June and continued to be relatively high from June to August. It then increased markedly in September, which is the typhoon season. Significant snowfall was recorded in

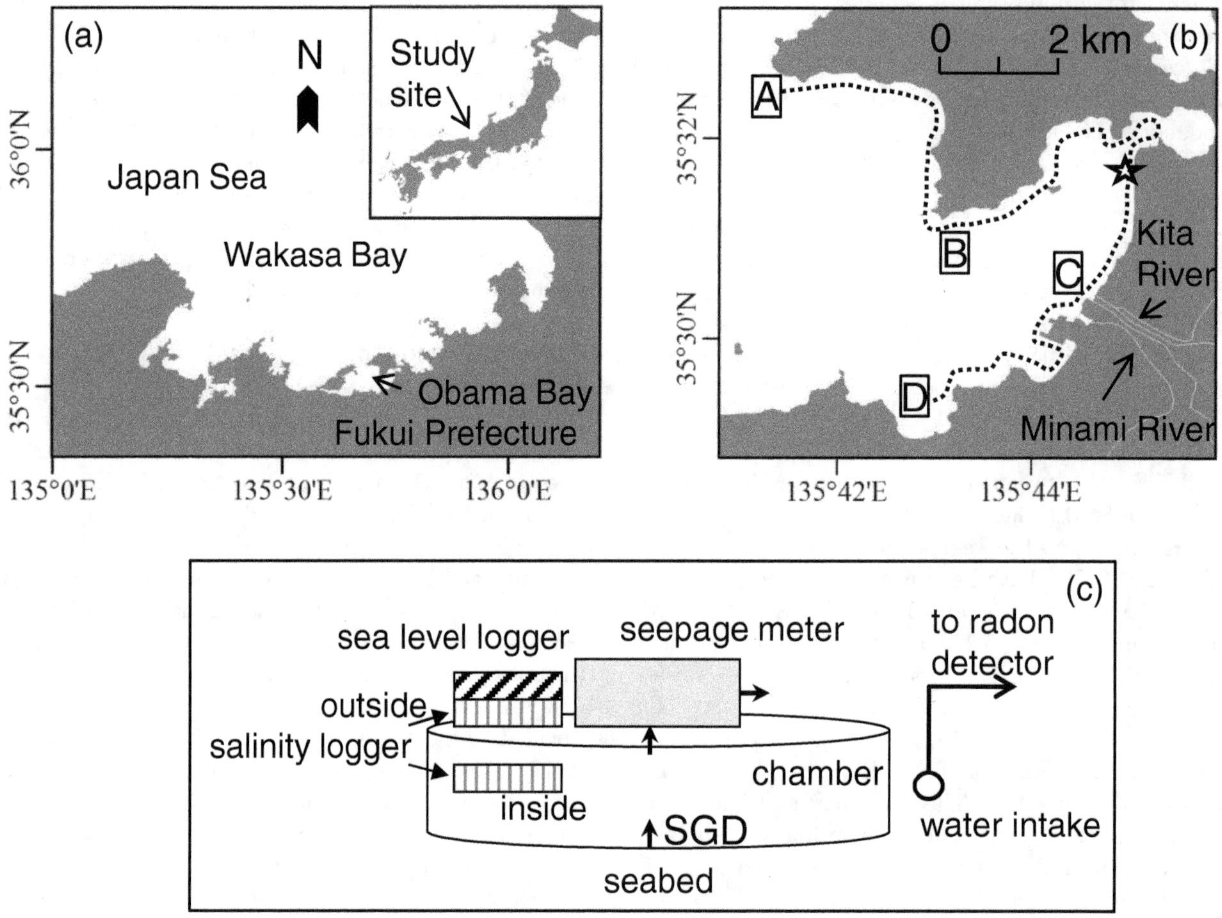

Fig. 1 **a** Location of the study site, Obama Bay. **b** Map showing the continuous ^{222}Rn monitoring track (*dotted line*), the station at which the time-series measurements of ^{222}Rn were made, and the seepage meter (*star mark*). Monitoring along the coastline from a boat was conducted from *A* to *D*. **c** Schematic illustration of the deployment of the seepage meters and loggers

winter (December to February). The mean monthly river discharge was relatively high from January to March, was lowest for the year in May, and increased slightly in June and July. It then increased markedly in September, when heavy rain fell in Obama City.

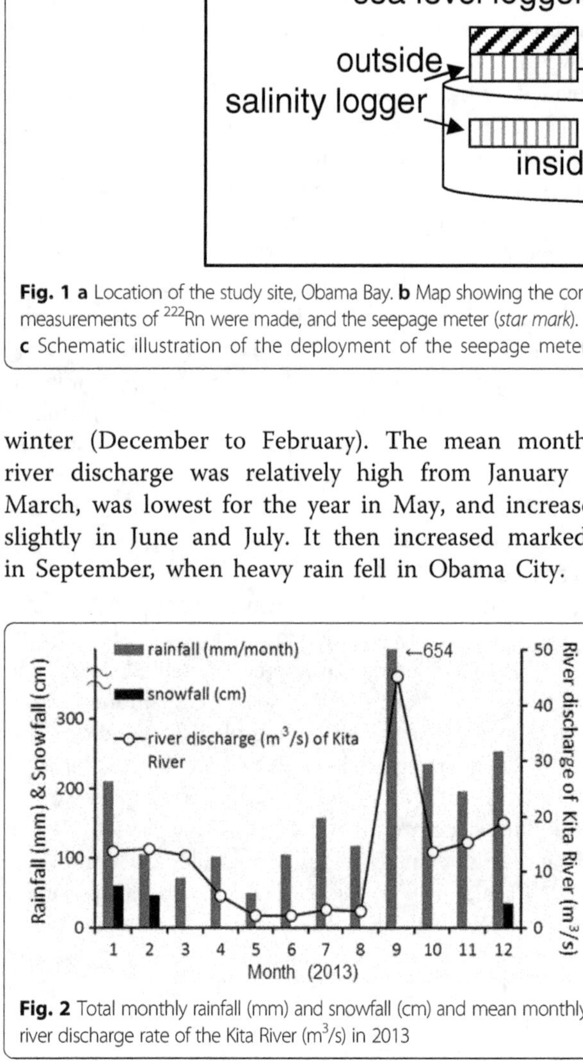

Fig. 2 Total monthly rainfall (mm) and snowfall (cm) and mean monthly river discharge rate of the Kita River (m^3/s) in 2013

Measurements and sampling

The distributions of the ^{222}Rn concentrations along the eastern coast of Obama Bay were measured on 13, 15, and 16 March, 7 June, 22 July, and 12 September, 2013, with electronic Rn detectors (RAD 7: Durridge) from a research boat belonging to Fukui Prefectural University. The boat ran close to the coast, within 10 m of land, where the depth of the water column was around 2 m, at a speed of 1–2 knots. The tracking line is shown in Fig. 1a. Monitoring was conducted from Tomari (A) to Seihama (D). The total distance of the tracking line was about 15 km.

Seawater was pumped at 5–10 L/min from 0.5 m below the surface, and the temperature and salinity of the water were measured every 1 min using a CTD instrument (AAQ1183: JFE-Advantech). The ^{222}Rn in the seawater was measured and analyzed according to Stieglitz et al. (2010), who developed an Rn detector system which consists of three Rn detectors connected in parallel and interfaced with an air–water exchanger (Dulaiova et al. 2005). We used two sets of Rn detector systems, with a measurement interval for each system of 10 min, so the average values and standard deviations of the ^{222}Rn concentrations were obtained every 5 min. Seawater samples for analyses of nutrient concentrations and Chl-a were collected every 10 min and every 20 min, respectively. However, seawater sampling for analyses of Chl-a failed at some points in March.

At a station on the tracking line (star in Fig. 1b), we measured the temporal variations in the ^{222}Rn concentrations in addition to the SGD fluxes, using a seepage meter. The water depth at the station was around 1 m. Divers deployed the seepage chamber (base radius, 15 cm) on the seabed at the station on 18 April 2013. A diagram of the seepage chambers is shown in Fig. 1c. Salinity and temperature loggers (Infinity-CT or MDS-CT, JFE-Advantec) were set inside and outside the chambers, whereas sea-level loggers (DIK-613A, Daikirika) were set only on the outsides of the chambers. The details of the seepage meters have been described by Taniguchi and Iwakawa (2004). We calibrated the seepage meters in the laboratory with a Coriolis flowmeter (FD-S, Keyence) before and after the field measurements were made. The seepage fluxes were determined in mL/min from each calibration curve and then converted to cm/d using the area of the chambers (cm^2). Seawater was pumped from the sea bottom at 5–10 L/min next to the chamber (Fig. 1c) from 20 April to 15 May in 2013. The ^{222}Rn concentrations in the seawater were measured every 10 min using a ^{222}Rn detector (RAD 7: Durridge) interfaced with an air–water exchanger located on land. The seawater samples for nutrient analysis were collected once a day from 20 April to 12 May in 2013. The times of water sampling are shown with the data on nutrient concentrations (Fig. 10).

Samples and data analysis

The seawater samples were filtered through GF/F 0.7 μm glass filters, frozen, and stored in the freezer. The concentrations of nitrate (NO_3), nitrite (NO_2), phosphate (PO_4), and dissolved silicate (DSi) were measured with an auto-analyzer (QuAAtro, BL-Tech). The concentration of ammonium (NH_4) was measured with a fluorometer (Trilogy, Turner Design) with a CDOM/NH4 module (Model 7200-041, Turner Design), using the ortho-phthaldialdehyde (OPA) method (Holmes et al. 1999). In this study, we defined dissolved inorganic nitrogen (DIN) as the sum of NO_3 and NO_2 because almost all the concentrations of NH_4 at the surface along the coastline were below the detection limit (0.1 μM). Dissolved inorganic phosphorous (DIP) is defined as PO_4. The concentration of Chl-a on the GF/F filters was quantified with a calibrated fluorometer (Trilogy, Turner Design). The analytical errors were within 10%, suggesting that the values analyzed were sufficiently accurate to draw conclusions from the data.

For the data analysis, we used the ^{222}Rn concentrations in the river water flowing into the bay, which had been measured monthly in 2013 by Sugimoto et al. (2016). Because approximately 80% of the riverine ^{222}Rn entering the bay is contributed by the two major rivers (Kita and Minami rivers), the mean ^{222}Rn concentrations in the two rivers were assumed to represent the value for the surface river waters (Sugimoto et al. 2016). We also used the ^{222}Rn concentrations and salinity in the terrestrial groundwater measured in March 2013 by Sugimoto et al. (2016).

The temporal changes in the tidal level from 20 April to 15 May 2013 were also obtained from the Japan Meteorological Agency. We used fast Fourier transforms (FFTs), providing power spectra, to evaluate the fluctuation cycles in the time series for wind velocity, sea level, SGD flux, and ^{222}Rn concentration.

We evaluated the factors limiting primary productivity using the equations of Steel (1962) for temperature (F_T) and nutrients (F_N), as follows:

$$F_T = T/T_{opt} \times \exp(1 - T/T_{opt}) \tag{1}$$

$$F_N = \mathrm{MIN}(\mathrm{DIN}/(K_N + \mathrm{DIN}), \mathrm{DIP}/(K_P + \mathrm{DIP})) \tag{2}$$

where T_{opt} (25 °C) is the optimum temperature for phytoplankton growth, and K_N (1.7 μM) and K_P (0.19 μM) are the half-saturation constants for DIN and DIP, respectively. These parameters were obtained from studies of the Japanese coast (Yanagi and Onitsuka 1999). We set T, DIN, and DIP to the maximum and minimum values extracted from the temperatures, DIN concentrations, and DIP concentrations observed in each month, respectively.

Results and Discussion

Distributions of the concentrations of ^{222}Rn, nutrients, and Chl-a along the coastline

High-resolution mapping of ^{222}Rn was performed in March, June, July, and September of 2013. The concentrations of ^{222}Rn, salinity, and Chl-a along the coastline in each month are shown in Figs. 3, 4, and 5, respectively. For the purpose of illustration, the tracking line was divided into three zones by four points (A–D, shown in Fig. 1b). The results of ^{222}Rn mapping show common patterns in the spatial distributions during June, July, and September: the concentrations were relatively high in the zone between B and D but low in the zone between A and B (Fig. 3). The concentration in the zone between C and D was highest in July. The average concentrations of ^{222}Rn on the tracking line (from A to D) in June, July, and September were nearly equal, at 2.8, 3.3, and 2.6 dpm/L, respectively. The average concentration of ^{222}Rn in March was 9.6 dpm/L, much higher than in the other 3 months. The distribution of

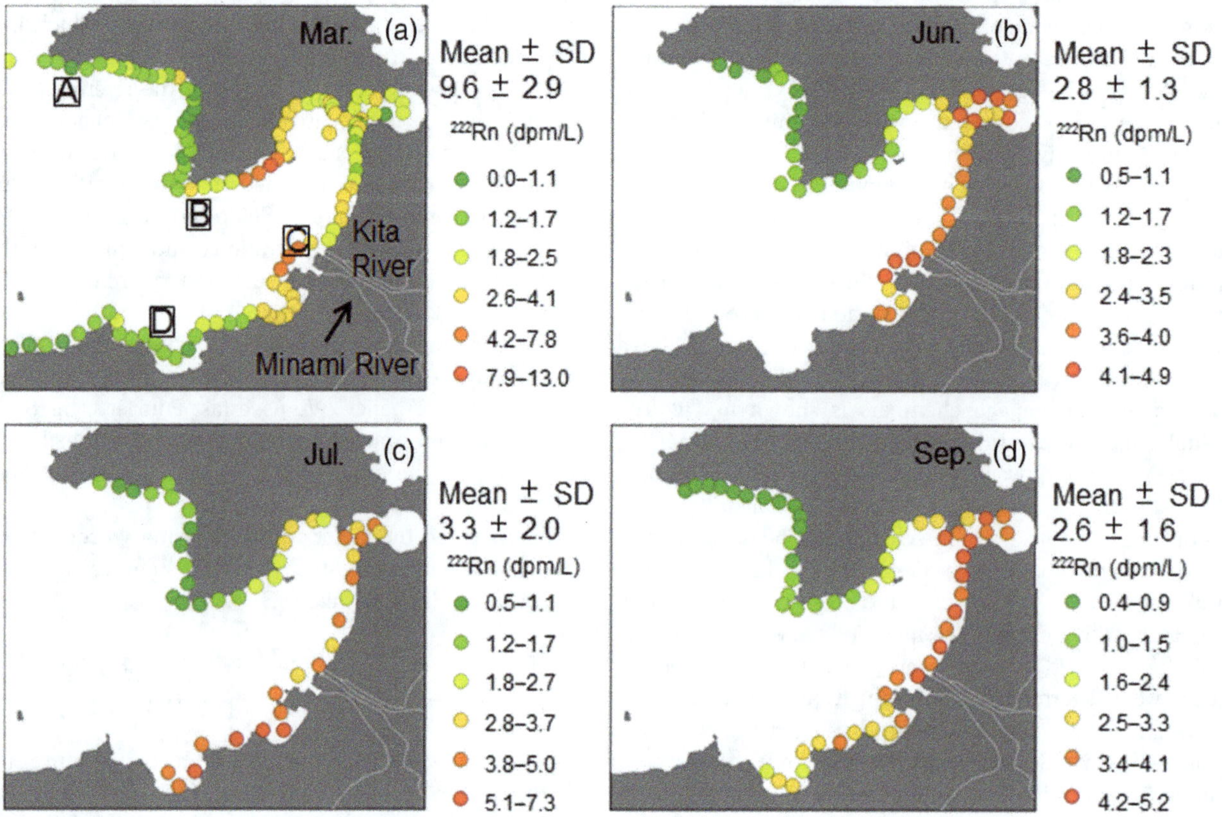

Fig. 3 Distributions of ^{222}Rn (dpm/L) in **a** March, **b** June, **c** July, and **d** September, determined by monitoring the coastline from a boat. All the data in March are shown in the same panel, although each datum was obtained on a different day

the ^{222}Rn concentration in March differed from that in the other months. The ^{222}Rn concentration was highest around point B but relatively low in the zone between B and C.

In June, July, and September, salinity was relatively low in the zone between B and D but high in the zone between A and B (Fig. 4). The distributions of low-salinity water corresponded strongly to the zones in which the ^{222}Rn concentrations were relatively high in all surveys. The distribution of Chl-a was constant throughout the survey, including in March. The Chl-a concentrations were relatively high in the zone between B and D but low in the zone between A and B (Fig. 5). The distribution of high-Chl-a water corresponded strongly to the zones in which the ^{222}Rn concentration was relatively high.

The correspondence between the distributions of ^{222}Rn and Chl-a is shown in Fig. 6. The distributions of both parameters corresponded well in June, July, and September, but not in March. The locations of the peak values for the two parameters differed in March. The correlation coefficients indicated significant correlations between the concentrations of ^{222}Rn and Chl-a in June, July, and September. Significant correlations were also obtained between the concentrations of ^{222}Rn and PO_4 in June and between salinity and NO_3 in March, June, and September (Table 1).

We created the zones shown in Fig. 1b using points A–D to investigate the differences in the biogeochemical properties of each zone along the coastline. The datasets obtained from all the observations made in March, June, July, and September were integrated, and the average concentrations of ^{222}Rn, salinity, NO_3, and PO_4 in each zone are shown in Fig. 7. In the zone between A and B, salinity was relatively high and the concentrations of ^{222}Rn, NO_3, and PO_4 were low. Salinity was lowest and the concentration of NO_3 was highest in the zone between B and C, whereas PO_4 was not high in this zone. The concentrations of ^{222}Rn and PO_4 were highest in the zone between C and D.

Temporal changes in the concentration of ^{222}Rn, SGD flux, and biogeochemical properties

Figure 8 shows temporal variations in the ^{222}Rn concentration measured with a Radon detector; the SGD flux measured with a seepage meter, the wind velocity (a parameter of the physical environment), and salinity levels inside and outside the seepage chamber at the point at

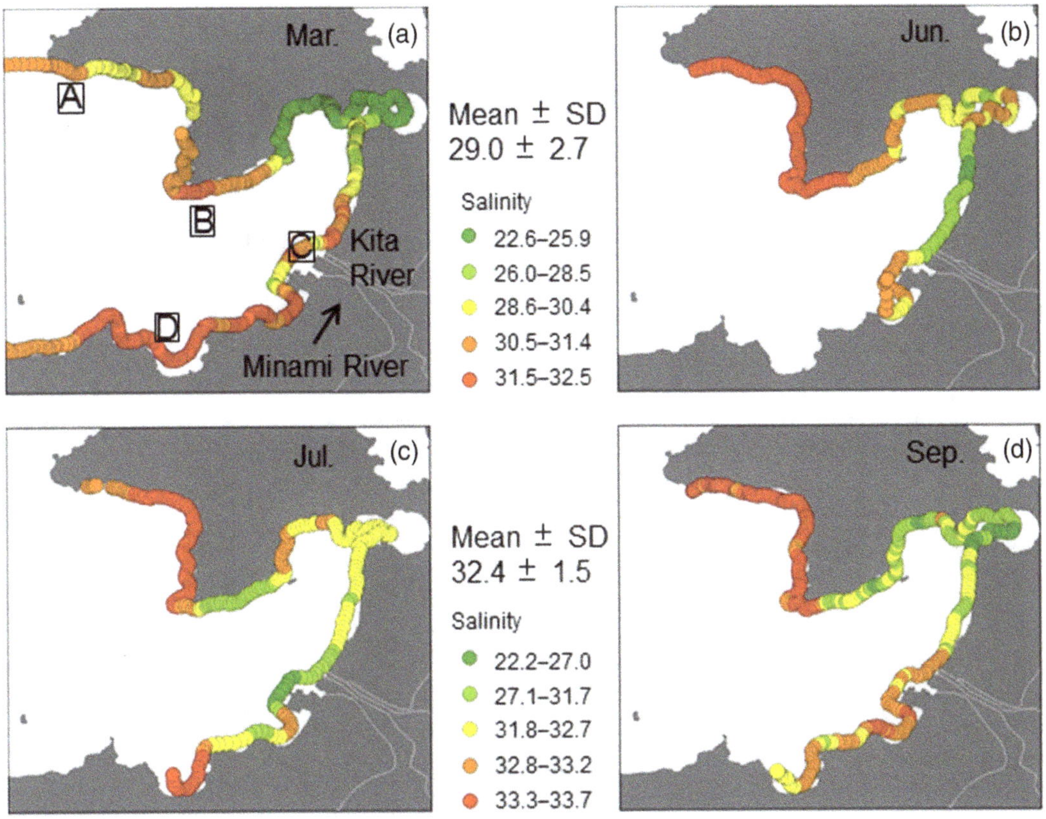

Fig. 4 Distributions of salinity in **a** March, **b** June, **c** July, and **d** September, determined by monitoring the coastline from a boat. All the data in March are shown in the same panel, although each datum was obtained on a different day

which the ^{222}Rn concentration was relatively high (star; location shown in Fig. 1b). The sea level varied from 0.7 to 1.2 m. The concentrations of ^{222}Rn increased and decreased over a 5–6 day cycle, although the SGD flux showed no such cycle (Fig. 8a). The minimum values of the cycle were nearly stable from April to May, whereas the SGD flux dropped sharply to almost zero in May. Figure 9 shows the FFT power spectra for the SGD flux, the sea level at the station, the concentration of ^{222}Rn, and the wind velocity and salinity inside and outside the chamber. The power spectrum for the SGD flux shows variations with diurnal (25 h) and semidiurnal (12.5 h) cycles, with characteristics similar to those of the power spectrum for sea level. The temporal variations in the ^{222}Rn concentrations throughout the monitoring period showed a 5–6 day cycle and were coincident with those in the wind velocity; ^{222}Rn increased when the northward wind became strong but decreased when the southward wind became dominant (Fig. 8b). The results of the FFT analysis also revealed that ^{222}Rn and wind velocity peaked with a periodicity of ~100 h (Fig. 9). As for the shorter cycles, the power spectra for wind, salinity, and the concentration of ^{222}Rn displayed peaks with periodicities of 24, 20–24, and 20 h, respectively.

Figure 8c shows the temporal variations in salinity inside and outside the chamber. The salinity outside the chamber shows negative peaks, with the period of low salinity ranging from 1 to 7 h. Peaks were observed almost daily, whereas no peaks occurred between 24 and 26 April, from 30 April to 3 May, or between 6 and 9 May. Several peaks coalesced into a single peak on 26–27 April. The salinity inside the chamber also showed negative peaks 6–12 h after the negative peaks in the outside salinity, although the magnitude of the declines was much smaller than the declines in the outside salinity. The inside salinity also displayed a declining trend in April and an increasing trend in May.

The temporal variations in salinity outside the chamber are presented with the wind velocity data (Fig. 8d). The northward wind showed peaks over a 5–6 day cycle. The negative peaks in the outside salinity appeared 5–6 h after the northward wind strengthened. The outside salinity remained high during periods when the southward wind was dominant.

Temporal variations in the concentrations of nutrients (NO_3, PO_4, DSi) and associated biogeochemical properties at the station described above are shown in Fig. 10. The plots in Fig. 10a indicate the times of water

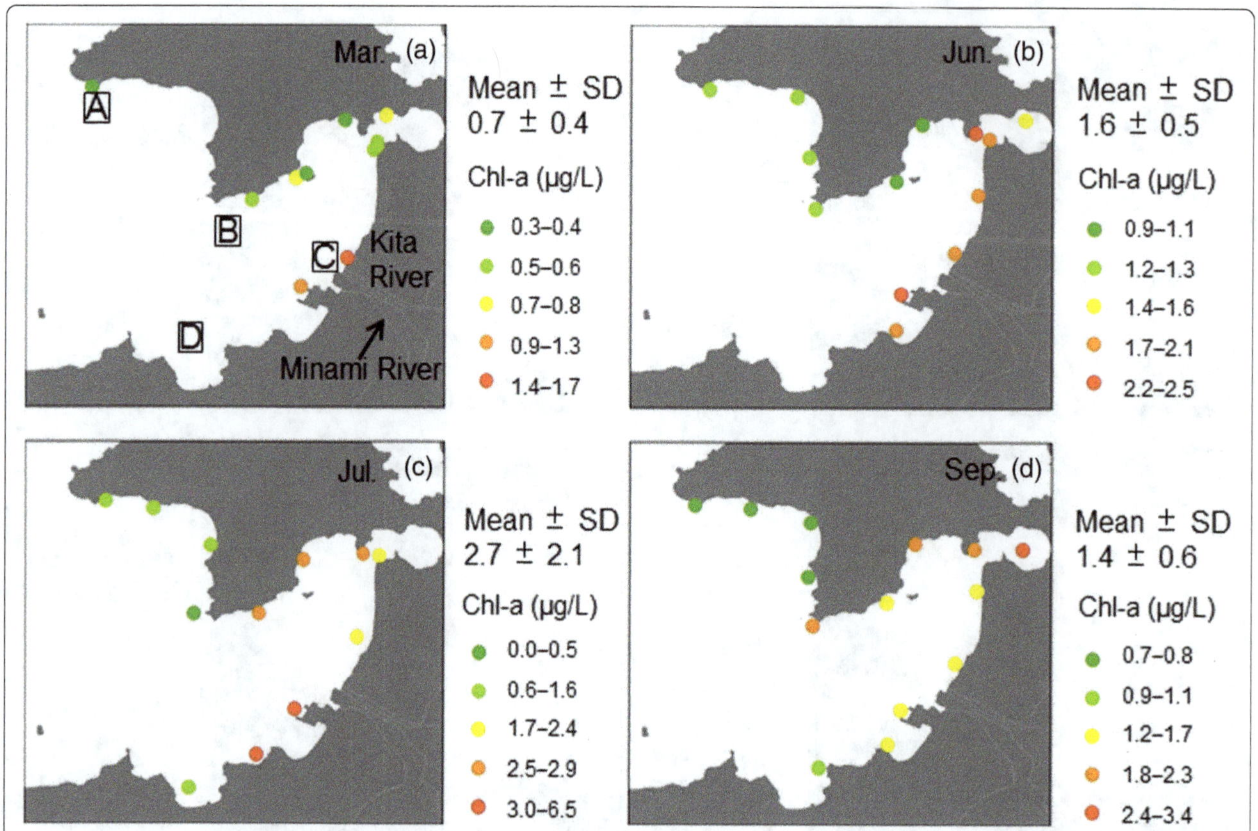

Fig. 5 Distributions of chlorophyll-a (µg/L) in **a** March, **b** June, **c** July, and **d** September, determined by monitoring the coastline from a boat. All the data in March are shown in the same panel, although each datum was obtained on a different day

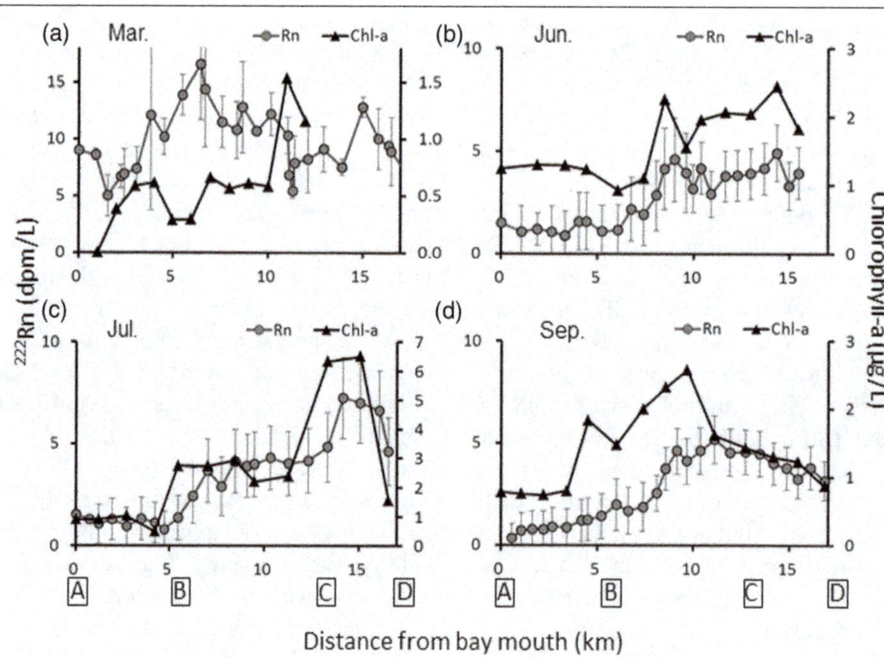

Fig. 6 Line plots of ^{222}Rn (dpm/L) and chlorophyll-a (µg/L) in **a** March, **b** June, **c** July, and **d** September, observed by monitoring the coastline from a boat. Letters *A* to *D* in **c** and **d** correspond to those in Fig. 1b

Table 1 Correlation coefficients for salinity and nutrients, ^{222}Rn and nutrients, and ^{222}Rn and chlorophyll-a

	Sal. vs. NO_3	Sal. vs. PO_4	^{222}Rn vs. NO_3	^{222}Rn vs. PO_4	^{222}Rn vs. Chl-a
March, 2013	−0.76**	0.12	0.37*	0.18	−0.17
June, 2013	−0.66**	−0.15	0.14	0.54**	0.91**
July, 2013	−0.35	−0.11	−0.12	0.37	0.79**
September, 2013	−0.82**	0.10	0.38	0.29	0.52*

Significant correlations are marked with asterisks
**Significant ($p < 0.01$) *significant ($p < 0.05$) *no mark* not significant
Sal. salinity, *vs.* versus

sampling for nutrient analysis. The ^{222}Rn concentrations at the daily sampling times were extracted from Fig. 8a and are plotted in Fig. 10b. The tide level shows diurnal (25 h) and semidiurnal (12.5 h) cycles, but the timing of water sampling was not related to the tidal cycle. The concentrations of PO_4 varied between 0.1 and 0.2 μM and increased and decreased over a 5–6 day cycle. The concentration of DSi ranged between 0.0 and 6.4 μM and showed several relatively large peaks that lasted for 1–2 days during the monitoring period. The concentrations of NO_3 ranged between 0.0 and 1.0 μM and peaked on the days when DSi increased. The temporal variations in the concentrations of PO_4 throughout the monitoring period showed a weak relationship with the daily average concentrations of ^{222}Rn ($r = 0.27$, $p = 0.1$), whereas the changes in NO_3 and DSi showed no relationship to the changes in ^{222}Rn.

Distribution of ^{222}Rn concentrations and its source

The concentrations of ^{222}Rn at the northern end of the tracking line, the point closest to the outside of the bay, were less than 1.5 dpm/L throughout the sampling period (Fig. 3), confirming that the influence of SGD was infrequent. In contrast, high ^{222}Rn concentrations (up to 15 dpm/L) were observed inside the bay throughout the monitoring period, suggesting the influence of SGD in Obama Bay.

The average ^{222}Rn concentration was highest in March (9.6 ± 2.9 dpm/L), whereas the values in June, July, and September (2.8 ± 1.3, 3.3 ± 2.0, and 2.6 ± 1.6 dpm/L, respectively) were relatively low. The average salinity in March was 29.0 ± 2.7, much lower than outside the bay at the end of February (33.3; Sugimoto et al. 2016). The average salinity was lowest in September (26.5 ± 2.6) in

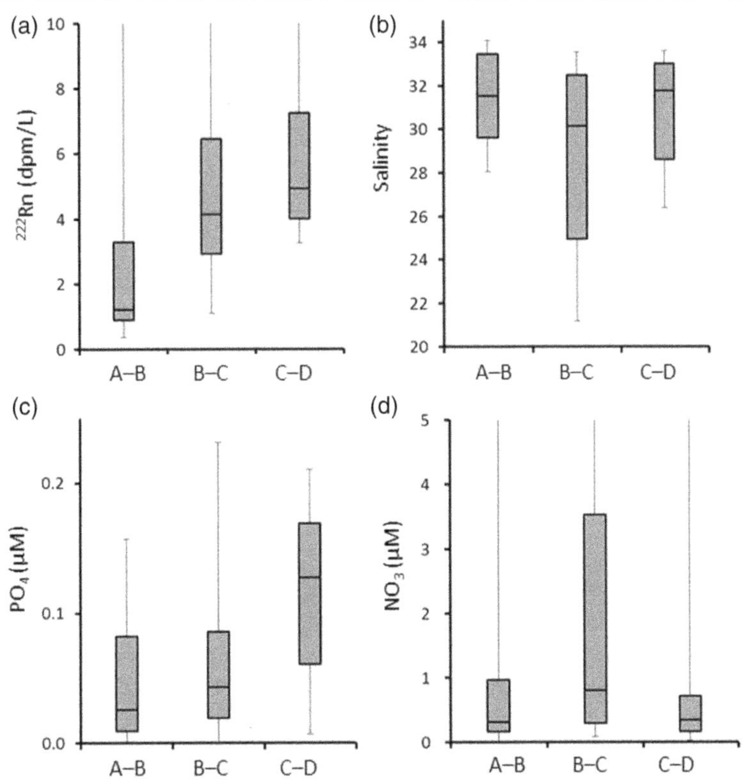

Fig. 7 Box plots of **a** ^{222}Rn (dpm/L), **b** salinity, **c** phosphorous (μM), and **d** nitrate (μM), showing the minimum and maximum values, 25th and 75th percentiles, and median values, from data obtained by monitoring the coastline from a boat in March, June, July, and September

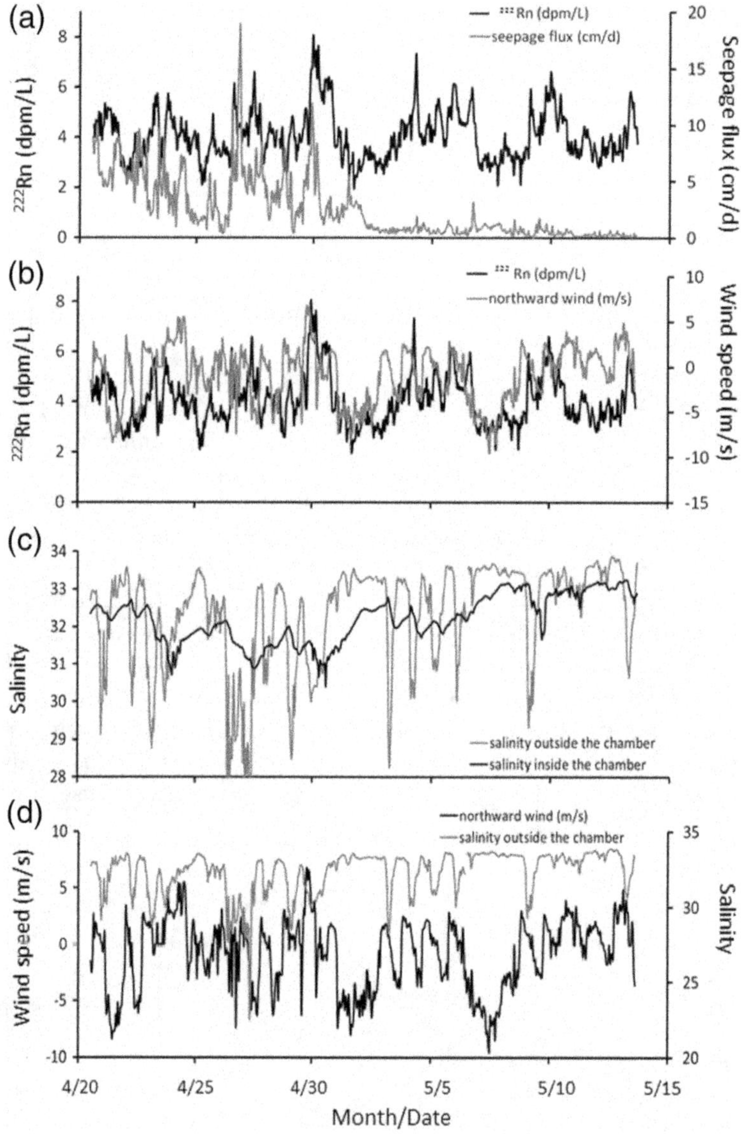

Fig. 8 Time series of **a** ^{222}Rn (dpm/L) and SGD fluxes measured with a seepage meter (cm/d), **b** ^{222}Rn (dpm/L) and northward wind (m/s), **c** salinity inside and outside the chamber, and **d** northward wind (m/s) and salinity outside the chamber, observed at a fixed station (*star*, shown in Fig. 1b) from 20 April to 15 May, 2013. The wind velocity data were obtained from AMeDAS

response to heavy typhoon rain. The river discharge was relatively high and largely stable from January to March, suggesting the influence of snowmelt water after significant snowfall in January and February (Fig. 2). These results are consistent with those of Sugimoto et al. (2016), who reported that the SGD rates flowing into Obama Bay, estimated from monthly ^{222}Rn data with a steady-state mass balance model, were relatively high in spring (March–April) when snowmelt water was the highest.

The ^{222}Rn concentration was measured along a few kilometers of the coastline every 5 min in this study, using two sets of a three-radon-detector system at a boat speed of 1–2 knots. The spatial resolution was slightly higher than that in previous pioneering works (Stieglitz et al. 2010; Hosono et al. 2012), and the distribution of the ^{222}Rn concentration along the coast of the semi-enclosed bay was determined in detail. Similar distributions were observed in June, July, and September, when the ^{222}Rn concentration was highest around the mouths of the Kita and Minami rivers, and the high-^{222}Rn water dispersed along the eastern coast toward the north (Fig. 3).

The first candidate source of high-^{222}Rn water is the river water. The ^{222}Rn concentrations in the river water collected at the river mouths in Obama Bay were relatively high (maximum, 234 dpm/L; Sugimoto et al. 2016) because groundwater seeps from the river bed and then flows rapidly down to the river mouth. These concentrations are consistent with those at other river mouths in

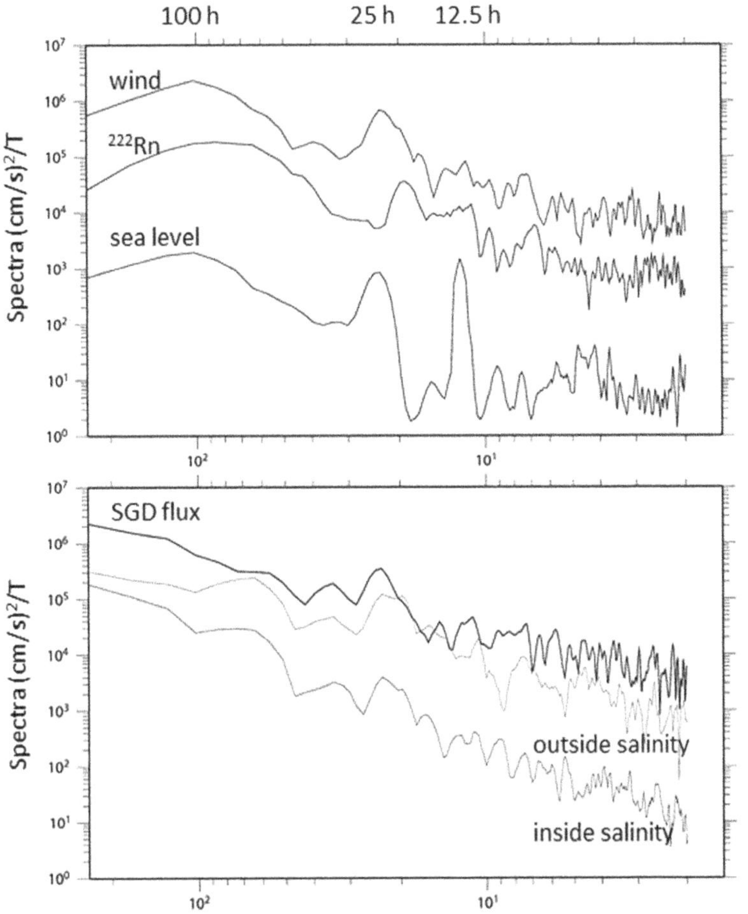

Fig. 9 Spectra obtained with fast Fourier transforms (FFTs) for the time series of ^{222}Rn, SGD flux, sea level, and salinity outside and inside the chamber, observed at a fixed station (*star* in Fig. 1b) from 20 April to 15 May 2013. The northward wind data were obtained from AMeDAS

Japan (e.g., Hosono et al. 2012; Shiokawa et al. 2013), whereas those at the mouths of continental rivers are much lower (e.g., Cable et al. 1996). The total river discharge was large enough (approximately 1.0–2.0 × 10^6 m^3/day; Sugimoto et al. 2016), and the axis of the bay was small enough (<4 km) for the river water to disperse throughout the bay within the period of ^{222}Rn degradation (half-life, 3.83 days).

The second candidate source of high-^{222}Rn water is the relatively large SGD offshore from the mouths of the Kita and Minami rivers, as has been mentioned in other surveys conducted by Obama City and models of the flow streams of terrestrial groundwater in the watersheds. Interviews with local residents confirmed the presence of fresh SGD at the point. However, we could not access the sites at the mouths of the Kita and Minami rivers to deploy seepage meters at that time. The concentrations of PO_4 and ^{222}Rn were both highest along the coastline and correlated with one another (Fig. 7; Table 1) around this site (in the zone between C and D, see Fig. 1b), indicating the presence of significant SGD.

Temporal changes in ^{222}Rn concentration and SGD flux

We continuously monitored the concentration of ^{222}Rn and measured the SGD flux using a seepage meter at a specific station (location shown in Fig. 1b) for a month, to determine the major sources of ^{222}Rn. The seepage meter located on the sea bottom indicated the presence of local SGD at the station (Fig. 8a). SGD includes the submarine fresh groundwater discharge (SFGD), originating from the freshwater on land, and the recirculated submarine groundwater discharge (RSGD), originating from the seawater that infiltrates the seabed under tidal influences (Taniguchi 2002a; Taniguchi et al. 2002b). The SFGD and RSGD generally show large differences in salinity, so they can be separated. We calculated them using the salinity inside and outside the chamber, as shown in Fig. 8c. The contribution rate of fresh SGD (SFGD/SGD × 100 [%]) was calculated as follows (Ishitobi et al. 2007):

$$SGD = SFGD + RSGD \tag{3}$$

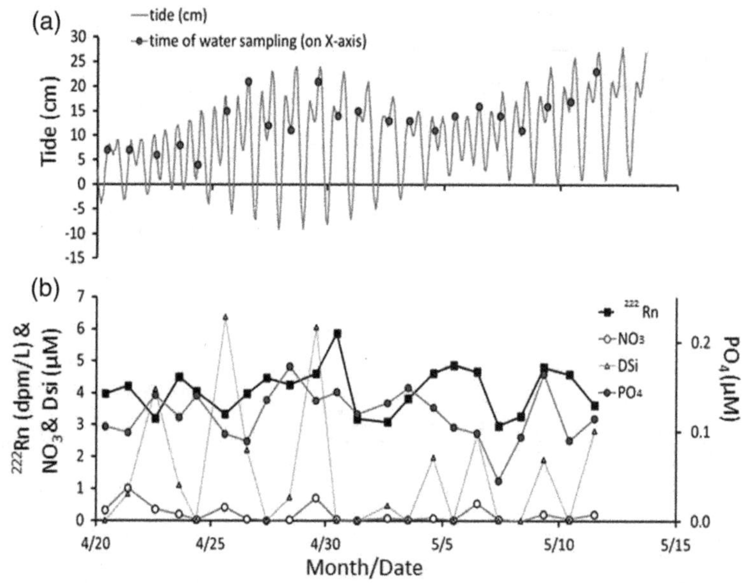

Fig. 10 As in Fig. 8 but for **a** tidal level (m) and the concentrations of **b** ^{222}Rn (dpm/L), phosphorous (μM), nitrate (μM), and silicate (μM). Values on the *X*-axes of the plots in **a** indicate the time (h) of water sampling for nutrient analysis, and those on the *Y*-axes of the plots are the tide levels at those times

$$SGD \times C_{sgd} = (SFGD \times C_{sfgd}) + (RSGD \times C_{rsgd}) \quad (4)$$

where SGD is the seepage flux determined with a seepage meter, C_{sgd} is the salinity inside the chamber, C_{rsgd} is the salinity outside the chamber, and C_{sfgd} is the salinity of the fresh groundwater on land. In this study, C_{sfgd}, which is the salinity (calculated from conductivity measurements) of the terrestrial groundwater calculated by Sugimoto et al. (2016), was assumed to be zero. The level of SGD flux changed sharply on 1 May, so we divided the datasets for SGD flux and salinity into the periods 20 April–1 May and 2–15 May. C_{rsgd} was set to the salinity averaged over each period, and C_{sgd} was set to the salinity averaged over the 1 h before the end of each period. We then solved the system of equations to calculate SFGD and RSGD. When C_{sgd} was larger than C_{rsgd}, the contribution of SFGD was assumed to be zero.

The contribution rates of SFGD (%) varied among months at the station: 1.4% in April but 0.0% in May 2013. The average SGD flux was 4.6 cm/day in April and 0.8 cm/day in May, with minimum and maximum values for the whole monitoring period of 0.1 and 18.9 cm/day, respectively. The average SGD fluxes were smaller than the results in the Chokai area (average SGD flux = 38.9 cm/day; Hosono et al. 2012), where the tidal range is as small as that in Obama Bay, but large SFGD fluxes were observed. The average SFGD flux, 0.41 cm/day, was consistent with that obtained with the ^{222}Rn mass balance model throughout the bay (0.62 cm/day; Sugimoto et al. 2016).

Temporal changes in the SGD flux and sea level showed variations with diurnal (25 h) and semidiurnal (12.5 h) cycles (Fig. 9), suggesting the influences of diurnal and semidiurnal tides. This indicates that the SGD at the site consisted mainly of tidally influenced RSGD. The observed SGD flux dropped sharply to almost zero, and the salinity inside the chamber increased in May, while the seasonal change in the SGD flux was consistent with that obtained with the ^{222}Rn mass balance model (Sugimoto et al. 2016). These trends indicated that the reduced SGD flux in May was partly attributable to the disappearance of the snowmelt water and partly to the rise in sea level from May, when SGD was suppressed. The seasonal changes in snowmelt water and sea level are mainly attributed to the seasonal change in air temperature, which varies little from year to year. These factors may explain the reduction in SGD observed in this study.

The minimum values of the 5–6 day cycle of the ^{222}Rn concentration were almost constant from April to May, whereas the local SGD flux observed at the station dropped sharply in May (Fig. 8), suggesting the influence of the advection of high-^{222}Rn water. The observed distribution of the ^{222}Rn concentration along the coast suggested that high-^{222}Rn water was present around the mouths of the Kita and Minami rivers during the observation period (Fig. 3). The results of the FFT analysis (i.e., that both the ^{222}Rn concentration and wind velocity showed peaked with a periodicity of ~100 h; Fig. 9) indicated that wind played an important role in the advection of high-^{222}Rn water.

We estimated the time (T_i) required for the high-^{222}Rn water near the mouths of the Kita and Minami rivers to reach the station. As an example, for the northward wind velocity of ~2 m/s that prevailed for 6 h daily between 2 and 13 May (Fig. 8), the velocity of the wind-driven flow, V (m/s), was calculated as follows:

$$V = \mathrm{ro\_a} \times C_d \times \mathrm{W\_y} \times \sqrt{\mathrm{W\_x}^2 + \mathrm{W\_y}^2 / \mathrm{ro\_s} \times \mathrm{dt/dz}} \quad (5)$$

where ro_a is the density of air, C_d is the friction coefficient at the sea surface, ro_s is the density of seawater, and W_x and W_y are the wind velocities along the X- and Y-axes, respectively. We set the Y- and X-axes to north–south and west–east, respectively; dt is the time scale on which the wind blows in a particular direction and dz is the friction depth. The friction depth (D) was defined as follows (Officer 1976):

$$D = \pi \sqrt{2 * N_z / f} \quad (6)$$

where π is 3.14, f is the Coriolis coefficient (7.3×10^{-5}) at latitude 35° north, and N_z is the eddy viscosity coefficient. D ranges from 5 to 50 m, assuming N_z ranges from 1 to 100 cm^2/s. The minimum value of D exceeds the water depth of the coastal zone of the bay, so we set dz to equal the water depth of 1–5 m. We assumed the following values: ro_a, 1.205 kg/m^3; C_d, 0.0013; ro_s, 1022 kg/m^3; W_x, zero; W_y, 2 m/s; and dt, 6 × 3600 s.

Using equation (5), the resulting V value is 0.03–0.13 m/s. We would expect the actual flow speed to be higher than this value because of the additional influence of the density-driven flow. Because the distances between the river mouths and the station were ~2 km, T_i is less than 1.0 day, which is much shorter than the half-life of ^{222}Rn (3.83 days). Therefore, the high-^{222}Rn water near the mouths of the Kita and Minami rivers is a potential source of ^{222}Rn in the area of the station.

Influence of the advection of high-^{222}Rn water on ^{222}Rn concentrations

The results of the FFT analysis indicated that both the ^{222}Rn concentration and the wind velocity peaked with a periodicity of ~100 h (Fig. 9), suggesting that wind was strongly related to the long-term temporal changes in the ^{222}Rn concentration at the observation site. In this section, we examined the influences of the advection of high-^{222}Rn water, caused by wind, on temporal changes in the concentrations of ^{222}Rn in more detail.

In the short term, the temporal changes in the wind, salinity, and ^{222}Rn concentration displayed variations with periods of around 24, 20–24, and 20 h, respectively (Fig. 9). Therefore, the cycle of the ^{222}Rn concentration was not perfectly coincident with the variations in wind velocity. This is considered to be partly because the advection of the low-salinity high-^{222}Rn water within a day was controlled not only by the wind but also by the tidal currents, the density flow (such as the estuarine circulation), vertical mixing, and the interactions of all these factors, and partly by the technological limitations of the FFT analysis, which were attributed to the spike-like changes in salinity and the ^{222}Rn concentration. Therefore, the temporal variation in salinity was mainly controlled by wind, in addition to many physical factors, and the temporal variation in the ^{222}Rn concentration could be synchronized to that in salinity.

The salinity at the station increased during the periods when the southward wind was dominant (Fig. 8d), suggesting that the northward dispersion of the low-salinity high-^{222}Rn water flowing from the mouths of Kita and Minami rivers was suppressed. The 1 h averaged ^{222}Rn concentrations, measured every 10 min at the station (location indicated by a star in Fig. 1b), are plotted versus salinity outside the chamber in Fig. 11, to allow a discussion of the influence of wind.

The northward wind was dominant on 30 April and then weakened on the next day. The southward wind became strong on 2 May (Fig. 8d), so we used the datasets for the ^{222}Rn concentration and salinity in the period 30 April–2 May. The endmembers and mixing curves are shown in Fig. 11a, and the data for the 3 days in this period, which were measured from noon on 1 day to noon on the next day, are plotted in Fig. 11b, along with a part of each mixing curve. The ^{222}Rn concentration and salinity of seawater outside the bay, measured on 26 April 2013, were 0.9 dpm/L and 34.0, respectively. The ^{222}Rn concentrations measured on the same date in the two major rivers (Kita and Minami rivers) were 72.5 and 73.6 dpm/L, respectively, so we used the mean value of 73.1 dpm/L as the endmember for river water. The mean concentrations of ^{222}Rn in the terrestrial groundwater and the seawater outside the bay were 660 dpm/L. The salinity of the river water and terrestrial groundwater was assumed to be zero, all according to Sugimoto et al. (2016). The ^{222}Rn concentration and salinity at 0.9 m under the seabed at the station were determined to be 54.2 ± 23.5 dpm/L and 31.3 by direct measurement (Sugimoto et al. 2017, unpublished data), so we used a value of 54 dpm/L as the endmember for RSGD.

Half the plots in Fig. 11b are located above the mixing curve between seawater outside the bay (SW) and river water (RW), suggesting that the local SGD contributed to the ^{222}Rn concentration, whereas the major component of SGD, either terrestrial groundwater (tGW) or recirculated submarine groundwater (RSGD), could not be determined. In contrast, the daily changes in the ^{222}Rn concentration were in response to changes in wind direction. The concentration of ^{222}Rn was relatively high on 30 April–1 May, the period in which northward wind was dominant, whereas the ^{222}Rn concentration gradually

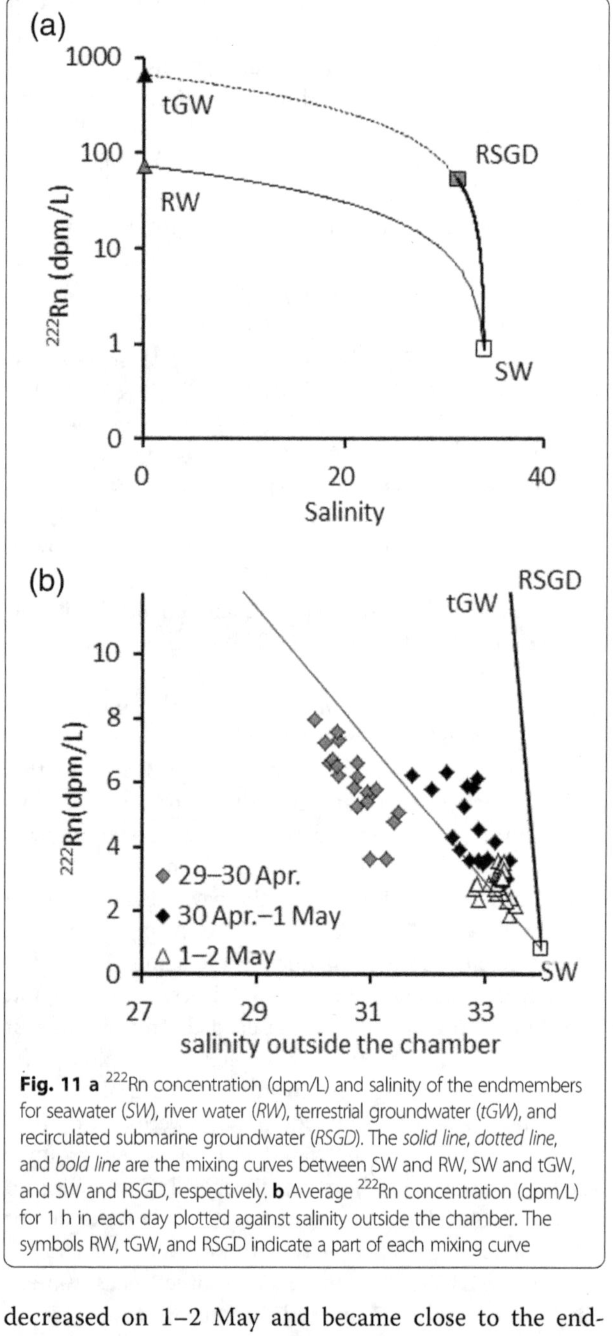

Fig. 11 a ^{222}Rn concentration (dpm/L) and salinity of the endmembers for seawater (*SW*), river water (*RW*), terrestrial groundwater (*tGW*), and recirculated submarine groundwater (*RSGD*). The *solid line*, *dotted line*, and *bold line* are the mixing curves between SW and RW, SW and tGW, and SW and RSGD, respectively. **b** Average ^{222}Rn concentration (dpm/L) for 1 h in each day plotted against salinity outside the chamber. The symbols RW, tGW, and RSGD indicate a part of each mixing curve

decreased on 1–2 May and became close to the endmember of SW on 2–3 May, the period in which the southward wind was dominant (Fig. 11b). These results provide evidence of the advection of high-^{222}Rn water, which was controlled mainly by the wind during the observation period. Moreover, as shown in Fig. 7, high-^{222}Rn and high-PO_4 seawater was present around the mouths of the Kita and Minami rivers (the zone between C and D, shown in Fig. 1b). This water moved northward when the northern wind became strong, supplying ^{222}Rn and PO_4 to the station at which the temporal variations in these parameters were measured.

The results of our various observations, including those made with seepage meters, the continuous monitoring of ^{222}Rn concentrations at a single station, and the monitoring survey of ^{222}Rn concentrations from a boat, together with measurements of the biogeochemical properties, suggest that the advection of high-^{222}Rn water plays an important role in controlling the temporal changes in the ^{222}Rn concentration at the station in the zone between B and C, whereas the local SGD might be another source of ^{222}Rn in the period when significant SGD flux was observed. The mass balance model (Burnett et al. 2008), which estimates SGD fluxes using ^{222}Rn records from monitoring surveys, could not be applied because the temporal changes in the advection of high-^{222}Rn water were significant.

Influence of SGD on the biogeochemical environment along the coastline

The results of the monitoring survey along the coastline suggest that groundwater, which is characterized by high ^{222}Rn, contributes to the nutrient supply in the bay. The N:P:Si molar ratio of the fresh groundwater on land was approximately 26:1:27, whereas that in the river water was 82:1:150 (Sugimoto et al. 2016), suggesting the groundwater is rich in PO_4. The concentrations of PO_4 and ^{222}Rn were both highest along the coastline in the zone between C and D (Fig. 7) and varied simultaneously (Fig. 10). This mutual correlation suggests that nutrients are supplied by SGD to the surface layer along the coastline.

The factors limiting primary productivity are listed in Table 2. The limitation factors for the maximum and minimum temperatures at each observation point (F_T) were nearly 1.0 in June, July, and September, suggesting that temperature was not a limiting factor in these seasons. F_T in March ranged from 0.64 to 0.81, indicating that temperature was one of the factors limiting primary productivity. Of the factors limiting primary productivity

Table 2 Maximum and minimum temperatures and concentrations of nutrients along the monitoring line shown by the dotted line in Fig. 1b in each month and limiting factors (F_T, T_N) for primary productivity

		Temp. (°C)	F_T	NO_{2+3} (μM)	DIP (μM)	F_N
March	Max.	11.9	0.80	20.7	0.18	0.49
	Min.	8.1	0.64	0.4	0.00	0.00
June	Max.	23.3	1.00	1.9	0.21	0.53
	Min.	20.6	0.98	0.1	0.00	0.00
July	Max.	29.8	0.98	8.3	0.23	0.55
	Min.	27.5	1.00	0.1	0.08	0.05
September	Max.	28.9	0.99	3.8	0.18	0.49
	Min.	26.3	1.00	0.0	0.00	0.00

Max. maximum, *Min.* minimum, *Temp.* temperature

in the sea, light intensity was not considered in this study because of a lack of data. However, light intensity was assumed to rarely limit primary production at the surface, as inferred from the limitation factor for light measured at the bottom in the shallow zone of Obama Bay (Sugimoto et al. 2017). F_N values for the minimum concentrations of nutrients at each observation point were all nearly 0.0, and F_N values for the maximum concentrations were around 0.5, suggesting that nutrients were a major factor limiting primary production at all the observation points.

The correspondence and correlation between the concentrations of ^{222}Rn and Chl-a at the surface in June, July, and September (Fig. 6, Table 1) suggest that the nutrients supplied by SGD influenced the phytoplankton biomass. The correlation of these factors in March was not significant, mainly because not only nutrients limited primary production but also temperature and probably light intensity. Of course, it is important to consider the influence of the seawater residence time on the concentrations of both ^{222}Rn and Chl-a. Although the links among ^{222}Rn, nutrient supply, and primary production must be more fully clarified in future studies, the results of this survey provide persuasive evidence that SGD influences the phytoplankton biomass in Obama Bay, whereas the nutrient supply pathways are not limited to the local SGD but are also influenced by its advection.

Conclusions

In this study, we conducted high-resolution mapping and made time-series measurements of ^{222}Rn and the biogeochemical properties along the coastline, combined with SGD measurements made with seepage meters, to examine the influence of SGD on the biogeochemical environment in Obama Bay. Our results show that (1) the observed concentrations of ^{222}Rn along the coast of the bay indicate that groundwater affected the biogeochemical properties of the bay and was not limited to the local SGD; (2) ^{222}Rn flowing into the bay from rivers (in which groundwater seeps from the river bed) had a strong effect on the distribution of ^{222}Rn concentrations along the coast; (3) high-^{222}Rn water was always present around the river mouths, and the northward advection of the water affected the distribution of ^{222}Rn concentrations; and (4) the southward wind suppressed the advection of high-^{222}Rn water and was the main control on temporal variations in the ^{222}Rn concentration at a station located to the north of the Kita and Minami rivers. Although the effect of the residence time of the seawater on the correlation between the concentrations of ^{222}Rn and Chl-a must be clarified in a future study, the results of this study provide persuasive evidence that the nutrients supplied by SGD influence the phytoplankton biomass in Obama Bay.

Abbreviations

AMeDAS: Automated Meteorological Data Acquisition System; Chl-a: Chlorophyll-a; SGD: Submarine groundwater discharge

Acknowledgements

The observations made with seepage meters along a line extending from onshore to offshore were made in collaboration with Wakasa High School. We are indebted to the members of the diving club and their teachers, Dr. Yasuyuki Kosaka and Mr. Hiroaki Hirayama, and the headmaster of Wakasa High School for their assistance. We sincerely thank Mr. Teruhiko Nakajima and Mr. Tomohiro Kawashiro of Fukui Prefectural University for supporting the development and deployment of the seepage chambers and the ^{222}Rn monitoring at the fixed stations. We are grateful to the journal editors and three anonymous reviewers for their helpful comments and suggestions.

Funding

This work was performed with the support of the Research Project Human–Environmental Security in Asia-Pacific Ring of Fire: Water–Energy–Food Nexus (R-08-Init) at the Research Institute for Humanity and Nature (RHIN).

Authors' contributions

SK, RS, DT, OT, and JS conceived and designed the study. HH, YM, and MY performed the field observations and data analysis. SN collaborated with the corresponding author in the data analysis and discussion. MT collaborated with the authors in the planning of the field observations. All the authors have read and approved the final manuscript.

Competing interests

The authors declare that they have no competing interests.

Author details

[1]Field Science Education and Research Center, Kyoto University, Kitashirakawaoiwake, Sakyo-ku, Kyoto 606-8502, Japan. [2]Research Center for Marine Bioresources, Fukui Prefectural University, 49-8-2, Katsumi, Obama, Fukui 917-0116, Japan. [3]Research Institute for Humanity and Nature, 457-4, Kamigamo Motoyama, Kita-ku, Kyoto 603-8047, Japan. [4]Idea Consultants Co., 2-2-2, Hayabuchi, Tsuzuki-ku, Yokohama, Kanagawa 224-0025, Japan. [5]Graduate School of Biosphere Science, Hiroshima University, 1-4-4 Kagamiyama, Higashi-hiroshima, Hiroshima 739-8528, Japan. [6]Graduate School of Maritime Sciences, Kobe University, 5-1-1 Fukae-minami, Higashi-nada-ku, Kobe 658-0022, Japan.

References

Bruce JR (1925) The metabolism of shore-living dinoflagellates. Br J Exp Biol 2: 413–426

Burnett WC, Dulaiova H (2003) Estimating the dynamics of groundwater input into the coastal zone via continuous radon-222 measurements. J Environ Radioact 69:21–35

Burnett WC, Kim G, Lane-Smith DA (2001) Continuous monitor for assessment of 222Rn in the coastal ocean. J Radioanal Nucl Chem 249:167–172

Burnett WC, Aggarwal PK, Aureli A, Bokuniewicz H, Cable JE, Charette MA, Kontar E, Krupa S, Kulkarni KM, Loveless A, Moore WS, Oberdorfer JA, Oliveira J, Ozyurt N, Povince P, Privitera AMG, Rajar R, Ramessur RT, Scholten J, Stieglitz J, Taniguchi M, Turner JV (2006) Quantifying submarine groundwater discharge in the coastal zone via multiple methods. Sci Total Environ 367:498–543

Burnett WC, Peterson R, Moore WS, de Oliveira J (2008) Radon and radium isotopes as racers of submarine groundwater discharge—results from the Ubatuba, Brazil SGD assessment intercomparison. Estuarine Coastal Shelf Res 76:501–511

Cable JE, Burnett WC, Chanton JP, Weatherly GL (1996) Estimating groundwater discharge into the northeastern Gulf of Mexico using radon-222. Earth Planet Sci Lett 144:591–604

Campbell EE, Bate GC (1996) Groundwater as a possible controller of surf diatom biomass. Revista Chilena De Historia Natural 69(4):503–510

Dulaiova H, Peterson R, Burnett WC, Lane-Smith DA (2005) A multi-detector continuous monitor for assessment of Rn-222 in the coastal ocean. J Radioanal Nucl Chem 263:361–365

Ellins KK, Roman-Mas A, Lee R (1990) Using 222Rn to examine groundwater/surface discharge interaction in the Rio Grade, DeManati, Puerto Rico. J Hydrol 115:319

Gobler CJ, Sanudo-Wilhelmy SA (2001) Temporal variability of groundwater seepage and brown tide blooms in a Long Island embayment. Mar Ecol Prog Ser 217:299–309

Holmes RM, Aminot A, Kerouel R, Hooker BA, Peterson BJ (1999) A simple and precise method for measuring ammonium in marine and freshwater ecosystems. Can J Fish Aquat 56(10):1801–1808

Honda H, Sugimoto R, Kobayashi S, Tahara D, Tominaga O (2016) Temporal and spatial variation in primary production in Obama Bay. Bull Jpn Soc Fish Oceanogr 80(4):269–282

Hosono T, Ono M, Burnett WC, Tokunaga T, Taniguchi M, Akimichi T (2012) Spatial distribution of submarine groundwater discharge and associated nutrients within a local coastal area. Environ Sci Technol 46:5319–5326

Ishitobi T, Taniguchi M, Shimada J (2007) Estimations of groundwater discharge and changes in fresh-salt water interface by measurements of submarine groundwater discharge in the coastal zone. Ground Water 49(3):191–204 (in Japanese with English abstract)

Johannes RE (1980) The ecological significance of the submarine discharge of groundwater. Mar Ecol Prog Ser 3:365–373

Kohout FA, Kolipinski MC (1967) Biological zonation related to groundwater discharge along the shore of Biscayne Bay. Miami, Florida. In: Lauff G (ed) Estuaries, vol 83. AAAS Publ., Washington, D.C, pp 488–499

Laroche J, Nuzzi R, Waters R, Wyman K, Falkowski P, Wallace D (1997) Brown Tide blooms in Long Island's coastal waters linked to interannual variability in groundwater flow. Glob Chang Biol 3(5):397–410

Lee DR (1977) A device for measuring seepage flux in lakes and estuaries. Limnol Oceanogr 22:140–147

Miller D, Ullman C, William J (2004) Ecological consequences of ground water discharge to Delaware Bay, United States. Ground Water 42(7):959–970

Moore WS (1996) Large groundwater inputs to coastal waters by 226Ra enrichments. Nature 380:612–614

Nestler J (1977) A preliminary study of the sediment hydrology of a Georgia salt marsh using Rhodamine WT as a tracer. Southeastern Geol 18:265–271

Officer CB (1976) Physical oceanography in estuaries. John Wiley & Sons, New York, p 465

Rutkowski CM, Burnett WC, Iverson RL, Chanton JP (1999) The effect of groundwater seepage on nutrient delivery and seagrass distribution in the Northeastern Gulf of Mexico. Estuaries 22:1033–1040

Sanders JG (1979) The importance of salinity in determining the morphology and composition of algal mats. Bot Mar 22:159–162

Santos IR, Niencheski F, Burnett WC, Peterson R, Chanton JP, Andrade CFF, Milani IB, Schmidt A, Knoeller K (2008) Tracing anthropogenically driven groundwater discharge into a coastal lagoon from southern Brazil. J Hydrol 353:275–293

Sasajima S, Sakamoto K (1962) Subsurface geology and groundwater of Obama Plain, Fukui pref., central Japan. Part 2: groundwater of Obama Plain. Memoirs of the Faculty of Liberal Arts, University of Fukui. Ser. II Natural science (in Japanese with English abstract)

Shiokawa M, Yamaguchi A, Umezawa Y (2013) Estimation of groundwater-derived nutrient inputs into the west coast of Ariake Bay. Bull Coastal Oceanography 50(2):157–167 (in Japanese with English abstract)

Steel JH (1962) Environmental control of photosynthesis in the sea. Limnol Oceanogr 7:137–150

Stieglitz TC, Cook PG, Burnett WC (2010) Inferring coastal processes from regional-scale mapping of 222Radon and salinity: examples from the Great Barrier Reef, Australia. J Environ Radioact 101:544–552

Sugimoto R, Honda H, Kobayashi S, Takao Y, Tahara D, Tominaga O, Taniguchi M (2016) Seasonal changes in submarine groundwater discharge and associated nutrient transport into a tideless semi-enclosed embayment (Obama Bay, Japan). Estuar Coasts 39:13–26

Sugimoto R, Kitagawa K, Nishi S, Honda H, Yamada M, Kobayashi S, Shoji J, Ohsawa S, Taniguchi M, Tominaga O (2017) Phytoplankton primary productivity around submarine groundwater discharge in nearshore coasts. Mar Ecol Prog Ser 563: 25–33

Taniguchi M (2002a) Tidal effects on submarine groundwater discharge into the ocean. Geophysical Res Lett 29(12). doi:10.1029/2002GL014987

Taniguchi M, Iwakawa H (2004) Submarine groundwater discharge in Osaka Bay, Japan. Limnology 5:25–32

Taniguchi M, Burnett WC, Cable JE, Turner JV (2002) Investigation of submarine groundwater discharge. Hydrol Process 16:2115–2129

Valiela I, Costa JE (1988) Eutrophication of Buttermilk Bay, a cape cod coastal embayment: concentrations of nutrients and watershed nutrient budgets. Environ Manag 12(4):539–553

Valiela I, Foreman K, LaMontagne M, Hersh D, Costa J, Peckol P, DeMeo-Andreson B, D'Avanzo C, Babione M, Sham C, Brawley J, Lajtha K (1992) Couplings of watersheds and coastal waters: sources and consequences of nutrients enrichment in Waquoit Bay, Massachusetts. Estuaries 15:443–457

Waska H (2011) Submarine groundwater discharge (SGD) as a main nutrient source for benthic and water-column primary production in a large intertidal environment of the Yellow Sea. J Sea Res 65(1):103–113

Waska H, Kim G (2010) Differences in microphytobenthos and macrofaunal abundances associated with groundwater discharge in the intertidal zone. Mar Ecol Prog Ser 407:159–172

Yanagi T, Onitsuka G (1999) Numerical model on the lower trophic level ecosystem in Hakata Bay. Umi-no-Kenkyu 8:245–251 (in Japanese with English abstract)

14

The role of snowmelt runoff on the ocean environment and scallop production in Funka Bay, Japan

Satoshi Nakada[1,6*], Katsuhisa Baba[2], Masatoshi Sato[2], Masafumi Natsuike[3], Yoichi Ishikawa[4], Toshiyuki Awaji[5], Koji Koyamada[1] and Sei-Ichi Saitoh[3]

Abstract

This study investigated the role of snowmelt runoff on water circulation, water mass formation, and the production of cultured scallop larvae, as a part of a land-sea linkage, by analyzing hydrological data in conjunction with nutrient data and by conducting sensitivity experiments based on a coupled land-sea model of Funka Bay, Japan, a typical semi-enclosed bay. A comparison between observed data and the simulated runoff showed that, using newly compiled datasets of nutrient concentrations in rivers and groundwater, the model was sufficiently accurate to estimate the terrestrial dissolved inorganic nitrogen (DIN) flux from the river and submarine groundwater discharges (SGDs). The average volume flux from the SGDs accounted for 26% of the riverine runoff flux. The DIN flux from SGDs accounted for a maximum of 40% of the total DIN loading to the bay before the snowmelt period. Sensitivity experiments using an ocean simulation indicated that the freshwater flux supplied by snowmelt runoff not only enhances clockwise circulations along with upwelling along the coast, but also modifies the distributions of wintertime water masses in the bay. However, the snowmelt runoff has little effect on larvae transport since wind forcing, rather than riverine buoyancy, dominates the circulation patterns. The annual density of scallop spat was highly correlated with snowmelt runoffs associated with high DIN concentrations, which suggested that riverine nutrients can increase the biomass of phytoplankton in near-shore seas and improve food availability for scallop spawners, resulting in increased egg production in March to April. Therefore, the nutrient flux from agricultural source areas through the large snowmelt runoff has an important role in larvae production. Land-sea linkages need to be identified to design sustainable and synergetic systems of aquaculture and agriculture for the integrated management of coastal regions.

Keywords: Snowmelt runoff; Scallop aquaculture; Land-sea linkage; OGCM; Nutrient flux; Submarine groundwater discharge

Background

Snowmelt runoff can greatly affect not only the ocean's physical environments, through volume and buoyancy fluxes, but also oceanic ecosystems, through nutrient fluxes in coastal seas (Uiboupin and Arino 2010). Figure 1 displays a map of months showing seasonal river runoff maxima in coastal regions, climatological areas of snow cover, and central lines of intertropical convergence zones (ITCZs). Many rivers located around the ITCZ during June to August had their maximum runoff during the same rainy period, although there were also maximum runoff periods during September and October, with time lags of a few months from the rainy period. The same tendency is predominant for many of the rivers located around the ITCZ during October to December and December to February. Areas of deep snow cover were mainly found in the northern hemisphere and rarely found in the southern hemisphere. The maximum seasonal runoffs occurred between December and August at middle latitudes (35° to 50°N) and between May and October at latitudes higher

* Correspondence: snakada@maritime.kobe-u.ac.jp
[1]Institute for Liberal Arts and Sciences, Kyoto University, Kitashirakawa-Oiwake Cho, Sakyo-ku, Kyoto 606-8502, Japan
[6]Graduate School of Maritime Sciences, Kobe University, 5-1-1 Fukae-minami, Higashi-nada-ku, Kobe 658-0022, Japan
Full list of author information is available at the end of the article

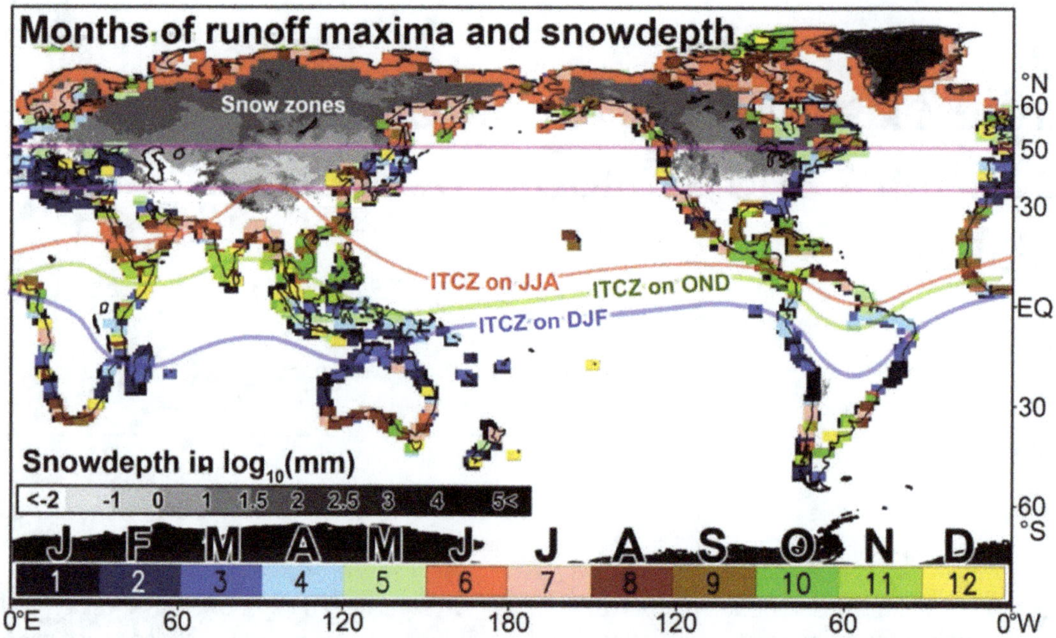

Figure 1 Climatological mean global snow cover and map of months showing seasonal runoff maxima. Chart of climatological mean global snow cover derived from datasets of the terrestrial water budget data archive (Willmott and Matsuura 2001) and map of months showing seasonal runoff maxima based on the climatological runoff dataset (Dai and Trenberth 2002). Red, green, and blue lines indicate the intertropical convergence zones (ITCZ) in JJA (June, July, and August), in OND (October, November, and December), and in DJF (December, January, and February), respectively. Purple lines indicate the latitudes 35°N and 50°N.

than 50°N. The maximum seasonal runoffs occurred during March to April and March to June (snowmelt seasons) for middle and higher latitudes, respectively, which could be due to the large-volume snowmelt runoffs that flow into coastal oceans. Therefore, coastal seas, over approximately two-thirds of the northern hemisphere latitude band, can experience snowmelt runoff between spring and early summer.

Fisheries and aquafarms located in coastal seas can be influenced by terrestrial inputs, such as freshwater, nutrients, and momentum through inflow from rivers (Kimmerer 2002). Aquacultures operate mainly in coastal oceans and currently contribute approximately 48% of the aquatic animal food for human consumption (FAO 2011; Bondad-Reantaso et al. 2012). The fish and shellfish populations of coastal seas are also influenced by the input of terrestrial nutrients through submarine groundwater discharge (SGD) (Liu et al. 2011; Nakada et al. 2011). Large snowmelt runoffs, along with high nutrient fluxes, can affect fishery production in coastal oceans. However, no study has yet investigated how snowmelt runoff influences the aquaculture production in coastal oceans. Estimation of nutrient loading through rivers and SGDs around the snowmelt season is essential to investigate the role of snowmelt runoff in fishery production.

A map of the maximum snow depth in Japan (Figure 2) shows that the northern half of Japan is snow-covered during the winter; in particular, the western coastal areas of Honshu and Hokkaido islands are in the snowbelt (maximum snow depth >200 cm). River runoffs from snow-covered areas reached their seasonal maxima between March and May, with the majority of maxima reached in April, excluding local areas around peninsulas and islands (e.g., Noto Peninsula and Sado Island), and the Pacific side. On the Pacific side, runoffs reached maximum levels in April to May and August to September. The relationship between snow-covered areas and runoff maxima indicates that coastal seas may be affected by snowmelt runoffs in Japan. In Funka Bay (northern Japan; Figure 3), the large amount of snowmelt runoff can affect the bay waters, supporting vast primary production through terrestrial and oceanic nutrients (Yoshimura and Kudo 2003). The Japanese scallop, which represents an important fishery resource in the bay (Minoda and Fuji 1985), is cultivated in aquafarms (Figure 3) along the coast (10% of the area of Funka Bay). The scallops feed on diatoms to spawn eggs in the spring and during the snowmelt period (Shimada et al. 2000). Larvae that hatch from the eggs are released from the aquafarms and transported by clockwise circulation (Takahashi et al. 2010). Aquaculturists must effectively collect larvae or spats released from spawners to achieve high spat densities (Baba et al. 2009), since the spat density affects the annual production of cultured scallops.

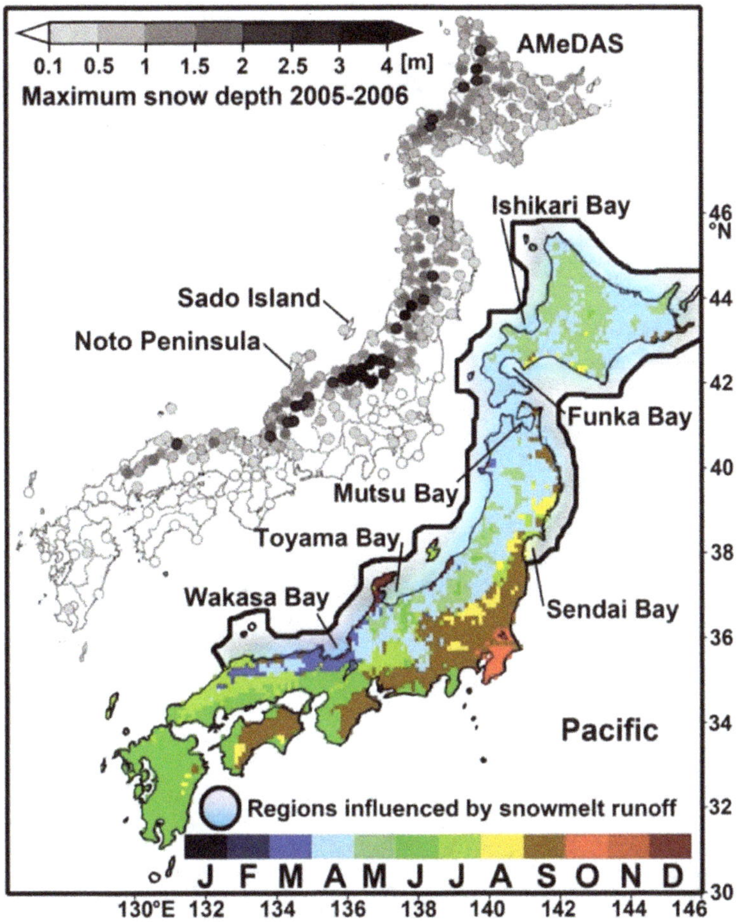

Figure 2 Snow-cover chart and map of months of runoff maxima. Snow-cover chart estimated using the AMeDAS dataset (JMA) and a map of months of runoff maxima in Japan based on simulated data (Yoshimura et al. 2008). The capital letters along the color bars are abbreviations for each month of the year.

Funka Bay is a typical semi-enclosed bay. The subarctic Oyashio water (OW) and subtropical Tsugaru water (TW), originating from the Kuroshio, alternately flow into Funka Bay from the North Pacific (Figure 4a) and mix over various timescales (Nakada et al. 2013a). This paper does not distinguish the coastal Oyashio water from the Oyashio water when examining the role of riverine water in the Funka Bay water masses. These unique oceanographic features are crucial for fishery resource management (e.g., skipjack, salmon, and kelp) and ocean biodiversity (Ohtani and Kido 1980). The majority of rivers that flow into the bay (Figure 3) have small watersheds (up to approximately 500 km^2) with steep gradients (1/25). The mountains surrounding Funka Bay generally grow beech forests (*Fagus crenata*). The watersheds include mostly agricultural lands, but it is unclear whether they are nutrient sources for the primary production around aquafarms. Figure 4b highlights the high concentration of chlorophyll *a* (chl *a*) distributed along the coast, although the presence of colored dissolved organic material (CDOM) and total suspended matter (TSM) in coastal waters may cause errors in chl *a* maps (Darecki and Stramski 2004). The map of chl *a* implies that terrestrial runoff can supply abundant nutrients and sediments, including CDOM and TSM, around aquafarms through snowmelt events and thereby greatly influences coastal fishery productivity. Terrestrial water can remain in the bay for more than a month owing to weak water exchange (Miyake et al. 1988).

Recently, a coupled land-sea model was developed to generate an operational ocean prediction system in the bay to inform fishermen of the ocean state in real time (Nakada et al. 2012a). The model reproduced realistic river runoffs and oceanographic features, such as the clockwise circulation, that agreed quantitatively with observations. New comprehensive analyses can expect to use high-resolution datasets produced by coupled land-sea simulations that are validated by observational datasets. Several studies have employed this approach, using the coupled model to investigate the oceanic dynamical behaviors of sub-mesoscale eddies in coastal oceans (e.g., Zhao et al. 2013).

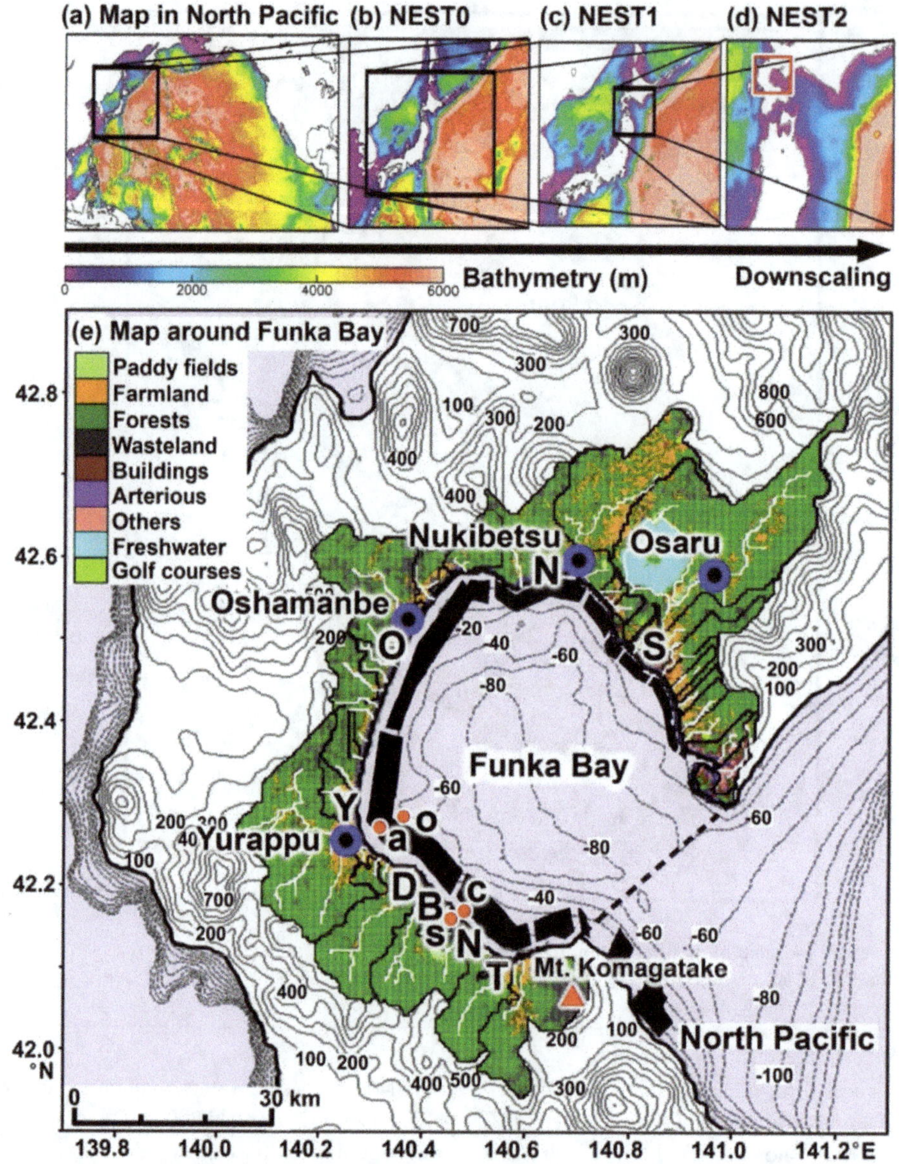

Figure 3 Bathymetry maps of numerical domains in the North Pacific and map of the study area. (a) Bathymetry map of the North Pacific showing the three-step nesting approach from **(b)** the northwestern North Pacific model (NEST0) to **(c)** the medium-scale model (NEST1) to **(d)** the coastal-scale model (NEST2). Colors indicate the depth in meters. The red square denotes the study area. **(e)** Map of the study area showing watersheds and aquafarms (black blocks) around Funka Bay. The colors indicate land use in the watersheds. The white lines and colored areas indicate river paths and watersheds, respectively. Blue circles with black dots represent river observation stations with an emphasis on the four major watersheds (solid line). Gray lines indicate isobaths and topographic contours in meters. The dashed line defines the mouth of the bay. The eight capital letters (T, N, B, D, Y, O, N, and S) are the abbreviations for the major rivers (Torizaki, Nigori, Otoshibe, Nodaoi, Yurappu, Oshamanbe, Nukibetsu, and Osaru rivers, respectively). Red dots denote water sampling sites for water in streams (s), near-shore (c), aquafarms (a), and offshore seas (o).

In this study, we investigated the role of snowmelt runoff in water circulation, water mass formation, and the production of scallop larvae in the bay to reveal a land-sea linkage using hydrological data in conjunction with nutrient data and by conducting sensitivity experiments based on the coupled land-sea model. The paper is organized as follows: the second section describes the details of numerical methods and nutrient analyses of sampled waters. The third section presents the simulated results and their validations and describes sensitivity experiments for the snowmelt runoff. The fourth section discusses the relationships between the terrestrial runoff and oceanic environment, in terms of nutrient supply and larval transport related to scallop production. The fifth section provides the conclusions.

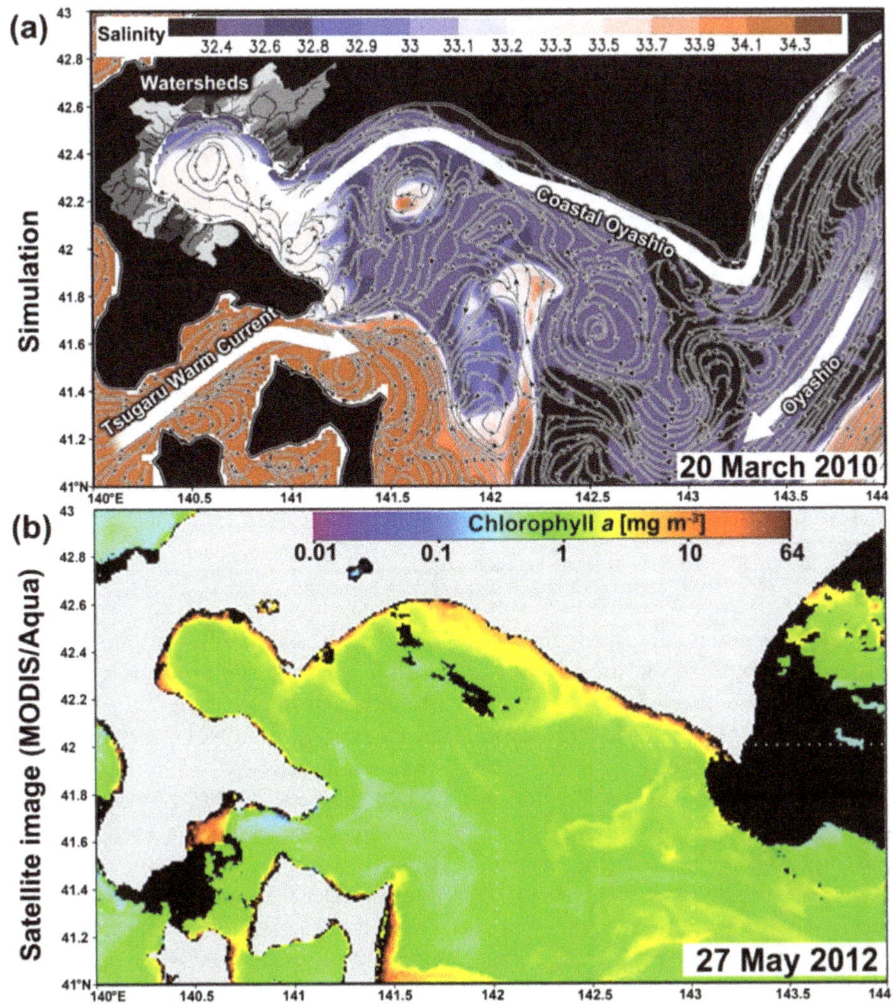

Figure 4 Distributions of simulated salinity and current, and satellite image of chl *a* concentrations. (a) Horizontal distributions of the simulated salinity (colors) and current at a depth of −12 m on 20 March 2010. The white stream lines with arrows indicate the flow direction estimated by the simulated velocity field. **(b)** Satellite image of chlorophyll *a* concentrations (mg m^{-3}) estimated from MODIS/Aqua ocean color remote sensing data around Funka Bay, southwestern Hokkaido, on 27 May 2012. The land and cloud data are represented by gray and black, respectively.

Methods

Overview of the coupled land-sea model

To simulate ocean states, land-surface conditions, and river runoff, a coupled land-sea model (Figure 5) was generated using a three-step nesting method with four-dimensional variational data assimilation. The model consisted of the hydrometeorological and multi-runoff utility model (HaRUM) (Nakada et al. 2012a) and a three-dimensional ocean general circulation model (OGCM) developed at Kyoto University (Ishikawa et al. 2009). The HaRUM was designed to minimize the calculation cost for ocean forecasts and to calculate the SGD.

River discharges were predicted using a distributed-tank model, based on radiation, heat, and water mass budgets, by calculating physical components including the solar insolation ($S\downarrow$), downward longwave radiation flux from the atmosphere ($L_a\downarrow$), upward longwave radiation flux from the ground surface ($L_g\uparrow$), sensible heat flux (H), and latent heat flux (lE), which determines rainfall and snowfall (T_r), precipitation (P), evapotranspiration (E), and snowmelt (M) (see Kondo and Watanabe (1992) and Kondo and Yamazaki (1990) for detailed formulae). These budgets can be calculated using the five meteorological properties (rainfall, air temperature [T_a], wind speed, cloud cover, and relative humidity) from the grid point value datasets of the meso-scale model (GPV-MSM) provided by the Japan Meteorological Agency on an hourly basis. The river discharge, Q_t, generated by the net water input ($P + M - E$) is predicted for each grid of the distributed-tank model (Figure 5). The model has three serial tanks, represented as n = 1 to 3, ten model coefficients (A_n, B_n, and S_n) for the side outlets (Q_n) and bottom outlets (L_n), and specified depth thresholds of water levels (Z_n) in each tank. If snow covers the ground surface (Figure 5) and the

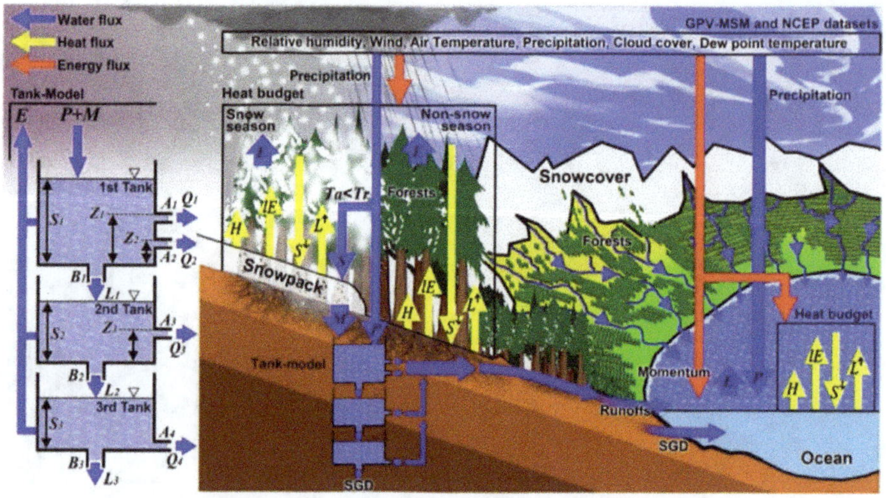

Figure 5 Flowchart to estimate river runoff and structure of the tank model with three serial tanks. The physical quantities depicted in the diagram are described in the text.

snow surface temperature satisfies $T_s < T_a$ (where T_s denotes the ground surface temperature), the snowmelt, M, is calculated by a radiation and heat balance.

A three-dimensional OGCM developed at Kyoto University was employed using the hydrostatic and Boussinesq approximations and a realistic bottom topography. The OGCM implemented a hybrid σ-z vertical coordinate system to improve simulation of the free surface motion of the ocean. To enhance the representation of upper ocean circulation further, this model adopted several sophisticated parameterizations, including a turbulence closure scheme for the mixed layer parameterization (Noh 2004), UTOPIA (Leonard et al. 1993) for horizontal advection, and an isopycnal mixing scheme (Gent and McWilliams 1990; Griffies 1998). To improve the realistic reproduction of salinity distributions in the coastal ocean or around the aquafarms in Funka Bay, we employed a three-step nesting or downscaling method (Figure 3) that implemented a large-scale, four-dimensional variational data assimilation model (Ishikawa et al. 2009). The assimilation model (NEST0) covered the western half of the North Pacific at horizontal resolutions of (1/6)° and (1/8)° for longitude and latitude, respectively, and provided accurate boundary conditions for two nested models embedded in the western region. The first nested model (NEST1) without assimilation covered the northwestern North Pacific and centered on the mixed water region at medium resolutions of (1/18)° and (1/24)° for longitude and latitude, respectively. The second nested model (NEST2) without assimilation was resolved with the finest horizontal resolutions of (1/54)° and (1/72)° for longitude and latitude, respectively, to reproduce the small-scale physical features within coastal zones. Note that 78 vertical levels were spaced between 4 m apart near the sea surface and 500 m apart at the bottom. The boundary conditions used in each nested model were provided from upper-level (larger-scale) modeling results using the nesting technique by Oey and Chen (1992). Assimilated elements in our study were satellite-derived sea surface temperature, sea surface height data, and *in situ* observation data on temperature and salinity (see Ishikawa et al. (2009) for more details). The river runoff derived from the HaRUM was converted to freshwater flux and input into the meshes closest to the coast of Funka Bay. The freshwater flux was reflected by a decrease in salinity as water was elevated by freshwater inputs into the water column at each surface grid. The high-resolution ocean simulation in NEST2 was spun up

Table 1 Dates of water sampling observations from 2011 to 2013

Year	Date	Water sampling
2011	May 18	
	June 20	
	August 29	
	October 11	
2012	March 16	
	April 12	River water (Yurappu) and seawater (stations a and o in Figure 3)
	April 25	
	May 14	
	June 25	
	August 14	
	December 14	
2013	April 15	Stream water station s, seawater, around station a, c, and o in Figure 3
	April 16	River water (Yurappu)

using initial conditions following the methods of Ishikawa et al. (2009).

Reanalysis and predicted meteorological GPV-MSM datasets and the National Centers for Environmental Prediction reanalysis datasets were input into the coupled model using the mean hourly air temperature, precipitation, cloud cover, relative humidity, dew-point temperature, and wind speed. The coupled model was validated using observational data and showed quantitative reproducibility for temporal variations in the bay after several calibrations (Nakada et al. 2012a, 2013a).

We conducted sensitivity experiments by changing the runoffs in the OGCM to evaluate the influence of snowmelt runoff on circulation and water mass formation in the bay. Two numerical runs were conducted: 1) a case using the terrestrial runoff into the bay as the control run (CR) and 2) a case without the total runoff and SGDs (NF) during spring to summer (March to July).

Water sampling and measurements for nutrient concentrations

The datasets of nutrient concentrations observed at the mouths of eight major rivers (Figure 3) were used to calculate riverine nutrient fluxes (Yoshimura and Kudo 2003). To derive further concentration data for the Yurappu River and other streams, additional water samples were collected at river mouths between April and August in 2011 and 2012 (Table 1). In our study, the streams were defined by the small areas of the watersheds (<10 km^2). Seawater was also sampled as a control. The water samples, obtained using acid-cleaned buckets, were collected from the surface (approximately 1 m deep). Nutrients (nitrates plus nitrites) were determined using continuous-flow automated analyzers (QuAAtro, BL TEC K.K., Osaka, Japan; AutoAnalyzer II, Technicon Co., tarry town, New York, the U.S.) following the manufacturers' protocols. The observation dates for the riverine, ground, and stream waters are described in Table 1.

Groundwater data, including nutrient concentrations measured at observational wells, were downloaded from the website of the National Land Information Division, National and Regional Policy Bureau, Ministry of Land, Infrastructure, Transport and Tourism (http://nrb-www.mlit.go.jp/kokjo/inspect/inspect.html), which provided historical observational datasets of groundwater quality. We removed data that lacked information about the elevation at the well or screen depths, because observation depths at the wells could not be calculated without this information. Data without observed water levels were removed as it was not known whether the screens had been submerged in well water. Furthermore, data with observation depths lower than −70 m or higher than 35 m were removed because groundwater located outside these depths is located far from the coast and contributes little to the SGD. A total of 24 (out of 89) wells were chosen for the period of 1980 to 2006.

Results

Simulated terrestrial volume fluxes to the bay

Figure 6a,b,c,d shows riverine and submarine groundwater discharges based on the HaRUM outputs and runoff observations. Large discharges arising from snowmelt were visible in the simulated and observed runoffs of four major rivers for late April; the discharges had seasonal maxima and displayed interannual variability. The low observed water runoffs associated with the baseflow were reproduced successfully, along with local peaks generated by intense rainfall after the snowmelt period. The temporal tendency after each peak agreed well with the observations and was mainly composed of the interflow discharged from the third outlet. The high degree of consistency between the observed and estimated runoffs for the four major rivers during the observation period resulted in high correlations (approximately 0.8) at significance levels similar to previous studies (Nakada et al. 2012a). The Nash-Sutcliffe efficiency coefficient (Nash and Sutcliffe 1970) ranged between 0.6 and 0.7, except for the Yurappu River coefficient (0.4), which fell into the range (0.36 to 0.75) of 'satisfactory' performance (Motovilov et al. 1999). The relative root-mean-square errors (RMSEs), or the fractional RMSEs, for climatologically averaged runoffs were 21.8% for the Yurappu River (14.5 ± 4.6 m^3 s^{-1}), 12.4% for the Osaru River (20.5 ± 7.9 m^3 s^{-1}), 31.4% for the Oshamanbe River (7.4 ± 2.9 m^3 s^{-1}), and 40.9% for the Nukibetsu River (8.3 ± 3.4 m^3 s^{-1}). The average of the fractions was approximately 27%.

The tank model without the SGD effect (L_3; the bottom outlet in the third tank) displayed a positive interannual runoff trend, leading to a decrease in the model's reproducibility (not shown). Meanwhile, the tank model applied for large watersheds (3,000 km^2) did not need the bottom outlet in the lowest tank (e.g., Iwanami et al. 2013): no SGD condition was required. Watersheds in our study area, which are largely characterized by Neogene or Paleogene volcanic rocks (andesite), retained their high permeability. These facts suggested that the SGD effect is essential for realistic simulations of volume and nutrient fluxes in small watersheds (<500 km^2) with high permeability.

Time series of all runoffs were estimated using the coefficients determined by Nakada et al. (2012a) and summed to derive the total discharge from all watersheds. The robust time series of the total discharge (Figure 6e) clearly showed seasonal maxima around late April, in the snowmelt period. The interannual variations of snowmelt runoffs were also visible and indicated that the largest snowmelt runoffs occurred in 2010. The maximum

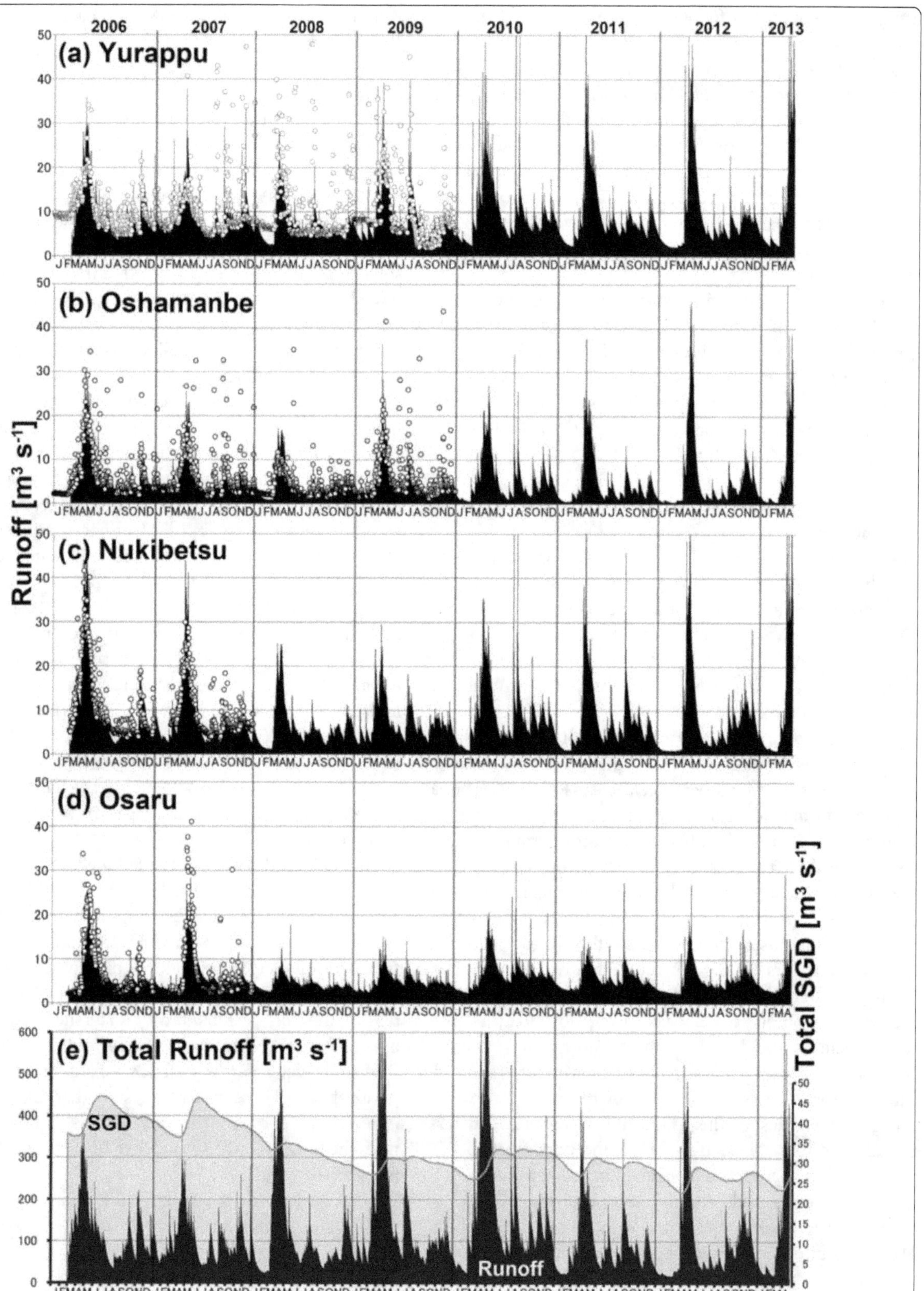

Figure 6 Riverine and submarine groundwater discharges and time series of the simulated volume flux of the SGD. (a-d) Temporal variations in observed (open circles) and simulated (bars) daily runoffs for four major rivers from January 2006 to April 2013. **(e)** Estimated volume transports of the total discharge from all rivers (black) and submarine groundwater discharges (SGDs) (gray). All units are $m^3\ s^{-1}$.

freshwater loading to Funka Bay occurred when the discharge was 300 to 600 $m^3 s^{-1}$ around late April.

Figure 6e shows the time series of the simulated volume flux of the SGD, which indicates marked increases in the flux after the seasonal maxima of snowmelt runoff from the rivers. The seasonal maxima of SGDs were 28 to 46 $m^3 s^{-1}$, occurring around June. The time lag between the seasonal maxima of SGDs and the snowmelt runoff was approximately 1 month. During the snowmelt period, the water level in the third tank seasonally increased owing to the generation of snowmelt water, which led to a flux maximum that was mainly caused by leakage from the third tank. The flux of SGD after the seasonal maximum gradually decreased from summer to fall.

Simulated terrestrial nutrient fluxes to the bay

The nutrient concentrations of terrestrial waters were estimated based on our field surveys, in addition to the observational dataset of riverine concentrations (Yoshimura and Kudo 2003). Observed data were mainly derived in spring to early summer (March to July). Figure 7a shows seasonal variations in nutrient concentrations (DIN) for riverine waters sampled at river mouths. Concentrations averaged over all months were used to substitute missing data from summer to winter (August to February), allowing derivation of the time series of the DIN flux from watersheds (except for the Yurappu River). Seasonal peaks in seasonal variations of DIN occurred in March or April, except for the Osaru River. After a seasonal peak, concentrations decreased during the period in April when the snowmelt runoff increased, and became lower than average for each river except the Osaru River. The time lag between DIN and river runoff peaks was approximately 1 month, indicating that nutrient concentrations in rivers were attenuated by the large volume of snowmelt water.

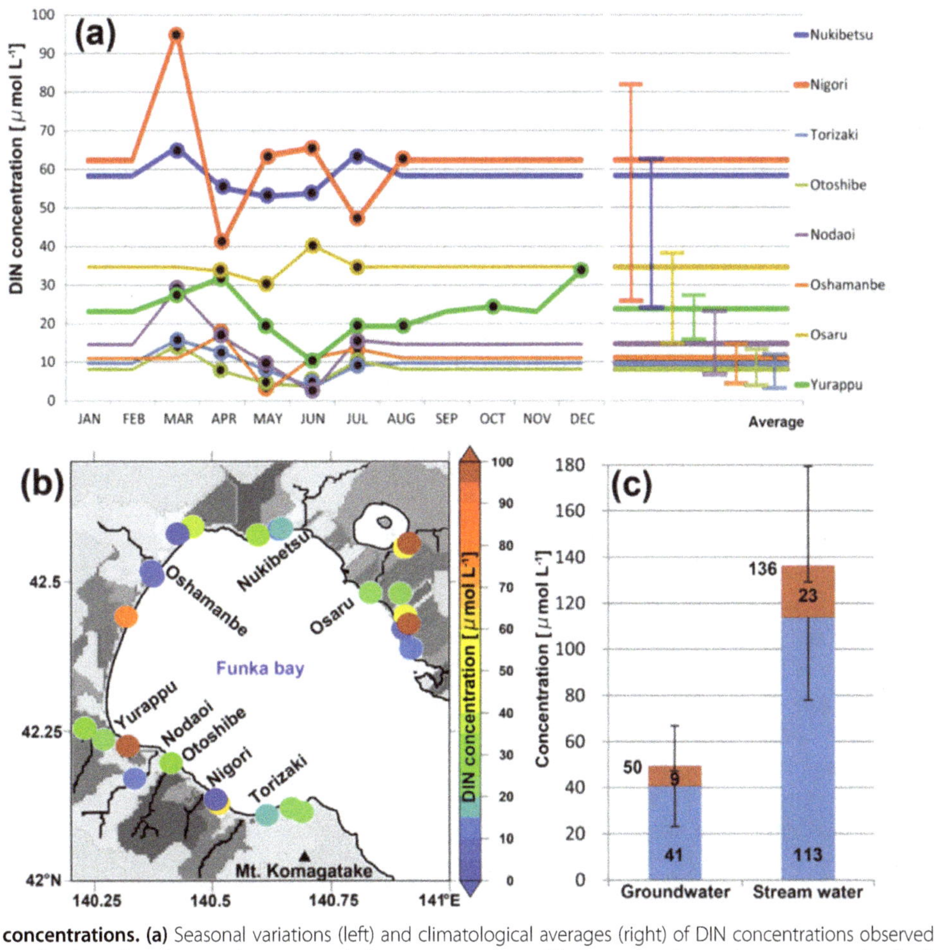

Figure 7 Nutrient concentrations. (a) Seasonal variations (left) and climatological averages (right) of DIN concentrations observed at the mouths of the eight major rivers. Error bars on the averages indicate errors estimated by assuming the nutrient concentration for the missing data period lies between zero and the maximum values. Black dots with each river color indicate values calculated by observational datasets. **(b)** DIN concentrations of groundwater measured at observation wells. Shaded areas indicate each watershed along the bay. Black lines denote the paths of eight major rivers. **(c)** The average concentrations of NH_4 (red), $NO_2 + NO_3$ (blue), and DIN (red + blue) for groundwater and stream waters. Error bars indicate the standard errors of the concentrations for NH_4 and $NO_2 + NO_3$ (downward bars) and DIN (upward bars).

Table 2 Climatologically averaged DIN riverine fluxes from the eight major rivers

Rivers	DIN (mol s^{-1})	
	Averages	Errors
Torizaki	0.04	0.02
Nigori	0.10	0.04
Otoshibe	0.04	0.02
Nodaoi	0.10	0.05
Yurappu	0.37	0.11
Oshamanbe	0.09	0.04
Nukibetsu	0.56	0.28
Osaru	0.74	0.24
Average	0.26	0.14

It is essential to estimate errors due to lack of data. Lower and upper error limits were estimated by assuming the nutrient concentration for the period in which data were missing to be minimum (zero) and the maximum values, respectively. The lower limit of error bars, with DIN averages, in the right panel in Figure 7a shows that the DIN in the missing data period can be less than half the averaged DIN value, except at the Yurappu River. The climatologically averaged concentration over all riverine waters (28 μmol L^{-1}) includes errors of −20 and +8 μmol L^{-1}. The averaged error value is ±11 μmol L^{-1} or 39% as a fraction of the averaged value. The quantity of data observed at the Yurappu River was greater than for other rivers, suggesting that further observation will reduce these errors. The missing data period did not occur during the snowmelt period, although the error magnitudes were comparable to those for seasonal variations. Therefore, these errors were mostly insensitive to our analyses, but as shown later, errors can influence the climatologically averaged DIN riverine fluxes (Table 2).

A total of 24 wells were located near the ocean to monitor groundwater quality, distributed fairly evenly across the coastal areas around the bay (Figure 7b). DIN concentrations in the groundwater from the wells were 1.3 to 418.3 μmol L^{-1}, with an average of 50 ± 18 μmol L^{-1} (Figure 7c), approximately twice the climatologically averaged concentration over all riverine waters (28 ± 8 μmol L^{-1}). A total of 15 sites (63%) showed higher concentrations than the lower limit of the averaged riverine concentrations (20 μmol L^{-1}), and these sites were largely located in or around the watersheds of the Osaru, Yurappu, and Nukibetsu rivers and at the base of Mt. Komagatake. This finding indicated that SGDs from these watersheds can increase the DIN concentration in the coastal bottom water. The average groundwater concentration of NH_4 (9 ± 2 μmol L^{-1}) was much lower than the concentration of $NO_2 + NO_3$ (41 ± 18 μmol L^{-1}). The average concentration of DIN in stream water (NH_4: 23 ± 7 plus $NO_2 + NO_3$: 113 ± 36 μmol L^{-1}) was 136 ± 43 μmol L^{-1} (approximately five times the concentration in riverine water). High-concentration DIN water could flow directly into the coastal water around aquafarms through streams and SGDs from narrow watersheds (<10 km^2) in the mountains.

Average DIN concentrations in the Nigori and Nukibetsu rivers were two times higher than in other rivers (Figure 7a). These high concentrations could be affected by nutrient runoff from agricultural fields, because paddy fields, farmlands, and golf courses account for 23% to 30% of the land use in the watersheds of both rivers (Figure 8). The fractions of forests in both watersheds ranged from 62% to 74%, lower than the average fraction (75%) for all watersheds. In contrast, agricultural fields accounted for 6% to 17% for the other watersheds, leading to low concentrations of DIN in the associated rivers. Fractions of forests and agricultural fields indicate that nutrient loading from forests does not contribute to the increases in DIN concentrations in rivers. The DIN concentration in the near-shore water, measured around the Nigori River (Figure 3) (4.6 ± 0.2 μmol L^{-1}), was higher than the area-averaged concentration in the aquafarm and near-shore areas (2.0 ± 1.1 μmol L^{-1}) and at offshore observation sites (0.4 ± 0.2 μmol L^{-1}), for April 2011 to 2013.

The total DIN flux from all rivers and SGDs was calculated by multiplying the river, stream, and submarine groundwater discharges by DIN concentrations in those waters. Figure 9a shows that temporal variations of total DIN fluxes were similar to total volume fluxes for both rivers and SGDs (Figure 6). Variations in volume fluxes determine the variations in DIN fluxes. Since volume fluxes from SGDs were lower than those from rivers, higher DIN concentrations in SGDs had a comparably minor influence on the total DIN flux.

Figure 9b shows temporal variations in the percentages of volume and DIN fluxes from SGDs in total fluxes. Variations in the percentages of the volume and DIN were similar, but the percentages of the volume flux were generally higher than those of the DIN flux and often exceeded 50% around February. The percentages of the DIN flux from the SGD in the total terrestrial DIN flux increased from 10% to 15% in the snow-cover season (or winter) and reached its seasonal maximum (30% to 40%) around February for each year except 2007 and 2009. Local maxima also occurred between summer and fall, when both the riverine DIN and volume fluxes were low. Seasonal minima were reached during the snowmelt periods, after seasonal maxima, because of the large DIN flux generated by snowmelt runoff from rivers.

Climatologically averaged DIN fluxes from eight major rivers and their errors (Table 2) were calculated using averaged DIN concentrations (with associated errors) for each river (Figure 7) and averaged volume fluxes and errors (RMSEs), with error calculations based on error

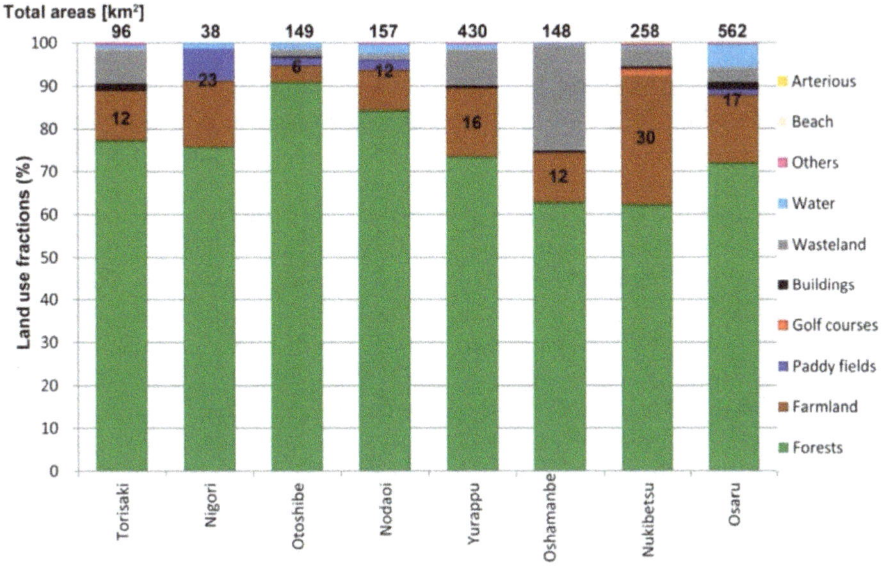

Figure 8 Percentages of land use in the seven major watersheds. See Figure 3 for the map of land use in each watershed. The numerals on the bars represent the summed percentages of paddy fields, farmlands, and golf courses in the watersheds. The numerals at the top of the bars indicate the areas of each watershed.

propagation. Errors in DIN fluxes from each river ranged from 0.02 to 0.28 mol s^{-1}. The average value was 0.26 ± 0.14 mol s^{-1} or 54% of the averaged value. The error was affected by errors in the averaged DIN concentration (39%), rather than those in the volume flux (27%). The DIN flux from Osaru River was the highest and corresponded to the highest volume flux. In contrast, the Nigori River hardly affected the total DIN flux as it showed the lowest volume flux (1.6 m^3 s^{-1}), although the averaged DIN concentration of the Nigori River was the largest. The Yurappu, Nukibetsu, and Osaru rivers contributed 82% of the total DIN flux because their

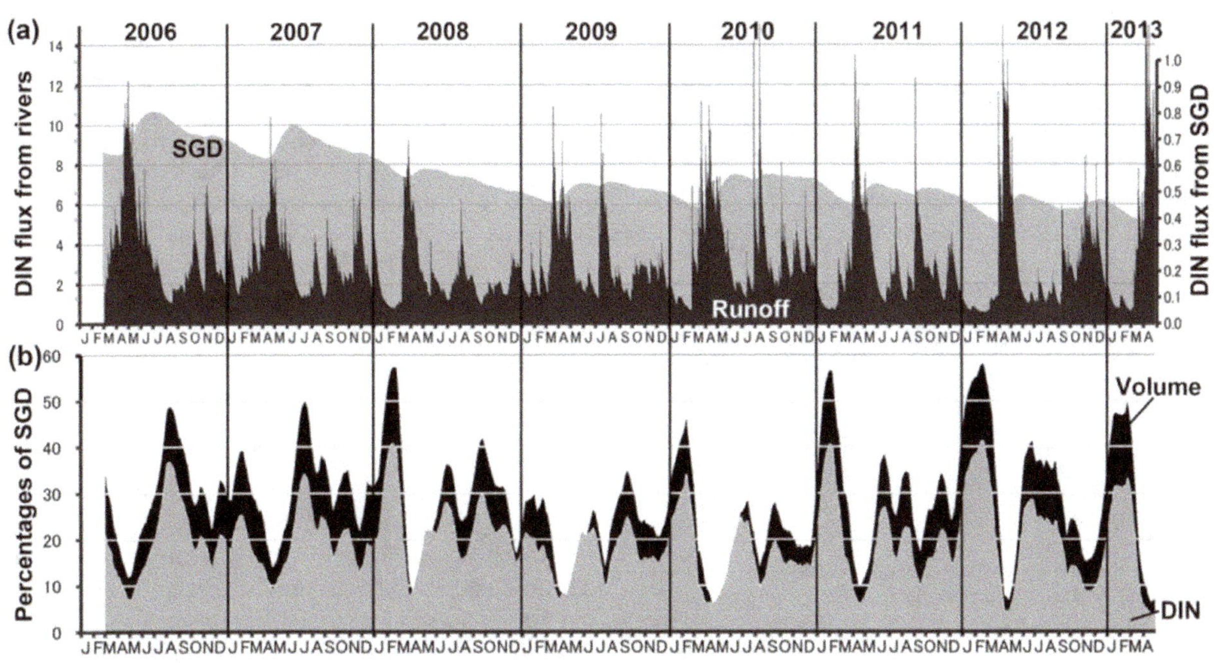

Figure 9 Temporal variations of DIN fluxes. (a) Temporal variations in estimated daily mean DIN fluxes (mol s^{-1}) from all rivers (black) and all submarine groundwater discharges (SGDs) (gray) between January 2006 and April 2013. **(b)** Temporal variations in percentages of the estimated volume (black) and DIN (gray) fluxes from SGDs divided by the total flux (all fluxes from the rivers plus all fluxes from the SGDs).

volume fluxes and DIN concentrations were higher than those for other rivers (Figure 7); however, errors were two to ten times higher than those for other rivers. Therefore, more observations of DIN concentrations in these rivers could decrease errors in future studies.

Figure 10 summarizes climatologically averaged contributions to total volume and DIN fluxes from the major, small (<10 km^2), and other watersheds, corresponding to eight major rivers, streams, and other minor rivers, respectively. Volume fluxes from SGDs accounted for 26% of riverine volume fluxes (123 m^3 s^{-1}). The estimated riverine volume flux accounted for 79% of the total flux (155 m^3 s^{-1}), which was comparable to a previous estimate (150 m^3 s^{-1}; Hasegawa and Isoda 1997). Volume fluxes from all SGDs (32 m^3 s^{-1}: 21%), and the small and other rivers (37 m^3 s^{-1}: 24%), accounted for 45% of the total flux, which could be a second source of terrestrial freshwater. DIN fluxes from all SGDs accounted for 16% of the riverine DIN flux (3.29 mol s^{-1}), which represents a typical fraction (e.g., Taniguchi et al. 2002). The calculated DIN flux from the eight major rivers (2.04 mol s^{-1}) was comparable to a previous estimate (2.42 mol s^{-1}; Yoshimura and Kudo 2003) and accounted for 53% of the total DIN flux (3.84 mol s^{-1}). The remaining 47% is provided by all SGDs (0.55 mol s^{-1}: 14%), and the small and other rivers (1.25 mol s^{-1}: 33%), which indicated that they can supply nutrient fluxes comparable to riverine fluxes from the major rivers. Currently, no study has evaluated these fluxes, but our approach successfully considered each of the fluxes in detail. Streams (narrow watersheds) contributed 5% (6 m^3 s^{-1}) of the riverine volume flux, but contributed 15% (0.50 mol s^{-1}) of the riverine DIN flux owing to high DIN concentrations in the steam water. DIN fluxes of SGDs from small and other watersheds (0.04 and 0.08 mol s^{-1}, respectively) were negligible (1% and 2%, respectively) compared to the total DIN flux. Seasonal differences in each contribution or percentage were rarely observed in volume and DIN fluxes. It is important to note that these respective values can often include errors of (typically) 27% in volume fluxes, an average derived from model errors, and 41% to 54% for the DIN fluxes propagated from model and observational errors.

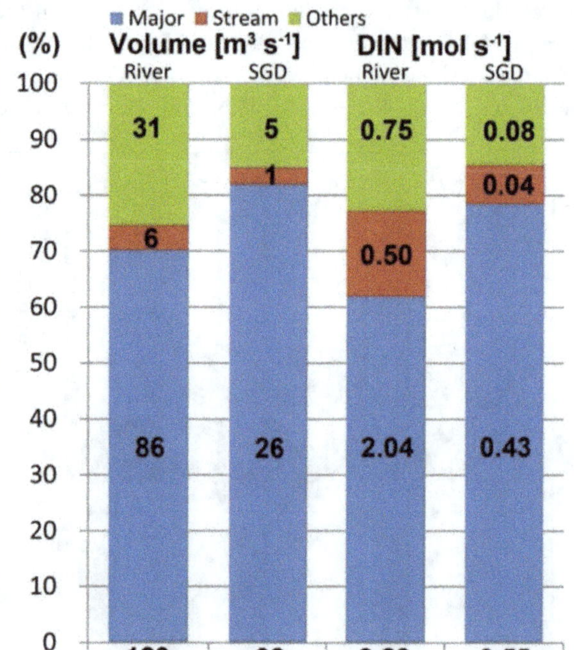

Figure 10 Contributions to total volume and DIN fluxes. Fractions of volume and DIN fluxes from the watershed groups of major rivers (blue), streams (red), and other small rivers (green) averaged over 2006 to 2013. Figures on the color bars denote the climatological volume and DIN fluxes of rivers and SGDs in each watershed group. The values at the bottom of each bar indicate total volume and DIN fluxes from the all rivers and SGDs.

Results simulated using the coupled land-sea model

The coupled model correctly simulated the typical ocean phenomena that occurred during spring and summer, such as the Oyashio inflow during the spring (Ohtani 1981), the summertime clockwise circulation induced by terrestrial freshwater (Satoh et al. 2003), and the local wind in the bay (Takahashi et al. 2007). These phenomena were often captured by conductivity-temperature-depth (CTD) sensor and acoustic Doppler current profiler (ADCP) observations (e.g., Takahashi et al. 2010), which were used to validate the simulated results. The simulated results were validated using observational datasets (salinity and temperature) at satisfactory model skill scores (Murphy and Epstein 1989) ranging from 0.5 to 0.8 (see Nakada et al. 2013a).

Figure 11 concurrently shows the horizontal distributions of the simulated salinity, velocity, and terrestrial runoffs. Each distribution of the terrestrial runoff was calculated by multiplying the runoffs by the stream orders at each grid, normalized to the maximum orders in each watershed. We highlight the maps for March and July in 2008 and 2010, which indicate a clear transition from the Oyashio inflow pattern to clockwise circulation. The maps for March (both years) show the anticlockwise circulation generated by the inflow of low-salinity Oyashio water. The large number of visible river paths indicates that large runoffs occurred in March because of snowmelt runoff; their number decreased in July, when predominantly major rivers were found in both years. The low-salinity water distributed along the coast of the bay was formed by snowmelt runoff in March for both years and was most dominant in July. Low-salinity water originating from terrestrial discharge in July formed around the estuaries of the major rivers, indicating the regions of freshwater influence (Simpson 1997). Similar

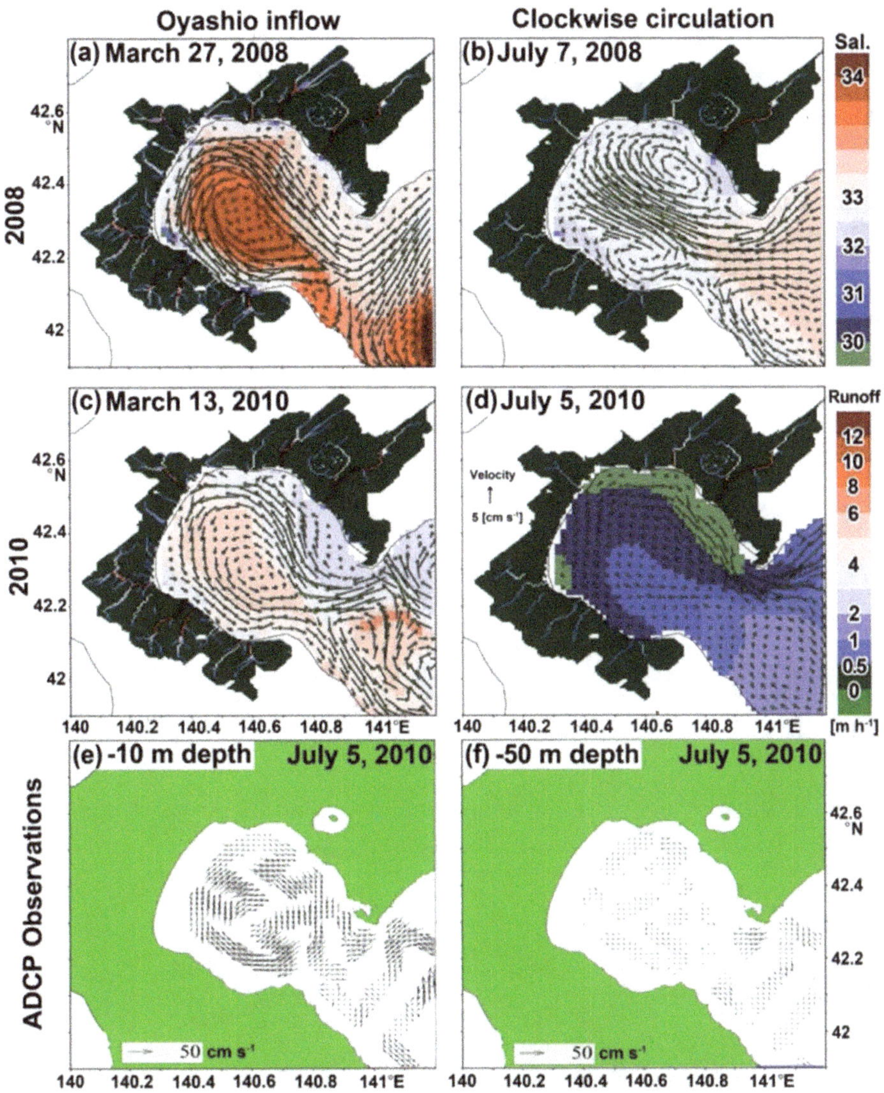

Figure 11 Horizontal distributions of the simulated salinity, velocity, and terrestrial runoffs. Daily mean maps of the simulated salinity (colors on the ocean), current (vectors) at the surface layer (−2 m depth), and terrestrial runoff (colors on the land) **(a)** on 27 March 2008, **(b)** on 7 July 2008, **(c)** on 13 March 2010, and **(d)** on 5 July 2010. The levels of runoff volume transport are depicted by red to blue to white colors along the river paths. The lower panels show snapshots of the observed velocity field at depths of **(e)** -5 m and **(f)** −50 m on 5 July 2010. Velocity fields were measured using the acoustic Doppler current profiler installed on the R/V *Kinsei-maru* of the HFRI.

anticlockwise and clockwise circulations also formed during other years (not shown).

The observed velocity field at the surface (−10 m depth) on July 5 (Figure 11e) shows a clockwise circulation that is similar to the simulation, although current speeds for the circulation were generally five times greater than simulated speeds. The underestimation of the surface velocity field can be attributed to the model's vertical resolution; the layer thickness around the surface layer, from 0 to 20 m, is coarse (4 m), which may insufficiently resolve the surface boundary layer. Local inconsistencies between the simulation and observations were found to correspond to detailed flow patterns around the bay mouth and along the northern coast in the outer bay. The inconsistency around the bay mouth can be attributed to tidal components, for example, the 10 cm s^{-1} velocity observed at the bay mouth (Ohtani 1981) can be attributed to the M2 tidal component. The local wind pattern can induce the observed inconsistency in the outer bay, because the horizontal resolution of the wind (GPV/MSM) was originally 5 km, insufficient to resolve the 10 km spatial pattern of the wind. The observed speed at a depth of −50 m (Figure 11f) was comparable to the simulated speeds (not shown), which also suggests that the simulation underestimated the surface velocity field.

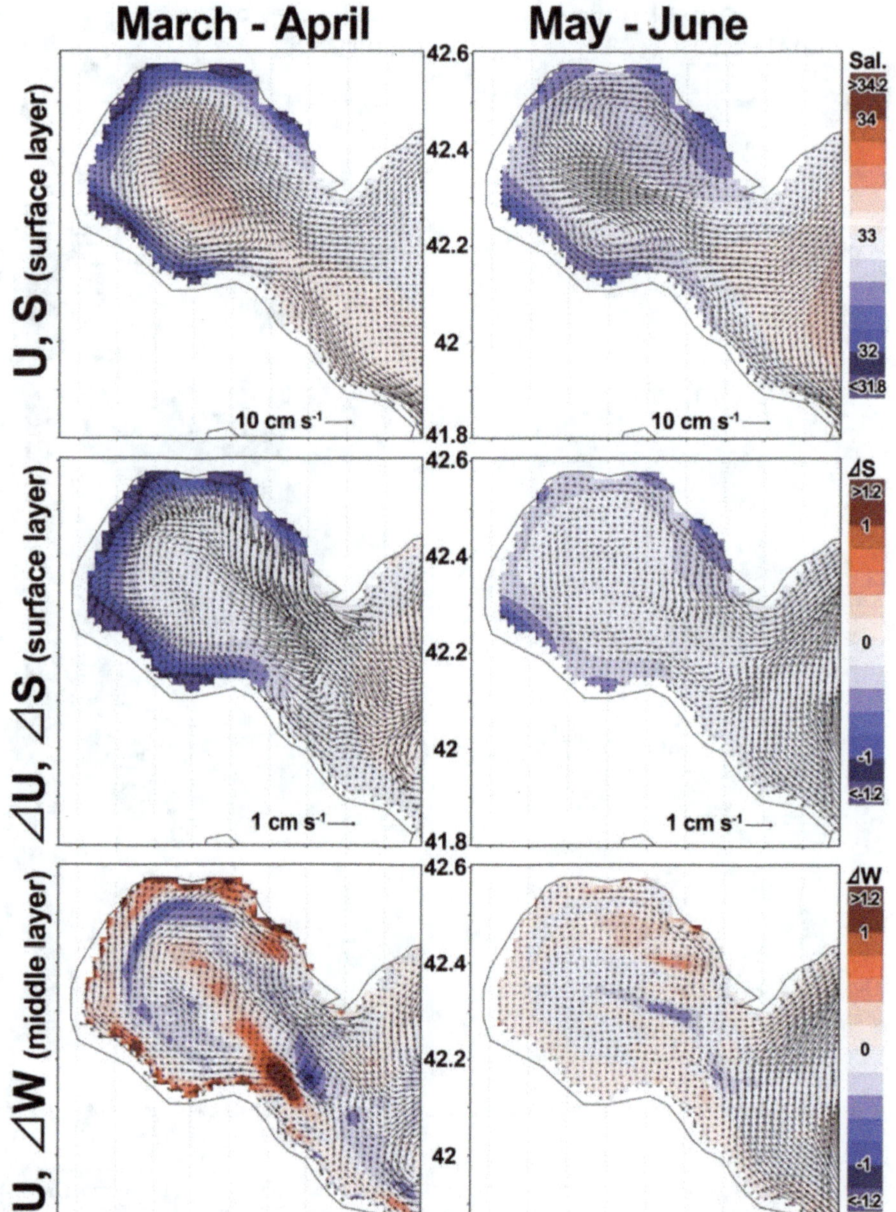

Figure 12 Bi-monthly mean maps. Upper panels: bi-monthly mean maps of the simulated salinity (colors) and current (vectors) at the surface layer (−2 m depth) in (left) March to April and (right) May to June for the case including terrestrial runoff. Middle panels: bi-monthly mean maps of the differences in simulated salinity and current at the surface layer between the cases with and without terrestrial runoff, over the same period as the upper panels. Lower panels: differences in the vertical velocity (colors) and horizontal current (vectors) averaged over the middle layer (depths from −10 to −30 m) between the cases with and without terrestrial runoff, over the same period as the upper panels.

Sensitivity experiments for snowmelt runoff

Influence of runoff on circulation

Sensitivity experiments were conducted for 1 year, 2008, because this year showed typical snowmelt runoffs (Figure 6) and oceanographic events, including a typical Oyashio and Tsugaru warm water inflow (Figure 11). The salinity and velocity fields, averaged over March to April, in the control run (CR) case (upper panels of Figure 12) represented the average pattern of the Oyashio inflow that is associated with a low-salinity flow, forming the anticlockwise velocity field. The low-salinity areas formed along the coast of the bay and were less saline during the snowmelt period (March to April) than during early summer (May to June). The velocity field, averaged over May to June, represented the large clockwise circulation.

The difference between surface salinity maps for the CR case and the NF case, which is without terrestrial runoffs during March to July (middle panels of Figure 12), indicated that the terrestrial discharge associated with snowmelt decreased the salinity field (by approximately −1.1) along the coast. The salinity difference became small during May to June; the areas of marked salinity difference (less than −0.6) in May to June were confined to the areas around major river estuaries.

Differences in surface velocity fields during the period of snowmelt showed that offshore flows from rivers formed along the coast. These seaward flows deflected to the left, leading to a clockwise velocity field around the center of the bay. Maps of the vertical velocity difference (lower panels of Figure 12) at depths from −10 to −30 m showed upwelling in the near-shore area along the coast of the bay and downwelling in the offshore area. The flow pattern characterized by seaward flows and upwelling can be regarded as the estuarine circulation along the coast of the bay. This arrangement suggests that the terrestrial discharge driven by snowmelt could generate the estuarine circulation along the entire coast of the bay. The clockwise velocity field, enhanced by snowmelt runoff, is significant in March to April; however, the estuarine circulation diminished in May to June, and the clockwise flows weakened and moved closer to the coast.

The temporal variation of the vorticity difference between the CR and NF cases (CR minus NF) (Figure 13) showed that the vorticity difference decreased from early March to the beginning of April during the snowmelt period. The sign of the vorticity difference changed from positive to negative, indicating the transition process from the anticlockwise flow pattern associated with the Oyashio inflow to clockwise circulation in the bay. The decrease in the vorticity difference was highly correlated with the terrestrial discharge (correlation coefficient = 0.92), with a time lag of 10 days. The time lag may indicate the period needed to reach the stable negative vorticity generated by buoyancy due to the terrestrial runoff through the geostrophic adjustment. Once the clockwise velocity field was established until the peak of the snowmelt runoff in the middle of April, the negative vorticity was retained by the small runoff although the clockwise velocity field gradually weakened because of the spin down. Our sensitivity experiment clearly showed that the terrestrial buoyancy flux enhanced the clockwise velocity field associated with estuarine circulation.

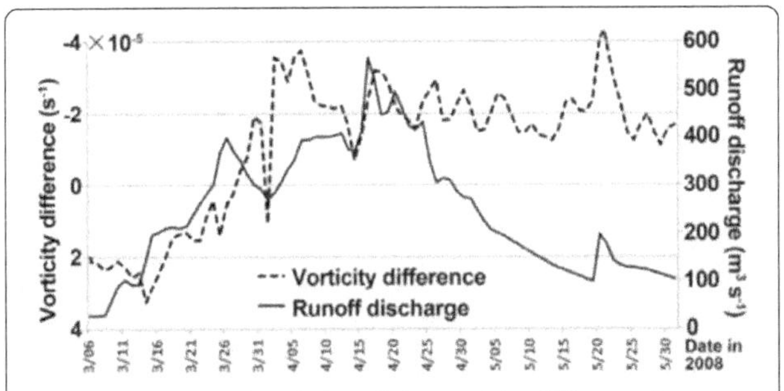

Figure 13 Temporal variation of the vorticity difference between the CR and NF cases. Temporal variation in the daily mean difference of the simulated vorticity (CR minus NF) averaged over the surface layer (−2 m depth) of the bay (dashed line), and estimated daily mean volume flux from the terrestrial runoff (black line) from early March to late May 2008.

Influence of runoff on water mass formation

The experiment indicated that the low-salinity Funka Bay surface water (FS) was reduced in the NF case (without snowmelt runoffs) compared with the CR case (Figure 14a). The FS was predominantly formed by riverine water flowing into the bay rather than rainfall on the bay. The mixed water (MW), which is a mixture of the FS, the summertime OW (S-OW), and the Tsugaru Water (TW) largely formed at depths of −50 to −70 m in both cases. The amount of MW increased from June to November in the CR case (Figure 14b). In contrast, the amount of MW decreased from October in the NF case and was much smaller than for the CR case. The lower salinity water formed by terrestrial freshwater is mixed with the TW by vertical convection from the beginning of the surface cooling period in September through the winter, leading to the emergent MW in winter. The terrestrial freshwater input yields wintertime MW that is less saline than the TW, and can suppress the increase of the TW volume in the CR case, and salinization of the bay water. The TW, at a depth of −40 to −80 m, is an apparent intrusion due to the emergent MW. The bottom intrusions in August, September, and November were visible at depths below −80 m, flushing out the oxygen-deficient bottom water and suppressing the mortality rate of the cultured scallop. Meanwhile, the volume of the TW in the NF case continuously increased from September. The wintertime TW in the CR case ranged from 4×10^{10} to 9×10^{10} m^3. However, the amount of wintertime TW (approximately 14×10^{10} m^3) in the NF case was 1.4 to 3 times that in the CR case, where it occupied almost all of the bay water. The amount of OW in the NF case (maximum volume of 4×10^{10} m^3) was 1.2 times that in the CR case (maximum volume of 3.5×10^{10} m^3), which suggests that the clockwise vorticity, enhanced by terrestrial discharge, can suppress the anticlockwise inflow of the Oyashio water (Figure 12). Highly saline Funka Bay water formed during the winter (FW) in the NF case (1.9×10^{10} m^3); it was also 1.2 times higher than for the CR case (1.6×10^{10} m^3) during spring of the following year (2009; not shown).

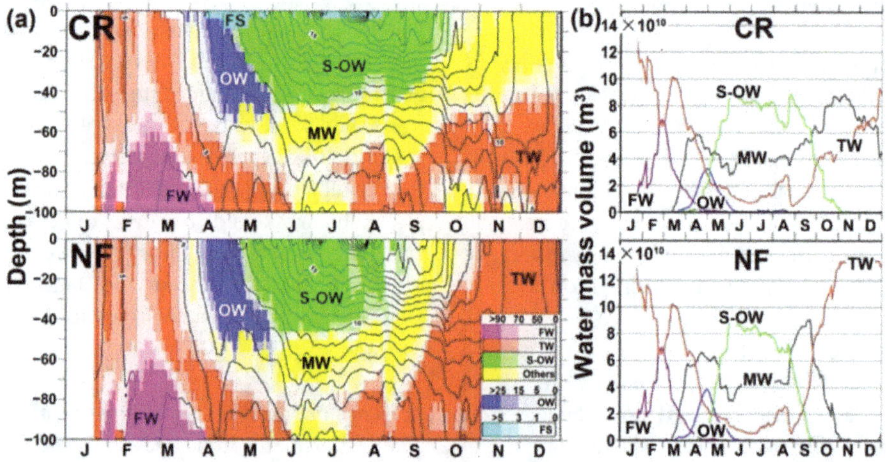

Figure 14 Seasonal and temporal variations of water masses. (a) Seasonal variation in the vertical distribution of water masses and area-averaged simulated temperatures over Funka Bay. The temperature contour interval is 1°C. The letters indicate the six water masses: the Tsugaru warm current water system (TW), the Oyashio water system (OW), the warm low-salinity Funka Bay water that formed during the summer (FS), the cold dense Funka Bay water that formed during the winter (FW) (Hanawa and Mitsudera 1986; Ohtani 1981), the OW surface waters that were warmed mainly by sea-surface heating (S-OW; Rosa et al. 2007), and all other water, which is defined as mixed water (MW). These water masses were classified using the T-S diagram for this study (see Figure Three of Nakada et al. 2013a). The colors indicate percentages of each water mass occupying the water volume at each depth. **(b)** Temporal variations of water mass volumes in the bay.

Discussion

Relationship between terrestrial discharges, nutrients, and larval production

After the winter cooling, nutrients in the bay are vertically homogeneous in February (Lee et al. 2002). A massive spring phytoplankton bloom, characterized by a high chl *a* concentration (>20 μg L^{-1}), is then induced by high-nutrient waters, increasing solar radiation, and water column stability during March to April (Kudo and Matsunaga 1999) and is terminated by nitrate (NO_3) depletion (Kudo et al. 2000). The riverine nutrients can contribute to the summertime primary production in the entire bay (Yoshimura and Kudo 2003). This study found that terrestrial nutrient loading in the coastal ocean is driven predominantly not only by riverine flux, but also by the nutrient flux through the SGD that accounts for a maximum of 40% of the total flux before the snowmelt period. This DIN flux also contributed to the primary production in near-shore seas.

Our results revealed that DIN concentrations were higher in the near-shore and aquafarm areas than in offshore observation sites during the snowmelt period. The concentration of chl *a* in near-shore seas (Figure 4) is strongly associated with terrestrial nutrients. The concentrations of riverine nutrients were much higher than concentrations in the mixed layer of the bay (9 to 12 μmol L^{-1}), even before the spring bloom (Kudo et al. 2000). This suggests that the high DIN flux from rivers can significantly contribute to the increase in DIN concentration, primary production, and chl *a* in the shallow coastal seas during the winter.

The *in situ* chl *a* concentration in February is highly correlated with the density of scallop spat because the scallop spawners can improve food availability in years of high chl *a* and allocate enough energy to gonadal development (Baba et al. 2009). The correlation between the simulated volume flux and the observed spat density (Figure 15) was also highest in spring for each year except 2011. The remarkably high density of the spat in 2011 (Figure 15) might have been influenced by the Great East Japan Earthquake on March 11, because the tsunami that surged through the coastal area of the bay disrupted the sea floor sediment, leading to high concentrations of nutrients in the surface layer. Higher correlations were generally found in March to April than in February to April, which suggests that the terrestrial nutrient flux generated by the large snowmelt runoff in March to April played a more significant role in food availability than the flux in February. Spawners utilized the food or chl *a* produced by the terrestrial nutrients in March to April before the scallops' spawning season from late April until late May (Kosaka and Ito 2006). The eggs and larvae can float in the surface ocean without feeding on phytoplankton until they become young scallops and attach to aquaculture facilities for approximately 40 days. Therefore, eggs and larvae should not be dependent on the phytoplankton supplied by the bloom or the Oyashio inflow in March to April. The nutrients or chl *a* affected by oceanic events could have a significant role in food availability for the young scallops feeding on the phytoplankton.

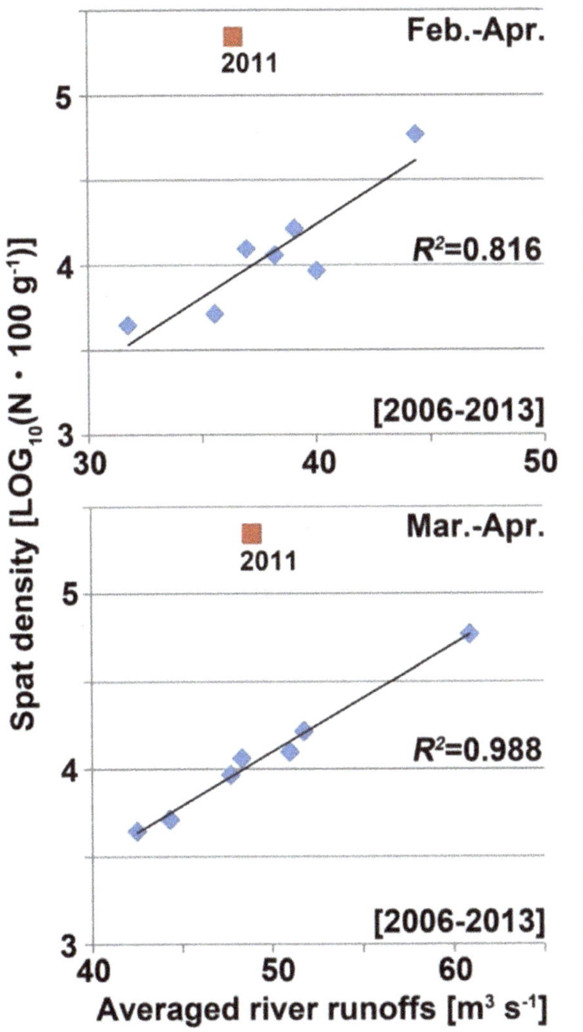

Figure 15 Scatter plots of spat densities. For scallops and simulated volume fluxes averaged over February to April and March to April of 2006 to 2013 (see Baba et al. 2009 for the sampling and calculation procedures used to derive the spat density data). R^2 denotes the squared correlation coefficient.

The late inflow of cold Oyashio water, from late April through to May, can decrease the egg and larval production because the low temperature stunts gonad growth (e.g., Sastry and BlakeSource 1971). Sensitivity experiments showed that the clockwise velocity modified by terrestrial discharge could diminish the anticlockwise inflow of cold Oyashio water (Figure 14), which can suppress the decrease of the water temperature in the bay. The high correlation between the simulated volume flux and the observed spat density (Figure 15) can secondarily explain how the relationship between larval production and Oyashio water volume is affected by the terrestrial discharge.

The observed chl *a* in February around the aquafarm is considered to be an index of the food availability for scallop spawners and is used for yearly predictions of spat density (Baba et al. 2009). The 2013 prediction, using the observed chl *a*, was inconsistent with the observed spat density because the observed chl *a* can reflect spatial and temporal variations, or inhomogeneity, of chl *a* concentration affected by intermittent inputs of both terrestrial and oceanic nutrients through the biochemical mechanism of phytoplankton (Yoshimura and Kudo 2011). The prediction's representativeness may be limited by the location of the observation site. This study emphasized that predictions of spat density based on chl *a* concentration could be improved by including information about terrestrial runoff or nutrient flux, instead of examining the space-time inhomogeneity of chl *a*.

DIN nutrient concentrations from the Nigori and Nukibetsu rivers were twice the climatologically averaged concentration of all riverine waters (28 ± 8 μmol L^{-1}), owing to the supply from agricultural lands in the watersheds. The proportion of agricultural areas in these watersheds determines the baseline concentration of terrestrial nutrients (Ileva et al. 2009). The nutrient concentration during the snowmelt period was diluted by the large volume of snowmelt water. However, the temporal variation in the nutrient flux was controlled by the volume flux rather than the nutrient concentration. Similar results have been suggested by previous studies (Iwanami et al. 2013). This tendency for dilution during the snowmelt period may be a general feature of snow-covered coastal areas around the subarctic zone. The spawning density was affected by riverine nutrients in near-shore areas, which suggests that nutrient loading from agricultural lands played an essential role in food availability for spawners. However, additional studies are needed to determine whether the current loading of terrestrial nutrients is appropriate for scallop aquaculture. The increase in the nutrient flux can improve food availability, but it can also induce harmful algal blooms of organisms such as *Alexandrium tamarense* (Anderson et al. 2012).

Water circulation and larval migration

The large snowmelt runoff induces upwelling and seaward advection as components of estuarine circulation (Figure 12) and forms nutrient-rich low-salinity water in the surface layer of Funka Bay. The rich nutrients that upwell from the deeper layer may contribute to high chl *a* concentrations near the aquafarms. Terrestrial water accumulates in the bay because the clockwise circulation, driven by the local wind (Takahashi et al. 2010) and terrestrial runoff (Satoh et al. 2003), can restrict the exchange of bay water with external water (Isoda et al. 2000), resulting in a low-salinity surface water (Hasegawa and Isoda 1997). After the snowmelt period, larvae can be trapped along the coast and transported to the northern coast by clockwise circulation.

The clockwise circulation can be referred to as a type 2 buoyant plume (Garvine 2001), as the buoyant water at the river mouth turns left along the coast in the northern hemisphere. This water can be trapped on the coast and self-advected along the left-hand coast (Chapman and Lentz 1994). Satoh et al. (2003) interpreted the clockwise circulation as a geostrophic flow proceeding along the left-handed coast, induced by an internal Kelvin-wave propagation under the condition of a preexisting oceanic mixed layer, based on the theory proposed by McCreary et al. (1997). Sensitivity experiments on freshwater inputs in the ocean model indicated that the buoyancy flux supplied by the river runoff contributed secondarily to the velocity field or the clockwise circulation in the bay. The contribution of the riverine buoyancy to the circulation was estimated at approximately 20% (= $V_r(1\ [\mathrm{cm\ s^{-1}}])/V_s(5\ [\mathrm{cm\ s^{-1}}]) \times 100$) of the total momentum of the clockwise velocity field, where V_r is the typical velocity difference that is revealed by sensitivity experiments or induced by riverine runoffs (Figure 12) and V_s is the typical simulated velocity of the clockwise circulation (Figure 11); the remaining 80% was presumably supplied by the wind (Takahashi et al. 2007). Episodically, strong winds can form sub-mesoscale vortices in the bay and distort the clockwise circulation (Nakada et al. 2012b). Such vortices have often been observed in other coastal seas and can locally modify ocean currents (Nakada et al. 2013b).

Conclusions

This study comprehensively investigated the roles of snowmelt runoff in water circulation in terms of riverine buoyancy flux, in water mass formation in terms of freshwater volume flux, and in the production of scallop larvae in terms of riverine nutrient flux, for Funka Bay, based on hydrological and nutrient data. Sensitivity experiments using a high-resolution coupled land-sea simulation were also conducted. The terrestrial freshwater and nutrient (DIN) fluxes from rivers and SGDs into the bay were estimated using a hydrological model and climatological datasets compiled from nutrient concentration data for the rivers, streams, and groundwater. The comparison between observed and simulated runoffs indicated that the model accurately reproduced the terrestrial runoff, allowing for estimates of the terrestrial nutrient flux because variations in terrestrial volume fluxes controlled variations in the terrestrial DIN fluxes. The average volume flux from SGDs accounted for 26% of the flux for riverine runoff. The nutrient flux from SGDs accounted for a maximum of 40% of the total loading to the coastal ocean of the bay in February before the snowmelt period.

The impact of snowmelt runoff on water circulation and water mass formation in the bay was first evaluated by sensitivity experiments using the OGCM. This study revealed, as part of the land-sea linkage, that snowmelt runoff has three important roles in the ocean environment and scallop production: 1) Buoyancy fluxes supplied from the runoff can enhance the clockwise circulation in the bay, but rarely dominate the velocity field. 2) The terrestrial freshwater input can diminish the springtime inflow of the cold OW, leading to an increase in larval production and a decrease in the volume of wintertime TW in the bay, and can modify distributions of the wintertime water masses. 3) Fluxes of high-concentration nutrients from the runoff can form a favorable environment for spawners by improving food availability. Therefore, the snowmelt runoff hardly contributes to larvae transport but can increase egg production through the spring nutrient supply. In addition, upwelling may advect rich nutrients from deeper regions to the surface in the near-shore ocean, and the clockwise circulation may transport larvae northward along the coast.

Agricultural fields were important sources of nutrients for feeding scallop spawners. This study, therefore, demonstrated that the integrated management of agricultural farms and aquafarms (Iwasaki 2013) in the large bay is essential for controlling the nutrient flux from the land to the ocean. The volume, buoyancy, and nutrient fluxes supplied from terrestrial snowmelt runoff can be key factors in fishery or aquaculture production in subarctic coastal regions, alongside watersheds in the snow zone. Additional studies are needed to elucidate land-ocean linkages and to design sustainable and synergetic systems of aquaculture and agriculture.

Abbreviations

ADCP: acoustic Doppler current profilers; chl *a*: chlorophyll *a*; CR: the control run case that includes terrestrial runoffs into the bay; CTD: conductivity, temperature, depth (CTD) sensors; DIN: dissolved inorganic nitrogen; FS: the low-salinity Funka Bay surface water; FW: highly saline Funka Bay water formed in winter; GPV-MSM: grid point value datasets of the meso-scale model; HaRUM: hydrometeorological and multi-runoff utility model; ITCZs: intertropical convergence zones; MW: the mixed water (MW) that blends FS, the summertime OW (S-OW), and Tsugaru water (TW); NF: the case without snowmelt runoffs during spring to summer (March to July); OGCM: ocean general circulation model; OW: subarctic Oyashio water; SGD: submarine groundwater discharge; S-OW: the summertime OW; TW: Tsugaru water.

Competing interests

The authors declare that they have no competing interests.

Authors' contributions

SN designed this work; collected the data; carried out the water sampling; conducted the numerical simulation; analyzed, observed, and simulated the data; and prepared the manuscript. KB observed the larvae and spat density and helped in their interpretation. MS observed physical properties such as water temperature, salinity, and current velocity. MN carried out the water sampling and assayed the water to derive nutrient concentrations. YI prepared the ocean general circulation model and its data assimilation system. KK provided the high-end computational environment, visualization system, and super-computing resources. TA and SS proposed the topic, provided discussion, and collaborated with the corresponding author in the construction of the manuscript. All authors read and approved the final manuscript.

Authors' information

SN received a bachelor's degree in Fisheries Science and a master's degree in Fisheries Science (studying in the Faculty of Fisheries Sciences) from Hokkaido University, Hakodate, Japan, in 1999 and 2001, respectively. He received a PhD from Kyushu University (the Department of Earth System Science and Technology, Interdisciplinary Graduate School of Engineering Sciences), Ohnojo, Japan, in 2008. He was a researcher for the multidisciplinary consulting company Knowledge Consulting for Solution, Co., Ltd., Tokyo, Japan, during 2001 to 2003 and JAPAN NUS Co., Ltd., Tokyo, Japan, during 2003 to 2005, and a fellow of Japan Society for the Promotion of Science during 2006 to 2008, at Yonsei University, Seoul, Korea, in 2008, and at the Research Institute of Humanity and Nature (RIHN) during 2009 to 2010. He was mainly engaged in the development of numerical models, analyses of their outputs, and design for visualization systems. He is currently a researcher at the Institute for Liberal Arts and Sciences, Kyoto University, Kyoto, Japan. His current research is predominantly aimed at developing the operational ocean forecast system based on a land-sea-coupled model using data assimilation methods and high-end computers, investigating land-ocean interactions based on the large amount of output datasets produced by high-resolution ocean simulations and high-density observations.

Dr. KB is a researcher in the Hokkaido Research Organization, Fisheries Research Department, Hakodate Fisheries Research Institute, Hakodate, Japan. His research interest is to study the biological mechanisms of scallops, such as spawning, and ecosystems favorable to fishery productions with the aim of predicting the annual spat density.

MS is a researcher in the Hokkaido Research Organization, Fisheries Research Department, Hakodate Fisheries Research Institute, Hakodate, Japan. His research interest is to investigate relationships among fishery productions such as scallops, ocean circulation, water masses, and oxygen-deficient water.

MN received BE and MS degrees (studying in Faculty of Agriculture) from Kyoto University, Kyoto, Japan. He is currently a student in the Graduate School of Fishery, Faculty of Fisheries Sciences, Hokkaido University, Japan. His research interest is to examine the mechanisms of harmful blooming algae, such as *Alexandrium tamarense*, in terms of nutrient fluxes, pertaining to marine biological systems in the coastal ocean.

YI received a BE degree and a PhD in Science from Kyoto University, Kyoto, Japan, in 1994 and 1999, respectively. He was an Assistant Professor in the Graduate School of Science in Kyoto University in 2001. Currently, he is a Principal Research Scientist in Data Research Center of Marine-Earth Science (DrC), Japan Agency for Marine-Earth Science and Technology (JAMSTEC), Japan, Yokohama (2012). His research interests are in physical oceanography and climate physics, especially numerical modeling and data assimilation systems.

TA received a BS degree (studying in the Faculty of Engineering) from Kyoto University in 1972 and received MS and Doctor of Science degrees at the Graduate School of Science, Kyoto University, Kyoto, Japan. He is currently an executive vice president for education at Kyoto University (from 2010).

KK received BE and ME degrees and a PhD in Electrical and Electronic Engineering (studying at the Faculty of Engineering) from Kyoto University, Kyoto, Japan, in 1983, 1985, and 1994, respectively. He is currently a professor at the Institute for Liberal Arts and Sciences, Kyoto University, Kyoto, Japan. From 1985 to 1998, he has a 14-year career in IBM Japan Ltd., Tokyo, Japan. From 1998 to 2001, he was an associate professor at Iwate Prefectual University. From 2001 to 2003, he was an associate professor at Kyoto University. His research interests include modeling, simulation, and visualization. He is a member of IEEE Computer Society, directors of Visualization Society Japan, and the Institute of Systems, Control and Information Engineers, and a president of Japan Society of Simulation Technology. He received the IEMT/IMC outstanding paper award in 1998, the VSJ contribution award in 2009, and the VSJ outstanding paper award in 2010.

SS is a professor in the Laboratory of Satellite Oceanography, Division of Marine Bio-resource and Environmental Science, Hokkaido University, Hakodate, Japan. He was a research engineer for the Japan Weather Association, Tokyo, Japan during 1984 to 1993. From 1993 to 2000, he was an associate professor at Hokkaido University. His research interests cover many fields, including fishery oceanography, satellite oceanography, marine ecosystems, and marine GIS.

Acknowledgements

We thank Mr. M. Watanobe, Mr. M. Kanamori, and the crew and officers of the R/V Kinsei-Maru of Hakodate Fisheries Research Institute for providing *in situ* data. We deeply thank Professor I. Kudo (Faculty of Fisheries Sciences, Hokkaido University) and Dr. T. Yoshimura (Central Research Institute of Electric Power Industry) for suggesting improvements to this paper and providing the nutrient datasets (Yoshimura and Kudo 2003), which greatly helped in the calculation of the terrestrial DIN flux. This work was partially supported by the 'Hakodate Marine Bio Cluster Project' in the knowledge Cluster Program from 2009, the Grant-in-Aid for University and Society Collaboration from the Ministry of Education, Culture, Sports, Science and Technology (MEXT), Japan, and the Grant-in-Aid for Research Programs on Climate Change Adaptation (RECCA). We acknowledge the National Aeronautics and Space Administration (NASA) for the MODIS/Aqua data and the Research Institute for Sustainable Humanosphere, Kyoto University, for the archived data of GPV-MSM. We deeply appreciate the constructive comments from two anonymous reviewers and editors.

Author details

[1]Institute for Liberal Arts and Sciences, Kyoto University, Kitashirakawa-Oiwake Cho, Sakyo-ku, Kyoto 606-8502, Japan. [2]Hokkaido Research Organization, Fisheries Research Department, Hakodate Fisheries Research Institute, 1-2-66 Yunokawa-cho, Hakodate, Hokkaido 042-0932, Japan. [3]Faculty of Fisheries Sciences, Hokkaido University, 3-1-1, Minato-cho, Hakodate, Hokkaido 041-8611, Japan. [4]Data Research Center for Marine-Earth Sciences, Japan Agency for Marine-Earth Science and Technology, Yokohama Institute for Earth Sciences, 3173-25 Showa-machi, Kanazawa-ku, Yokohama, Kanagawa 236-0001, Japan. [5]Kyoto University Headquarters, Yoshida-Honmachi, Sakyo-ku, Kyoto 606-8501, Japan. [6]Graduate School of Maritime Sciences, Kobe University, 5-1-1 Fukae-minami, Higashi-nada-ku, Kobe 658-0022, Japan.

References

Anderson DM, Alpermann TJ, Cembella AD, Collos Y, Masseret E, Montresor M (2012) The globally distributed genus *Alexandrium*: multifaceted roles in marine ecosystems and impacts on human health. Harmful Algae 14:10–35

Baba K, Sugawara R, Nitta H, Endou K, Miyazono A (2009) Relationship between spat density, food availability, and growth of spawners in cultured *Mizuhopecten yessoensis* in Funka Bay: concurrence with El Nino southern oscillation. Can J Fish Aquat Sci 66:6–17, doi:0.1139/F08-183

Bondad-Reantaso MG, Subasinghe RP, Josupeit H, Cai J, Zhou X (2012) The role of crustacean fisheries and aquaculture in global food security: past, present and future. J Invert Path 110(2):158–165, doi:org/10.1016/j.jip.2012.03.010

Chapman DC, Lentz SJ (1994) Trapping of a coastal density front by the bottom boundary layer. J Phys Oceanogr 24:1464–1479, doi: http://dx.doi.org/10.1175/1520-0485(1994)024<1464:TOACDF>2.0.CO;2

Dai A, Trenberth KE (2002) Estimates of freshwater discharge from continents: latitudinal and seasonal variations. J Hydro 3:660–87, doi:10.1175/1525-7541(2002)003<0660:EOFDFC>2.0.CO;2

Darecki M, Stramski D (2004) An evaluation of MODIS and SeaWiFS bio-optical algorithms in the Baltic Sea. Remote Sens Environ 89(3):326–350, doi:10.1016/j.rse.2003.10.012

FAO (2011) FAO yearbook: fishery and aquaculture statistics 2009. Food and Agriculture Organization of the United Nations, Rome, Italy, p 223

Garvine RW (2001) The impact of model configuration in studies of buoyant coastal discharge. J Mar Res 59:193–225, doi:10.1357/002224001762882637

Gent PR, McWilliams JC (1990) Isopycnal mixing in the ocean circulation model. J Phys Oceanogr 20:150–155, doi:10.1175/1520-0485(1990)020<0150:IMIOCM>2.0.CO;2

Griffies SM (1998) The Gent-McWilliams skew flux. J Phys Oceanogr 28:831–841, doi:10.1175/1520-0485(1998)028<0831:TGMSF>2.0.CO;2

Hanawa K, Mitsudera H (1986) Variations of water system distribution in the Sanriku coastal area. J Oceanogr Soc Japan 42:435–446

Hasegawa N, Isoda Y (1997) Fresh water budget of Funka Bay. Umi to Sora 73:113–121 (in Japanese with English abstract)

Ileva NY, Shibata H, Satoh F, Sasa K, Ueda H (2009) Relationship between the riverine nitrate-nitrogen concentration and the land use in the Teshio River watershed, North Japan. Sustain Sci 4:189–198, doi:10.1007/s11625-009-0081-5

Ishikawa Y, Awaji T, Toyoda T, In T, Nishina K, Nakayama T, Shima S, Masuda S (2009) High-resolution synthetic monitoring by a 4-dimensional variational data assimilation system in the northwestern North Pacific. J Mar Syst 78:237–248, doi:10.1016/j.jmarsys.2009.02.016

Isoda Y, Takeyama Y, Azumaya T (2000) Ocean structural change around the diatom bloom in Funka Bay. Bull Coast Oceanogr 38:3–14 (in Japanese with English abstract)

Iwanami H, Wada T, Sakamoto K, Kudo I, Chikita KA (2013) Mechanism of nutrients supply from the Tokachi River catchment considering characteristics of discharge and land use. J Jpn Assoc Hydro Sci 43(1):3–24

Iwasaki S (2013) Fishers-based watershed management in Lake Saroma, Japan. Ocean & Coastal Manag 81:58–65

Kimmerer WJ (2002) Effects of freshwater flow on abundance of estuarine organisms: physical effects or trophic linkages? Mar Eco Prog Series 243:39–55

Kondo J, Watanabe T (1992) Studies on the bulk transfer coefficients over a vegetated surface with a multilayer energy budget model. J Atmo Sci 49(23):2183–2199, doi:10.1175/1520-0469(1992)049<2183:SOTBTC>2.0.CO;2

Kondo J, Yamazaki T (1990) A prediction model for snowmelt, snow surface temperature and freezing depth using a heat balance method. J Appl Meteorol 29:375–384, doi:10.1175/1520-0450(1990)029<0375:APMFSS>2.0.CO;2

Kosaka Y, Ito H (2006) Japan. In: Shumway SE, Parsons GJ (eds) Scallops: biology, ecology and aquaculture. Elsevier, Amsterdam, the Netherlands, pp 1093–1141

Kudo I, Matsunaga K (1999) Environmental factors affecting the occurrence and production of the spring phytoplankton bloom in Funka Bay, Japan. J Oceanogr 55:505–513

Kudo I, Yoshimura T, Yanada M, Matsunaga K (2000) Exhaustion of nitrate terminates a phytoplankton bloom in Funka Bay, Japan: change in SiO_4:NO_3 consumption rate during the bloom. Mar Ecol Prog Ser 193:45–51

Lee CW, Kudo I, Yokokawa T, Yanada M, Maita Y (2002) Dynamics of bacterial respiration and related growth efficiency, dissolved nutrients and dissolved oxygen concentration in a subarctic coastal embayment. Mar Freshw Res 53:1–7

Leonard BP, MacVean MK, Lock AP (1993) Positivity-preserving numerical schemes for multidimensional advection. NASA Tech Memo 106055: 62 ICOMP-93-05, Washington, DC. http://ntrs.nasa.gov/archive/nasa/casi.ntrs.nasa.gov/19930017902.pdf

Liu SM, Li RH, Zhang GL, Wang DR, Du JZ, Herbeck LS (2011) The impact of anthropogenic activities on nutrient dynamics in the tropical Wenchanghe and Wenjiaohe estuary and lagoon system in East Hainan, China. Mar Chemist 125:49–68

McCreary JP, Zhang S, Shetye SR (1997) Coastal circulation driven by river outflow in a variable-density 1 1/2-layer model. J Geophys Res 102:15535–15554

Minoda T, Fuji A (1985) Biological environments in Funka Bay. Tokai University Press p 1106, (In Japanese), Coastal Oceanography of Japanese Islands

Miyake H, Tanka I, Murakami T (1988) Outflow of water from Funka Bay, Hokkaido, during early spring. J Oceanogr Soc Japan 44:163–170, doi:10.1007/BF02302640

Motovilov Y, Gottschalk GL, England K, Rodhe A (1999) Validation of distributed hydrological model against spatial observations. Agric Forest Meteo 98–99:257–277, doi:10.1016/S0168- 1923(99)00102-1

Murphy AH, Epstein ES (1989) Skill scores and correlation coefficients in model verification. Mon Weather Rev 117:572–581, doi:10.1175/1520-0493(1989)117<0572:SSACCI>2.0.CO;2

Nakada S, Yasumoto J, Taniguchi M, Ishitobi T (2011) Submarine groundwater discharge and seawater circulation in a subterranean estuarine beneath a tidal flat. Hydrol Process 25:2755–2763, doi:10.1002/hyp.8016

Nakada S, Ishikawa Y, Awaji T, In T, Shima S, Nakayama T, Isada T, Saitoh S (2012a) Modeling runoff into a region of freshwater influence for improved ocean prediction: an application in Funka Bay. Hydro Res Lett 6:47–52

Nakada S, Uenaka T, Ishikawa Y, Matsui K, Sakamoto N, Koyamada K, Awaji T, Saitoh S (2012b) A visualization study of vortexes extracted from the tremendous ocean simulation data: an application in Funka Bay. Jpn Soc Sim Tech 4(4):145–152, doi:org/10.1016/j.pocean.2013.10.008

Nakada S, Ishikawa Y, Awaji T, In T, Koyamada K, Watanobe M, Okumura H, Nishida Y, Saitoh S (2013a) An integrated approach to the heat and water mass dynamics of a large bay: high-resolution simulations of Funka Bay, Japan. J Geophys Res Ocean 118:1–18, doi:10.1002/jgrc.20262

Nakada S, Hirose N, Senjyu T, Fukudome K, Tsuji T, Okei N (2013b) Operational ocean prediction experiments for smart coastal fishing. Prog Oceanogr 121:125–140, doi.org/10.1016/j.pocean.2013.10.008

Nash JE, Sutcliffe JV (1970) River flow forecasting through conceptual models part I - a discussion of principles. J Hydro 10(3):282–290, doi:10.1016/0022-1694(70)90255-6

Noh Y (2004) Sensitivity to wave breaking and the Prandtl number in the ocean mixed layer model and its dependence on latitude. Geophys Res Lett 31: L23305, doi:10.1029/2004GL021289

Oey LY, Chen P (1992) A nested-grid ocean model - with application to the simulation of meanders and eddies in the Norwegian coastal current. J Geophys Res 97:20063–20086

Ohtani K (1981) Funka Bay (I) physical environments. Bull Coastal Oceanogr 19:68–80 (in Japanese)

Ohtani K, Kido K (1980) Oceanographic structure in Funka Bay, vol 31. Bull Fac Fish Hokkaido University, Hakodate, Japan, pp 84–114

Rosa AL, Isoda Y, Uehara K, Aiki T (2007) Seasonal variations of water system distribution and flow patterns in the southern sea area of Hokkaido, Japan. J Oceanogr 63:573–588, doi:10.1007/s10872-007-0051-4

Sastry AM, BlakeSource NJ (1971) Regulation of gonad development in the bay scallop, *Aequipecten irradians*. Lamarck Biol Bull 140(2):274–283

Satoh C, Isoda Y, Shimizu M (2003) Clockwise circulation in the upper layer of Funka Bay in summer. Bull Coast Oceanogr 40:181–188 (In Japanese with English Abstract)

Shimada H, Nishida Y, Ito Y, Mizushima T (2000) Relationship among growth and survival of cultured scallops (*Patinopecten yessoensis* JAY), and environmental condition in the coastal area off Yakumo, Funka Bay, Hokkaido, Japan. Sci Rep Hokkaido Fish Exp Stn 27:49–62 (In Japanese with English Abstract)

Simpson JH (1997) Physical processes in the ROFI regime. J Mar Syst 12:3–15, doi:10.1016/S0924-7963(96)00085-1

Takahashi D, Kido K, Nishida Y, Kobayashi N, Higaki N, Miyake H (2007) Dynamical structure and wind-driven upwelling in a summertime anticyclonic eddy within Funka Bay, Hokkaido, Japan. Conti Shelf Res 27:1928–1946

Takahashi D, Miyake H, Nakayama T, Kobayashi N, Kido K, Nishida Y (2010) Response of a summertime anticyclonic eddy to wind forcing in Funka Bay, Hokkaido, Japan. Conti Shelf Res 30:1435–1449, doi:10.1016/j.csr.2010.05.003

Taniguchi M, Burnett WC, Cable JE, Turner JV (2002) Investigations of submarine groundwater discharge. Hydrol Process 16:2115–2129, doi:10.1002/hyp.1145

Uiboupin R, Arino O (2010) Study of snowmelt impact on SST and TSM fields in the coastal zone of Barents Sea. IEEE International Geoscience and Remote Sensing Symposium, pp 4212–4215, doi:10.1109/IGARSS.2010.5654375

Willmott CJ, Matsuura K (2001) Terrestrial water budget data archive: monthly time series (1950-1999). http://climate.geog.udel.edu/~climate/html_pages/archive.html

Yoshimura T, Kudo I (2003) Riverine nutrient loadings and their impact on primary production in Funka Bay, Japan. Oceanogr Jpn 12:185–193 (in Japanese with English abstract)

Yoshimura T, Kudo I (2011) Seasonal phosphorus depletion and microbial responses to the change in phosphorus, availability in a subarctic coastal environment. Mar Chemi 126:182–192

Yoshimura K, Sakimura T, Oki T, Kanae S, Seto S (2008) Toward flood risk prediction: a statistical approach using a 29-year river discharge simulation over Japan. Hydro Res Lett 2:22–26, doi:10.3178/hrl.2.22

Zhao K, Nakada S, Sakamoto N, Koyamada K, Bajaj C, Ishikawa Y, Awaji T, In T, Saitoh S (2013) A visualization for the dynamic behaviors of the mixture of water mass for Northwestern Pacific near Japan. Int J Model Simul Sci Comput 4(1341002):18, doi:10.1142/S179396231341002X

PERMISSIONS

All chapters in this book were first published in PEPS, by Springer; hereby published with permission under the Creative Commons Attribution License or equivalent. Every chapter published in this book has been scrutinized by our experts. Their significance has been extensively debated. The topics covered herein carry significant findings which will fuel the growth of the discipline. They may even be implemented as practical applications or may be referred to as a beginning point for another development.

The contributors of this book come from diverse backgrounds, making this book a truly international effort. This book will bring forth new frontiers with its revolutionizing research information and detailed analysis of the nascent developments around the world.

We would like to thank all the contributing authors for lending their expertise to make the book truly unique. They have played a crucial role in the development of this book. Without their invaluable contributions this book wouldn't have been possible. They have made vital efforts to compile up to date information on the varied aspects of this subject to make this book a valuable addition to the collection of many professionals and students.

This book was conceptualized with the vision of imparting up-to-date information and advanced data in this field. To ensure the same, a matchless editorial board was set up. Every individual on the board went through rigorous rounds of assessment to prove their worth. After which they invested a large part of their time researching and compiling the most relevant data for our readers.

The editorial board has been involved in producing this book since its inception. They have spent rigorous hours researching and exploring the diverse topics which have resulted in the successful publishing of this book. They have passed on their knowledge of decades through this book. To expedite this challenging task, the publisher supported the team at every step. A small team of assistant editors was also appointed to further simplify the editing procedure and attain best results for the readers.

Apart from the editorial board, the designing team has also invested a significant amount of their time in understanding the subject and creating the most relevant covers. They scrutinized every image to scout for the most suitable representation of the subject and create an appropriate cover for the book.

The publishing team has been an ardent support to the editorial, designing and production team. Their endless efforts to recruit the best for this project, has resulted in the accomplishment of this book. They are a veteran in the field of academics and their pool of knowledge is as vast as their experience in printing. Their expertise and guidance has proved useful at every step. Their uncompromising quality standards have made this book an exceptional effort. Their encouragement from time to time has been an inspiration for everyone.

The publisher and the editorial board hope that this book will prove to be a valuable piece of knowledge for researchers, students, practitioners and scholars across the globe.

LIST OF CONTRIBUTORS

Kentaro Nakamura
Precambrian Ecosystem Laboratory (PEL), Japan Agency for Marine-Earth Science and Technology (JAMSTEC), 2-15 Natsushima-cho, Yokosuka, Kanagawa 237-0061, Japan
Department of Systems Innovation, School of Engineering, The University of Tokyo, 7-3-1 Hongo, Bunkyo-ku, Tokyo 113-8656, Japan

Ken Takai
Precambrian Ecosystem Laboratory (PEL), Japan Agency for Marine-Earth Science and Technology (JAMSTEC), 2-15 Natsushima-cho, Yokosuka, Kanagawa 237-0061, Japan
Subsurface Geobiology Advanced Research (SUGAR) Project, Japan Agency for Marine-Earth Science and Technology (JAMSTEC), 2-15 Natsushima-cho, Yokosuka, Kanagawa 237-0061, Japan

Makio C. Honda, Yoshikazu Sasai and Eko Siswanto
Japan Agency for Marine-Earth Science and Technology (JAMSTEC), Yokosuka, Kanagawa, Japan

Akira Kuwano-Yoshida
Disaster Protection Research Institution, Kyoto University, Shirahama, Wakayama, Japan

Hidenori Aiki
Institute for Space-Earth Environmental Research, Nagoya University, Nagoya, Aichi, Japan

Meghan F. Cronin
National Ocean and Atmosphere Administration (NOAA), Pacific Marine Environment Laboratory (PMEL), Seattle, WA, USA

Takazo Shibuya
Laboratory of Ocean-Earth Life Evolution Research (OELE), Japan Agency for Marine-Earth Science and Technology (JAMSTEC), 2-15 Natsushima-cho, Yokosuka 237-0061, Japan

Masafumi Saitoh
Laboratory of Ocean-Earth Life Evolution Research (OELE), Japan Agency for Marine-Earth Science and Technology (JAMSTEC), 2-15 Natsushima-cho, Yokosuka 237-0061, Japan
Department of Earth and Planetary Sciences, Tokyo Institute of Technology, 2-12-1 Ookayama, Meguro, Tokyo 152-8551, Japan
Earth-Life Science Institute, Tokyo Institute of Technology, 2-12-1 Ookayama, Meguro, Tokyo 152-8551, Japan
Research and Development (R&D) Center for Submarine Resources, Japan Agency for Marine-Earth Science and Technology (JAMSTEC), 2-15 Natsushima-cho, Yokosuka 237-0061, Japan

Yuichiro Ueno
Laboratory of Ocean-Earth Life Evolution Research (OELE), Japan Agency for Marine-Earth Science and Technology (JAMSTEC), 2-15 Natsushima-cho, Yokosuka 237-0061, Japan
Department of Earth and Planetary Sciences, Tokyo Institute of Technology, 2-12-1 Ookayama, Meguro, Tokyo 152-8551, Japan
Earth-Life Science Institute, Tokyo Institute of Technology, 2-12-1 Ookayama, Meguro, Tokyo 152-8551, Japan

Naohiro Yoshida
Earth-Life Science Institute, Tokyo Institute of Technology, 2-12-1 Ookayama, Meguro, Tokyo 152-8551, Japan
Department of Environmental Chemistry and Engineering, Tokyo Institute of Technology, 4259 Nagatsuda, Midori, Yokohama 226-8502, Japan

Yukio Isozaki
Department of Earth Science and Astronomy, The University of Tokyo, 3-8-1 Komaba, Meguro, Tokyo 153-8902, Japan

Jianxin Yao and Zhansheng Ji
Institute of Geology, Chinese Academy of Geological Science, Baiwanzhuang Road 26, Beijing 100037, China

Katsumi Shozugawa and Motoyuki Matsuo
Department of Chemistry, The University of Tokyo, 3-8-1 Komaba, Meguro, Tokyo 153-8902, Japan

John Wakabayashi
Department of Earth and Environmental Sciences, California State University Fresno, 2576 E. San Ramon Avenue, Mail Stop ST-24, Fresno, CA 93740, USA

Tomohisa Irino
Faculty of Environmental Earth Science, Hokkaido University, N10-W5 Kita-ku, Sapporo 060-0810, Japan

Ryuji Tada, Akinori Karasuda, Shunsuke Kurokawa and Arisa Seki
Graduate School of Science, the University of Tokyo, 7-3-1 Hongo, Bunkyo-ku, Tokyo 113-0033, Japan

Ken Ikehara
Geological Survey of Japan, National Institute of Advanced Industrial Science and Technology, Tsukuba Central 7, 1-1-1 Higashi, Tsukuba, Ibaraki 305-8567, Japan

Takuya Sagawa
Institute of Science and Engineering, Kanazawa University, Kakuma-machi, Kanazawa, Ishikawa 920-1192, Japan

Song Lu
Graduate School of Environmental Science, Hokkaido University, N10-W5 Kita-ku, Sapporo 060-0810, Japan

Takashi Tonegawa, Yoshio Fukao, Gou Fujie, Tsutomu Takahashi and Shuichi Kodaira
Research and Development Center for Earthquake and Tsunami, Japan Agency for Marine-Earth Science and Technology, 3173-25, Showa-machi, Kanazawa-ku, Yokohama 236-0001, Japan

Shunsuke Takemura
Department of Material System Science, Graduate School of Nanobioscience, Yokohama City University, 22-2, Seto, Kanazawa-ku, Yokohama 236-0027, Japan
Earthquake Research Institute, University of Tokyo, 1-1-1, Yayoi, 30 Bunkyo-ku, Tokyo 113-0032, Japan

Ryuji Tada
Department of Earth and Planetary Science, Graduate School of Science, The University of Tokyo, 7-3-1 Hongo, Bunkyo-ku, Tokyo 113-0033, Japan

Kenji M. Matsuzaki
Department of Earth and Planetary Science, Graduate School of Science, The University of Tokyo, 7-3-1 Hongo, Bunkyo-ku, Tokyo 113-0033, Japan
Atmosphere and Ocean Research Institute, The University of Tokyo, 5-1-5 Kashiwanoha, Kashiwa, Chiba 277-8564, Japan

Takuya Itaki
Geological Survey of Japan, AIST, Marine Geology Research Group/Institute of Geology and Geoinformation, Tsukuba Central 7, 1-1-1 Higashi, Tsukuba, Ibaraki 305-8567, Japan

Shin-ichi Kamikuri
Faculty of Education, Ibaraki University, 2-1-1 Bunkyo, Ibaraki 310-8512, Japan

Ryuji Tada, Akinori Karasuda and Arisa Seki
Graduate School of Science, The University of Tokyo, Tokyo, Japan

Tomohisa Irino and Song Lu
Faculty of Environmental Earth Science, Hokkaido University, Sapporo, Japan

Ken Ikehara, Saiko Sugisaki and Takuya Itaki
Research Institute of Geology and Geoinformation, Geological Survey of Japan, AIST, Tsukuba, Japan

Chuang Xuan
National Oceanography Centre Southampton, Southampton, UK

Takuya Sagawa
Institute of Science and Engineering, Kanazawa University, Kanazawa, Japan

Yoshimi Kubota
Department of Geology and Paleontology, National Museum of Nature and Science, Tsukuba, Japan

Richard W. Murray
Department of Earth and Environment, Boston University, Boston, USA

Carlos Alvarez-Zarikian
International Ocean Descovery Program, Texas A&M University, College Station, USA

William T. Anderson Jr
Marine Sciences, Florida International University, Miami, USA

Maria-Angela Bassetti
Laboratoire CEFREM, Université de Perpignan, Perpignan, France

Bobbi J. Brace
Department of Earth and Atmospheric Sciences, University of Nebraska, Lincoln, USA

Steven C. Clemens
Earth, Environmental, and Planetary Sciences, Brown University, Providence, USA

Marcio H. da Costa Gurgel
Escola de Artes, Ciencias e Humanidades (EACH), Universidade de São Paulo, São Paulo, Brazil

Gerald R. Dickens
Department of Earth, Environmental, and Planetary Science, Rice University, Houston, USA

Stephen J. Gallagher
School of Earth Sciences, University of Melbourne, Melbourne, Australia

Ann G. Dunlea and Liviu Giosan
Department of Geology and Geophysics, Woods Hole Oceanographic Institution, Woods Hole, USA

Andrew C. G. Henderson
School of Geography, Politics and Sociology, University of Newcastle Upon Tyne, Newcastle Upon Tyne, UK

Ann E. Holbourn
Institute for Geosciences, Christian-Albrechts-Universitat zu Kiel (IFM), Kiel, Germany

Christopher W. Kinsley
Department of Earth, Atmospheric and Planetary Sciences, Massachusetts Institute of Technology, Cambridge, USA

Gwang Soo Lee
Petroleum and Marine Research Division, Korea Institute of Geoscience and Mineral Resources, Daejeon, South Korea

Kyung Eun Lee
Department of Ocean Science, Korea Maritime and Ocean University, Busan, South Korea

Johanna Lofi
Géosciences Montpellier, Université Montpellier II, Montpellier, France

Christina I. C. D. Lopes
IPMA, Lisbon, Portugal

Mariem Saavedra-Pellitero
Department of Geosciences, FB5, University of Bremen, Bremen, Germany

Larry C. Peterson
Rosenstiel School of Marine and Atmospheric Science, University of Miami, Miami, USA

Raj K. Singh
School of Earth, Ocean and Climate Sciences, Indian Institute of Technology Bhubaneshwar, Bhubaneshwar, India

Samuel Toucanne
Centre Bretagne, IFREMER, Plouzané cedex, France

Shiming Wan
Institute of Oceanology, Chinese Academy of Sciences, Beijing, China

Hongbo Zheng
Research Center for Earth System Science, Yunnan University, Kunming, China

Martin Ziegler
Department of Earth Sciences, Utrecht University, Utrecht, Netherlands

Takashi Nakagawa
Department of Mathematical Science and Advanced Technology, Japan Agency for Marine-Earth Science and Technology, 3173-25, Showa-machi, Yokohama 236-0001, Japan
Department of Earth Sciences, University of Hong Kong, Pokfulam Road, Hong Kong, Hong Kong

Hikaru Iwamori
Department of Solid Earth Geochemistry, Japan Agency for Marine-Earth Science and Technology, 2-15, Natsushima-cho, Yokosuka 237-0061, Japan
Earthquake Research Institute, The University of Tokyo, 1-1-1, Yayoi, Bunkyo, Tokyo 113-0032, Japan

Ryunosuke Yanagi
Department of Earth and Planetary Sciences, Tokyo Institute of Technology, 2-12-1, Ookayama, Meguro, Tokyo 152-8551, Japan

Atsushi Nakao
Earthquake Research Institute, The University of Tokyo, 1-1-1, Yayoi, Bunkyo, Tokyo 113-0032, Japan

Takazo Shibuya
Laboratory of Ocean-Earth Life Evolution Research (OELE), Japan Agency for Marine-Earth Science and Technology (JAMSTEC), 2-15 Natsushima-cho, Yokosuka 237-0061, Japan
Research and Development (R&D) Center for Submarine Resources, Japan Agency for Marine-Earth Science and Technology (JAMSTEC), 2-15 Natsushima-cho, Yokosuka 237-0061, Japan
Project Team for Development of New-generation Research Protocol for Submarine Resources, Japan Agency for Marine-Earth Science and Technology (JAMSTEC), 2-15 Natsushima-cho, Yokosuka 237-0061, Japan

Ken Takai
Laboratory of Ocean-Earth Life Evolution Research (OELE), Japan Agency for Marine-Earth Science and Technology (JAMSTEC), 2-15 Natsushima-cho, Yokosuka 237-0061, Japan
Research and Development (R&D) Center for Submarine Resources, Japan Agency for Marine-Earth Science and Technology (JAMSTEC), 2-15 Natsushima-cho, Yokosuka 237-0061, Japan
Project Team for Development of New-generation Research Protocol for Submarine Resources, Japan Agency for Marine-Earth Science and Technology (JAMSTEC), 2-15 Natsushima-cho, Yokosuka 237-0061, Japan
Department of Subsurface Geobiological Analysis and Research (SUGAR), Japan Agency for Marine-Earth Science and Technology (JAMSTEC), 2-15 Natsushima-cho, Yokosuka 237-0061, Japan

Katsuhiko Suzuki
Research and Development (R&D) Center for Submarine Resources, Japan Agency for Marine-Earth Science and Technology (JAMSTEC), 2-15 Natsushima-cho, Yokosuka 237-0061, Japan
Project Team for Development of New-generation Research Protocol for Submarine Resources, Japan Agency for Marine-Earth Science and Technology (JAMSTEC), 2-15 Natsushima-cho, Yokosuka 237-0061, Japan

Kenji Shimizu
Research and Development (R&D) Center for Submarine Resources, Japan Agency for Marine-Earth Science and Technology (JAMSTEC), 2-15 Natsushima-cho, Yokosuka 237-0061, Japan
Kochi Institute for Core Sample Research, Japan Agency for Marine-Earth Science and Technology (JAMSTEC), 200 Monobe, Nankoku, Kochi 783-8502, Japan

Motoko Yoshizaki and Hideo Tsunakawa
Department of Earth and Planetary Sciences, Tokyo Institute of Technology, 2-12-1 Ookayama, Meguro-ku, Tokyo 152-8551, Japan

Shigenori Maruyama
Department of Earth and Planetary Sciences, Tokyo Institute of Technology, 2-12-1 Ookayama, Meguro-ku, Tokyo 152-8551, Japan
Earth-Life Science Institute, Tokyo Institute of Technology, 2-12-1 Ookayama, Meguro-ku, Tokyo 152-8551, Japan

Masahiko Sato
Geological Survey of Japan, National Institute of Advanced Industrial Science and Technology, 1-1-1 Higashi, Tsukuba 305-8567, Japan

Kentaro Nakamura
Department of Systems Innovation, The University of Tokyo, 7-3-1 Hongo, Bunkyo-ku, Tokyo 113-8656, Japan

Soichi Omori
The Open University of Japan, 2-11 Wakaba, Mihama-ku, Chiba 261-8586, Japan

Takashi Furumura
Earthquake Research Institute, The University of Tokyo, 1-1-1 Yayoi, Bunkyo-ku, Tokyo 113-0032, Japan

Tae-Kyung Hong
Department of Earth System Sciences, Yonsei University, 50 Yonsei-ro, Seodaemun-gu, Seoul 120-749, South Korea

Brian LN Kennett
Research School of Earth Sciences, The Australian National University, Canberra ACT 0200, Australia

Shuhei Masuda, Satoshi Osafune and Tadashi Hemmi
Research and Development Center for Global Change, Japan Agency for Marine-Earth Science and Technology (JAMSTEC), Yokosuka, Japan

Shiho Kobayashi
Field Science Education and Research Center, Kyoto University, Kitashirakawaoiwake, Sakyo-ku, Kyoto 606-8502, Japan

Ryo Sugimoto, Daisuke Tahara and Osamu Tominaga
Research Center for Marine Bioresources, Fukui Prefectural University, 49-8-2, Katsumi, Obama, Fukui 917-0116, Japan

Hisami Honda, Makoto Yamada and Makoto Taniguchi
Research Institute for Humanity and Nature, 457-4, Kamigamo Motoyama, Kita-ku, Kyoto 603-8047, Japan

Yoji Miyata
Idea Consultants Co., 2-2-2, Hayabuchi, Tsuzuki-ku, Yokohama, Kanagawa 224-0025, Japan

Jun Shoji
Graduate School of Biosphere Science, Hiroshima University, 1-4-4 Kagamiyama, Higashi-hiroshima, Hiroshima 739-8528, Japan

Satoshi Nakada
Graduate School of Maritime Sciences, Kobe University, 5-1-1 Fukae-minami, Higashi-nada-ku, Kobe 658-0022, Japan

Koji Koyamada
Institute for Liberal Arts and Sciences, Kyoto University, Kitashirakawa-Oiwake Cho, Sakyo-ku, Kyoto 606-8502, Japan

Katsuhisa Baba and Masatoshi Sato
Hokkaido Research Organization, Fisheries Research Department, Hakodate Fisheries Research Institute, 1-2-66 Yunokawa-cho, Hakodate, Hokkaido 042-0932, Japan

Masafumi Natsuike and Sei-Ichi Saitoh
Faculty of Fisheries Sciences, Hokkaido University, 3-1-1, Minato-cho, Hakodate, Hokkaido 041-8611, Japan

Yoichi Ishikawa
Data Research Center for Marine-Earth Sciences, Japan Agency for Marine-Earth Science and Technology, Yokohama Institute for Earth Sciences, 3173-25 Showa-machi, Kanazawa-ku, Yokohama, Kanagawa 236-0001, Japan

Toshiyuki Awaji
Kyoto University Headquarters, Yoshida-Honmachi, Sakyo-ku, Kyoto 606-8501, Japan

Satoshi Nakada
Institute for Liberal Arts and Sciences, Kyoto University, Kitashirakawa-Oiwake Cho, Sakyo-ku, Kyoto 606-8502, Japan
Graduate School of Maritime Sciences, Kobe University, 5-1-1 Fukae-minami, Higashi-nada-ku, Kobe 658-0022, Japan

Index

www.ingramcontent.com/pod-product-compliance
Lightning Source LLC
LaVergne TN
LVHW080922210926
848485LV00004B/710